Lecture Notes in Computer Science 718
Edited by G. Goos and J. Hartmanis

Jennifer Seberry Yuliang Zheng (Eds.)

Advances in Cryptology – AUSCRYPT '92

Workshop on the Theory and Application
of Cryptographic Techniques
Gold Coast, Queensland, Australia
December 13-16, 1992
Proceedings

Springer-Verlag
Berlin Heidelberg New York
London Paris Tokyo
Hong Kong Barcelona
Budapest

Series Editors

Gerhard Goos
Universität Karlsruhe
Postfach 69 80
Vincenz-Priessnitz-Straße 1
D-76131 Karlsruhe, Germany

Juris Hartmanis
Cornell University
Department of Computer Science
4130 Upson Hall
Ithaca, NY 14853, USA

Volume Editors

Jennifer Seberry
Yuliang Zheng
Department of Computer Science, University of Wollongong
Northfields Avenue, Wollongong NSW 2522, Australia

CR Subject Classification (1991): E.3-4, D.4.6, G.2.1, C.2.0, K.6.5

ISBN 3-540-57220-1 Springer-Verlag Berlin Heidelberg New York
ISBN 0-387-57220-1 Springer-Verlag New York Berlin Heidelberg

Printed in Germany

Typesetting: Camera-ready by authors
Printing and binding: Druckhaus Beltz, Hemsbach/Bergstr.
45/3140-543210 - Printed on acid-free paper

Preface

The AUSCRYPT'92 conference held on the Gold Coast, Queensland, Australia, 13-16 December, 1992 is the second conference held in the Southern Hemisphere in cooperation with the International Association for Cryptologic Research. The conference was very enjoyable and ran very smoothly, largely due to the efforts of the General Chair, Professor Bill Caelli of the Queensland University of Technology and his colleagues Ed Dawson, Barry Arnison, Helen Bergen, Eleanor Crosby, Diane Donovan, Ian Graham, Helen Gustafson, and Lauren Nielson. There were 114 attendees from 18 countries and 5 continents.

This is the third conference held outside the EUROCRYPT series, held in European countries each northern spring, and the CRYPTO series held in Santa Barbara, California, USA each August. The other two were AUSCRYPT'90 held in Sydney, New South Wales, Australia in January 1990 and ASIACRYPT'91 held in Fujiyoshida, Japan in December 1992.

There were 77 submissions from 18 countries and 55 were accepted from 15 countries. Thirty were submitted from Asia and 15 accepted, 17 from Europe and 13 accepted, 12 from North America and 8 accepted and 18 from Australia of which 9 were accepted. In addition there were 7 presentations representing 5 countries at the rump sessions. After refereeing, 3 of them were selected to be published in these proceedings. All refereeing was carried out blind: no names were attached to papers. Programme Committee members' submissions were anonymous and went through exactly the same refereeing procedure as all other papers except that they were always sent to referees not in their own country. In addition the Committee chose four invited speakers: Yvo Desmedt from University of Wisconsin-Milwaukee, USA, Peter Landrock, the IACR President from Denmark, Valery Korzhik from Russia and John Snare, Australia's representative on the International Standards Committees. Please remember that these invited talks were not refereed and the authors bear full responsibility for their contents.

It is our pleasure to acknowledge the efforts of all those who contributed to making the conference a success. We especially wish to thank the members of the Programme Committee: Mike Burmester (RHNBC, University of London, UK), Yvo Desmedt (University of Wisconsin-Milwaukee, USA), Hideki Imai (University of Tokyo but formerly of Yokohama National University, Japan), Svein Knapskog (University of Trondheim, Norway), Rudi Lidl (University of Tasmania, Hobart, Australia), John Loxton (Macquarie University, Sydney, Australia), Tsutomu Matsumoto (Yokohama National University, Japan), Josef Pieprzyk (University of Wollongong, New South Wales, Australia), Rei Safavi-Naini, (University of Wollongong, New South Wales, Australia) and the Programme Chair Jennifer Seberry (University of Wollongong, New South Wales, Australia). Many of these referees will have used other persons to advise and evaluate and we sincerely thank those anonymous persons. Josef Pieprzyk ably organized the Rump Session.

We must thank my two valuable helpers Tor Jomar Nordhagen (Norway) and Marc Gysin (Switzerland) who helped so much with the electronic processing, entering into the computer all referees' comments that came by snail mail, then helped to email and print letters for the acceptances and rejections so we could get comments to authors as speedily as possible.

We wish to thank all the authors for sending their submissions (even ones that were unsuccessful), the speakers, and all the participants of this and other IACR conferences. We have established a tradition for high quality research and we hope this continues.

Wollongong, New South Wales, Australia
July 1993

Jennifer Seberry
Yuliang Zheng

General Chair

Bill Caelli (Queensland University of Technology, Australia)

Program Chair

Jennifer Seberry (University of Wollongong, Australia)

Program Committee

Mike Burmester	(RHBNC, University of London, UK)
Yvo Desmedt	(University of Wisconsin-Milwaukee, USA)
Hideki Imai	(University of Tokyo, Japan)
Svein Knapskog	(University of Trondheim, Norway)
Rudi Lidl	(University of Tasmania, Australia)
John Loxton	(Macquarie University, Australia)
Tsutomu Matsumoto	(Yokohama National University, Japan)
Josef Pieprzyk	(University of Wollongong, Australia)
Rei Safavi-Naini	(University of Wollongong, Australia)

In cooperation with

The International Association for Cryptologic Research (IACR)

and

The Centre for Computer Security Research, University of Wollongong

and

The Information Security Research Centre, Queensland University of Technology

Table of Contents

Table of Contents

Session 4: THEORY OF S-BOXES
Chair: Tsutomu Matsumoto

Session 5: CRYPTANALYSIS
Chair: Peter Landrock

Session 6: PROTOCOLS I
Chair: Rei Safavi-Naini

Session 7: PROTOCOLS II
Chair: Ed Dawson

Session 8: SEQUENCES
Chair: Rudi Lidl

Session 9: PSEUDORANDOMNESS
Chair: Bill Caelli

Session 10: ODDS AND ENDS
Chair: Valery Korzhik

Session 11: PUBLIC KEY CRYPTOGRAPHY I
Chair: Jennifer Seberry

Session 12: PUBLIC KEY CRYPTOGRAPHY II
Chair: John Snare

Rump Session
Chair: Josef Pieprzyk

Session 1

AUTHENTICATION AND SECRET SHARING I

Chair: Diane Donovan
(Queensland University of Technology, Australia)

Threshold Cryptosystems

Yvo Desmedt*

EE & CS Department, University of Wisconsin–
Milwaukee, WI 53201, U.S.A.

Abstract. Often the power to use a cryptosystem has to be shared. In threshold schemes, t-out-of-l have the power to regenerate a secret key (while less than t have not). However threshold schemes cannot be used directly in many applications, such as threshold signatures in which t-out-of-l have to co-sign a message. A normal threshold scheme would require the shareholders to send their shares to a trusted person who would sign for them. But the use of such a trusted person violates the main point of threshold signatures!
We first overview the research in the field and then discuss a threshold decryption/signature scheme which is as secure as RSA. We conclude by giving a list of open problems.

1 Introduction

1.1 The Concept and its Importance

In the classical mechanical example used to illustrate sharing two bank clerks are necessary to open the bank vault. While sharing schemes are useful for such control purposes, more is required when wholesale transactions are co-signed by two members of the bank. Indeed, when one requires that two bank clerks authorize (or sign) a large transaction (appending an authenticator), it is required that the authenticator is message dependent. The classical way to achieve this is to use first a sharing secret scheme which outputs the secret key. This secret key is then given to an authentication scheme. The first solution has the obvious disadvantage that each of the clerks knows the secret key and could perform (now or later) a fraud with it, *e.g.*, by authenticating a different message. In the second solution the shareholders transfer their complete power to a third party. A better solution would be that the bank clerks calculate the authenticator (or signature) all together such that above fraud is impossible, *i.e.*, while *not* divulging their shares. This last solution is precisely what *threshold authentication* (or signature) is about. In general, the goal of threshold cryptography is to design cryptosystems in which the power to perform a certain operation is shared. A further generalization is to allow that we have a more general access structure [29] than just a threshold.

The importance of threshold signature is clear when one considers that majority is often a threshold too and that in a democracy no member of the congress

* A part of this work has been supported by NSF Grant NCR-9106327.

(parliament) has the power himself to make a law, but that a majority is required. Even the concept of the president having a veto power can be included when using general access structures instead of threshold schemes.

Threshold cryptography can contribute in the fight against abuse of power by the government or any individual inside an organization. To give another illustration besides threshold signatures consider threshold decryption. In the United States a proposal that would allow the government to eavesdrop encrypted messages has been heavily debated [36, p. 46] and [11]. Micali [31] proposed to use threshold schemes to achieve this goal. In his scheme each user of a cryptosystem has to deposit shares of his secret key in banks and other respectful organizations. When a court order is provided shareholders are forced to reveal their shares to, let us say, the Federal Bureau of Investigation (FBI). A major disadvantage of his scheme is that once these shares have been provided to the FBI this organization could abuse the obtained secret key to decrypt any message, even those sent after the court order was successfully appealed. So once such a court order has been successfully appealed, the receiver has to publish a new public key. Threshold decryption allows for a scenario in which the court specifies which messages have to be decrypted. Under the assumption that no number of shareholders larger or equal than the threshold will ever become corrupt, threshold decryption allows a threshold of shareholders to decrypt specific messages without leaking during this process anything about other plaintexts or the secret key. So there is even no need to publish a new public key after a successful appeal.

Another important property of threshold cryptography is its reliability. This reliability is needed in many organizations that must provide their services (*e.g.*, sign and validate documents) even when the head is on vacation or sick.

1.2 Its Origin

Let us now briefly overview the origin of threshold cryptography. The first threshold cryptosystems were independently developed by Boyd [3], Croft-Harris [8] and Desmedt [16].

A 2-out-of-l RSA (signature) scheme [37] has been presented by Boyd [3]. It is not clear whether this scheme is as secure as RSA. Shares of the secret exponent d are given such that any two shareholders can sign a message. The final output is a normal RSA signature. So, to verify the signature an outsider does not need to know who the shareholders are. In this approach it is sufficient that all shareholders as a group have one public key and the shareholders themselves could be ***anonymous***. Examples of such groups are organizations we are familiar with in our society, such as banks, companies, etc. The generation of the signature in this scheme is serial (one shareholder sends his partial result to the other to obtain the RSA signature).

A scheme for t-out-of-l "authorizations" has been proposed by Croft and Harris [8]. They also attempted to present a threshold key exchange protocol. Their scheme uses a prime order subgroup of $Z_p(*)$ (predating [38]). Because the authorizations are independent of the message the scheme is insecure for many applications. The threshold key exchange protocol suffers from a security bug (less than the threshold of shareholders can break the system).

The author introduced the concept of society oriented cryptography [16]. Sender and receiver are then organizations instead of individuals. The author described (what he would now call) threshold decryption, in which a threshold of shareholders are needed to decipher a message. To motivate the need for threshold decryption the author cites that in some modern software companies two supervisors are responsible for software developing projects. Continuity is then guaranteed when one of supervisors leaves — a not unlikely event in this business. Different problems are also introduced, many of which are still unsolved. The solution presented for threshold decryption (in which the threshold is 50%) is entirely based on mental games [24].

Although mental games can easily provide threshold cryptosystems they are unsatisfactory because modern implementations are completely impractical by requiring the shareholders to heavily interact. The *challenge* is the development of ***non-interactive*** threshold cryptosystems. In Section 2 we overview the results, achieved so far, in attaining this goal. Seen above we further ignore in our discussions solutions that need interaction between shareholders.

1.3 Definition

Let us avoid to be too formal. Instead we will focus on the essential aspects of threshold cryptography. Moreover readers for whom the examples above are sufficient to grasp the meaning of threshold cryptography can skip the rest of this section.

Let us suppose that we have a cryptosystem A which is used in a cryptographic protocol (A, R), where R is a collection of machines, so that we have a robust (workable) and secure system. Let t and l be fixed integers. If we now replace the machine A by a collection of machines $\mathbf{A} = (A_1, A_2, \ldots, A_l, C)$ such that:

1. (***Reliability***) $\forall \{i_1, \ldots, i_t\} \subset \{1, \ldots, l\}$: the cryptographic protocol $((A_{i_1}, A_{i_2}, \ldots, A_{i_t}, C), R)$, in which $(A_{i_1}, A_{i_2}, \ldots, A_{i_t}, C)$ behave as an entity, satisfies the former robustness and security condition,
2. (***Threshold Security***) $\forall t' \leq t-1 : \forall \{i_1, \ldots, i_{t'}\} \subset \{1, \ldots, l\} : (\tilde{A}_{i_1}, \tilde{A}_{i_2}, \ldots, \tilde{A}_{i_{t'}}, \tilde{C})$ cannot give more power to an adversary by trying to help the adversary, where $\tilde{A}_{i_j}$ private information (tape) is identical to the one of A_{i_j} and $\tilde{A}_{i_j}$ has eavesdropped in previous executions of the scheme the output of each shareholder, where, in the case of unconditional security $\tilde{A}_{i_j}$ and $\tilde{C}$ are any machines, while for conditional Threshold Security their computer power is (polynomially) bounded,

then we call $(A_1, A_2, \ldots, A_l, C)$ a t-out-of-l ***threshold cryptosystem***. If A, used in the protocol (A, R), has been called an S-system, then we call $(A_1, A_2, \ldots, A_l, C)$ a threshold S-system.

To generalize the definition to include general sharing, we modify the Reliability Condition (1.) to require that $\{i_1, \ldots, i_t\} \in \Gamma_{\{1,\ldots,l\}}$ and we modify the Threshold Security Condition (2.) so that $\{i_1, \ldots, i_{t'}\} \notin \Gamma_{\{1,\ldots,l\}}$, where $\Gamma_{\{1,\ldots,l\}}$ is the access structure. We say that the entity **A** has *anonymous members* [16] if the machines R above do not "see" the difference between the execution of (A, R) and $((A_{i_1}, A_{i_2}, \ldots, A_{i_t}, C), R)$. To achieve this "anonymity" C will perform the external communications with R, receiving "partial results" from $A_{i_1}, A_{i_2}, \ldots, A_{i_t}$. We call C the *combiner*.

Threshold schemes could be considered to be special cases of threshold cryptography in which A's output is equal to the key.

1.4 Background

We assume that the reader is familiar with the concept of threshold schemes [2, 39] and general sharing schemes [29] allowing a specific access structure to reconstruct a secret key. For a bibliography on sharing schemes consult [41].

An important tool in modern threshold cryptography is the concept of homomorphic sharing scheme [1]. Informally, a sharing scheme is homomorphic if the product of the shares is a share of the product of the secret keys (in general the first operator could be different from the second, moreover the product could be a sum for example). A homomorphic sharing scheme is composite [1] if revealing all the shares of the key $k_1 * k_2$ does not leak anything additional about (k_1, k_2) besides what $k_1 * k_2$ does. So a composite sharing scheme is "leak free".

2 Overview of Existing Schemes

We overview research on threshold cryptography and its generalization towards sharing cryptographic power in a general access structure setting. Papers on related topics are discussed jointly. (So, papers that are multi disciplinary will be cited more than once.) We do not rediscuss research already cited in Section 1.2. We avoid to discuss schemes that require internal interaction between the shareholders, to avoid overlap with mental games (secure distributed computation).

2.1 Threshold Decryption

In Sections 1.1 and 1.2 we already discussed two applications of threshold decryption.

A scheme allowing l-out-of-l shareholders to decrypt incoming ciphertexts encrypted by the El Gamal [19] public key encryption scheme has been presented by Frankel [20]. This was generalized towards t-out-of-l decryption of El Gamal type ciphertexts in [13]. The Threshold Security of this scheme is easy to prove. In other words it satisfies the requirement that whatever they attempt $t-1$ shareholders cannot help an adversary to eavesdrop. So it is as secure as the original El Gamal scheme. Pedersen observed that there is no need for a trusted key distributor in these schemes [35].

t-out-of-l decryption of RSA [37] type ciphertexts has been presented in [14]. Laih and Harn [30] presented a shared decryption scheme based on RSA that generalizes the threshold to any access structure. For some access structures the shareholders need exponential computer power. The Threshold Security of both schemes seems heuristic, because it is not clear how to prove (or disprove) that $t-1$ shareholders are of no help to an eavesdropper. For threshold decryption this problem has been solved in [21]. Because RSA is not a proven secure public key scheme, only the Threshold Security fact has been proven, evidently. We briefly discuss the scheme in further detail in Section 3. In [21] it has also been observed that using the general sharing scheme of Simmons-Jackson-Martin [42], as adapted in [22], it is easy to make shared (in the sense of a general access) decryption of RSA with proven Threshold Security. For some access structure Simmons-Jackson-Martin generates exponentially large shares, requiring those access structures to be excluded when the number of shareholders is not very small (otherwise the scheme is impractical, moreover it would require the shareholders a computer power that would allow them to break RSA anyway).

A proven secure public key threshold decryption scheme is presented in [9]. While all above schemes require that no shareholder will jam the computation (*e.g.*, by outputting random) this restriction has been removed as far as possible.

All the above shared decryption schemes are for an anonymous membership scenario. Hwang [28] presented a discrete log based decryption scheme in which the sender knows the shareholders.

Clearly the security of all these schemes is conditional. Unconditionally secure threshold decryption is easy to achieve by combining the one-time-pad [43, 40] with any threshold (general sharing) scheme.

2.2 Threshold Pseudorandom

The generation of a pseudorandom string by a threshold of anonymous shareholders has been studied in [5, 9]. The first scheme is based on RSA, while the second is based on secure one way functions with homomorphic trapdoor (such as RSA).

2.3 Threshold Encryption

The concept of sharing the power to encrypt a message makes no sense in a public key context, but does for symmetric cryptosystems. De Santis, Desmedt, Frankel and Yung present a proven secure threshold encryption scheme in [9] which is based on a threshold pseudorandom generator.

Unconditionally secure threshold encryption is similar as decryption.

2.4 Threshold Authentication and Signatures

Unconditional Security Threshold generation of authenticators by a set with anonymous members has been discussed in [14] (for a variant see also [18]).

This has been extended to include general access structures in [12]. As demonstrated in [12], the shared generation of authenticators can be achieved using any composite sharing scheme [1]. It was also observed that the Simmons-Jackson-Martin [42] scheme, as adapted in [22], is a composite sharing scheme.

Threshold verification of user identifiers has been proposed in [10].

Signatures First we observe that threshold signatures are somewhat related to multisignatures (*e.g.*, [33, 32]) and group signatures [7]. In multisignatures a multiple of individuals are all going to sign a message. One could consider multisignatures as l-out-of-l threshold signatures with known members, where l is not necessarily fixed. Group signatures could be viewed as 1-out-of-l threshold "signatures" with anonymous members in which the anonymity has a trapdoor. To avoid to be sidetracked we do no longer discuss multisignatures and group signatures.

The shared generation by anonymous members of an RSA signature was discussed in [14, 30, 21]. These schemes are very similar to the threshold (shared) RSA decryption, discussed in Section 2.1, and have similar security.

Harn and Yang [27] discussed the threshold generation by anonymous members of undeniable signatures [6].

De Santis, Desmedt, Frankel and Yung present a proven secure threshold signature scheme with anonymous members in [9].

In undeniable signatures verifiers need the help of the signer to verify the validity of the signature. Transferring this power to a threshold of known shareholders has been discussed in [34]. It is important to observe that although any threshold of shareholders can help in the verification, the reliability of this scheme is only average. The scheme, based on a zero-knowledge protocol (see also Section 2.5), requires that the shareholder's computer does not go down immediately after having giving an output, but remains up for a certain time. We call a scheme *very reliable* when one assumes that the number of shareholders that must be up is larger or equal than a certain threshold but the set could change dynamically at any moment during the execution of the external (the combiner with the receivers R) protocol. If the scheme under consideration requires no external interaction it is automatically very reliable. But the verification process in [34] requires external interaction and has only an average reliability.

Other Authentication Schemes A proven secure threshold authentication scheme with anonymous co-authenticators was presented in [14]. The underlying authentication scheme is an adaptation [15] of the Goldwasser-Micali-Rivest scheme [26]. While the authentication tree is no longer needed, a zero-knowledge protocol is used (see also Section 2.5). The scheme only provides average reliability.

2.5 Threshold zero-knowledge proofs

A zero-knowledge protocol in which a threshold of known provers prove the simultaneous discrete log has been proposed [34]. For the case that shareholders are anonymous a threshold of shareholders can in perfect zero-knowledge prove the knowledge of a square root, as briefly explained in [14]. Using [4, 17] this protocol can easily be generalized to a whole class of zero-knowledge proofs for different languages, as we further explain in Appendix A. Both protocols have only an average reliability. How to achieve very reliable computationally zero-knowledge proofs with a threshold of provers is explained in [9].

3 Threshold RSA as Secure as RSA

To illustrate the concept of threshold cryptosystems and combiner, we now describe in more detail one threshold cryptosystem. Desmedt and Frankel announced in [14, p. 460] a t-out-of-l threshold RSA scheme as secure as RSA. We here overview this Frankel-Desmedt scheme. We omit all proofs, which are given in [21].

Let $P(z) = \sum_{j=0}^{Q-1} z^j$, where Q is a prime greater than or equal to $l+1$. Let $Z[u] \cong Z[z]/(P(z))$, *i.e.*, the ring of integer polynomials modulo $P(z)$. Let the public identity of shareholder i be: $x_i = \sum_{j=0}^{i-1} u^j$. Finally let us define the function $F_0 : Z[u] \to Z; b_0 + b_1 u + \cdots + b_{Q-2} u^{Q-2} \to b_0$.

Key Initialization The trusted key distributor[2]:

Step 1 chooses $n = pq$ and e and d as described in the RSA algorithm and publishes the public key (e, n) of the organization,

Step 2 chooses with uniform probability distribution a random polynomial $f(x)$ of degree $t-1$ over $Z_\gamma[u]$ such that $f(0) = d \bmod \gamma$ where $\gamma = \lfloor n^2/\lambda(n) \rfloor \cdot \lambda(n)$,

Step 3 privately transmits to each co-signer i the share $K_i = f(x_i)$ evaluated in $Z_\gamma[u]$.

Observe that $\lfloor n^2/\lambda(n) \rfloor \cdot \lambda(n)$ is not revealed in the above protocol. So a shareholder considers K_i as belonging to $Z[u]$.

Co-Signing Assume that the set of co-signers B ($|B| = t$), who will co-sign, is known to all co-signers. Each shareholder $i \in B$:

Step 0 *precomputes* in $Z[u]$ (so modulo $P(z)$) $\alpha_{i,B} = K_i \cdot y_{i,B}$ where

$$y_{i,B} = \prod_{\substack{j \in B \\ j \neq i}} \frac{(0 - x_j)}{(x_i - x_j)}$$

to obtain the integer $a_{i,B} = F_0(\alpha_{i,B})$.

Step 1 co-signs a (hashed) message M by sending $S_{M,i,B} = M^{a_{i,B}}$ to the combiner, C.

---

[2] Using mental games only during the initialization phase there is no need for a key distributor when t is small enough.

The Combiner sends the message and the signature $S_M = \prod_{i \in B} S_{M,i,B} \bmod n$ to the receiver.

One demonstrates in [21] that after precomputation the scheme is roughly as efficient as RSA when $t < \log_2 n$. It is also proven that the scheme is Reliable and Threshold Secure, *i.e.*, as secure as RSA. A reader could object to call this scheme non-interactive. Indeed shareholders must know who is active (which set B will co-sign), which normally requires some interaction. This problem can be solved by combining the scheme with the homomorphic sharing scheme presented in [17], as is discussed in detail in [9].

4 Conclusion

We have discussed the purpose and the need of threshold cryptosystems. We have seen that a threshold scheme on itself does not provide the required security. Our overview has demonstrated that this area has attracted a lot of research. While secure distributed computation has a more general scope (for an overview consult [23]), many of these schemes are not practical. Threshold cryptography is primarily interested in non-interactive solutions. Moreover, to be as realistic as possible, many schemes do not require that shareholders send in a given order their output to the combiner.

The attentive reader can see that some cryptosystems have no threshold equivalent yet. For example no threshold perfect zero-knowledge proof for graph isomorphism has been presented so far. Moreover no non-interactive threshold generation of DSS signatures has been found. May be some of these open problems can be proven to have no non-interactive solution. Other open problems are to improve the performance of many schemes and/or to prove their limitations. Although many threshold cryptosystems are based on some homomorphic property it has not been proven that such a property is necessary to achieve non-interactive threshold cryptosystems.

A Threshold zero-knowledge proofs

We sketch how to generalize the threshold zero-knowledge proof of [14].

We follow the notations of [4]. So let us briefly overview it. To simplify the discussion we focus on the case that $\mathcal{G} = \mathcal{G}' = \mathcal{G}''$ is an Abelian group (although some zero-knowledge protocols do not satisfy exactly this description they can trivially be modified to fit it). To have easy notations we assume $m = 1$. f is homomorphism and $f(\mathcal{G}) = \mathcal{H} \subset \mathcal{H}'$. $I \in \mathcal{H}$ is the "public number" which is a part of the input x and the "secret number" S belongs to a set $\mathcal{G}$.

The protocol in which A is the prover and R the verifier, is as follows:
First the verifier checks that $I \in \mathcal{H}'$. Then repeat sufficiently many times:

Step 1 A selects a random $X \in \mathcal{G}$ and sends R: $Z = f(X)$ (A's cover).
Step 2 R sends A a random $q \in Q$ (R's query).
Step 3 When $q \in Q$, A sends R: $Y = X \cdot S^q$ (A's answer).

Step 4 R verifies that $Y \in \mathcal{G}$ and that $f(Y) = Z \cdot I^q$ (R's verification).

We now assume that multiplying and calculating inverses in $G(\cdot)$ can be done in polynomial time and that G is sampleable.

Let us introduce the reader to techniques we use. In zero-knowledge sharing schemes [17] the (view of the) key distribution phase can be simulated [25]. All *existing* homomorphic sharing schemes have the property that the key $k \in G$ can be written as:

$$k = \prod_{i \in B} \psi_{i,B}(s_i)\ , \tag{1}$$

where B is in the access structure, s_i is i's share, and $\psi_{i,B}$ is a function from the set from which i's shares are chosen to G. When a sharing scheme satisfies (1) we call it *multiplicative*. It was proven [22] that when the set of keys is an Abelian group all ideal homomorphic sharing schemes are multiplicative. (It is an open problem whether any non-ideal homomorphic sharing scheme is multiplicative.) So, the homomorphic schemes of [17] for $G(\cdot)$ being any Abelian group are multiplicative, because they are ideal for a superset of G. Moreover they are zero-knowledge. Also the Simmons-Jackson-Martin [42] general sharing scheme, as adapted in [22], is multiplicative if $G(\cdot)$ is an Abelian group. If the shares are polynomially bounded it is also a zero-knowledge sharing scheme.

We now present the generalization. We assume that the set of co-provers B involved in the proof is in the access structure and is known to all co-provers. For a zero-knowledge interactive proof which can be described as above we explain how to modify it to obtain a zero-knowledge scheme with shared provers. We assume that a key distributor has given each shareholder its share s_i of the secret key S using a zero-knowledge multiplicative sharing scheme. Replace Steps 1–3 by:

Step 1 Each shareholder $i \in B$ selects a random $X_i \in \mathcal{G}$ and sends: $Z_i = f(X_i)$ to C, the combiner. C calculates $Z = \prod_{i \in B} Z_i$ and sends Z to R.
Step 2 R sends C a random $q \in Q$ which C broadcasts to all shareholders.
Step 3 When $q \in Q$, each shareholder $i \in B$ sends C: $Y_i = X_i \cdot (\psi_{i,B}(s_i))^q$. Then C calculates $Y = \prod_{i \in B} Y_i$ and sends it to R.

The resulting protocol has the Threshold Security property. In other words, the joint view of $B \cup \{C\} \cup \{R\}$, where B is *not* in the access structure, is simulatable (approximable). As in Section 3 the reader could object to call this scheme non-interactive, because shareholders must know who is active. This can easily be solved when the Simmons-Jackson-Martin [42] general sharing scheme as adapted in [22] is used. It can also be solved using [17] based on a similar technique as in [9].

References

1. Benaloh, J. C.: Secret sharing homomorphisms: Keeping shares of a secret secret. In Advances in Cryptology, Proc. of Crypto '86 (Lecture Notes in Computer Science 263) (1987) A. Odlyzko, Ed. Springer-Verlag pp. 251–260

2. Blakley, G. R.: Safeguarding cryptographic keys. In Proc. Nat. Computer Conf. AFIPS Conf. Proc. (1979) pp. 313–317
3. Boyd, C.: Digital multisignatures. In Cryptography and coding (1989) H. Beker and F. Piper, Eds. Clarendon Press pp. 241–246
4. Burmester, M. V. D., Desmedt, Y. G., Piper, F., Walker, M.: A general zero-knowledge scheme. In Advances in Cryptology, Proc. of Eurocrypt '89 (Lecture Notes in Computer Science 434) (1990) J.-J. Quisquater and J. Vandewalle, Eds. Springer-Verlag pp. 122–133
5. Cerecedo, M., Matsumoto, T., Imai, H.: Non-interactive generation of shared pseudorandom sequences. Presented at Auscrypt'92, Mudgeeraba, Queensland, Australia, to appear in the proceedings (Lecture Notes in Computer Science), Springer-Verlag December 13–16, 1992
6. Chaum, D., van Antwerpen, H.: Undeniable signatures. In Advances in Cryptology — Crypto '89, Proceedings (Lecture Notes in Computer Science 435) (1990) G. Brassard, Ed. Springer-Verlag pp. 212–216
7. Chaum, D., van Heyst, E.: Group signatures. In Advances in Cryptology, Proc. of Eurocrypt '91 (Lecture Notes in Computer Science 547) (April 1991) D. W. Davies, Ed. Springer-Verlag pp. 257–265
8. Croft, R. A., Harris, S. P.: Public-key cryptography and re-usable shared secrets. In Cryptography and coding (1989) H. Beker and F. Piper, Eds. Clarendon Press pp. 189–201
9. De Santis, A., Desmedt, Y., Frankel, Y., Yung, M.: Quorum cryptography and non-interactive protocols. In preparation (Available from authors when completed)
10. De Soete, M., Quisquater, J.-J., Vedder, K.: A signature with shared verification scheme. In Advances in Cryptology — Crypto '89, Proceedings (Lecture Notes in Computer Science 435) (1990) G. Brassard, Ed. Springer-Verlag pp. 253–262
11. Denning, D. E.: Panel: Digital telephony October 13–16, 1992. National Computer Security Conference, Baltimore
12. Desmedt, Y.: Threshold cryptography. Invited paper, to be presented at SPRC '93, 3rd Symposium on State and Progress of Research in Cryptography, Roma, Italy February 15–16, 1993
13. Desmedt, Y., Frankel, Y.: Threshold cryptosystems. In Advances in Cryptology — Crypto '89, Proceedings (Lecture Notes in Computer Science 435) (1990) G. Brassard, Ed. Springer-Verlag pp. 307–315
14. Desmedt, Y., Frankel, Y.: Shared generation of authenticators and signatures. In Advances in Cryptology — Crypto '91, Proceedings (Lecture Notes in Computer Science 576) (1992) J. Feigenbaum, Ed. Springer-Verlag pp. 457–469
15. Desmedt, Y. G.: Abuse-free cryptosystems: Particularly subliminal-free authentication and signature. Submitted to the Journal of Cryptology, under revision April 1989
16. Desmedt, Y.: Society and group oriented cryptography : a new concept. In Advances in Cryptology, Proc. of Crypto '87 (Lecture Notes in Computer Science 293) (1988) C. Pomerance, Ed. Springer-Verlag pp. 120–127
17. Desmedt, Y., Frankel, Y.: Perfect zero-knowledge sharing schemes over any finite Abelian group. Presented at Sequences '91, June 17–22, 1991, Positano, Italy, to appear in: the Proceedings, Springer-Verlag 1991
18. Desmedt, Y., Frankel, Y., Yung, M.: Multi-receiver / multi-sender network security: efficient authenticated multicast/ feedback. In IEEE INFOCOM '92, Eleventh Annual Joint Conference of the IEEE Computer and Communications Societies (Florence, Italy, May 4–8, 1992) pp. 2045–2054

19. ElGamal, T.: A public key cryptosystem and a signature scheme based on discrete logarithms. IEEE Trans. Inform. Theory **31** (1985) 469–472
20. Frankel, Y.: A practical protocol for large group oriented networks. In Advances in Cryptology, Proc. of Eurocrypt '89 (Lecture Notes in Computer Science 434) (1990) J.-J. Quisquater and J. Vandewalle, Eds. Springer-Verlag pp. 56–61
21. Frankel, Y., Desmedt, Y.: Parallel reliable threshold multisignature. Tech. Report TR–92–04–02 Dept. of EE & CS, Univ. of Wisconsin-Milwaukee April 1992. Submitted to ACM Transactions on Computer Systems with title: Distributed reliable threshold multisignatures
22. Frankel, Y., Desmedt, Y.: Classification of ideal homomorphic threshold schemes over finite Abelian groups. Presented at Eurocrypt '92, Balatonfüred, Hungary, to appear in Advances in Cryptology, Proc. of Eurocrypt '92, (Lecture Notes in Computer Science), Springer-Verlag May 24–28, 1992
23. Franklin, M., Yung, M.: Varieties of secure distributed computing. Presented at the Second Advanced Workshop on Sequences: Combinatorics, Compression, Security and Transmission, June 17–22, 1991, Positano, Italy, to appear in: Sequences, Springer-Verlag
24. Goldreich, O., Micali, S., Wigderson, A.: How to play any mental game. In Proceedings of the Nineteenth annual ACM Symp. Theory of Computing, STOC (May 25–27, 1987) pp. 218–229
25. Goldwasser, S., Micali, S., Rackoff, C.: The knowledge complexity of interactive proof systems. Siam J. Comput. **18** (1989) 186–208
26. Goldwasser, S., Micali, S., Rivest, R.: A digital signature scheme secure against adaptive chosen-message attacks. Siam J. Comput. **17** (1988) 281–308
27. Harn, L., Yang, S.: Group-oriented undeniable signature schemes without the assistance of a mutually trusted party. Presented at Auscrypt'92, Mudgeeraba, Queensland, Australia, to appear in the proceedings (Lecture Notes in Computer Science), Springer-Verlag December 13–16, 1992
28. Hwang, T.: Cryptosystems for group oriented cryptography. In Advances in Cryptology, Proc. of Eurocrypt '90 (Lecture Notes in Computer Science 473) (1991) I. Damgård, Ed. Springer-Verlag pp. 352–360
29. Ito, M., Saito, A., Nishizeki, T.: Secret sharing schemes realizing general access structures. In Proc. IEEE Global Telecommunications Conf., Globecom'87 (1987) IEEE Communications Soc. Press pp. 99–102
30. Laih, C.-S., Harn, L.: Generalized threshold cryptosystems. Presented at Asiacrypt'91, November 11–14, 1991, Fujiyoshida, Yamanashi, Japan, to appear in: Advances in Cryptology. Proc. of Asiacrypt'91 (Lecture Notes in Computer Science), Springer-Verlag
31. Micali, S.: Fair public-key cryptosystems. Presented at Crypto '92, Santa Barbara, California, U.S.A., to appear in Advances in Cryptology — Crypto '92, Proceedings (Lecture Notes in Computer Science), Springer-Verlag Augustus 16–20, 1992
32. Ohta, K., Okamoto, T.: A digital multisignature scheme based on the Fiat-Shamir scheme. Presented at Asiacrypt'91, November 11–14, 1991, Fujiyoshida, Yamanashi, Japan, to appear in: Advances in Cryptology. Proc. of Asiacrypt'91 (Lecture Notes in Computer Science), Springer-Verlag 1991
33. Okamoto, T.: A digital multisignature scheme using bijective public-key cryptosystems. ACM Trans. on Computer Systems **6** (1988) 432–441
34. Pedersen, T. P.: Distributed provers with applications to undeniable signatures. In Advances in Cryptology, Proc. of Eurocrypt '91 (Lecture Notes in Computer Science 547) (April 1991) D. W. Davies, Ed. Springer-Verlag pp. 221–242

35. Pedersen, T. P.: A threshold cryptosystem without a trusted party. In Advances in Cryptology, Proc. of Eurocrypt '91 (Lecture Notes in Computer Science 547) (April 1991) D. W. Davies, Ed. Springer-Verlag pp. 522–526
36. Rivest, R. L., Hellman, M. E., Anderson, J. C.: Responses to NIST's proposal. Commun. ACM **35** (1992) 41–54
37. Rivest, R. L., Shamir, A., Adleman, L.: A method for obtaining digital signatures and public key cryptosystems. Commun. ACM **21** (1978) 294–299
38. Schnorr, C. P.: Efficient identification and signatures for smart cards. In Advances in Cryptology — Crypto '89, Proceedings (Lecture Notes in Computer Science 435) (1990) G. Brassard, Ed. Springer-Verlag pp. 239–252
39. Shamir, A.: How to share a secret. Commun. ACM **22** (1979) 612–613
40. Shannon, C. E.: Communication theory of secrecy systems. Bell System Techn. Jour. **28** (1949) 656–715
41. Simmons, G. J.: An introduction to shared secret and/or shared control schemes and their application. In Contemporary Cryptology, G. J. Simmons, Ed. IEEE Press 1992 pp. 441–497
42. Simmons, G. J., Jackson, W., Martin, K.: The geometry of shared secret schemes. Bulletin of the Institute of Combinatorics and its Applications **1** (1991) 71–88
43. Vernam, G. S.: Cipher printing telegraph systems for secret wire and radio telegraphic communications. Journal American Institute of Electrical Engineers **XLV** (1926) 109–115

Authentication Codes with Perfect Protection

L. Tombak *
R. Safavi-Naini **

Department of Computer Science University of Wollongong
Northfields Ave., Wollongong 2522, AUSTRALIA

Abstract. In this paper we prove new results on authentication codes with perfect protection. We will prove that perfect protection for impersonation follows from perfect protection for substitution only if the source is uniform and derive a necessary and sufficient condition for codes that provide perfect protection for both types of attack. We prove a new lower bound on the probability of deception in substitution for codes with perfect protection and characterize the codes that satisfy the bound with equality.

1 Introduction

In this paper we prove some new results on authentication codes (A-codes). A highly desirable property of an A-code is to ensure minimum probability of deception. Such codes are said to provide perfect protection. In this paper we consider codes that provide such protection for impersonation and substitution simultaneously. Our first major theorem generalizes a result due to Massey. We prove that codes that provide perfect protection for substitution provide perfect protection for impersonation if the source is uniform. This was known for the case that the communicants use uniform strategy. We show that this is not a necessary condition and the result holds for a uniform source independent of communicants' strategy. We will also demonstrate that for a non-uniform source perfect protection for impersonation does not always follow from perfect protection for substitution. Next we prove a necessary and sufficient condition for A-codes to provide perfect protection for impersonation *and* substitution. This is a powerful condition that allows us to give an alternative proof of Stinson's general characterization theorem ([4], theorems 4.1 and 2.5).

In the second part of this paper we consider a subclass of enemy's strategies: those that are determined by a probability distribution on the set of *valid keys* when a cryptogram is intercepted. This is an important class of strategy which the enemy uses when the encoding process is algorithmic (such as using cryptographic algorithms). This will allow us to prove a new lower bound on the probability of deception in substitution for codes with perfect protection for

* Support for this work was provided by Australian Research Council grant A49030136.
** Support for this work was provided in part by Australian Research Council grant A49030136.

impersonation and characterize A-codes that satisfy the bound. This completes Stinson's characterization theorem in the following way: in an A-code with k source state, E encoding rules and M cryptograms, if the code provides perfect protection for impersonation, probability of deception, is lower bounded by the maximum of $((k-1)/(M-1), M/(kE))$. If $(k-1)/(M-1) > M/kE$, the code must have at least E_0 encoding rules and codes with minimum number of encoding rules correspond to BIBDs. If $(k-1)/(M-1) \leq M/kE$ then the code has at most E_0 encoding rules and cannot provide perfect protection for substitution (other than the case $E = E_0$). The least achievable probability of deception in this case is M/kE. We give a characterization of codes with $P_0 = k/M$ and $P_1 = M/kE$.

2 Preliminaries

In the model of authentication system introduced by Simmons [5, 6, 7], there are three participants. A *transmitter* who wants to send the state of a source S to a *receiver* and an *opponent* whose aim is to deceive the receiver into accepting a fraudulent message he/she has devised. To protect against the opponent's aim the communicants use an A-code which is a collection of mappings from the set $\mathcal{S}$, $|\mathcal{S}| = S$, of plaintext into the set $\mathcal{M}$ of cryptogram, $|\mathcal{M}| = M$. The collection is indexed by keys which are elements of a set $\mathcal{E}$ of size E. The incidence matrix A of an A-code is an $E \times M$ zero-one matrix of size whose rows are indexed by the elements of $\mathcal{E}$ and columns by those of $\mathcal{M}$, and $a_{ij} = 1$ if $m_j \in M(e_i)$ where $M(e_i)$ is the set of cryptograms that are authentic under e_i. Matrix A has exactly k ones in each row. We assume a source state and an encoding rule uniquely determine a cryptogram (no splitting) and every cryptogram has a non-zero probability.

The communicants use a probability distribution π on $\mathcal{E}$, which is their *strategy*, to choose a key e, hence determining the set $M(e)$ of authentic cryptograms under e. The enemy may use an *impersonation* attack in which he/she fabricates a message m and initiate communication, or a *substitution* attack in which he sees a message m and replaces it with m'. His/Her attack is successful if the fraudulent message is in $M(e)$. Probability of success for the best enemy's strategy for the two types of attack is denoted by P_0 and P_1 respectively.

An A-code is said to provide *perfect protection* if the enemy's best strategy is random selection with uniform distribution from the set of allowable cryptograms. The probability of success in this case can be proved [1, 2, 4, 5] to be k/M and $(k-1)/(M-1)$ for impersonation and substitution respectively. The following theorems give necessary and sufficient conditions for A-codes to provide prefect protection.

Theorem 1 ([5, 4]). *$P_0 \geq k/M$ and equality holds if and only if,*

$$\sum_{\{e_i \in \mathcal{E}: m \in M(e_i)\}} \pi_i = k/M,$$

for every message m.

Let $P_S(e, m)$ be the probability of the source state that is mapped onto m by the encoding rule e.

Theorem 2 ([1, 2, 4]). *In any A-code $P_1 \geq (k-1)/(M-1)$ and equality holds if and only if,*

$$\frac{\sum_{\{e_i: m, m' \in M(e_i)\}} \pi_i P_S(e_i, m)}{\sum_{\{e_i: m \in M(e_i)\}} \pi_i P_S(e, m)} = \frac{k-1}{M-1}$$

for all $m, m' \in \mathcal{M}$, $m \neq m'$.

Stinson characterized A-codes that provide perfect protection for impersonation and substitution and have minimum possible number of encoding rules.

Theorem 3 ([4], theorem 4.1). *Suppose we have an A-code with $P_0 = k/M$ and $P_1 = (k-1)/(M-1)$. Then $E \geq E_0 = (M^2 - M)/(k^2 - k)$ and equality occurs if and only if the incidence matrix A corresponds to that of a $(M, k, 1) - BIBD$, and both the source states and encoding rules are equiprobable.*

In this paper we will be mainly concerned with substitution attack, i.e., when a message m is received. We allow the enemy to choose his/her fraudulent cryptogram m' by one of the following ways;

- use a probability distribution q^m on the set $\mathcal{M}_m = \mathcal{M} \setminus \{m\}$ and choose m' according to this distribution;
- use a probability distribution p^m on the set $E(m)$ of keys that are incident with m to select a key e and then randomly select $m' \in M(e)$.

Enemy's strategy in the first case is completely determined by the probability distributions $q^m, \forall m \in \mathcal{M}$, and in the second case by the distributions $p^m, \forall m \in \mathcal{M}$. We call these, class S_M (for message) and S_K (for key) strategies respectively.

Substitution strategy in Simmons' model of authentication corresponds to class S_M. We will prove that class S_K is a proper subclass of S_M. However its independent investigation will be instrumental in proving our results.

3 New Results for Class $\mathcal{S}_M$

We will use incidence matrix of the A-code to derive an expression for the probability of success, p_1, when the communicants are using a strategy π and the enemy's strategy is given by $\{q^m : m \in \mathcal{M}\}$, where q^m is defined as above.

Proposition 4.

$$p_1 = \sum_{m \in \mathcal{M}} \sum_{m' \in \mathcal{M}_m} \sum_{j=1}^{E} \pi_j a_{jm} a_{jm'} P_S(e_j, m) q^m_{m'}.$$

Proof: Let $P(m)$ denote the probability of message m occurring in the channel,

$$P(m) = \sum_{j=1}^{E} \pi_j a_{jm} P_S(e_j, m).$$

π_j is the probability that the communicants choose e_j and we have $\sum_m P(m) = 1$. The probability of success if m is replaced by m', with the given strategies, is denoted by $p_1(m, m')$ and is equal to,

$$\begin{aligned} p_1(m, m') &= P(m' \text{ valid}|m \text{ received}), \\ &= \sum_j \pi_j a_{jm} a_{jm'} P_S(e_j, m)/P(m), \qquad (1) \\ &= \frac{\sum_{j=1}^{E} \pi_j a_{jm} a_{jm'} P_S(e_j, m)}{\sum_j \pi_j a_{jm} P_S(e_j, m)}. \qquad (2) \end{aligned}$$

This is the same as payoff(m, m') in the game theory model and can be used to find probability of success,

$$\begin{aligned} p_1 &= \sum_{m \in \mathcal{M}} P(m) \sum_{m \neq m'} p_1(m, m') q'^{\,m}_{m} \qquad (3) \\ &= \sum_m \sum_{m' \neq m} \sum_{j=1}^{E} \pi_j a_{jm} a_{jm'} P_S(e_j, m) q'^{\,m}_{m}. \qquad (4) \end{aligned}$$

□

Let Π denote an $M \times M$ matrix such that,

$$\begin{aligned} \Pi_{mm'} = \Pi_{m'm} &= \sum_{j=1}^{E} \pi_j a_{jm} a_{jm'}, \quad m \neq m' \qquad (5) \\ &= 0, \quad m = m'. \end{aligned}$$

Proposition 4 gives an alternative proof for theorem 2.

Proposition 5. *In an A-code $P_1 \geq (k-1)/(M-1)$ and equality holds if and only if $p_1(m, m') = (k-1)/(M-1)$ for all $m, m' \in \mathcal{M}$, $m \neq m'$.*

Proof: We have ,

$$\sum_{m' \in \mathcal{M}_m} p_1(m, m') = k - 1,$$

and so for every m there exists an m_0 with $p_1(m, m_0) \geq (k-1)/(M-1)$. If for every m the enemy uses a distribution given by $q^m_{m_0} = 1$ and zero otherwise, we have,

$$p_1 = \sum_{m \in \mathcal{M}} P(m) \sum_{m_0 \in \mathcal{M}_m} p_1(m, m_0) q^m_{m_0} \geq \frac{k-1}{M-1} \sum_m P(m) = \frac{k-1}{M-1},$$

which proves the first part.
If $P_1 = (k-1)/(M-1)$ then $p_1(m, m_0) = (k-1/(M-1), \forall m, m_0 \in \mathcal{M}, \ m \neq m_0$.

Conversely, let $p_1(m, m_0) = (k-1)/(M-1)$ for all $m, m_0 \in \mathcal{M}, \ m \neq m_0$. Using (3) we have,

$$p_1 = \frac{k-1}{M-1} \sum_{m \in \mathcal{M}} P(m) \sum_{m' \in \mathcal{M}_m} q^m_{m'} = \frac{k-1}{M-1} = P_1.$$

□

If the source is uniform, perfect protection for substitution can be stated as a condition on the matrix Π.

Proposition 6. *For a uniform source $P_1 = (k-1)/(M-1)$ if and only if $\Pi_{mm'} = k(k-1)/(M(M-1)$, for all $m, m' \in \mathcal{M}, \ m \neq m'$.*

Proof: Let,

$$\Pi_{mm'} = \sum_{j=1}^{E} \pi_j a_{jm} a_{jm'} = \frac{k(k-1)}{M(M-1)}.$$

Then from proposition 4,

$$p_1 = \frac{1}{k} \times \frac{k(k-1)}{M(M-1)} \times \sum_{m \in \mathcal{M}} \sum_{m' \in \mathcal{M}_m} q^m_{m'} = \frac{k-1}{M-1} = P_1.$$

Conversely, let $P_1 = (k-1)/(M-1)$. Using uniform source and expression (1) results in,

$$\Pi_{mm'} = kP(m)p(m, m'),$$

which by proposition 5 gives,

$$\Pi_{mm'} = \frac{k(k-1)}{M-1} \times P(m), \ \forall m, m' \neq m.$$

Since Π is symmetric we have,

$$P(m) = P(m'), \ \forall m, m', m' \neq m.$$

That is $P(m) = 1/M$,

$$\Pi_{mm'} = \frac{k(k-1)}{M(M-1)}.$$

□

Proposition 6 is used to prove theorem 7 which is an extension of Massey's result ([1], proposition 2).

Theorem 7. *For a uniform source if $P_1 = (k-1)/(M-1)$ then $P_0 = k/M$.*

Proof: Let $P_1 = (k-1)/(M-1)$. From proposition 6 we have, $\Pi_{mm'} = k(k-1)/M(M-1)$, and so,

$$\sum_{m' \in \mathcal{M}_m} \Pi_{mm'} = (M-1) \times \frac{k(k-1)}{M(M-1)} = \frac{k(k-1)}{M}.$$

Also,

$$\sum_{m' \in \mathcal{M}_m} \Pi_{mm'} = \sum_{m' \in \mathcal{M}_m} \sum_{j=1}^{E} \pi_j a_{jm} a_{jm'} = (k-1) \sum_{j=1}^{E} \pi_j a_{jm}.$$

Hence,

$$\sum_{j=1}^{E} \pi_j a_{jm} = \frac{k}{M}, \ \forall m \in \mathcal{M}.$$

That is $P_0 = k/M$.

□

We will show in example 1 below, that the uniform source is *a necessary* condition in this theorem in the sense that for non-uniform source $P_1 = (k-1)/(M-1)$ does not necessarily imply $P_0 = k/M$.

Theorem 8 gives a necessary and sufficient condition for an A-code to provide perfect protection for substitution and impersonation.

Theorem 8. *An A-code provides perfect protection for impersonation and substitution if and only if,*

$$\sum_{j=1}^{E} \pi_j a_{jm} a_{jm'} P_S(e_j, m) = \frac{k-1}{M(M-1)}$$

for all $m, m' \in \mathcal{M}, m \neq m'$. These requires that cryptograms occur with the same probability in the channel.

Proof:
Necessity: Let

$$\sum_{j=1}^{E} \pi_j a_{jm} a_{jm'} P_S(e_j, m) = \frac{k-1}{M(M-1)}. \tag{6}$$

Then,

$$p_1 = \sum_{m} \sum_{m' \neq m} \sum_{j=1}^{E} \pi_j a_{jm} a_{jm'} P_S(e_j, m) q^{m}_{m'} = \frac{k-1}{M-1} = P_1,$$

and the code provides perfect protection for substitution.

Hence from proposition 4 we have,

$$p_1(m, m') = \frac{\sum_j \pi_j a_{jm} a_{jm'} P_S(e_j, m)}{P(m)} = \frac{k-1}{M-1},$$

which by using (6) gives $P(m) = 1/M$ and the probability distribution on the cryptogram space is uniform. We note that,

$$\sum_{m \in \mathcal{M}} a_{jm} P_S(e_j, m) = 1,$$

and so we can write,

$$\pi_j = \sum_m \pi_j a_{jm} P_S(e_j, m). \tag{7}$$

Since for all m,

$$\begin{aligned}
\sum_j \pi_j a_{jm} &= \sum_{m' \in \mathcal{M}} \sum_j \pi_j a_{jm'} a_{jm} P_S(e_j, m'), \\
&= \sum_{m' \in \mathcal{M}_m} \sum_j \pi_j a_{jm} a_{jm'} P_S(e_j, m') + \sum_j \pi_j a_{jm}^2 P_S(e_j, m), \\
&= \sum_{m' \in \mathcal{M}_m} \frac{k-1}{M(M-1)} + P(m) = \frac{k}{M},
\end{aligned}$$

We have,

$$P_0 = \frac{k}{M}.$$

Sufficiency:
Let $P_1 = (k-1)/(M-1)$ and $P_0 = k/M$. Using proposition 5, for all $m, m' \in \mathcal{M}$, $m \neq m'$,

$$\sum_{j=1}^{E} \pi_j a_{jm} a_{jm'} P_S(j, m) = \frac{k-1}{M-1} P(m).$$

Since $P_0 = k/M$, using (7) results in,

$$\begin{aligned}
P_0 = \sum_{j=1}^{E} \pi_j a_{jm} &= \sum_{j=1}^{E} \sum_{m' \in \mathcal{M}} \pi_j a_{jm'} a_{jm} P_S(e_j, m'), \\
&= \sum_{m' \in \mathcal{M}_m} \sum_{j=1}^{E} \pi_j a_{jm} a_{jm'} P_S(e_j, m') + \sum_{j=1}^{E} \pi_j a_{jm}^2 P_S(e_j, m), \\
&= P(m) + \sum_{m' \in \mathcal{M}_m} \frac{k-1}{M-1} P(m') = \frac{k-1}{M-1}(1 - P(m)) + P(m), \\
&= k/M.
\end{aligned}$$

That is $P(m) = 1/M$. But $P_1 = (k-1)/(M-1)$ implies $p(m, m') = (k-1)/(M-1)$ (proposition 5), which by using (1) gives,

$$\sum_{j=1}^{E} \pi_j a_{jm} a_{jm'} P_S(e_j, m) = \frac{k-1}{M(M-1)}.$$

□

Proposition 9 allows us to give an alternativee proof of Stinson's theorem 3.

Proposition 9. *Let $P_1 = (k-1)/(M-1)$ and $P(m) > 0, \forall m \in \mathcal{M}$. Then $E \geq M(M-1)/(k(k-1))$ and equality is obtained if the incidence matrix of the code is that of a BIBD.*

Proof: Because of perfect protection $P(m'|m) > 0$ and every pair of cryptograms should occur together at least once. So $E \times k(k-1) \geq M(M-1)$ and the result follows.

□

Theorem 3 is readily obtained by combining proposition 9 and theorem 8. This is true because the incidence matrix of an A-code that satisfies both theorem 8 and proposition 9 is that of a BIBD and so $\sum_j a_{jm}a_{jm'} = 1$. This gives,

$$\pi_\ell P_S(e_\ell, m) = \frac{k(k-1)}{M(M-1)} = \frac{1}{kE}. \tag{8}$$

But for a given ℓ, we have $\sum_m P_S(e_\ell, m)a_{\ell m} = 1$ and hence $\pi_\ell = \sum_m \pi_\ell P_S(e_\ell, m)a_{\ell m} = 1/(kE)\sum a_{\ell m} = 1/E$. From (8) it follows that $P_S = 1/k$ which completes the theorem.

It is important to note that $P_0 = k/M$ does not always follow from $P_1 = (k-1)/(M-1)$. In the following example $P_1 = (k-1)/(M-1)$, and $P_0 = k/M$ require uniform distribution on the source. However there exists a communicants strategy that results in $P_1 = (k-1)/(M-1)$ (without $P_0 = k/M$) for a non-uniform source.

Example 1. Let A be an A-code whose encoding matrix S and its corresponding incident matrix A are,

$$S = \begin{bmatrix} s_1 & s_2 & 0 \\ s_2 & 0 & s_1 \\ 0 & s_1 & s_2 \\ s_2 & s_1 & 0 \\ s_1 & 0 & s_2 \end{bmatrix},$$

$$A = \begin{bmatrix} 1 & 1 & 0 \\ 1 & 0 & 1 \\ 0 & 1 & 1 \\ 1 & 1 & 0 \\ 1 & 0 & 1 \end{bmatrix}.$$

Rows of S and A are indexed by keys e_i, $1 \leq i \leq 5$ and columns by m_i, $1 \leq i \leq 3$.

If $P_0 = k/M$ and $P_1 = (k-1)/(M-1)$ then by theorem 8 we have,

$$\sum_{j=1}^{E} \pi_j a_{jm} a_{jm'} P_S(e_j, m) = (k-1)/(M(M-1)) = 1/6, \; \forall m, m' \neq m. \tag{9}$$

Let m_3 be intercepted and m_2 is to be replaced. Then expression 9 reduces to $\pi_3 P_S(e_3, m_3) = 1/6$. Similarly if m_2 is intercepted and m_3 is to be replaced then $\pi_3 P_S(e_3, m_2) = 1/6$ and hence $P_S(e_3, m_2) = P_S(s_1) = P_S(e_3, m_3) = P_S(s_2)$.

But $P_S(s_1) + P_S(s_2) = 1$. Hence $P_S(s_1) = P_S(s_2) = 1/2$ and the code provides perfect protection for impersonation and substitution simultaneously only if the source is uniform. The communicants' strategy given by,

$$\pi_1 = \pi_2 = \pi_4 = \pi_5 = 1/6,\ \pi_3 = 1/3,$$

is a strategy that ensures perfect protection for impersonation and substitution.

Let the source be non-uniform and $P_S(s_1) = 1/3$, $P_S(s_2) = 2/3$. Then the probability distribution,

$$\pi_1 = 1/28, \pi_2 = 4/28, \pi_3 = 9/28, \pi_4 = \pi_5 = 7/28,$$

gives $P_1 = (k-1)/(M-1)$ but $P_0 \neq k/M$. This is because we have, $P(m_1) = 30/84$, $P(m_2) = 18/84$, $P(m_3) = 36/84$ and it is easy to verify,

$$\sum_{j=1}^{E} \pi_j a_{jm} a_{jm'} P_S(e_j, m) = \frac{k-1}{M-1} P(m),\ \forall m, m' \neq m.$$

However in this case the A-code does not provide perfect protection for impersonation because,

$$\begin{aligned}\sum_{j=1}^{E} \pi_j a_{j\ell} &= 19/28,\ \ell = 1,\\ &= 17/28,\ \ell = 2,\\ &= 20/28,\ \ell = 3.\end{aligned}$$

4 Strategies of class S_K

In this class of strategies, enemy chooses a probability distribution p^m on the set $E(m)$ of keys that are incident with the intercepted cryptogram m and chooses a key $e_i \in E(m)$ according to this distribution. Next the enemy selects a cryptogram $m' \in M(e_i)$ randomly (with uniform distribution). This is equivalent to randomly choosing a source state and finding its corresponding cryptogram using key e_i. Probability of success in substiting m by m' is p_1^K (superscript K denotes class S_K strategy).

Proposition 10.

$$p_1^K = \frac{1}{k-1} \sum_{m \in \mathcal{M}} \sum_{i=1}^{E} \sum_{j=1}^{E} \pi_j p_i^m a_{jm} P_S(e_j, m) \sum_{m' \in \mathcal{M}_m} a_{im'} a_{jm'}.$$

Proof: p_1^K is calculated by averaging probability of success when communicants use encoding rule e_j, enemy uses encoding rule e_i and selects cryptogram m' from $M(e_i)$ given m is received. This is,

$$\begin{aligned}&P(\text{ communicants } e_j, \text{ enemy } e_i, \text{ enemy select } m',\\ &\qquad m' \in M(e_i) \bigcap M(e_j)\ |m \text{ intercepted}),\end{aligned}$$

which because of the independence of events reduces to,

$$P(\text{communicants } e_j \mid m) \times p_i^m \times \frac{1}{k-1} \times a_{im'}a_{jm'}.$$

But,

$$P(\text{communicants } e_j \mid m) = \frac{P(\text{communicants } e_j, m)}{P(m)} = \frac{P_S(e_j, m)\pi_j a_{jm}}{P(m)}$$

and so,

$$\begin{aligned} p_1^K &= \frac{1}{k-1} \sum_{m \in \mathcal{M}} P(m) \frac{\sum_{m' \in \mathcal{M}_m} \sum_{i,j=1}^{E} \pi_j p_i^m a_{jm} P_S(e_j, m) a_{im'} a_{jm'}}{P(m)}, \\ &= \frac{1}{k-1} \sum_{m \in \mathcal{M}} \sum_{i,j=1}^{E} \pi_j p_i^m a_{jm} P_S(e_j, m) \sum_{m' \in \mathcal{M}_m} a_{im'} a_{jm'}. \qquad (10) \end{aligned}$$

□

Proposition 11. *Class S_K is a subclass of S_M. That is, for every strategy $\{p^m : m \in \mathcal{M}\}$ of class S_K there is a corresponding strategy $\{q^m : m \in \mathcal{M}\}$ of class S_M which gives the same probability of success.*

Proof: For a strategy of class S_K defined by $\{p^m,\ m \in \mathcal{M}\}$ we construct a strategy of class S_M given by

$$q_{m'}^m = \frac{1}{k-1} \sum_{i=1}^{E} p_i^m a_{im'},\quad m' \in \mathcal{M}_m.$$

Substituting $q_{m'}^m$ in proposition 4 results in 10 and $p_1^K = p_1$.

□

We require some notation which has been used by Stinson [2]. Given any encoding rule e_i and given any $m, m' \in M(e_i)$, define

$$\delta(e', m, m') = \frac{\sum_{j=1}^{E} \pi_j a_{jm} a_{jm'} P_S(e_j, m)}{\pi_i a_{im} a_{im'} P_S(e_i, m)}$$

Then, let $\delta = min(\delta(e_i, m, m'))$. Note that $\delta \geq 1$ and $\delta = 1$ if and only if for any m, m', $m \neq m'$ there is at most one e such that $m, m' \in M(e)$. Let P_1^K denote the probability of success for class S_K when the enemy and communicants use their best strategies.

Theorem 12. *Let $P_0 = k/M$. Then $P_1^K \geq \delta M/(kE)$ and equality holds if the incidence matrix of an A-code has equal number of ones in each column, $\delta(e_i, m, m') = \delta$ and the intersection of any two nonequal columns is equal to 0 or δ.*

Proof: From proposition 10

$$p_1^K = \frac{1}{k-1} \sum_{m \in \mathcal{M}} \sum_{i=1}^{E} \sum_{j=1}^{E} \pi_j p_i^m a_{jm} P_S(e_j, m) \sum_{m' \in \mathcal{M}_m} a_{im'} a_{jm'} \tag{11}$$

$$= \frac{1}{k-1} \sum_{m \in \mathcal{M}} \sum_{m' \in \mathcal{M}_m} \sum_{i=1}^{E} p_i^m a_{im'} \sum_{j=1}^{E} \pi_j a_{jm} a_{jm'} P_S(e_j, m). \tag{12}$$

Using definition of $\delta(e_i, m, m')$ and δ we have

$$\sum_{j=1}^{E} \pi_j a_{jm} a_{jm'} P_S(e_j, m) = \delta(e_i, m, m') \pi_i a_{im} a_{im'} P_S(e_i, m)$$

$$\geq \delta \pi_i a_{im} a_{im'} P_S(e_i, m)$$

Hence,

$$p_1^K \geq \delta \frac{1}{k-1} \sum_{m \in \mathcal{M}} \sum_{m' \in \mathcal{M}_m} \sum_{i=1}^{E} p_i^m a_{im'} \pi_i a_{im} a_{im'} P_S(e_i, m) \tag{13}$$

A possible enemy's strategy is,

$$p_i^m = \frac{\pi_i a_{im}}{\sum_{j=1}^{E} \pi_j a_{jm}}, \ i \in E(m).$$

This is a valid strategy because,

$$p_i^m > 0, \ i \in E(m), \ \text{and} \ \sum_{e_i \in E(m)} p_i^m = \sum_{i=1}^{E} p_i^m a_{im} = 1.$$

Using this strategy in (13) and noting that, $P_0 = k/M$ implies $\sum_j \pi_j a_{jm} = k/M, \ m \in \mathcal{M}$, we have,

$$P_1^K \geq \delta \frac{1}{k-1} \sum_{m \in \mathcal{M}} \sum_{m' \in \mathcal{M}_m} \sum_{i=1}^{E} p_i^m a_{im'} \pi_i a_{im} a_{im'} P_S(e_i, m),$$

$$\geq \delta \frac{M}{k(k-1)} \sum_{i=1}^{E} \sum_{m \in \mathcal{M}} \sum_{m' \in \mathcal{M}_m} \pi_i^2 a_{im} a_{im'} P_S(e_i, m),$$

where we have used $P_1^K \geq p_1^K$. This means,

$$P_1^K \geq \delta \frac{M}{k} \sum_{i=1}^{E} \pi_i^2 \geq \delta \frac{M}{kE}. \tag{14}$$

The second inequality is obtained because $\sum_{i=1}^{E} \pi_i^2$ is a concave function and its minimum is obtained for $\pi_i = 1/E, \ i = 1, 2, ..., E$. Equality in (14) implies the communicants best strategy is uniform, $\delta(e_i, m, m') = \delta$ which means that product of two unequal columns in the incidence matrix A is equal to 0 or δ. It is easy to see that uniform strategy for communicants and $P_0 = k/M$ implies that the columns of the incidence matrix A have equal number of ones which is equal to kE/M.

□

Combining theorem 12 and proposition 4 we have the following theorem.

Theorem 13. *In an A-code if $P_0 = k/M$ we have*

$$P_1 \geq max(\frac{k-1}{M-1}, \frac{M}{kE}).$$

5 Conclusion

We have studied the best possible performance of an A-code under substitution attack and its relation to the protection provided by the code for impersonation. We have proved that codes with best protection for impersonation, at best, limit probability of success of the enemy in substitution to one of the two distinct values depending on the number of encoding rules. If this number is higher than a threshold E_0 the code may provide perfect protection for substitution but for less encoding rules the best protection offered by the code is $M/(kE)$ and it is impossible to obtain perfect protection. We have derived a necessary and sufficient condition for A-codes that have perfect protection for impersonation and substitution simultaneously and have used it to give an alternative proof of Stinson's characterization theorem. We have also given a generalization of Massey's result for code with perfect protection for substitution.

References

1. J.L. Massey, *Cryptography, a selective survey*, Digital Communications, ed.E. Biglieri and G. Pratti, Elsvier Science Publ., North-Holland, (1986), 3-25.
2. D.R. Stinson, *Some constructions and bounds for authentication codes*, Journal Cryptology **1**, (1988), 37-51
3. D.R. Stinson, *The combinatorics of authentication and secrecy codes* , Journal Cryptology **2**, (1990), 23-49
4. D.R. Stinson, *Combinatorial characterization of authentication codes* , Proceedings Crypto 91, Lecture Notes in Computer Science **576**, (1992), 62-72
5. G.J. Simmons, *Message authentication: a game on gypergraphs*, Congressus Numerantium **45**, (1984), 161-192
6. G.J. Simmons, *Authentication theory/coding Theory*, Proc. of Crypto '84, Lect. Notes in Comp. Science **196**, (1985), 411-432
7. G.J. Simmons, *A game theory model of digital message authentication*, Congressus Numerantium **34**, (1982), 413-424
8. E.Brickell, *A Few Results in Message Authentication* ,Congressus Numerantium **43**, (1984), 141-154

Practical Proven Secure Authentication with Arbitration

Yvo Desmedt[1* **] and Jennifer Seberry[2*** †]

[1] EE & CS Department, University of Wisconsin–Milwaukee, WI 53201, U.S.A.

[2] Department of Computer Science, The University of Wollongong Wollongong, NSW, 2522, Australia

Abstract. Proven secure signature schemes and unconditionally secure authentication schemes with arbiter have been proposed. The former are not practical (too slow) and the latter cannot be reused. All these limitations are solved in this paper by presenting a resuable conditionally secure authentication scheme with arbiter. The scheme is unconditionally secure against denial by the sender of having sent a message (which signatures do *not* have) and conditionally secure against a receiver impersonating the sender or substituting a message and conditionally secure against a similar fraud by the arbiter.

1 Introduction

One can make a proven secure signature scheme [9, 11] based on any one way function. Unfortunately all proven secure signature schemes [7, 9, 1, 11, 2] are very impractical (to make some of them more practical the authentication tree could be used instead of pseudo random functions but this approach requires a lot of memory). So from a practical viewpoint it could be advantageous to use symmetric authentication schemes, however one then loses the signature property. In the classical notion of arbiter [8, p. 409] the arbiter has to be active when messages are transmitted.

Simmons [14] introduced unconditionally secure authentication schemes with arbitration. From a functional viewpoint the arbiter is not active, in Simmons'

* Part of this work was done while he was visiting professor at University of New South Wales, Department of Computer Science, ADFA, Australia, part while he was visiting the Center for Communication and Information Science, University of Nebraska–Lincoln and part while visiting the Department of Computer Science at the University of Wollongong, Australia.

** A part of this work has been supported by NSF Grant NCR-9106327 and Telecom Project 7027, Australia, 1991.

*** Written while on faculty of the Department of Electrical Engineering and the Department of Computer Science and Engineering, and Head of the Center for Communication and Information Science, University of Nebraska–Lincoln, NE 68588, USA.

† Research funded by Telecom grant 7027, ARC grant A49130102 and an ATERB grant.

scheme, during the transmission of the authenticated message while in the classical notion the arbiter must be active. Desmedt and Yung [6] (see also Brickell and Stinson [4]) improved Simmons' scheme by protecting the receiver against an impersonation (substitution) attack by the arbiter. Unfortunately all these schemes can only be used once (because otherwise they lose their security) and hence new keys have to be distributed for each new message as in a one time pad.

The purpose of this paper is to develop a practical proven secure conditional authentication scheme with arbitration. Our scheme has some similarities with [10], however our scheme is non-interactive and the keys can be re-used.

2 Definitions

Let us call S the sender, R the receiver, A the arbiter, and O the outside opponent. We can distinguish three stages in Simmons' solution [14]. The three stages are:

The key initialization phase in which S, R and A interact to come up with the necessary keys.

The authentication phase in which R receives a message and wants to ascertain that the message is authentic. A does *not* interact in this stage.

The dispute phase in which A is requested to resolve a dispute between S and R. Using some information gathered by A during the initialization phase A solves the dispute.

Our scheme contains these three stages as well.

Let us describe more precisely the threats with which we are faced. We follow closely Simmons's description of such threats (for the first three threats see [14]).

The outside opponent. The outsider, O, can try to impersonate the sender and/or substitute some message(s) for one sent from S to R, but which O has intercepted (actively eavesdropped).
The attack is said to be successful if and only if R accepts the message as authentic when it is not.

The sender. A dishonest $\tilde{S}$ can attempt to cheat by sending a message which R will accept as authentic, but which he can later deny having sent.
The attack is successful if and only if the following two conditions hold. First, R accepts the fraudulent message, and second, in a dispute A will decide that the message is *not* authentic.

The receiver. A dishonest $\tilde{R}$ can falsely claim to have received the message M from S. Two subcases can be distinguished: $\tilde{R}$ never received a message at all, or $\tilde{R}$ has received some authentic message(s) from S which he tries to alter.
The attack is successful if and only if in a dispute A certifies the message as being authentic.

The arbiter. A dishonest $\tilde{A}$ can send a message to R which R will accept as authentic. As in the case of the opponent's attack the arbiter can either choose an impersonation or a substitution attack.
The attack is successful if and only if the message originating at $\tilde{A}$ will be accepted by R.

We remark that it is not A's task *to force* R to accept messages originating from S.

The reader who is interested in formalizing the above informal definitions is referred to [6]. Although these definitions have been given for unconditionally secure schemes, they can very easily be adapted for conditionally secure ones.

3 The Scheme

We use S for the sender, R for the receiver and A for the arbiter. We assume the existence of a conditionally proven secure authentication scheme. When we mention keys we assume that these (symmetric) keys were chosen according to a prescribed algorithm and belong to the set K.

3.1 Distribution Phase

Step 1 A sends S an ordered tuple $(k_1, k_2, \ldots, k_n)$ of random, independently chosen, keys privately.

Step 2 A chooses with uniform probability distribution a random subset, I, of $\lfloor n/2 \rfloor$ indices between 1 and n and privately sends to R the tuple $(k'_1, k'_2, \ldots, k'_n)$ where $k'_i = k_i$ if $i \in I$, otherwise $k'_i = \epsilon$, where $\epsilon \notin K$, (for example ϵ may be the empty string).

Step 3 S privately sends a key, k_{n+1}, to R.

3.2 Authentication Phase

To send a message M the sender S forms $n+1$ message authentication codes ($MACs$) by processing the message with each of the $n+1$ keys, n provided by the arbiter and one by himself, using a proven secure authentication scheme. Call these $MACs$ $MAC_1, MAC_2, \ldots, MAC_{n+1}$. The sender sends $(M, MAC_1, MAC_2, \ldots, MAC_{n+1})$ where MAC_i is generated using the key k_i, the message M and the agreed authentication algorithm to R.

To verify whether R should accept M as (probably) being authentic R proceeds as follows: if $k'_i \neq \epsilon$ then R checks that MAC_i is correct, and does this for all i, $1 \leq i \leq n$, and additionally checks if MAC_{n+1} also matches. If these $\lfloor n/2 \rfloor + 1$ $MACs$ are correct R accepts M as authentic, otherwise R rejects. In the case R rejects R erases his keys and requests new keys unless all the $MACs$ were wrong.

3.3 Dispute Phase

If a dispute occurs the receiver presents $(M, MAC_1, MAC_2, \dots, MAC_n)$ to the arbiter. The arbiter will accept the message plus the MACs as correct if among $(M, MAC_1, MAC_2, \dots, MAC_n)$ all those MACs that R should have known were correct are indeed correct plus at least one more MAC is correct.

4 Proof of Security

We will use h as a security parameter so that the complexity of performing an attack on the underlying authentication scheme is bounded above by $1/p(h)$ where p is any polynomial. We assume S receives feedback from R whether he has accepted the message M or not.

Theorem 1. *Let n in our scheme be chosen linear in h the security parameter. Now if a conditionally proven secure authentication scheme exists then our scheme is secure against a denial attack by the sender, conditionally secure against an attack in which the receiver, the arbiter or an outsider attempts to modify the message or impersonate the sender.*

Proof. The receiver's attack will not succeed as he does not know enough keys and the authentication scheme was assumed to be secure. A similar proof holds for the arbiter's and an outsider's attacks. We now consider denial by the sender. We do not consider MAC_{n+1} as this was used only to protect against the arbiter.

If the sender wishes to have a false message accepted and then deny sending the message he optimizes his chance of winning by adopting a game plan. He wants R *to accept* and A *to reject.* Now if S has sent i $\{i : 0, \dots, \lfloor (n-2)/2 \rfloor\}$ correct MACs and $n-i$ incorrect MACs then R will reject the message and so S *loses.* If S sends i $\{i : \lfloor (n+2)/2 \rfloor, \dots, n\}$ correct MACs and $n-i$ incorrect MACs then if R accepts the message as genuine then so will A and again S *loses.* If S sends $\lfloor n/2 \rfloor$ correct MACs and $n - \lfloor n/2 \rfloor$ incorrect MACs, then S can *win* if he has chosen exactly the $\lfloor n/2 \rfloor$ MACs that R has, R will accept but the arbiter will reject. There are exactly

$$\binom{n}{\lfloor n/2 \rfloor}$$

ways of choosing subsets of the indices of the MACs with $\lfloor n/2 \rfloor$ elements .

Our assumption that S receives feedback from R whether a message M was accepted or not implies S can *win next time* if he guesses all the indices of the keys, $k'_j = \epsilon$, and sends the MAC_js of these $n - \lfloor n/2 \rfloor$ keys correctly and all other MACs incorrectly. In this case R will reject the message but not erase his keys (so R will not ask for new keys). The probability of this succeeding without detection is also

$$\frac{1}{\binom{n}{\lfloor n/2 \rfloor}} .$$

So the probability is negligible of an attack succeeding (even if repeated[3] polynomially many times). □

5 Conclusions

We have presented an authentication scheme with arbiter which is unconditionally secure against denial by the sender of having sent a message and conditionally secure against a receiver impersonating the sender or substituting a message and conditionally secure against a similar fraud by the arbiter.

The security obtained is the same as for the symmetric authentication scheme on which it is based. We observe that making practical proven secure authentication schemes is easy to achieve starting from pseudo-noise generators [12, 13] and unconditionally secure authentication schemes [5].

It is clear that the scheme presented in Section 3, can be adapted for DES. We remind the reader that DES is not a proven secure scheme and that some weaknesses have been found in the protocol for generating *MAC*s [3].

Acknowledgement

The authors thank Bart Preneel of the University of Louvain, Belgium for bringing [10] to their attention.

References

1. Bellare, M., Goldwasser, S.: New paradigms for digital signatures and message authentication based on non-interactive zero-knowledge proofs. In Advances in Cryptology — Crypto '89, Proceedings (Lecture Notes in Computer Science 435) (1990) G. Brassard, Ed. Springer-Verlag pp. 194–211
2. Bellare, M., Micali, S.: How to sign given any trapdoor function. Journal of the ACM **39** (1992) 214–233
3. Bird, R., Gopal, I., A.Herzberg, Jansen, P., Kutten, S., Molva, R., Yung, M.: Systematic design of two-party authentication protocols. In Advances in Cryptology — Crypto '91, Proceedings (Lecture Notes in Computer Science 576) (1992) J. Feigenbaum, Ed. Springer-Verlag pp. 44–61

[3] For the first type of attack, if the sender successfully modified i_1 *MAC*s the first time, i_2 different *MAC*s the second time and so on. Then observing

$$\frac{\binom{n-\lfloor n/2\rfloor}{i_1}}{\binom{n}{i_1}} \cdot \frac{\binom{n-\lfloor n/2\rfloor - i_1}{i_2}}{\binom{n-i_1}{i_2}} = \frac{\binom{n-\lfloor n/2\rfloor}{i_1+i_2}}{\binom{n}{i_1+i_2}}$$

the numerate reader can show that the probability of successful attack remains the same as above. The same can be said for the second type of attack or a combination of both types.

4. Brickell, E. F., Stinson, D. R.: Authentication codes with multiple arbiters. In Advances in Cryptology, Proc. of Eurocrypt '88 (Lecture Notes in Computer Science 330) (May 1988) C. G. Günther, Ed. Springer-Verlag pp. 51–55
5. den Boer, B.: A simple and key-economical authentication scheme, March 30–April 3, 1992. Presented at System Security, Dagstuhl, Germany
6. Desmedt, Y., Yung, M.: Arbitrated unconditionally secure authentication can be unconditionally protected against arbiter's attacks. In Advances in Cryptology — Crypto '90, Proceedings (Lecture Notes in Computer Science 537) (1991) A. J. Menezes and S. A. Vanstone, Eds. Springer-Verlag pp. 177–188
7. Goldwasser, S., Micali, S., Rivest, R.: A digital signature scheme secure against adaptive chosen-message attacks. Siam J. Comput. **17** (1988) 281–308
8. Meyer, C. H., Matyas, S. M.: Cryptography: A New Dimension in Computer Data Security. J. Wiley New York 1982
9. Naor, M., Yung, M.: Universal one-way hash functions and their cryptographic applications. In Proceedings of the twenty first annual ACM Symp. Theory of Computing, STOC (May 15–17, 1989) pp. 33–43
10. Rabin, M. O.: Digitized signatures. In Foundations of Secure Computation (New York, 1978) R. A. DeMillo, D. P. Dobkin, A. K. Jones, and R. J. Lipton, Eds. Academic Press pp. 155–168
11. Rompel, J.: One-way functions are necessary and sufficient for secure signatures. In Proceedings of the twenty second annual ACM Symp. Theory of Computing, STOC (May 14–16, 1990) pp. 387–394
12. Rueppel, R. A.: Stream ciphers. In Contemporary Cryptology, G. J. Simmons, Ed. IEEE Press 1992 pp. 65–134
13. Schrift, A. W., Shamir, A.: The discrete log is very discreet. In Proceedings of the twenty second annual ACM Symp. Theory of Computing, STOC (May 14–16, 1990) pp. 405–415
14. Simmons, G. J.: A Cartesian product construction for unconditionally secure authentication codes that permit arbitration. Journal of Cryptology **2** (1990) 77–104

Session 2

AUTHENTICATION AND SECRET SHARING II

Chair: Josef Pieprzyk
(University of Wollongong, Australia)

Authentication Codes under Impersonation Attack

R. Safavi-Naini *
L. Tombak **

Department of Computer Science, University of Wollongong
Northfields Ave, Wollongong 2522, AUSTRALIA

Abstract. Performance of authentication codes under impersonation attack is considered. We assume two classes of strategies for the enemy and give expressions for the probability of success in each case and study the relationship of the two classes. We prove a new lower bound on the probability of deception which points out the importance of the average distance between the encoding rules of the code in the protection provided by it. Codes with perfect protection for each class are characterized and some constructions based on error correcting codes are proposed.

1 Introduction

Authentication codes (A-codes) provide protection against an active spoofer whose aim is to modify the message or fabricate an acceptable message. An A-code is a collection of mappings, indexed by keys, from the set of source states into the set of cryptogram space. The communicants secretly choose a key, hence specifying the set of authentic cryptograms for their communication. The enemy, uncertain about the key, wants to choose an authentic cryptogram and deceive the receiver. A *perfect authentication system*, as defined by Simmons [2], is optimum in the use of the uncertainty introduced in the encoding process, i.e., is a system in which the probability of enemy's success is equal to the uncertainty due to the encoding. An A-code provides *perfect protection* if it forces the enemy to random selection. Characterization and construction of perfect systems and perfect protection codes are studied by a number of authors including [4, 7, 9, 6]. Most of these results are based on the model of system proposed by Simmons which specifies communicants' and enemy's action. For example in *impersonation attack* the enemy's action is to select a cryptogram using a probability distribution. However it is natural to consider other possible actions for the enemy, such as trying to guess the correct key. This is especially appealing if the size of key space is smaller than the cryptogram space, and the question remains as to whether he/she gains anything by doing this. In this paper we look at this two possible enemy's actions when the attack is limited to impersonation.

* Support for this work was provided in part by Australian Research Council grant A49030136.

** Support for this work was provided by Australian Research Council grant A49030136.

Our concern will be determining the best strategies of the communicants and the enemy in each case, finding their relationship and characterizing codes that result in total uncertainty of the enemy in his/her attack.

We assume the communicants have a strategy which is a probability distribution on the key space and use it to select the key for their communication. The enemy might take two types of action: choosing a cryptogram from cryptogram space using a probability distribution on the cryptograms (class $\mathcal{M}$ strategy) or selecting a key, using a probability distribution on the key space (class $\mathcal{K}$ strategy). The probability distribution in each case is the enemy's strategy. We will derive expressions for the probability of deception in terms of the incidence matrix of the code and the strategies of the communicants and enemy which results in lower bounds on the probability of deception and shows that for both types of action finding the best strategies amounts to solving a linear programming problem whose solution in practice becomes unwieldy. We prove that the enemy's best strategy always belong to class $\mathcal{M}$ but if his/her best strategy in this class is random selection, the same lower bound on the probability of deception is obtained by random strategy of class $\mathcal{K}$. We show that class $\mathcal{K}$ is properly contained in class $\mathcal{M}$ but study of the code under class $\mathcal{K}$ strategies is illuminating as it points out the importance of the average distance between the encoding rules in the performance of an A-code. Class $\mathcal{K}$ is also practically important in cases when the enemy cannot easily have access to all possible cryptograms but he/she can generate a cryptogram which is valid under an assumed key. This average distance can be regarded as a measure of protection provided by the authentication codes the same way that minimum distance of an error correcting codes indicates the error correcting/detecting capability of such codes. This becomes more evident if either the enemy or the communicant use random strategy. It can be shown that in this case the code with maximum average distance are the ones with constant weight column. Moreover if communicants' strategy is random lower bound on the probability of success for class $\mathcal{K}$ will be the same as class $\mathcal{M}$. In randomly selecting a key we expect to have a chance of success inversely proportional to the size of key space but this result shows that for random strategy of communicants this bound is the same as that obtained by random selection from cryptogram space.

For each class of strategy, codes that provide perfect protection are the ones that force the best strategy of the enemy to be random selection. We propose constructions for codes that provide perfect protection of each class using error correcting codes but characterization of incidence matrices of such codes remain an open problem. In the next section we provide the required definitions for the rest of the paper. Sections 3, 4, and 5 contain the main results while the section 6 gives some concluding remarks.

2 Preliminaries

We consider an authentication scenario with three participants: a transmitter and receiver (*communicants*) who want to communicate over a publicly exposed channel and an *enemy* who tries to deceive the receiver in accepting a fraudulent message as genuine. We are only concerned with honest communicants. An *authentication code* (A-code) is a collection $\mathcal{E}$, $|\mathcal{E}| = E$ of mappings from the set $\mathcal{S}$, $|\mathcal{S}| = k$, of the source states into the set $\mathcal{M}$, $|\mathcal{M}| = M$, of codewords. We use $M(e)$ to denote the subset of codewords that are authentic under the key $e \in \mathcal{E}$ and $E(m)$ to denote the subset of encoding rules that are incident with $m \in \mathcal{M}$. The code provides protection only if $k < M$. The *incidence matrix* of an A-code is a zero-one matrix of size $E \times M$ in which $A_{em} = 1$ only if $m \in M(e)$. A has exactly k ones in each row. *Hamming weight* of a binary vector e, denoted by $wt(e)$, is defined as the number of nonzero coordinates of e. *Hamming distance* between two binary vectors e_1, e_2 of a binary vector space, denoted by $d(e_1, e_2)$, is equal to $wt(e_1 + e_2)$ where addition is bitwise and binary. The distance between two encoding rules is the Hamming distance between the corresponding rows of A.

The communicants use a probability distribution $\pi = (\pi_1, \cdots, \pi_E)$ on the key space as their *strategy* and choose an encoding rule e using the distribution. In an *impersonation attack*, the enemy's aim is to construct a codeword in $M(e)$. Simmons [3] gave a game theoretic formulation of this scenario in which the objective of the communicants is to minimize the value of the game and showed that

$$P_d \geq 2^{-I(M;E)}, \tag{1}$$

where P_d is the probability of deception when the communicants and the enemy use their best strategies.

An authentication system is called *perfect* if it satisfies bound (1) with equality. It can be seen that perfectness is a property of the system and depends on the source, A-code and the communicants' strategy. Codes with *perfect protection of zero order* are those for which the best strategy of the enemy in impersonation attack is random selection from cryptogram space [1] and this results in the best probability of deception to be equal to k/M. Perfect protection for impersonation is independent of the source and as shown by Stinson [8] depends only on the strategy of the communicants and the A-code. In fact protection properties of A-code for impersonation is solely determined by the incidence matrix of the A-code and the communicants' strategy. Using this matrix the best strategy of the communicants can be determined which ensures the minimum probability of success for the enemy (independent of the source). When the communicants' strategy is random protection properties of the code will depend only on the structure of the incidence matrix.

3 Bound on the probability of success in impersonation

We consider two types of possible action by the enemy in an impersonation attack and in each case calculate probability of success of the enemy. We assume the strategy of the communicants is given by a probability distribution π on $\mathcal{E}$.

3.1 Class $\mathcal{K}$

In class $\mathcal{K}$ the enemy chooses a probability distribution p on $\mathcal{E}$ and selects an encoding rule e_i using p. Next the enemy randomly chooses an element of $m_k \in M(e_i)$ for impersonation.

Theorem 1. *Let A, p and π be defined as above and P_I denote the best probability of success in impersonation. Then we have the following equivalent bounds:*

$$P_I \geq \frac{1}{k} pAA^t\pi; \tag{2}$$

$$P_I \geq \frac{1}{k} \sum_{j,k} p_j \pi_k (e_j . e_k); \tag{3}$$

$$P_I \geq 1 - (1/2k) \sum_{l=0}^{M} lB_l; \tag{4}$$

where

$$B_l = \sum_{\substack{e_i + e_j = t, \\ wt(t) = l}} p_i \pi_j,$$

and $e_j.e_k$ denotes the inner product of the vectors e_i and e_j. The equality is obtained if and only if the code provide perfect protection for class $\mathcal{M}$.

Proof: Let the enemy choose e_j (using his/her strategy) and $P^K_{i,j}$ denote his/her probability of success in this case (superscript shows class $\mathcal{K}$ is used). $P^K_{i,j}$ is the probability that $m_k \in M(e_j) \bigcap M(e_i)$, i.e., the message chosen by the enemy using rule e_j, is in $M(e_i)$. Let $d_{i,j}$ denote the Hamming distance between e_i and e_j, then

$$P^K_{i,j} = \frac{k - d_{i,j}/2}{k}.$$

Finding the average of this value gives the probability of success,

$$P^K_I = \sum_{i,j} p_j \pi_i P^K_{i,j}. \tag{5}$$

It can be verified that

$$e_i.e_j = [AA^t]_{i,j} = k - d_{i,j}/2,$$

and hence

$$P^K_I = 1/k(pAA^t\pi^t), \tag{6}$$

$$P^K_I = 1/k \sum_{i,j} p_i \pi_j e_i.e_j. \tag{7}$$

To prove bound (4) we note that the equation (5) can be written as

$$P_I^K = \sum_{i,j} p_j \pi_i (1 - d_{ij}/2k),$$

$$= 1 - (1/2k) \sum_{l=0}^{M} l B_l. \quad (8)$$

where equation (8) is obtained by grouping all the encoding rules that have the same inner product (same Hamming distance).

In all cases the bound follows because $P_I \geq P_I^K$. Equality in (2), (3), (4) is obtained if and only if the code provide perfect protection in class $\mathcal{M}$. This will be shown in corollary 6. □

Bounds (3) states that the probability of deception is lower bounded by the average inner product of the encoding rules which is written in terms of the average distance between the encoding rules in 4. The expression $\sum_{l=0}^{M} l B_l$ can be regarded as a generalized average distance when compared to error correcting codes.

Simmons [2] pointed out an interesting duality between design criterion of error correcting codes and A-codes: the objective of an A-code is to spread evenly the acceptable substitutes of a codeword and in error correcting code the aim is to cluster the most likely substitutes. Bound 4 suggests another duality between these codes, that is, *good A-codes should have high average distance,* while good error correcting codes have high minimum distance.

Generalized average distance depends on the strategy of the communicants and the enemy. For a given communicant strategy, enemy's objective is to choose a strategy that minimizes this average and so the communicants best strategy is obtained by finding π that maximizes this minimum. When the enemy and communicants use random strategies, this average distance coincides with the one given by MacWilliams et al. [5] for error correcting codes.

We use expression (7) to find the best strategy of class $\mathcal{K}$ for the communicants and/or the enemy.

Theorem 2. *The best class $\mathcal{K}$ strategy of the communicants (or the enemy) for a given code A can be obtained by solving a linear programming problem.*

Proof: We have

$$P_I^K = \frac{1}{k} p A A^t \pi^t.$$

If the enemy knows π, then his/her best strategy is to calculate $b = AA^t\pi^t$, find j corresponding to the maximum component of b and choose p with $p_j = 1$ and zero elsewhere.

Assuming the enemy uses his/her best strategy, the communicants best strategy is to find a distribution π for which the maximum element of $AA^t\pi^t$ is minimum. So finding the communicants' best strategy is by solving the following minimax problem:

$$\text{minimize over }_{\pi}\text{ (maximum over }_{j})\, b_j,\ \ 1 \leq j \leq E,$$
$$\sum_i \pi_i = 1 \qquad , \pi_j \geq 0\ .$$

which is equivalent to the following linear programming problem

$$\text{minimize } x$$
$$\text{subject to } b_i \leq x, \quad 1 \leq i \leq E,$$
$$\sum_i \pi_i = 1, \quad \pi_i \geq 0, 1 \leq i \leq E.$$

which proves the theorem.

We note that, for a given code, if the enemy only knows the fact that the communicants will use their best strategy, without actually knowing what the strategy is, then he/she can solve the same linear programming problem to find the communicants strategy and then calculate his/her best strategy. □

This can be interpreted geometrically. Columns of AA^t are vectors of an E dimensional vector space and $b = AA^t\pi^t$ is a vector in the convex region $\mathcal{B}$ defined by them. The enemy is in the worst situation if all components of b are equal in which case the best strategy of the enemy is random selection which gives him a probability of success equal to the common value of b's components. In section 4 we will show that if π is uniform then this common value, which is equal to the probability of success, is at least k/M.

3.2 Class $\mathcal{M}$

We define a second type of action in which the enemy uses a probability distribution on the set of possible cryptograms and chooses the fake message, to impersonate the transmitter, according to this distribution. Enemy's best chance of success is denoted by P_I^M. Class $\mathcal{M}$ strategies correspond to what is considered as enemy's strategies by other authors [3, 8, 7, 6] and hence $P_I^M = P_I$.

Proposition 3. *The best strategy of communicants for class $\mathcal{M}$ is obtained by solving a linear programming problem.*

Proof:
Probability of a message m_l being valid is

$$P(m_l \text{ valid}) = \sum_{i \in E(m_l)} \pi_i$$

which gives the probability of success for class $\mathcal{M}$

$$P_I^M = \sum_l q_l \sum_{i \in E(m_l)} \pi_i = q(\pi A)^t, \tag{9}$$

where q is the enemy's strategy. Again if the enemy knows π, his/her best strategy is to find the j for which $P(m_j \text{ valid}) = \sum_{i=1}^{E} \pi_i a_{ij}$ is maximum and choose m_j for impersonation.

The communicants best strategy is to choose the distribution π for which this maximum is smallest. So the communicants must solve the following problem to find their best strategies:

$$\text{minimize}_{\text{ over } \pi} \text{ (maximum}_{\text{ over } j}) \ P(m_i \text{ valid}), \quad 1 \leq i \leq E$$
$$\sum_i \pi_i = 1, \qquad \pi_j \geq 0.$$

which can be written as a linear programming problem. □

Corollary 4. $P_I \geq k/M$.

Proof: $P_I \geq P_I^M \geq k/M$ where the last inequality is obtained when the enemy uses the random distribution on cryptogram space. □

The two types of enemy's actions are related as the following proposition shows.

Proposition 5. *Class $\mathcal{K}$ is properly contained in $\mathcal{M}$. That is, for a given strategy p of $\mathcal{K}$, there exists a corresponding strategy q of $\mathcal{M}$ with the same probability of success.*

Proof: Let p be an enemy strategy for class $\mathcal{K}$. Now $q = pA/k$ is a strategy of class $\mathcal{M}$ with the probability of deception given by

$$P_I^M = q(\pi A)^t = \frac{1}{k} pA(\pi A)^t = P_I^K.$$

□

It is interesting to note that if the best strategy of the enemy in class $\mathcal{M}$ is random selection from cryptogram space, then his/her best strategy for class $\mathcal{K}$ is random selection from key space. This is true because if there exist a π for which πA is a vector of equal components then πAA^t has equal components too. However the inverse, in general, is not true as the following example shows.

Example 1. Let

$$A = \begin{bmatrix} 1\,1\,0\,1 \\ 1\,1\,1\,0 \\ 1\,0\,1\,1 \end{bmatrix}.$$

Then

$$A.A^t = \begin{bmatrix} 3\,2\,2 \\ 2\,3\,2 \\ 2\,2\,3 \end{bmatrix}$$

and $\pi = [1/3, 1/3, 1/3]$ gives a vector $AA^t\pi$ with equal components but πA does not have this property.

In fact this is the only case that the enemy's best strategy can be obtained from class $\mathcal{K}$ because in all other cases the best strategy of class $\mathcal{M}$ is a vector with exactly one nonzero component (equal to one).

Corollary 6. *In an A-code that provides perfect protection against impersonation the enemy's probability of success will remain the same if he/she uses random selection of class $\mathcal{K}$ and this is the only case that the enemy's best strategy belong to class $\mathcal{K}$.*

4 Random Strategies

As noted in the previous section, finding the best communicants' strategy in practice is computationally expensive. This makes random strategy a good candidate for many applications. Using this strategy for the communicants simplifies expressions (2),(3) and (4) and allows us to relate performance of the authentication system directly to the structure of the incidence matrix.

Proposition 7. *The probability of success for class $\mathcal{K}$ is bounded as*

$$1-\frac{d_{max}(E-1)}{kE} \leq P_I^K \leq 1-\frac{d_{min}(E-1)}{kE}, \tag{10}$$

where $2d_{min}$ and $2d_{max}$ are the minimum and maximum distance between the encoding rules. The equality is obtained when the codewords are equidistant.

Proof: From (7) we have

$$P_I^K = \frac{1}{k}\sum_{\substack{j,k \\ i\neq j}} p_j\pi_k \sum_u e_{ju}e_{ku} + k\sum_j p_j\pi_j. \tag{11}$$

If the minimum distance between the encoding rules is $2d$ we have

$$\sum_u e_{ju}e_{ku} \leq k - d_{min},$$

which gives

$$P_I^K \leq 1/k[(k-d_{min})\sum_{\substack{j,k \\ i\neq j}} p_j\pi_k + k\sum_j p_j\pi_j],$$

and

$$P_I^K \leq 1-\frac{d_{min}(E-1)}{kE}. \tag{12}$$

Similar argument results in the lower bound in (10). □

Hence increasing minimum distance will decrease probability of success for strategies of class $\mathcal{K}$.

Johnson's bound gives the maximum number of binary vectors of weight k which are at least d apart [5, page 525]. Using this bound we have the following upper bound on the minimum distance ($2d_{min}$) of a code,

$$d_{min} \leq \frac{Ek(M-k)}{M(E-1)}. \tag{13}$$

Equality is obtained if columns of A have the same number of ones (kE/M).

For an A-code whose minimum distance satisfy (13) with equality the upper bound in (10) will be k/M., i.e.,

$$1 - \frac{d_{max}(E-1)}{kE} \leq P_I^K \leq k/M,$$

which means that the best strategy of class $\mathcal{K}$ is at most as good as random selection from cryptogram space and gives a probability of deception equal to k/M. We note that using the maximum value of the minimum distance results in the weakest upperbound. We will see in theorem 9 that this upperbound coincides with the lower bound of class $\mathcal{K}$ for codes with perfect protection. In other cases the upperbound gives the best chance of success for class $\mathcal{K}$ strategies.

Corollary 8. *If the distance between the encoding rules is constant and equal to*

$$2d = \frac{kE(M-k)}{M(E-1)},$$

the probability of deception (using strategies of class $\mathcal{K}$) is k/M.

Example 2. The following code has the maximum possible distance $d_{min} = 6$:

$$A = \begin{bmatrix} 1\,1\,1\,1\,0\,0\,0\,0\,0 \\ 1\,0\,0\,0\,1\,1\,1\,0\,0 \\ 0\,1\,0\,0\,1\,0\,0\,1\,1 \end{bmatrix}.$$

We have

$$\pi A = \begin{bmatrix} \pi_1 + \pi_2 & \pi_1 + \pi_3 & \pi_1 & \pi_1 & \pi_2 + \pi_3 & \pi_2 & \pi_2 & \pi_3 & \pi_3 \end{bmatrix}$$

which because of symmetry with respect to π_1, π_2, and π_3 results in the best communicants' strategy to be uniform. This gives a probability of success equal to $P_I = 2/3$ which is bigger than $k/M = 4/9$ and shows that the code does not provide prefect protection in class $\mathcal{M}$. However

$$AA^t = \begin{bmatrix} 4\,1\,1 \\ 1\,4\,1 \\ 1\,1\,4 \end{bmatrix}$$

and we have $P_I^K = 1/2 < 2/3 = P_I$. The lower bound of (10) in this case becomes the trivial one, that is, $P_I^K \geq 0$.

We noted that, for a given code, the probability of deception for strategies of class $\mathcal{K}$ is lower bounded by the average intersection of the encoding rules. For a given k, M, E, a minimum value for this bound can be obtained which is, in fact, a lower bound for all codes.

Theorem 9. *The probability of deception for class $\mathcal{K}$ is lower bounded by k/M and equality is obtained for codes that have equal number of ones in each column of A.*

Proof: Let x_i denote the number of ones in the ith column of A. Using (3) and uniform distribution for π we have

$$P_I^K \geq \frac{1}{k} \sum_{j+1}^{E} \sum_{k=1}^{E} \sum_{s=1}^{M} \pi_j a_{js} a_{ks} p_k.$$

The enemy's chance of success is minimum if he/she uses uniform strategy; hence

$$P_I^K \geq \frac{1}{kE^2} \sum_j \sum_k \sum_s a_{js} a_{ks} = \frac{1}{kE^2} \sum_i x_i^2.$$

The lower bound is minimum if $\sum_i x_i^2$ is minimum and because $\sum_i x_i = kE$ the minimum is when $x_i = x_j$, for all i, j. In this case $x_i = kE/M$, $1 \leq i \leq M$, and

$$P_0^K \geq k/M \tag{14}$$

□

This theorem shows that the lower bound on the probability of success for class $\mathcal{K}$ and $\mathcal{M}$ are the same and this lower bound is obtained when enemy uses random strategy of each class.

It is interesting to note that random strategy of class $\mathcal{K}$ has always a minimum chance of success equal to $1/E$. It is also known that minimum chance of success for impersonation (using class $\mathcal{M}$ strategies) is k/M. If $1/E < k/M$ it seems that random selection from key space is worse than random selection from cryptogram space. Theorem 9 shows that this is not the case and the two strategies provide the same minimum probability of success.

5 Perfect Protection

An A-code provides perfect protection for a class of strategies if the enemy's best strategy in that class is the uniform strategy of that class.

5.1 Class $\mathcal{K}$

Proposition 10. *An A-code provides perfect protection for class $\mathcal{K}$ if and only if πAA^t is a vector of equal components.*

Proof: This follows from expression (7). □

An A-code is called *distance invariant* if the number of encoding rules that are at distance i, $0 < i \leq 2k$ from a given rule e is independent of e.

Proposition 11. *A distance invariant code provides perfect protection for class $\mathcal{K}$.*

Proof:
$B = AA^t$ and hence $B_{ij} = e_i.e_j$. The result follows as $wt(e_i) = wt(e_j) = k$ and using a uniform communicant's strategy gives a probability of deception equal to $1/E \times \sum_i B_{ij}$ which is independent of j. □

Example 3. The following code is distance invariant with two possible distances between the encoding rules:

$$A = \begin{bmatrix} 1111110000 \\ 0011111100 \\ 0011001111 \\ 1111000011 \end{bmatrix},$$

and

$$AA^t = \begin{bmatrix} 6424 \\ 4642 \\ 2464 \\ 4246 \end{bmatrix}.$$

The code provides perfect protection for class $\mathcal{K}$ when communicants use a uniform strategy. It cannot provide perfect protection for class $\mathcal{M}$ as two cryptograms are authentic under all keys.

The above proposition gives some sufficient condition for A-codes that provide perfect protection in class $\mathcal{K}$ but in general characterizing A for which is it is possible for the communicants to choose a strategy which forces the enemy to randomly select a key is an open problem.

5.2 Class $\mathcal{M}$

The necessary and sufficient condition for a code to provide perfect protection for class $\mathcal{M}$ is that πA has equal components [8].

Proposition 12. *An A-code provides perfect protection for class $\mathcal{M}$ if either of the following hold:*

– *the incidence matrix of the code has equal number of ones in each column. This corresponds to 1-designs.*
– *There exist a submatrix of A consisting of a subset of rows of A with the same number of ones in each column.*

Proof:
In the first case π is uniform and in the second case has equal components on the rows of the submatrix and zero for its other components. □

There are many ways of constructing 1-designs. Vectors of constant weight in an error correcting code whose number of non-zero weights is less than its dual distance is an example of such constructions.

Proposition 12 only gives sufficient condition. For all codes that conform with the conditions of this proposition, communicants strategy is uniform on the nonzero components of π. In fact if the second condition holds the code can be optimized by leaving all the encoding rules which have zero probability in the best strategy of the communicants and retain the same probability of success. However it is possible to have codes with perfect protection and communicants' strategy not uniform.

Example 4. Let

$$A = \begin{bmatrix} 1\,1\,1\,0\,0 \\ 0\,1\,0\,1\,1 \\ 0\,0\,1\,1\,1 \\ 1\,0\,0\,1\,1 \end{bmatrix}.$$

For this code $\pi = [2/5\ 1/5\ 1/5\ 1/5]$ is the optimum strategy and the code provides perfect protection.

6 Concluding Remarks

We have studied performance of an authentication code under impersonation attack when the enemy can choose the fraudulent message either directly from the cryptogram space or by selecting a key and then choosing a cryptogram which is authentic under that key. This approach has allowed us to derive a new bound on the probability of success in terms of the average distance of the A-code. We showed that codes with highest average distance have columns of equal weight and provide perfect protection for class $\mathcal{M}$.

Codes with perfect protection in each case are characterized and sufficient conditions on the incidence matrices of such codes are given. However the characterization of such matrices in general remain an open problem.

References

1. J.L. Massey, *Cryptography, a selective survey*, Digital Communications, ed.E. Biglieri and G. Pratti, Elsvier Science Publ., North-Holland, (1986), 3-25.

2. G.J. Simmons, *Authentication theory/coding Theory*, Proc. of Crypto '84, Lect. Notes in Comp. Science **196**, (1985), 411-432
3. G.J. Simmons, *A game theory model of digital message authentication*, Congressus Numerantium **34**, (1982), 413-424
4. E.G. Gilbert, F.J. MacWilliams, N.J. Sloane, *Codes which Detect Deception*, Bell Sys. Tech. J., **53-3**, (1974), 1-19
5. F.J. MacWilliams and N.J. Sloane, *The Theory of Error-Correcting Codes*, North-Holland Publishing Company, 1978
6. M. De Soete, Some constructions for authentication-secrecy codes, Advances in Cryptology, Eurocrypt '88, (1988), 51-55
7. E.F. Brickell, *A few results on Message Authentication*, Proc. of the 15th Southeastern Conf. on Combinatorics, Graph Theory and Computing, Boca raton LA, (1984), 141-145
8. D.R. Stinson, *Combinatorial characterization of authentication codes* , Proceedings Crypto 91, Lecture Notes in Computer Science **576**, (1992), 62-72
9. D.R. Stinson, *The combinatorics of authentication and secrecy codes* , Journal Cryptology **2**, (1990), 23-49

Cumulative Arrays and Geometric Secret Sharing Schemes

Wen-Ai Jackson[1]* and Keith M. Martin[2]**

[1] Department of Mathematics, Royal Holloway, Egham Hill, Egham, Surrey, TW20 OEX, United Kingdom
[2] Department of Pure Mathematics, The University of Adelaide, GPO Box 498, Adelaide SA 5001, Australia

Abstract. Cumulative secret sharing schemes were introduced by Simmons et al (1991) based on the generalised secret sharing scheme of Ito et al (1987). A given monotone access structure together with a security level is associated with a unique cumulative scheme. Geometric secret sharing schemes form a wide class of secret sharing schemes which have many desirable properties including good information rates. We show that every non-degenerate geometric secret sharing scheme is 'contained' in the corresponding cumulative scheme. As there is no known practical algorithm for constructing efficient secret sharing schemes, the significance of this result is that, at least theoretically, a geometric scheme can be constructed from the corresponding cumulative scheme.

1 Introduction

A *secret sharing scheme* is a system by which a *secret* is protected among a group of *participants* in such a way that only specified groups of these participants (the *access structure*) can reconstruct the secret. Each participant is distributed a piece of information relating to the secret (a *share*) and groups of participants in the access structure can reconstruct the secret by pooling their shares. If groups not in the access structure cannot gain any information relating to the secret by pooling their shares then the secret sharing scheme is described as being *perfect*.

The most basic model for a secret sharing scheme was first proposed by Brickell and Davenport [2] and takes the form of a matrix. The columns of this matrix are indexed by the secret and the participants. The rows of this matrix relate to all the possible allocations of shares to participants and the corresponding values of the secret. For further details of this model see [2].

Secret sharing schemes were first discussed by Blakley [1] and Shamir [9]. They both dealt with schemes with an access structure comprising of all groups of participants of at least some fixed size (*threshold schemes*). Simmons [10] was the first to model a wider variety of access structures. This model used

* This work was supported by the Science and Engineering Research Council Grant GR/G 03359

** This work was supported by the Australian Research Council

finite geometry and thus we call such schemes *geometric schemes.* A number of authors ([3, 8]) have proposed measures of efficiency of a secret sharing scheme known as *information rates.* For more details of these, see Stinson [12]. Although other methods exist for constructing secret sharing schemes (see for example [3]), geometric schemes have proved to be a good source of schemes with high information rates. Other advantages of geometric schemes are given in Jackson and Martin [7].

Constructing geometric schemes has thus far been a fairly ad hoc process. It would be highly desirable to try and find some algorithmic process which could be used to find geometric schemes with good information rates. The contribution of this paper towards that aim is to show that any geometric scheme is in fact contained in a structure known as a cumulative array (or several copies of this array). Cumulative arrays will be discussed in the next section and are based on a method for constructing secret sharing schemes first proposed by Ito et al [5] and first applied to geometric schemes in Simmons et al [11].

The paper is structured as follows. Section 2 discusses access structures and their relationship to cumulative arrays. Section 3 introduces geometric secret sharing schemes. The main result is presented for the simplest case in Sect. 4 and generalised in Sect. 5. To conclude the paper, some consequences of the main result are considered in Sect. 6 with relation to the problem of developing an algorithm for producing 'good' geometric schemes.

2 Monotone Access Structures and Cumulative Arrays

Let the set of participants be $\mathcal{P} = \{p_1, \ldots, p_n\}$. An access structure Γ defined on $\mathcal{P}$ comprises of a collection of subsets of $\mathcal{P}$. We say that Γ is a *monotone access structure* if whenever $A \in \Gamma$ and $B \supseteq A$, then $B \in \Gamma$ $(A, B \subseteq \mathcal{P})$. For the remainder of the paper let Γ be a monotone access structure. The sets $C \in \Gamma$ with $C \backslash c \notin \Gamma$ for all $c \in C$ are called the *minimal sets* of Γ, the collection of which is denoted by Γ^-. This set Γ^- is uniquely determined by Γ. Let $\Gamma^- = \{C_1, \ldots, C_c\}$. We say Γ is *connected* if $\mathcal{P} = \bigcup_{i=1}^{c} C_i$. Note that if Γ is connected then Γ^- uniquely determines Γ. For the remainder of this paper we will assume that Γ is connected. The monotone access structure of the form

$$\Gamma = \{A \subseteq \mathcal{P} \mid |A| \geq k\}, \quad k \geq 1$$

is called the (k, n)*–threshold* access structure. Secret sharing schemes with this access structure are called (k, n)*–threshold* schemes.

The monotone access structure Γ^* called the *dual* of Γ can be defined in a variety of equivalent ways. We will define Γ^* by $\Gamma^* = \{S \subseteq \mathcal{P} \mid S \cap C \neq \emptyset$ for all $C \in \Gamma\}$. Let $\Gamma^{*-} = \{S_1, \ldots, S_s\}$. A set $B \subseteq \mathcal{P}$ is *unauthorised* if $B \notin \Gamma$. If $B \cup p \in \Gamma$ for all $p \in \bar{B} = \mathcal{P} \backslash B$, then B is called a *maximal unauthorised* set of Γ. Let Γ^+ be the collection of the maximal unauthorised sets of Γ. The significance of Γ^* is given by the following result [11, Lemma 3].

Result 1. *Let* $\Gamma^{*-} = \{S_1, \ldots, S_s\}$, *and let* $\bar{S}_i = \mathcal{P} \backslash S_i$ $(1 \leq i \leq s)$. *Then* $\Gamma^+ = \{\bar{S}_1, \ldots, \bar{S}_s\}$. *So if* $B \subseteq \mathcal{P}$, $B \notin \Gamma$ *then* $\bar{B} \in \Gamma^*$.

Cumulative arrays were first defined in [11] but the idea behind them was first exhibited in [5]. Let Γ be a monotone access structure as defined above. We recall the definition of a cumulative map from [11]. A *cumulative map* (α, S) for Γ is a finite set S accompanied by a mapping $\alpha: \mathcal{P} \to 2^S$ (where 2^S is the collection of all subsets of S) such that for $Q \subseteq \mathcal{P}$,

$$\bigcup_{a \in Q} a^\alpha = S \iff Q \in \Gamma.$$

Example 1. Let $\Gamma = ab+bc+cd$ defined on set $\mathcal{P} = \{a, b, c, d\}$. Then a cumulative map for Γ is given by

$$S = \{s_1, s_2, s_3\}, \quad a^\alpha = s_1, \; b^\alpha = \{s_2, s_3\}, \; c^\alpha = \{s_1, s_2\}, \; d^\alpha = s_3.$$

We have the following result from [11].

Result 2. *Let Γ be a monotone access structure defined on set $\mathcal{P}$. If (α, S) is a cumulative map for Γ then $|S| \geq |\Gamma^+|$.*

We refer to an (α_m, S) for Γ with $|S| = |\Gamma^+|$ as a *minimal* cumulative map. In [5] a minimal cumulative map was constructed as follows:

Let $S = \{S_1, S_2, \ldots, S_s\}$ (the sets of Γ^{*-}). For each $p \in \mathcal{P}$ let

$$p^{\alpha_m} = \{S_i \mid p \in S_i, 1 \leq i \leq s\}.$$

This was shown to be a cumulative map in [5] and is minimal by Result 1.

We have the following result.

Theorem 3. *For a monotone access structure Γ on a set $\mathcal{P}$, there exists a unique minimal cumulative map (α_m, S). Further, for any cumulative map (β, T) for Γ, there exists $T' \subseteq T$ such that (β', T') is a minimal cumulative map where $\beta': \mathcal{P} \to 2^{T'}$ is such that for all $p \in \mathcal{P}$, $p^{\beta'} = p^\beta \cap T'$.*

Theorem 3 says that α_m is the only minimal cumulative map for Γ and that any cumulative map for Γ essentially contains α_m. Hence without loss of generality we refer to α_m as *the* cumulative map for Γ. As in [11] we can represent the cumulative map for Γ as an $n \times s$ array with columns indexed by the sets of Γ^{*-}, rows indexed by elements of $\mathcal{P}$ and each entry either 0 or 1 such that

$$(p_i, S_j)\text{th entry} = 1 \iff p_i \in S_j.$$

We will add an extra initial row of all 1's, indexed by p_0, to this matrix and call the resulting $(n+1) \times s$ matrix *the cumulative array* for Γ. We denote this array by CA(Γ).

Example 2. For $\Gamma = ab + bc + cd$ as in Example 1, we have that

$$\mathrm{CA}(\Gamma) = \begin{array}{c} \\ p_0 \\ a \\ b \\ c \\ d \end{array} \begin{array}{c} \begin{array}{ccc} ac & bc & bd \end{array} \\ \begin{pmatrix} 1 & 1 & 1 \\ 1 & 0 & 0 \\ 0 & 1 & 1 \\ 1 & 1 & 0 \\ 0 & 0 & 1 \end{pmatrix} \end{array}.$$

The reason that the cumulative map is of interest is as follows. Let (α, S) be the cumulative map for Γ. Take an $(|S|, |S|)$–threshold scheme $\mathcal{X}$ ((n, n)–threshold schemes are known to exist for all $n \geq 1$; see for instance [9]), defined on set S, and then use the cumulative map to distribute sets of shares of $\mathcal{X}$ to each participant in $\mathcal{P}$. In other words, give the shares of $\mathcal{X}$ corresponding to p^α to participant p. If we then operate $\mathcal{X}$ in the normal way, only participants belonging to sets in Γ will be able to accumulate all the shares of $\mathcal{X}$ and hence reconstruct the secret in $\mathcal{X}$. This was the method used in [5] to show that there exists a perfect secret sharing scheme for every monotone access structure.

The disadvantages of the above system is that each participant in $\mathcal{P}$ tends to get many shares of scheme $\mathcal{X}$ and consequently the information rates of schemes formed in this way are low. However, we show here that the cumulative array is in fact a very significant structure in the theory of geometric secret sharing schemes. In [11] it was shown how to use the cumulative array to construct a geometric secret sharing scheme. As this produced a scheme with the same information rates as that of [5] this is not a very useful scheme in itself. However, rather than being used as a scheme itself, the cumulative array for Γ would appear to contain useful information when it comes to considering constructing other geometric schemes. We show in this paper that in fact any perfect geometric scheme is in some sense 'contained' in one or more copies of the cumulative array. The significance of this is that it may lead to more systematic methods for constructing geometric schemes as opposed to the somewhat ad hoc methods currently being used.

3 Geometric Secret Sharing Schemes

For a background on projective geometry over finite fields, see [4]. Let $\Sigma = \mathrm{PG}(d, q)$, the projective space of dimension d over the Galois field F_q, where q is a prime power. We represent points of Σ in homogeneous co-ordinates by $(d+1)$–tuples over F_q, and similarly for hyperplanes. Thus the concept of independence for both points and hyperplanes is inherited from the vector space of dimension $d+1$ over F_q.

Let Γ be a connected monotone access structure on $\mathcal{P}$. A *geometric secret sharing scheme* for Γ is a function ω which assigns to each $p \in \mathcal{P} \cup p_0$ a subspace p^ω of Σ, such that for $A \subseteq \mathcal{P}$,

$$p_0^\omega \subseteq A^\omega \text{ if and only if } A \in \Gamma,$$

where $A^\omega = \langle p^\omega \,|\, p \in A \rangle$, the subspace spanned by the subspaces p^ω, $p \in A$. We say that ω is *perfect* if for every $A \notin \Gamma$, $A \subseteq \mathcal{P}$, then $p_0^\omega \cap A^\omega = \emptyset$ (see [7] for more detail). If ω is perfect then we say that ω is a t–GS(Γ, q), where $t = \dim p_0^\omega + 1$, $(t \geq 1)$, where dim is the projective dimension.

The most general model for secret sharing is a matrix M whose columns are indexed by p_0 and the participants in $\mathcal{P}$. To implement such a scheme, a row r is chosen at random and the (r, p)th entry of M given to participant p. The secret is the (r, p_0)th entry. Various conditions are imposed on M for it to be a perfect secret sharing scheme for a monotone access structure. For more details, see [2]. It can be shown [7] that given a t–GS(Γ, q) we can obtain from it a perfect secret sharing matrix for Γ.

Example 3. Let $\mathcal{P} = \{a,b,c,d\}$ and $\Gamma^- = \{ab, bc, cd\}$. Then ω is a 1–GS(Γ, q), where $p_0^\omega = (1,1,1)$, $a^\omega = (1,0,0)$, $b^\omega = \langle(0,1,0),(0,0,1)\rangle$, $c^\omega = (1,1,0)$, $d^\omega = (0,0,1)$.

Let ω be τ be two t–GS(Γ, q) in $\Sigma_1 = \mathrm{PG}(d_1, q)$ and $\Sigma_2 = \mathrm{PG}(d_2,q)$ respectively. Suppose $d_1 \geq d_2$. Embed Σ_2 in the natural way into Σ_1, so we can consider τ to be a geometric scheme in Σ_1. We say that τ is ***contained in*** ω if for every $p \in \mathcal{P} \cup p_0$, $p^\tau \subseteq p^\omega$. Notice that $p_0^\tau = p_0^\omega$, since $p_0^\tau \subseteq p_0^\omega$ and they are both $t-1$ dimensional subspaces.

4 Embedding Geometric Schemes in Cumulative Schemes

Let Γ be a connected monotone access structure with $\Gamma^{*-} = S = \{S_1, \ldots, S_s\}$. Throughout this section let $\Theta = \mathrm{PG}(s-1, q)$ and let $F = \{f_1, \ldots, f_s\}$ be a hyperplane basis of Θ. We call the point or hyperplane basis

$$\{(1,0,\ldots,0),(0,1,0,\ldots,0),\ldots,(0,\ldots,0,1)\}$$

the ***standard*** basis.

We now use the cumulative map α_m for Γ defined on set S (as in Sect. 2) to construct a geometric scheme σ for Γ:

1. Let $p^\sigma = \bigcap_{S_i \notin p^{\alpha_m}} f_i = \bigcap_{p \notin S_i} f_i$, for all $p \in \mathcal{P}$.
2. Let p_0^σ be any point such that $p_0^\sigma \notin f_i$ for all i, $1 \leq i \leq s$.

In [11] this scheme is shown to be a 1–GS(Γ, q). Note that Part 2 above is equivalent to requiring $p_0^\sigma \notin \bar{S}_i^\sigma$, for all i, $(1 \leq i \leq s)$. We call σ a ***cumulative scheme*** (with basis F) for Γ in PG($s-1, q$) and denote it by 1–CS(Γ, q). Note that for $A \subseteq \mathcal{P}$, $A^\sigma = \langle p^\sigma \mid p \in A\rangle = \left\langle \bigcap_{p \notin S_i} f_i \mid p \in A \right\rangle = \bigcap_{A \cap S_i = \emptyset} f_i$. We will use the convention that $\bigcap_\emptyset f_i = \Theta$.

Example 4. Consider $\Gamma = ab + bc + cd$ once more. Let $\{e_1, e_2, e_3\}$ and $F = \{f_1, f_2, f_3\}$ be the standard point and hyperplane bases for $\Sigma = \mathrm{PG}(2, q)$, respectively. Then the cumulative array of Example 2 immediately gives a 1–CS(Γ, q) where $p_0^\omega = (1,1,1)$, $a^\omega = f_2 \cap f_3 = e_1$, $b^\omega = f_1 = \langle e_2, e_3\rangle$, $c^\omega = f_3 = \langle e_1, e_2\rangle$, $d^\omega = f_1 \cap f_2 = e_3$. This is the 1–GS($\Gamma, q$) (with basis F) exhibited in Example 3.

For the remainder of this paper we will assume a non-degeneracy condition on geometric schemes. This condition is somewhat technical (details can be found in the full paper) but is based on the fact that if a scheme is degenerate then we can easily obtain a scheme implemented in a smaller dimension and whose information rate is at least as good.

We present the main result of this section.

Theorem 4. *A* 1–GS(Γ, q) *is contained in a* 1–CS(Γ, q).

5 Embedding in the General Case

In this section we generalise Theorem 4 to a t–GS(Γ, q), $t > 1$. It is important to consider the generalisation since, for example, for the access structure $\Gamma^- = \{ab, bc, cd, de, ea\}$ on $\mathcal{P} = \{a, b, c, d, e\}$ we can find a 2–GS(Γ, q) which has a better information rate than any 1–GS(Γ, q) [13].

In the previous section we showed that a 1–GS(Γ, q) is contained in a cumulative geometric scheme 1–CS(Γ, q). In this section our aim is to show that a t–GS(Γ, q) is contained in the "product" of t cumulative geometric schemes.

Let Γ be a connected monotone access structure and let $\Gamma^{*-} = \{S_1, \ldots, S_s\}$. The ***t–cumulative array*** is an $(n+1) \times st$ array $X_t = [X{:}X{:}\cdots{:}X]$ consisting of t copies of the cumulative array X for Γ. Throughout this section let $\Theta = \mathrm{PG}(st-1, q)$ and let $F = \{f_1, \ldots, f_{st}\}$ be a hyperplane bases for Θ. Let $t_0 = t-1$. A ***t–cumulative scheme*** t–CS(Γ, q) (with basis F) for Γ in PG$(st-1, q)$, is a geometric scheme σ defined as follows.

1. Let $p^\sigma = \bigcap_{p \notin S_i} \bigcap_{j=0}^{t_0} f_{i+js}$ for all $p \in \mathcal{P}$.
2. Let p_0^σ be any subspace such that $\dim p_0^\sigma = t_0$ and $p_0^\sigma \cap \left(\bigcap_{j=0}^{t_0} f_{i+js}\right) = \emptyset$, $(1 \le i \le s)$.

So $A^\sigma = \bigcap_{A \cap S_i = \emptyset} \bigcap_{j=0}^{t_0} f_{i+js}$. Further, Part 2 is equivalent to requiring that $p_0^\sigma \notin \bar{S}_i^\sigma$, for all i, $1 \le i \le s$.

Example 5. Let $\Gamma = ab + bc + cd$. Using the cumulative array for Γ as given in Example 2 we obtain the 2–cumulative array X_2. Letting $\{e_1, \ldots, e_6\}$ and $\{f_1, \ldots, f_6\}$ be the standard point and hyperplane bases for $\Theta = \mathrm{PG}(5, q)$ we obtain σ the 2–cumulative scheme for σ in PG$(5, q)$:

$$X_2 = \begin{array}{c} \\ p_0 \\ a \\ b \\ c \\ d \end{array} \begin{array}{c} \begin{array}{cccccc} ac & cb & bd & ac & cb & bd \end{array} \\ \left(\begin{array}{ccc:ccc} 1 & 1 & 1 & 1 & 1 & 1 \\ 1 & 0 & 0 & 1 & 0 & 0 \\ 0 & 1 & 1 & 0 & 1 & 1 \\ 1 & 1 & 0 & 1 & 1 & 0 \\ 0 & 0 & 1 & 0 & 0 & 1 \end{array}\right) \end{array}$$

$$\begin{aligned} p_0^\sigma &= \langle (111000), (000111) \rangle \\ a^\sigma &= (f_2 \cap f_5) \cap (f_3 \cap f_6) = \langle e_1, e_4 \rangle \\ b^\sigma &= f_1 \cap f_4 = \langle e_2, e_3, e_5, e_6 \rangle \\ c^\sigma &= f_3 \cap f_6 = \langle e_1, e_2, e_4, e_5 \rangle \\ d^\sigma &= (f_1 \cap f_4) \cap (f_2 \cap f_5) = \langle e_3, e_6 \rangle \end{aligned}$$

Theorem 5. *A t–CS(Γ, q) is a t–GS(Γ, q).*

Proof. Let σ be a t–CS(Γ, q). If $A \in \Gamma$ then $S_i \cap A \neq \emptyset$ for all i $(1 \le i \le s)$, so $A^\sigma = \Theta$ and so $p_0^\sigma \subseteq A^\sigma$. If $A \notin \Gamma$ then A is unauthorised and so $A \subseteq \bar{S}_i$ for some $S_i \in \Gamma^{*-}$ (Result 1). So $A^\sigma \subseteq \bigcap_{j=0}^{t_0} f_{i+js}$ and as $p_0^\sigma \cap \left(\bigcap_{j=0}^{t_0} f_{i+js}\right) = \emptyset$ it follows that $p_0^\sigma \cap A^\sigma = \emptyset$, as required. Hence σ is a t–GS(Γ, q). □

We now generalize Theorem 4.

Theorem 6. *For $t \ge 1$, a t–GS(Γ, q) is contained in a t–CS(Γ, q).*

6 Consequences for Constructing Geometric Schemes

Throughout this section we continue to use the definitions and notation of Sects. 4 and 5. The main result of the previous section (Theorem 6) states that any t–geometric scheme for Γ can be embedded in the t–cumulative scheme for Γ in the appropriate finite field. This is a significant theoretical result in itself, however it would be very useful to be able to operate this process in 'reverse'. In other words, start with the t–cumulative scheme and then construct geometric schemes from it. This problem remains unsolved and it would be hoped that perhaps some practical algorithm could be developed to perform this task. Although we do not present such an algorithm here, we give some results which may prove useful in developing such a procedure.

Without loss of generality let $\{e_1, \ldots\}$ and $\{f_1, \ldots\}$ be the standard point and hyperplane bases.

The following theorem is a slightly stronger result than Theorem 6.

Theorem 7. *Let $t \geq 1$. Let ω be a t-GS(Γ, q) contained in σ, a t-CS(Γ, q) (with basis $\{f_1, \ldots, f_{st}\}$). Then $p^\omega \subseteq f_{i+js}$, $(0 \leq j \leq t_0)$ if and only if $p \notin S_i$.*

A consequence of Theorem 7 is the following result.

Corollary 8. *Suppose ω is a 1-GS(Γ, q) in $\Sigma = \mathcal{P}^\omega$ = PG(d, q). Then there exists an automorphism δ of Σ and $d+1$ columns of X = CA(Γ) forming a matrix $Y = [Y_{ij}]$ such that $(p_i^\omega)^\delta \subseteq \langle e_j \mid Y_{ij} \neq 0 \rangle$, where $\{e_1, \ldots, e_{d+1}\}$ is the standard point basis for Σ.*

The above corollary is saying that if a 1–GS(Γ, q) can be found in PG(d, q) then it can be "obtained" from $d+1$ columns of a cumulative array. From the proof of Corollary 8 it can be shown that the columns in X corresponding to any hyperplane basis $\{h_{i_1}, \ldots, h_{i_{d+1}}\}$ of Σ can be used.

Example 6. Let $\Gamma^- = \{ab, ac, ade, bde\}$ and $\Gamma^{*-} = \{ab, ad, bcd, ae, bcd\}$. We use the first four columns of the cumulative matrix to form ω, a 1–GS(Γ, q).

$$
\begin{array}{c} \\ p_0 \\ a \\ b \\ c \\ d \\ e \end{array}
\begin{array}{c} \begin{array}{ccccccc} ab & ad & bcd & ae & & bce \end{array} \\
\begin{pmatrix}
1 & 1 & 1 & 1 & \vdots & 1 \\
1 & 1 & 0 & 1 & \vdots & 0 \\
1 & 0 & 1 & 0 & \vdots & 1 \\
0 & 0 & 1 & 0 & \vdots & 1 \\
0 & 1 & 1 & 0 & \vdots & 0 \\
0 & 0 & 0 & 1 & \vdots & 1
\end{pmatrix} \end{array}
\quad
\begin{array}{l}
p_0^\omega = (1,1,1,1) \\
a^\omega = \langle (1,0,0,0), (1,1,0,1) \rangle \\
b^\omega = \langle (1,0,0,0), (0,0,1,0) \rangle \\
c^\omega = (0,0,1,0) \\
d^\omega = (0,1,1,0) \\
e^\omega = (0,0,0,1)
\end{array}
$$

Thus Corollary 8 provides some clues how to construct geometric schemes from cumulative schemes. We can also see from Theorem 6, every t–GS(Γ, q) can be obtained from σ a t–CS(Γ, q) in Θ = PG$(st-1, q)$, by reducing the size of each share p^σ to $p^{\sigma'}$ so that σ' is still a t–GS(Γ, q). However, by Corollary 8 we can separate this task into two smaller stages: firstly, find a subspace Σ (corresponding to $d+1$ columns of the cumulative array, for appropriate d) so that if $p^\omega = p^\sigma \cap \Sigma$ then ω is a t–GS(Γ, q); and secondly possibly reduce the size of each share p^ω to $p^{\omega'}$ so that ω' is still a t–GS(Γ, q).

By way of a conclusion we highlight some of the interesting avenues for further work. The main result of this paper shows that, at least in theory, it is possible to obtain geometric schemes from the columns of cumulative arrays. Is there a practical algorithm to obtain geometric schemes for a suitable choices of columns of the cumulative array? So far, the examples given have been obtained by a combination of exponential time calculations and inspired guesswork. How is the smallest number of columns necessarily determined, and how can such a number of columns be recognised? Or equivalently, is there a practical algorithm to show, given a geometric scheme, that there is no 'better' geometric scheme contained in it? It is hoped that some of the results here can be used to make further progress in answering some of these questions.

References

1. G. R. Blakley: Safeguarding cryptographic keys. Proceedings of AFIPS 1979 National Computer Conference **48** (1979) 313–317
2. E. F. Brickell and D. M. Davenport: On the Classification of Ideal Secret Sharing Schemes. J. Cryptology **2** (1991) 123–124
3. E. F. Brickell and D. R. Stinson: Some Improved Bounds on the Information Rate of Perfect Secret Sharing Schemes. To appear in J. Cryptology
4. J. W. P. Hirschfeld: Projective Geometries over Finite Fields. Clarendon Press, Oxford, 1979
5. M. Ito, A. Saito and T. Nishizeki: Secret Sharing Scheme Realizing General Access Structure. Proceedings IEEE Global Telecom. Conf., Globecom '87, IEEE Comm. Soc. Press (1987) 99–102
6. W.–A. Jackson and K. M. Martin: On Ideal Secret Sharing Schemes. Submitted to the J. Cryptology
7. W.–A. Jackson and 'K. M. Martin: Geometric Secret Sharing Schemes and their Duals. Submitted to Des. Codes Cryptogr.
8. K. M. Martin: New Secret Sharing Schemes from Old. Submitted to the J. Combin. Math. Combin. Comput.
9. A. Shamir: How to Share a Secret. Comm. ACM Vol 22 **11** (1979) 612–613
10. G. J. Simmons: How to (Really) Share a Secret. Advances in Cryptology – CRYPTO'88, Lecture Notes in Comput. Sci. **403** (1990) 390–448
11. G. J. Simmons, W.–A. Jackson and K. Martin: The Geometry of Shared Secret Schemes. Bull. Inst. Combin. Appl. **1** (1991) 71–88
12. D. R. Stinson: An Explication of Secret Sharing Schemes. Preprint, 1992
13. D. R. Stinson: Private communication, 1992.
14. T. Uehara, T. Nishizeki, E. Okamoto and K. Nakamura: A Secret Sharing System with Matroidal Access Structure. Trans. IECE Japan J69–A 9 (1986) 1124–1132

Nonperfect Secret Sharing Schemes

Wakaha Ogata, Kaoru Kurosawa and Shigeo Tsujii

Department of Electrical and Electronic Engineering,
Faculty of Engineering,
Tokyo Institute of Technology
2-12-1 O-okayama, Meguro-ku, Tokyo 152, Japan

E-mail: kkurosaw@ss.titech.ac.jp

Abstract. A nonperfect secret sharing scheme (NSS) consists of a family of access subsets Γ_1, a family of semi-access subsets Γ_2 and a family of non-access subsets Γ_3. In an NSS, it is possible that $|V_i| < |S|$, where $|V_i|$ is the size of the share and $|S|$ is the size of the secret. This paper characterizes nonperfect secret sharing schemes. First, we show that $(\Gamma_1, \Gamma_2, \Gamma_3)$ is realizable if and only if Γ_1 is monotone and $\Gamma_1 \cup \Gamma_2$ is monotone. Then, we derive a lower bound of $|V_i|$ in terms of a distance between Γ_1 and Γ_3. Finally, we show a condition for $(\Gamma_1, \Gamma_2, \Gamma_3)$ to achieve $|V_i| = |S|/2$ for all i.

1 Introduction

Secret sharing schemes permits a secret to be shared by among participants in such a way that only qualified subsets of participants (access subset) can recover the secret. Sharing schemes are useful in the management of cryptographic keys and in multiparty protocols.

"Perfect" sharing schemes have been studied extensively so far. In a perfect sharing scheme, any subset of participants is an access subset or a nonaccess subset which has absolutely no information on the secret. No subsets are allowed in between.

[1] and [2] introduced (k, n) threshold schemes. In such a scheme, the access subsets are all the subsets whose cardinality is more than $k - 1$. [3] described a more general structure. An access structure is a family consisting of all the access subsets. A family Δ is said to be monotone if

$$A \in \Delta, A \subseteq A' \Rightarrow A' \in \Delta \ .$$

They showed that a perfect secret sharing scheme exists if and only if the access structure is monotone. Subsequently, [4] gave a simpler and more efficient way to realize monotone access structures. If $|V_i| = |S|$ for all i, the secret sharing scheme is said to be ideal. [5] characterized ideal schemes in terms of matroid. They also analyzed the case when the access structure is the closure of the set of edges of a graph G. [6] showed that $|V_i| \geq |S|$ for all i in any monotone access structure by extending the technique of [7]. Lower bounds of $|V_i|$ which depend on the access structure have also been studied [6][9] [8][10][11]. We emphasize here that $|V_i| \geq |S|$ in any perfect secret sharing scheme.

Therefore, schemes which can achieve $|V_i| < |S|$ must be "nonperfect", where semi-access subsets should be allowed. A semi-access subset is a set of participants who can have some information on the secret but cannot recover the secret completely. (d, k, n) ramp schemes [12], which is an extension of (k, n) threshold schemes, are such an example.

This paper characterizes nonperfect secret sharing schemes. A nonperfect secret sharing scheme can be defined as $(\Gamma_1, \Gamma_2, \Gamma_3)$, where Γ_1 is a family of access subsets, Γ_2 is a family of semi-access subsets and Γ_3 is a family of non-access subsets. First, we show that $(\Gamma_1, \Gamma_2, \Gamma_3)$ is realizable if and only if Γ_1 is monotone and $\Gamma_1 \cup \Gamma_2$ is monotone. Then, we derive a lower bound of $|V_i|$ in terms of a distance between Γ_1 and Γ_3. Finally, we show a condition for $(\Gamma_1, \Gamma_2, \Gamma_3)$ to achieve $|V_i| = |S|/2$ for all i.

$\sharp X$ denotes the cardinality of a finite set X. $|X| \triangleq \log_2 \sharp X$. $A \setminus B = \{x | x \in A \textit{ but } x \notin B\}$. 2^P is the family of all subsets of P.

2 Some Lemmas on Entropy

Lemma 1. $H(S|XV) \geq H(S|X) - H(V)$.

Proof. $I(S, V|X) = H(S|X) - H(S|XV) = H(V|X) - H(V|SX)$
$\leq H(V|X) \leq H(V)$. □

Lemma 2. *If* $Y = X \cup V_1 \cup V_2 \cdots \cup V_k$, $H(S|Y) \geq H(S|X) - \sum_{i=1}^{k} H(V_i)$.

Proof. $H(S|Y) = H(S|XV_1 \cdots V_k) \geq H(S|X) - H(V_1 \cdots V_k)$ (from Lemma 1)

$$\geq H(S|X) - \sum_{i=1}^{k} H(V_i) .$$

□

Lemma 3.

1) $H(Z|Z') = 0 \Rightarrow H(Z|YZ') = 0 \Rightarrow H(X|YZ') \leq H(X|YZ)$.

2) $H(Z|Z') = H(Z'|Z) = 0 \Rightarrow H(Z|YZ') = H(Z'|YZ) = 0$
$\Rightarrow H(X|YZ') = H(X|YZ)$.

Proof. 1) $0 \leq H(Z|YZ') \leq H(Z|Z') = 0$. Hence, $H(Z|YZ') = 0$.

$$H(Z|XYZ') \leq H(Z|YZ') = 0 .$$
$$I(X; Z|YZ') = H(Z|YZ') - H(Z|XYZ') = 0 .$$
$$H(X|YZ') = H(X|YZZ') + I(X; Z|YZ')$$
$$= H(X|YZZ') \leq H(X|YZ) .$$

2) It is clear from 1) . □

Lemma 4. $H(Z'|Z) = 0 \Rightarrow H(X|YZZ') = H(X|YZ)$.

Proof. Since $0 \le H(Z'|XYZ) \le H(Z'|YZ) \le H(Z'|Z) = 0$, $H(Z'|YZ) = H(Z'|XYZ) = 0$. Then,

$$\begin{aligned} H(X|YZZ') &= H(X|YZ) - I(X,Z'|YZ) \\ &= H(X|YZ) - H(Z'|YZ) + H(Z'|XYZ) \\ &= H(X|YZ) . \end{aligned}$$

□

3 Nonperfect Secret Sharing Scheme

3.1 Definitions

- $P = \{P_1, \cdots, P_n\}$ is a set of participants.
- s is a secret uniformly distributed over a finite set S.
- v_i is the share of P_i distributed over a finite set V_i.
- $V \triangleq \{V_1, \cdots, V_n\}$.

$\tilde{\Gamma}_i$ denotes a subset of 2^P. Γ_i denotes a subset of 2^V. We use the convention for Γ_j and $\tilde{\Gamma}_j$ in such a way that $(V_{j1}, \cdots, V_{jk}) \in \Gamma_j$ iff $(P_{j1}, \cdots, P_{jk}) \in \tilde{\Gamma}_j$. (The index set in $\tilde{\Gamma}_j$ and that in Γ_j are the same.)

Definition 5. (Π, S, V) is a secret sharing scheme (SS) if Π is a mapping,

$$S \times R \rightarrow V_1 \times V_2 \times \cdots \times V_n$$

where R is a set of random inputs.

Definition 6. Let $\Gamma \subseteq 2^V$. We say that (Π, S, V, Γ) is a perfect secret sharing scheme (PSS) if (Π, S, V) is a secret sharing scheme and

1. $H(S|A) = 0$ for $\forall A \in \Gamma$.
2. $H(S|C) = H(S)$ for $\forall C \notin \Gamma$.

Definition 7. Suppose that $\Gamma_1 \cup \Gamma_2 \cup \Gamma_3 = 2^V, \Gamma_1 \cap \Gamma_2 = \Gamma_2 \cap \Gamma_3 = \Gamma_3 \cap \Gamma_1 = \phi, \Gamma_2 \neq \phi$. We say that $(\Pi, S, V, (\Gamma_1, \Gamma_2, \Gamma_3))$ is a nonperfect secret sharing scheme (NSS) if (Π, S, V) is a secret sharing scheme and

1. $H(S|A) = 0$ for $\forall A \in \Gamma_1$.
2. $0 < H(S|B) < H(S)$ for $\forall B \in \Gamma_2$.
3. $H(S|C) = H(S)$ for $\forall C \in \Gamma_3$.

3.2 Realizability

Lemma 8. *Suppose that $(\Pi, S, V, (\Gamma_1, \Gamma_2, \Gamma_3))$ is an NSS. Then,*

1. *If $A \in \Gamma_1, A' \supset A$, then $A' \in \Gamma_1$.*
2. *If $B \in \Gamma_2, B' \supset B$, then $B' \in \Gamma_1 \cup \Gamma_2$.*
3. *If $C \in \Gamma_3, C' \subset C$, then $C' \in \Gamma_3$.*

Definition 9. A family Γ is said to be monotone if $A \in \Gamma, A \subseteq A' \Rightarrow A' \in \Gamma$.

Definition 10. Let $\tilde{\Gamma}_1 \cup \tilde{\Gamma}_2 \cup \tilde{\Gamma}_3 = 2^P, \tilde{\Gamma}_2 \neq \phi$, We say that $(S, P, (\tilde{\Gamma}_1, \tilde{\Gamma}_2, \tilde{\Gamma}_3))$ is nonperfectly realizable if there exist Π and $V(|V| = |P|)$ such that $(\Pi, S, V, (\Gamma_1, \Gamma_2, \Gamma_3))$ is a nonperfect secret sharing scheme. "Perfectly realizable" is defined similarly.

Theorem 11. *Suppose that $\sharp S$ is not a prime. Let $\tilde{\Gamma}_1 \cup \tilde{\Gamma}_2 \cup \tilde{\Gamma}_3 = 2^P$, $\tilde{\Gamma}_2 \neq \phi$. $(S, P, (\tilde{\Gamma}_1, \tilde{\Gamma}_2, \tilde{\Gamma}_3))$ is nonperfectly realizable if and only if $\tilde{\Gamma}_1$ is monotone and $\tilde{\Gamma}_1 \cup \tilde{\Gamma}_2$ is monotone.*

Proof. From lemma 8, it is clear that, if $(S, P, (\tilde{\Gamma}_1, \tilde{\Gamma}_2, \tilde{\Gamma}_3))$ is nonperfectly realizable, then $\tilde{\Gamma}_1$ and $\tilde{\Gamma}_1 \cup \tilde{\Gamma}_2$ are monotone, respectively. Suppose that $\tilde{\Gamma}_1$ and $\tilde{\Gamma}_1 \cup \tilde{\Gamma}_2$ are monotone, respectively. Let $\sharp S = m_1 \times m_2$, where $m_1 > 1$ and $m_2 > 1$. Let $S_1 \triangleq \{0, \ldots, m_1 - 1\}, S_2 \triangleq \{0, \ldots, m_2 - 1\}$. Then, $s \in S$ can be expressed as (s_1, s_2) such that $s_1 \in S_1$ and $s_2 \in S_2$. Since $\tilde{\Gamma}_1$ is monotone, there exists Π_1 such that $(S, P, \tilde{\Gamma}_1)$ is perfectly realizable [3][4]. Similarly, there exists Π_2 such that $(S, P, \tilde{\Gamma}_1 \cup \tilde{\Gamma}_2)$ is perfectly realizable. To share $s = (s_1, s_2)$, apply Π_1 for s_1 and apply Π_2 for s_2, independently. Then, it is easy to see that $1 \sim 3$ of Def. 3 are satisfied. □

4 Lower Bound of the Size of the Shares

Suppose that $(\Pi, S, V, (\Gamma_1, \Gamma_2, \Gamma_3))$ is a nonperfect secret sharing scheme.

Lemma 12. *If $A \in \Gamma_1, C \in \Gamma_3, C \subset A$, then* $\sum_{V_i \in A \setminus C} H(V_i) \geq |S|$.

Proof. From lemma 2, $H(S|A) \geq H(S|C) - \sum_{V_i \in A \setminus C} H(V_i)$. From Def. 3 (1) and (2), $\sum_{V_i \in A \setminus C} H(V_i) \geq H(S|C) - H(S|A) = H(S) = |S|$. □

Theorem 13.

$$\max_i |V_i| \geq |S| / \sharp(A \setminus C), \quad \forall A \in \Gamma_1, \forall C \in \Gamma_3 .$$

Proof. First we assume $C \subset A$. Then, from lemma 12,

$$\sum_{V_i \in A \setminus C} H(V_i) \geq |S| \ . \tag{1}$$

On the other hand,

$$\sum_{p_i \in A \setminus C} H(V_i) \leq \sharp(A \setminus C) \max_i |V_i| \ . \tag{2}$$

From (1) and (2), we obtain

$$|S| \leq \sharp(A \setminus C) \max_i |V_i| \ . \tag{3}$$

Next, we assume $C \not\subset A$. Let $A' \triangleq C \cup A$. Since Γ_1 is monotone (from Theorem 1), $A' \in \Gamma_1$. Then, from (3), we have

$$|S| \leq \sharp(A' \setminus C) \max_i |V_i| \ . \tag{4}$$

It is clear that

$$\sharp(A \setminus C) = \sharp(A' \setminus C) \ . \tag{5}$$

From (4) and (5), we obtain

$$|S| \leq \sharp(A \setminus C) \max_i |V_i| \ . \tag{6}$$

□

5 $|V_i| = |S|/2$, When and How ?

This section characterizes nonperfect secret sharing schemes which achieve $|V_i| = |S|/2$ for all i.

Definition 14. $\tilde{\Gamma}^-$ denotes the family of minimal sets of a family $\tilde{\Gamma}$. $\tilde{\Gamma}^+$ denotes the family of maximal sets of $\tilde{\Gamma}$.

Definition 15. Let $\tilde{\Gamma} \subseteq 2^P$. $G(P, \tilde{\Gamma})$ denotes a hypergraph [13] such that P is a set of vertices and $\tilde{\Gamma}$ is a set of edges.

In this section, we suppose that

P1. $\tilde{\Gamma}_1 \cup \tilde{\Gamma}_2 \cup \tilde{\Gamma}_3 = 2^P, \tilde{\Gamma}_1 \cap \tilde{\Gamma}_2 = \tilde{\Gamma}_2 \cap \tilde{\Gamma}_3 = \tilde{\Gamma}_3 \cap \tilde{\Gamma}_1 = \phi, \tilde{\Gamma}_2 \neq \phi$.
$\tilde{\Gamma}_1$ and $\tilde{\Gamma}_1 \cup \tilde{\Gamma}_2$ are monotone, respectively.
P2. $\forall A \in \tilde{\Gamma}_1^-, \sharp A = 3$.
P3. $\forall P_i, \{P_i\} \in \tilde{\Gamma}_3$.
P4. $G(P, \tilde{\Gamma}_1^-)$ is connected.
P5. $\sharp S = q^2$, where q is a prime power.

From Theorem 2, (P2) and (P3) represent one of the simplest case which can achieve $|V_i| = |S|/2$.

Definition 16. $[\tilde{\Gamma}]_k \triangleq \{A | A \in \tilde{\Gamma}, \sharp A = k\}$.

Definition 17. Suppose $G(P, [\tilde{\Gamma}_3]_2)$ is a disjoint union of cliques $C_1, \ldots, C_{n'}$. We denote the vertex set of C_i by T_i. $T \triangleq \{T_1, \ldots, T_{n'}\}$. $\quad (n' \leq n)$

Definition 18. If $G(P, [\tilde{\Gamma}_3]_2)$ is a disjoint union of cliques, we define Δ_1, Δ_2 and Δ_3 as follows.

- $\{T_i, \ldots, T_k\} \in \Delta_1$ iff $\exists Q_i (\neq \phi) \subseteq T_i, \cdots, \exists Q_k (\neq \phi) \subseteq T_k, \quad Q_i \cup \cdots \cup Q_k \in \Gamma_1$.
- $\{T_i, \ldots, T_k\} \in \Delta_2$ iff $\exists Q_i (\neq \phi) \subseteq T_i, \cdots, \exists Q_k (\neq \phi) \subseteq T_k, \quad Q_i \cup \cdots \cup Q_k \in \Gamma_2$.
- $\Delta_3 = 2^T \setminus (\Delta_1 \cup \Delta_2)$.

Lemma 19. *Suppose that $G(P, [\tilde{\Gamma}_3]_2)$ is a disjoint union of cliques.*

(1) *If $T_i \neq T_j$, then, $\forall P_i \in T_i, \forall P_j \in T_j, \{P_i, P_j\} \in \tilde{\Gamma}_2$. (Therefore, $\{T_i, T_j\} \in \Delta_2$).*
(2) $\Delta_3 = \{\{T_1\}, \ldots, \{T_{n'}\}\}$.

Proof. From (P2), $\{P_i, P_j\} \in \tilde{\Gamma}_2$ or $\{P_i, P_j\} \in \tilde{\Gamma}_3$. If $\{P_i, P_j\} \in \tilde{\Gamma}_3$, then $T_i = T_j$ from the definition of T_i and T_j. (2) is clear. □

Definition 20. Suppose that $G(P, [\tilde{\Gamma}_3]_2)$ is a disjoint union of cliques. We say that $(\tilde{\Gamma}_1, \tilde{\Gamma}_2, \tilde{\Gamma}_3)$ is well defined if

(1) If $\{T_i, \ldots, T_k\} \in \Delta_1$, then $\forall P_i \in T_i, \cdots, \forall P_k \in T_k, \{P_i, \ldots, P_k\} \in \tilde{\Gamma}_1$.
(2) If $\{T_i, \ldots, T_k\} \in \Delta_2$, then $\forall P_i \in T_i, \cdots, \forall P_k \in T_k, \{P_i, \ldots, P_k\} \in \tilde{\Gamma}_2$.
(3) For $\forall T_i = \{P_{i1}, \ldots, P_{ik}\}$, any subset of $T_i \in \tilde{\Gamma}_3$.
(4) $\Delta_1 \cap \Delta_2 = \phi$.

Definition 21. An NSS $(\Pi, S, V, (\Gamma_1, \Gamma_2, \Gamma_3))$ has level 2 if $|V_i| = |S|/2$ for all i and $H(S|B) = H(S)/2$ for $\forall B \in \Gamma_2$.

Theorem 22. *If there exists an NSS $(\Pi, S, V, (\Gamma_1, \Gamma_2, \Gamma_3))$ with level 2, then*

C1. *$G(P, [\tilde{\Gamma}_3]_2)$ is a disjoint union of cliques.*
C2. *$(\tilde{\Gamma}_1, \tilde{\Gamma}_2, \tilde{\Gamma}_3)$ is well defined. (see Def. 20)*
C3. *For $\forall B_1 \in \Delta_2^+$ and $\forall B_2 \in \Delta_2^+$ such that $B_1 \neq B_2$, $\sharp(B_1 \cap B_2) \leq 1$.*

Theorem 23. *Let $AG(2, q)$ be a two dimensional Euclidean space over $GF(q)$. Suppose that*

- *$(C1) \sim (C3)$ are satisfied.*
- *There exists a mapping ϕ from T to $AG(2, q)$ such that*
  (1) *For $\forall B \in \Delta_2^+$, $\forall T_i \in B$, all $\phi(T_i)$ lie on a line.*
  (2) *For $\forall (T_i, T_j, T_k) \in \Delta_1^-$, $\phi(T_i)$, $\phi(T_j)$ and $\phi(T_k)$ don't lie on a line.*

Then, there exists an NSS $(\Pi, S, V, (\Gamma_1, \Gamma_2, \Gamma_3))$ with level 2.

5.1 Proof of Theorem 22

First, we prove (C1).

Lemma 24. *1.* $\{V_i, V_j, V_k\} \in \Gamma_1^- \Rightarrow H(V_iV_j|SV_k) = H(V_i|SV_k) = 0$.
2. $\forall V_i, \forall V_j, H(V_j|SV_i) = 0$.
3. $\{V_i, V_j\} \in \Gamma_3 \Rightarrow H(V_i|V_j) = H(V_j|V_i) = 0$.

Proof. **(1)** $I(V_iV_j, S|V_k) = H(V_iV_j|V_k) - H(V_iV_j|SV_k) = H(S|V_k) - H(S|V_iV_jV_k) = H(S)$.

$$0 \le H(V_i|SV_k) \le H(V_iV_j|SV_k) = H(V_iV_j|V_k) - H(S)$$
$$\le H(V_iV_j) - H(S) \le H(V_i) + H(V_j) - H(S) \le |V_i| + |V_j| - |S| = 0.$$

The last line comes from $|V_i| = |V_j| = |S|/2$.
(2) From (P4), there exists $A_1, \cdots, A_k \in \Gamma_1^-, V_1', \cdots, V_{k-1}'$ such that $(V_i, V_1' \in A_1), (V_1', V_2' \in A_2), \cdots, (V_{k-1}', V_j \in A_k)$. From (1) of this lemma,

$$H(V_j|SV_{k-1}') = H(V_{k-1}'|SV_j) = 0, \quad H(V_{k-1}'|SV_{k-2}') = H(V_{k-2}'|SV_{k-1}') = 0, \quad \cdots$$

Then, from Lemma 3 (2),

$$H(V_i|SV_j) = H(V_i|SV_{k-1}') = H(V_i|SV_{k-2}') = \cdots = H(V_i|SV_1') = 0 .$$

(3) Since $\{V_i, V_j\} \in \Gamma_3$, $I(S, V_i|V_j) = H(S|V_j) - H(S|V_iV_j) = H(S) - H(S) = 0$. Then from (2) of this lemma, $H(V_i|V_j) = H(V_i|SV_j) + I(V_i, S|V_j) = 0$. □

Lemma 25. $\{V_i, V_j\} \in \Gamma_3, \{V_i, V_k\} \in \Gamma_3 \Rightarrow \{V_j, V_k\} \in \Gamma_3$.

Proof. From Lemma 24 (3), $H(V_i|V_j) = H(V_j|V_i) = 0$. Then from Lemma 3 (2),

$$H(S|V_kV_j) = H(S|V_kV_i) = H(S) \quad (\textit{because } \{V_i, V_k\} \in \Gamma_3) \Rightarrow \{V_j, V_k\} \in \Gamma_3.$$

Thus (C1) is proved by Lemma 25.

Now, we prove (C2). In the following, W denotes a subset of 2^P. If there is no confusion, we use W also to denote the shares given to W. Q_i, U_i and B_i are similar. □

Lemma 26. $\forall W(\ne \phi) \subseteq T_i, \forall P_z \in T_i,\ H(S|XW) = H(S|XV_z)$.

Proof. Let $W = \{P_1, \ldots, P_k\}$. From Lemma 24 (3), $H(V_k|V_{k-1}) = 0$. Then, from Lemma 4, $H(S|XV_1 \cdots V_{k-1}V_k) = H(S|XV_1 \cdots V_{k-1})$. Similarly,

$$H(S|XV_1 \cdots V_{k-1}) = \cdots = H(S|XV_1) .$$

Again, from Lemma 24 (3), $H(V_1|V_z) = H(V_z|V_1) = 0$. Then, from Lemma 3 (2), $H(S|XV_1) = H(S|XV_z)$. □

Def. 20 (1). Suppose that $\{T_i,\ldots,T_k\} \in \Delta_1$. Then, from the definition of Δ_1,

$$\exists Q_i(\neq \phi) \subseteq T_i,\cdots,\exists Q_k(\neq \phi) \subseteq T_k, \quad Q_i \cup \cdots \cup Q_k \in \tilde{\Gamma}_1 \ .$$

For $\forall P_i \in T_i,\cdots,\forall P_k \in T_k$, let $W = Q_i, P_z = P_i$ in Lemma 26. Then $H(S|Q_i \cup \cdots \cup Q_k) = H(S|V_i \cdots Q_k)$. Similarly, $0 = H(S|Q_i \cup \cdots \cup Q_k) = H(S|V_i \cdots V_k)$.

Def. 20 (2). Similar to (1).

Def. 20 (3). Let $\{P_1,\ldots,P_k\}$ be a subset of T_i. In Lemma 26, let $W = \{P_1,\ldots, P_k\}, P_z = P_1$. Then $H(S|V_1 \cdots V_k) = H(S|V_1) = H(S)$.

Def. 20 (4). Suppose that $\Delta_1 \cap \Delta_2 \neq \phi$. Let $\{T_i,\ldots,T_k\} \in \Delta_1 \cap \Delta_2$, Then,

$$\exists Q_i \subseteq T_i,\cdots,\exists Q_k \subseteq T_k, \quad Q_i \cup \cdots \cup Q_k \in \tilde{\Gamma}_1,$$

$$\exists U_i \subseteq T_i,\cdots,\exists U_k \subseteq T_k, \quad U_i \cup \cdots \cup U_k \in \tilde{\Gamma}_2 \ .$$

From Lemma (26) and the properties of Def. (20)(1)(2), $\forall P_i' \in T_i,\cdots,\forall P_k' \in T_k$,

$$0 = H(S|Q_i \cdots Q_k) = H(S|V_i' \cdots V_k') \ .$$

$$H(S)/2 = H(S|U_i \cdots U_k) = H(S|V_i' \cdots V_k') \ .$$

This is a contradiction. □

Finally, we prove (C3).

Lemma 27. *If* $\{T_i,T_j,T_k\} \in \Delta_2$, *then* $\forall P_i \in T_i, \forall P_j \in T_j, \forall P_k \in T_k,\ H(V_i|V_jV_k) = 0$.

Proof. $H(V_i|V_jV_k) = H(V_i|SV_jV_k) + I(V_i,S|V_jV_k)$

$$\begin{aligned} &\leq H(V_i|SV_j) + H(S|V_jV_k) - H(S|V_iV_jV_k) \\ &= H(S|V_jV_k) - H(S|V_iV_jV_k) \quad \text{(from Lemma 24 (2))} \\ &= H(S)/2 - H(S)/2 = 0 \ . \quad \text{(from Lemma 19 and C2).} \end{aligned}$$

□

Suppose that $\sharp(B_1 \cap B_2) \geq 2$. Let $T_i \in B_1 \cap B_2, T_j \in B_1 \cap B_2$. From Lemma 19, $\{T_i, T_j\} \in \Delta_2$. Then it is easily proved that, $\forall T_k \in B_2 \backslash B_1$, $\{T_i,T_j,T_k\} \in \Delta_2$. From Lemma 27, $H(V_k|V_iV_j) = 0$ for $\forall P_i \in T_i, \forall P_j \in T_j, \forall P_k \in T_k$. Then, from Lemma 4, $H(S|B_1V_k) = H(S|B_1) = H(S)/2$. Therefore, $B_1 \cup P_k \in \Delta_2$. This is a contradiction. □

5.2 Proof of Theorem 23

Lemma 28. :

(1) If $\{T_i, \ldots, T_k\} \in \Delta_1$, *then* $\forall Q_i(\neq \phi) \subseteq T_i, \cdots, \forall Q_k(\neq \phi) \subseteq T_k, Q_i \cup \cdots \cup Q_k \in \Gamma_1$.

(2) If $\{T_i, \ldots, T_k\} \in \Delta_2$, *then* $\forall Q_i(\neq \phi) \subseteq T_i, \cdots, \forall Q_k(\neq \phi) \subseteq T_k, Q_i \cup \cdots \cup Q_k \in \Gamma_2$.

Proof. **(1)** From C2 (Def. 20(1)) and because $\tilde{\Gamma}_1$ is monotone.
(2) From C2 (Def. 20(2)) and because $\tilde{\Gamma}_1 \cup \tilde{\Gamma}_2$ is monotone, $Q_i \cup \cdots \cup Q_k \in \tilde{\Gamma}_1 \cup \tilde{\Gamma}_2$. If $Q_i \cup \cdots \cup Q_k \in \tilde{\Gamma}_1$, $\{T_i, \ldots, T_k\} \in \Delta_1$. This is against C2 (Def. 20(4)). □

Suppose that an NSS for $(\Delta_1, \Delta_2, \Delta_3)$, $\Sigma = (\hat{\Pi}, S, U, (\Delta_1, \Delta_2, \Delta_3))$ with level 2, has be found. Let u_i be the share given to T_i by this scheme. Then, we can obtain an NSS for $(\Gamma_1, \Gamma_2, \Gamma_3)$, $\hat{\Sigma} = (\Pi, S, V, (\Gamma_1, \Gamma_2, \Gamma_3))$ with level 2, by letting $v_j = u_i, \forall P_j \in T_i$. (C2) and Lemma 28 guarantee that $\hat{\Sigma}$ is really an NSS with level 2. We show Σ in what follows.

Lemma 29. $\forall A \in \Delta_1^-, \sharp A = 3$.

Proof. Let $A \in \Delta_1^-$. Then from (P2), it is clear that $\#A \leq 3$. From Lemma 19, if $\#A = 2, A \in \Delta_2$. Furthermore, $\Delta_1 \cap \Delta_2 = \phi$ from (C2). Therefore, $\#A = 3$. □

From (P5), we can express s as $s = s_1 \circ s_2$ such that $s_1 \in GF(q)$ and $s_2 \in GF(q)$. ($\circ$ means concatenation.) Let $\phi(T_i) = (x_i, y_i)$, where (x_i, y_i) is the coordinate of a point in $AG(2, q)$. The public information of each T_i is (x_i, y_i). The secret share u_i of T_i is given by

$$u_i = s_1 x_i + s_2 y_i + r \ .$$

over GF(q), where r is a random number. Then, it is clear that

(1) Any T_i alone has no information on S. (see Lemma 19)
(2) If $(T_i, T_j, T_k) \in \Delta_1^-$, they can recover S by solving simultaneous equations on (s_1, s_2, r) because they don't lie on a line. (see Lemma 29)
(3) $(\forall T_i, \forall T_j)$ have a half information on S. (see Lemma 19) $\forall B \in \Delta_2^+$ have also a half information on S. Because they lie on a line, the rank of simultaneous equations is 2.
(4) $|U_i| = |S|/2$ because $u_i \in GF(q)$ and $s_1 \in GF(q), s_2 \in GF(q)$.

Finally, the public information of $\forall P_j \in T_i$ is given by (x_i, y_i) and the secret share v_j of $\forall P_j \in T_i$ is given by $v_j = u_i$. It is easy to see that this scheme is an NSS with level 2.

5.3 Example

We show an example of Theorem 4. Let

- $P = \{P_1, \cdots, P_6\}$.
- $\tilde{\Gamma}_3 = \{\{P_1\}, \cdots, \{P_6\}, \{P_5, P_6\}\}$.
- $\tilde{\Gamma}_2 = (\{\{P_i, P_j\} | i \neq j\}) \backslash \{\{P_5, P_6\}\} \cup \{\{P_1, P_2, P_3\}, \{P_3, P_4, P_5\}, \{P_3, P_4, P_6\}\}$.
- $\tilde{\Gamma}_1 = 2^P \setminus (\tilde{\Gamma}_2 \cup \tilde{\Gamma}_3)$.
- $q = 3, \quad \sharp S = 3^2 = 9$.

Then, it is clear that

1. $P1 \sim P5$ are satisfied.
2. $G(P, [\tilde{\Gamma}_3]_2)$ is a disjoint union of cliques.

From Def. 9, we have

$$T_1 = \{P_1\}, \cdots, T_4 = \{P_4\}, T_5 = \{P_5, P_6\}, \quad T = \{T_1, \cdots, T_5\} .$$

From Def. 10 and Lemma 5. 1, we have

$$\begin{aligned} \Delta_3 &= \{\{T_1\}, \cdots, \{T_5\}\} , \\ \Delta_2 &= \{\{T_i, T_j\} | i \neq j\} \cup \{\{T_1, T_2, T_3\}, \{T_3, T_4, T_5\}\} , \\ \Delta_1 &= 2^T \setminus (\Delta_2 \cup \Delta_3) . \end{aligned}$$

It is easy to see that $C1 \sim C3$ are satisfied.

Now, we show an NSS with level 2. First, express the secret s as (s_1, s_2) such that $s_1 \in GF(3)$ and $s_2 \in GF(3)$. Because

$$\begin{aligned} \Delta_2^+ &= \{\{T_1, T_2, T_3\}, \{T_3, T_4, T_5\}, \{T_1, T_4\}, \{T_1, T_5\}, \{T_2, T_4\}, \{T_2, T_5\}\}, \\ \Delta_1^- &= \{\forall \{T_i, T_j, T_k\}\} \setminus \{\{T_1, T_2, T_3\}, \{T_3, T_4, T_5\}\}, \end{aligned}$$

we choose $(x_i, y_i) \in AG(2, 3)$ as follows.

$$\begin{aligned} (x_1, y_1) &= (0, 0) , \\ (x_2, y_2) &= (0, 1) , \\ (x_3, y_3) &= (0, 2) , \\ (x_4, y_4) &= (1, 2) , \\ (x_5, y_5) &= (2, 2) . \end{aligned}$$

It is easily verified that these points satisfy the condition of Theorem 4. Let

$$u_i = s_1 x_i + s_2 y_i + r \bmod 3 \qquad (i = 1, \cdots, 5),$$

where r is a random number. The public information of P_i is

- (x_i, y_i) for $i \neq 6$.
- (x_5, y_5) for $i = 6$.

The secret share v_i of P_i is

- $v_i = u_i$ for $i \neq 6$.
- $v_6 = u_5$ for $i = 6$.

5.4 Matroid

Consider the NSS scheme given in the proof of Theorem 23. Let the public information of P_i be (x_i, y_i) and let

$$M = \begin{pmatrix} x_1, y_1, 1 \\ x_2, y_2, 1 \\ \cdots \end{pmatrix} .$$

Then,

$$\begin{pmatrix} v_1 \\ v_2 \\ \vdots \end{pmatrix} = M \begin{pmatrix} s_1 \\ s_2 \\ r \end{pmatrix} . \tag{7}$$

Let

$$\mathcal{I} \triangleq \{independent\ rows\ of\ M\} .$$

Then, it is clear that $(P, \mathcal{I})$ is a representable matroid of rank 3 [14].

References

1. Blakley, G.R. : Safeguarding cryptographic keys. Proc. of the AFIPS 1979 National Computer Conference **48** (1979) 313-317
2. Shamir, A. : How to share a secret. Communications of the ACM **22** (1979) 612-613
3. Itoh, M., Saito, A., Nishizeki, T. : Secret sharing scheme realizing general access structure. Proc. of IEEE Globecom '87 Tokyo (1987) 99-102
4. Benaloh, J.C., Leichter, J. : Generalized secret sharing and monotone functions. Crypto'88 (1990) 27-36
5. Brickell, E.F., Davenport, D.M. : On the classification of ideal secret sharing schemes. Journal of Cryptology **4** (1991) 123-134
6. Capocelli, R.M., De Santis, A., Gargano, L., Vaccaro, U. : On the size of shares for secret sharing schemes. Crypto'91 (1991) 101-113
7. Karnin, E.D., Green, J.W., Hellman, M.E. : On secret sharing systems. IEEE Trans. IT-29 (1982) 35-41
8. Brickell, E.F., Stinson, D.R. : Some improved bounds on the information rate of perfect secret sharing schemes. Crypto'90 (1990) 242-252
9. Blund, C., De Santis, A., Stinson, D.R., Vaccaro, V. : Graph decomposition and secret sharing schemes. Eurocrypt'92 (1992) 1-20
10. Blund, C., De Santis, A., Gargano, L., Vaccaro, U. : On the information rate of secret sharing schemes. Crypto'92 (1992)
11. Stinson, D.R. : New general bounds on the information rate of secret sharing schemes. Crypto'92 (1992)
12. Blakeley, G.R., Meadows, C. : Security of ramp schemes. Crypto'84 242-268 (1984)
13. Berge, C. : Graphs and Hypergraphs. North Holland (1973)
14. Welsh, D.J.A. : Matroid Theory. Academic Press (1976)

A Construction of Practical Secret Sharing Schemes using Linear Block Codes

Michael Bertilsson and Ingemar Ingemarsson

Department of Electrical Engineering

Linköping University, 581 83 Linköping, Sweden

Abstract. In this paper we address the problem of constructing secret sharing schemes for general access structures. The construction is inspired by linear block codes. Already in the beginning of the eighties constructions of threshold schemes using linear block codes were presented in [6] and [7]. In this paper we generalize those results to construct secret sharing schemes for arbitrary access structure. We also present a solution to the problem of retrieving the secret.

1 Introduction

Secret sharing is a method of dividing a secret into shares that can be distributed among a set of participants. The shares are chosen in such a way that only some predefined subsets of participants can retrieve the secret. The idea of secret sharing was first presented independently in [2][3]. Since then several papers has been written on the subject, see [4] for a list of references.

We will use linear codes to do the construction of our secret sharing schemes. Linear error correcting block codes of length n, dimension k and alphabet q is a k-dimensional sub space of the n-dimensional space. For a thorough discussion of linear codes see [1]. Linear codes has been used earlier in some constructions of threshold schemes [6][7].

In [8] a construction using matroids is presented. Our construction can be seen as special case of that one, actually they show that most of the other known constructions can be seen as special cases of their construction. However that construction has drawbacks from a practical point of view. Assume that we have N participants and a secret of size q. They need to store a matrix of size $\sim q^N \times N$ with entries of size q, even if the matrix is public and only have to be stored in one place it requires far to much storage. Especially since q has to be large to give a secure system.

In our construction we only need a generator matrix of size $\sim N \times N$ with entries of size q. Note that the size of the matrix is independent of q. This matrix can be made public. We also give an algorithm that constructs our generator matrix for a given access structure. Finally, given the shares from a set of participants who should be able to retrieve the secret we show how to retrieve the secret. Furthermore, our construction will always give perfect schemes and in many cases the schemes will be ideal.

The basic properties of secret sharing schemes and the notation used are introduced in section 2. Our construction and the algorithm constructing the generator matrix is presented in section 3. Finally in section 4 we show how to retrieve the secret.

2 Notation and Basic results

2.1 Secret Sharing

Let $P = \{P_1, P_2, \ldots, P_N\}$ be the set of all participants. Then 2^P is the set of all subsets. The size of 2^P is 2^N.

The set of all subsets of users that can recover the secret will be called the *access structure* and will be denoted F, $F \subseteq 2^P$. The set of all subsets that can not recover the secret will be denoted $\bar{F}$, $\bar{F} \subseteq 2^P$.

Let us look at the following situation. We have a group of users that can recover the secret. What happens if we add more users to this group? It seams reasonable that this new group should still be able to recover the secret. This property means that F is a monotone set which can be written formally as follows

$$A, B \subseteq 2^P, A \in F, A \subseteq B \Rightarrow B \in F \tag{2.1}$$

For small sets P we can use F or $\bar{F}$ to describe the access structure. But the size of 2^P grows exponentially with the size of P. This means that in most cases both F and $\bar{F}$ will be large and it will be difficult to describe the access structure using neither F nor $\bar{F}$. Fortunately we can use the fact that F is monotone to give an easier description of the access structure. Since F is monotone it is described by a unique minimal subset in the following sense. A set A is *minimal* if the following condition is satisfied.

$$A, B \subseteq 2^P, A \in F, B \subseteq A \Rightarrow B \notin F \tag{2.2}$$

We denote the set of these sets Γ and use it to give a short description of F. Instead of writing Γ as a set, we will write it as a sum of products.

$$\Gamma = \sum_i \gamma_i \tag{2.3}$$

Each term γ_i is a product of P_i:s representing a minimal subset in F. In a similar way $\bar{F}$ can be represented by its maximal subsets denoted $\bar{\Gamma}$.

From now on we will refer to the access structure using either $F, \bar{F}, \Gamma, \bar{\Gamma}$. When we want to make it explicit that we refer to $\bar{F}$ or $\bar{\Gamma}$ we will call them the negative access structure.

We can regard (2.3) as a logical expression. By this we mean that + represents "or" and juxtaposition represents "and". Then we can read $\Gamma = AB + AC + AD$ as, (user A and B) or (user A and C) or (user A and D). This means that we can use relations and concepts from logic. Let us now introduce the dual of Γ.

Definition 1

The dual of an access structure Γ is constructed by replacing each sum in Γ by a product and each product by a sum. The dual is denoted by F^* or Γ^*.

As in the case of Γ we write the dual access structure as a sum of products

$$\Gamma^* = \sum_i \gamma_i^* . \tag{2.4}$$

A dual has the property that the dual of a dual is the primal, i.e. $\Gamma^{**} = \Gamma$. It should also be pointed out that an access structure can be the dual of itself, i.e. $\Gamma = \Gamma^*$.

Usually we describe an access structure using Γ and we would like an easy way of getting $\bar{\Gamma}$ from Γ. The dual can help us with this. The following theorem is a minor rewriting of some results from [5] and we state it without proof.

Theorem 1

Given an access structure Γ we can get $\bar{\Gamma}$ using the dual in the following way. For each term γ_i^* in Γ^* we get a term $\bar{\gamma}_i$ in $\bar{\Gamma}$ as

$$\bar{\gamma}_i = P \backslash \gamma_i^* \tag{2.5}$$

Finally we remind the reader of the following two definitions.

Definition 2

A secret sharing scheme is called *perfect* if each group of users not in F has the same knowledge about the secret.

Definition 3

The information *rate* of a secret sharing scheme is defined as the ratio between the size of the secret and the size of the shares. A perfect scheme with information rate one is called *ideal*.

3 The Construction

We assume that we have a secret that we want to distribute among a set of participants. To do this we have to make the following construction. The construction is divided into two steps. First we use the access structure to construct a generator matrix, G, over $GF(q)$. Then we use the secret to construct the information vector. We choose the information vector so the first component of the codeword will be the equal to the secret. Multiplying the generator matrix and the information vector we get a codeword. The components of the codeword are distributed as shares to the participants. The first component of the codeword can not be distributed to any participant since it is equal to the secret.

Definition 4

Denote a codeword $c = (c_0, c_i, \ldots, c_{n-1})$. The share of user P_i is the set of components $\{c_j\}$ for all j in the index set p_i. Let A be a subset of participants then

$$J_A = \bigcup_{P_i \in A} p_i \tag{3.1}$$

J_A is a union of index sets and is of course itself an index set.

The following theorem tells us how we should choose the columns of the generator matrix. We remind you that the secret is equal to c_0 and is thus associated with the index set $\{0\}$.

Theorem 2

A subset of users, A, can retrieve the secret, i.e. $A \in F$, if and only if the index sets J_A and $\{0\}$ fulfills the following requirement

$$rank(G(J_A \cup \{0\})) = rank(G(J_A)) \text{ for } A \in F. \tag{3.2}$$

A subset of users, B, that can not retrieve the secret, i.e. $B \in \overline{F}$, must fulfill the following requirement

$$rank(G(J_B \cup \{0\})) = rank(G(J_B)) + 1 \text{ for } B \in \overline{F}. \tag{3.3}$$

Proof

Equation (3.2) means that $G(\{0\})$ is linearly dependent of $G(J_A)$ and from Lemma 1 we know that this means that knowing $c(J_A)$ determines c_0 and we

2.2 Linear Codes

We use the following standard notation. Let G be the $k \times n$ generator matrix, H the $(n-k) \times n$ parity check matrix. G is a linear mapping from the information space U to the code space C. Both U and C are over $GF(q)$. As distance measure we use Hamming distance, denoted d. Hamming distance is defined as the number of positions in which two words differ. The weight, w, of a word is the number of non zero symbols in the word. d_{min} for a code is the minimum distance between any two code words. The minimum distance is equal to the weight of the code words with least weight.

H and G have the following relationship

$$GH^T = 0. \quad \textbf{(2.6)}$$

The parity check matrix also generates a code which is called the dual code. Let $c^{\perp}$ be a codeword in the dual code then

$$Gc^{\perp^T} = \bar{0} \quad \textbf{(2.7)}$$

for all codewords in the dual code. The same is also true for the parity check matrix and all codewords in the code.

We also need the following notation. Let A be an index set with values between 0 and $n-1$. $G(A)$ is a subset of the columns in G corresponding to A and $c(A)$ is a subset of the components in c corresponding to A.

We are interested in what happens when we add new columns from the generator matrix to our set of columns. Let A be an index set and let m be an index not in A. Let g be the column of G corresponding to index m. If $c_i(A) = c_j(A)$, what is the value of $c_i(m)$ and $c_j(m)$? There are two different cases to be studied, either g is linearly dependent of the columns in $G(A)$, or it is linearly independent. The following two lemmas are given without proof.

Lemma 1

Let g be linearly dependent of $G(A)$. If $c_i(A) = c_j(A)$ then $c_i(m) = c_j(m)$.

Lemma 2

Let g be linearly independent of $G(A)$. Choose a fix c_j. Then in the set $\{c_i, c_i(A) = c_j(A)\}$ there are exactly

$$q^{k - rank(G(A)) - 1} \quad \textbf{(2.8)}$$

codewords with $c_i(m) = a$ for every $a \in [0, \ldots, q-1]$.

know the secret. Equation (3.3) means that $G(\{0\})$ is linearly independent of $G(J_B)$ and from Lemma 2 we know that this means that for all possible values in $c(J_B)$ the value in c_0 takes on all values in $GF(q)$ with equal probability and we know nothing about the secret.

We have now chosen G and are ready to choose the information vector, u, using the secret, s. s must be an element of $GF(q)$. We choose the components of u such that

$$ug_o = s, \tag{3.4}$$

where g_0 is the first column in the generator matrix. This means that the first component of the codeword will be equal to the secret.

Now we have constructed both the generator matrix and the information vector. We multiply the information vector and the generator matrix to get a codeword.

$$c = uG \tag{3.5}$$

We can know distribute the different components of this codeword as shares to the users. The sets p_i determine which user should have which components.

Just as a remark, if G generates the access structure Γ, then the parity check matrix H of G generates the dual access structure Γ^* of Γ.

A few comments about this construction to make it clearer. If we are able to use only one column per user we get a $k \times (N+1)$ generator matrix and the system will be ideal. This is not possible for all access structures. It is important to notice that the shares of a subset $A \in F$ does not necessarily determine a single codeword. The shares determines a set of codewords that are equal in the positions $J_A \cup \{0\}$.

3.1 The Algorithm Constructing the Generator Matrix

In Theorem 2 we gave the requirements for the construction of the generator matrix. We will use this to give an algorithm that generates a generator matrix for a given access structure. Theorem 2 gives us relationships between the subsets of F and $\bar{F}$ and the column subsets of G. As stated in section 2 it is usually impractical to work with F and $\bar{F}$. Instead we would like to get conditions on G from Γ and $\bar{\Gamma}$. This is easily done. If (3.2) is true for an index set, J_A, it is also true if we add indices to J_A. So if we find a G that fulfills (3.2) for Γ, it will also fulfill it for F. In a similar way it is enough to use $\bar{\Gamma}$ instead of $\bar{F}$ in (3.3).

The algorithm is divided into two parts. First we use $\bar{\Gamma}$ to construct a generator matrix with some free parameters and then we use Γ to decide the values of these parameters.

Equation (3.3) says that if $B \in \bar{\Gamma}$, the first column of G is linearly independent of $G(J_B)$. This means that the null-space of $G(J_B)$ is q times larger then the null-space of $G(J_B \cup \{0\})$ and this can be written as

$$\exists u \text{ such that } uG(J_B) = \bar{0} \text{ and } uG(0) \neq 0. \tag{3.6}$$

If there are q^l codewords with zeros in the positions corresponding to J_B, q^{l-1} are equal to a in the first position for all $a \in GF(q)$.

From equation (3.8) and the comment following it we know that for each subset $B \in \bar{\Gamma}$ there should be at least one codeword with zeros in the positions corresponding to J_B and a non-zero symbol in the first position. (We actually get many such codewords.) One way to get such a codeword is to let it be a row in the generator matrix. This means that we will choose the entries of the generator matrix as follows.

Step 1 First we let the number of rows in G be equal to the number of terms in $\bar{\Gamma}$ and demand that all the entries in the first column of G are non-zero. Now we choose the rest of the entries in G using the terms in $\bar{\Gamma}$. Each term in $\bar{\Gamma}$ is associated with a row in G. Let g_i be the i:th row of G and let $g_i(J_B)$ be the index positions corresponding to the participants in B. Then we let $g_i(J_{\bar{\gamma}_i}) = 0$ and let the rest of the coefficients of G be undetermined, but non-zero.

Now we have a G with some unknown entries. This will be used in the next step where we will determine these unknown. Equation (3.2) says that if $A \in \Gamma$, the first column of G is linearly dependent of $G(J_A)$. This means that the null-space of $G(J_A)$ is equal to the null-space of $G(J_A \cup \{0\})$ and every codeword which is zero in the positions corresponding to J_A must also be zero in the first position. From this we can choose the values of the undecided g_{ij} as follows.

Step 2 Given the generator matrix constructed in the first step we will use Γ to choose the values of the undetermined coefficients of G. The following equation gives us demands on the undetermined coefficients of G.

$$\forall u, \text{ such that } uG(J_{\gamma_i}) = \bar{0}, uG(\{0\}) = 0 \tag{3.7}$$

3.2 Examples

We give three examples that are typical in that they show the problems that arise in the above algorithm. In the following examples we have chosen q small. Since q is the size of the secret q must in practice be made large. It should be large enough to make it improbable to be able to guess the value of the secret.

Example 3.1

Let $P = \{A, B, C\}$ and $\Gamma = A + BC$ then $\bar{\Gamma} = B + C$. From $\bar{\Gamma}$ we get $k = 2$ and

$$G = \begin{bmatrix} s_1 & a_1 & 0 & c_1 \\ s_2 & a_2 & b_2 & 0 \end{bmatrix} \tag{3.8}$$

Now we can use Γ. From $\gamma_1 = A$ and the first column we get

$$u_1 a_1 + u_2 a_2 = 0 \text{ and } u_1 s_1 + u_2 s_2 = 0. \tag{3.9}$$

Combining these we get

$$\left(s_1 - s_2 \frac{a_1}{a_2}\right) u_1 = 0 \Rightarrow s_1 - s_2 \frac{a_1}{a_2} = 0 \tag{3.10}$$

since u_1 could be non-zero. From $\gamma_2 = BC$ we get

$$u_2 b_2 = 0 \text{ and } u_1 c_1 = 0 \Rightarrow u_1 = u_2 = 0 \tag{3.11}$$

and we can choose b_2 and c_1 arbitrarily. We do not get any new demands on s_1 and s_2 either. In this case we can choose $q = 2$ and $s_1 = s_2 = a_1 = a_2 = b_2 = c_1 = 1$. We then get the following generator matrix

$$G = \begin{bmatrix} 1 & 1 & 0 & 1 \\ 1 & 1 & 1 & 0 \end{bmatrix}. \tag{3.12}$$

Example 3.2

Let $P = \{A, B, C\}$ and $\Gamma = AB + AC + BC$ then $\bar{\Gamma} = A + B + C$. From $\bar{\Gamma}$ we get $k = 3$ and

$$G = \begin{bmatrix} s_1 & 0 & b_1 & c_1 \\ s_2 & a_2 & 0 & c_2 \\ s_3 & a_3 & b_3 & 0 \end{bmatrix}. \tag{3.13}$$

We will not go through all the steps in this example. We just state that Γ gives us

$$-s_1\frac{b_3}{b_1}-s_2\frac{a_3}{a_2}+s_3 = 0 \tag{3.14}$$

$$-s_1\frac{c_2}{c_1}+s_2-s_3\frac{a_2}{a_3} = 0 \tag{3.15}$$

$$s_1-s_2\frac{c_1}{c_2}-s_3\frac{b_1}{b_3} = 0. \tag{3.16}$$

We now have three equations and nine unknowns. We start by choosing $s_i = 1$ and from equation (3.16) we see that we have to choose $a_2 \neq a_3$, otherwise $b_3 = 0$. This means that $q>2$. In a similar way $b_1 \neq b_3$ and $c_1 \neq c_2$. If we now choose $a_2 = b_1 = c_1 = 1$ and $a_3 = 2$ we get $b_3 = -1$ and $c_2 = 1-2^{-1}$. Finally we choose $q = 3$ and get

$$G = \begin{bmatrix} 1\ 0\ 1\ 1 \\ 1\ 1\ 0\ 2 \\ 1\ 2\ 2\ 0 \end{bmatrix}. \tag{3.17}$$

If we study this generator matrix we see that the last row is a linear combination of the first and the second row. This means that we can skip the last row and still get the same code. We can now write the generator matrix as

$$G = \begin{bmatrix} 1\ 0\ 1\ 1 \\ 1\ 1\ 0\ 2 \end{bmatrix}. \tag{3.18}$$

Example 3.3

Let $P = \{A, B, C, D\}$ and $\Gamma = AB+BC+CD$ then $\bar{\Gamma} = AC+AD+BD$. From $\bar{\Gamma}$ we get $k = 3$ and

$$G = \begin{bmatrix} s_1 & 0 & b_1 & 0 & d_1 \\ s_2 & 0 & b_2 & c_2 & 0 \\ s_3 & a_3 & 0 & c_3 & 0 \end{bmatrix}. \tag{3.19}$$

From Γ we get

$$\gamma_1 = AB \Rightarrow s_2 = s_1\frac{b_2}{b_1} \tag{3.20}$$

$$\gamma_2 = BC \Rightarrow -s_1\frac{b_2}{b_1} + s_2 - s_3\frac{c_2}{c_3} = 0 \tag{3.21}$$

$$\gamma_3 = CD \Rightarrow s_2 = s_3\frac{c_2}{c_3} \tag{3.22}$$

Combining these three equations gives

$$-s_2 + s_2 - s_2 = -s_2 = 0. \tag{3.23}$$

This means that $s_2 = 0$, but we have a demand that $s_i \neq 0$. The only solution to the three equations is not allowed. We can not construct a generator matrix with one column per user for this access structure. This means that we in our model can not create an ideal secret sharing scheme for the access structure $\Gamma = AB + BC + CD$. To find a solution to this access structure we must add another column to one of the participants. We choose to add a column to participant B and then $\bar{\Gamma}$ gives us the following generator matrix

$$G = \begin{bmatrix} s_1 & 0 & b_{11} & b_{12} & 0 & d_1 \\ s_2 & 0 & b_{21} & b_{22} & c_2 & 0 \\ s_3 & a_3 & 0 & 0 & c_3 & 0 \end{bmatrix}. \tag{3.24}$$

We must choose the b_{ij} so that the two columns are linearly independent otherwise it is useless to have two columns. This means that the only u that is orthogonal to the A column and the two B columns is $u = (0, 0, 0)$. We do not get any restrictions on the b_{ij} and a_i from $\gamma_1 = AB$. Neither does $\gamma_2 = BC$ give any restrictions on the b_{ij} and c_i. The only restrictions left are equation (3.24) and that the columns of B must be independent. For $q = 3$ we can choose

$$G = \begin{bmatrix} 1 & 0 & 1 & 1 & 0 & 1 \\ 1 & 0 & 1 & 2 & 1 & 0 \\ 1 & 1 & 0 & 0 & 1 & 0 \end{bmatrix}. \tag{3.25}$$

In the first example we get the generator matrix directly. In the second we can construct the generator matrix but some of the rows are linearly dependent and we can remove them from the generator matrix. In the last example we are not able to construct the generator matrix with only one column per participant. Instead we are forced to add extra column(s) to one or more of the participants. These three examples cover all possible cases.

4. Retrieving the Secret

We will regard the retrieval of the secret as a special case of erasure decoding of linear block codes. It is important to notice that the number of erased positions is usually more than $d_{min} - 1$ which means that we will not be able to determine all erased symbols. Due to the given construction we are always able to determine the secret, i.e. the first component of the codeword.

Let us look at the shares as a part of a received codeword that is erased in some of the positions. Use the received vector, v, and the error vector, e, as follows

$$v = (0, 0, a_i, \ldots) \tag{4.1}$$

$$e = (c_0, c_1, 0, \ldots). \tag{4.2}$$

v is zero in all positions where there are not any known shares and has values according to the shares in the others. e is zero in all positions corresponding to the known shares and is assumed to be equal to the codeword in the others. Let B denote the index set of the unknown shares. Observe especially that index 0 is in B. We can now set up the following equations

$$s = Hv^T = H(c+e)^T = He^T = H(B)e(B)^T, \tag{4.3}$$

since $Hc^T = 0$ by definition. Since both v and H are known s is also known. $H(B)$ is in general a rectangular matrix with more columns than rows. To be able to reconstruct *all* the unknown c_i, which are in $e(B)$, we have to find the pseudo-inverse of $H(B)$. I.e. find

$$H(B)^{-1}H(B) = I_{|B|}, \tag{4.4}$$

where $I_{|B|}$ is the identity matrix of sizc $|B|$. If we could find $H(B)^{-1}$ we could get the missing values in e from s in the following way

$$e(B)^T = H(B)^{-1}s. \tag{4.5}$$

The problem is that we can not always find a $H(B)^{-1}$ fulfilling the demands in equation (4.4). However we are in a better position in that we only have to find c_0, and it can be found if we are able to find the first row of $H(B)^{-1}$. Denote this first row of $H(B)^{-1}$ l then we have

$$c_0 = e_0 = -\sum_{i=0}^{k-1} l_i s_i. \tag{4.6}$$

l has the following properties, $lh_0 = 1$ and $lh_i = 0, i \in B$, where h_i are the columns of H.

This means that l is not orthogonal to the first column of $H(B)$, but is orthogonal to the rest of the columns. We can find such an l if and only if h_0 is linearly independent of the rest of the h_i, $i \in B$. To prove that this is true in our construction we first state the following lemma about G and H.

Lemma 3

Let A be an index set with values from the set $\{0, \ldots, n-1\}$ and let $\backslash A$ be all indices not in A. Then

$$rank(G(A)) = |A| \Rightarrow rank(H(\backslash A)) = n-k \quad \textbf{(4.7)}$$

$$rank(G(A)) < |A| \Rightarrow rank(H(\backslash A)) < n-k \quad \textbf{(4.8)}$$

Proof

Remember that $Gc^{\perp^T} = \bar{0}, \forall c^{\perp}$ in the dual code. We first prove (4.7). We take any codeword in the dual code such that $c^{\perp}(\backslash A) = \bar{0}$. Then $c^{\perp}(A)$ must be equal to $\bar{0}$ for all these codewords since the columns in $G(A)$ are linearly independent. So if $c^{\perp}(\backslash A) = \bar{0}$ we have the all zero codeword. This means that $uH(\backslash A) = \bar{0}$ if and only if $u = \bar{0}$ and $rank(H(\backslash A)) = n-k$.

Now we prove (4.8). Since the columns in $G(A)$ are linearly dependent there exist codewords in the dual code, such that $c^{\perp}(\backslash A) = \bar{0}$ and $c^{\perp}(A)$ are not all zero. This means that for some vector $u \neq \bar{0}$, $uH(\backslash A) = \bar{0}$ and the rank of $H(\backslash A)$ must be less than the number of rows.

Theorem 3

h_0 is linearly independent of all h_i, $i \in B$.

Proof

Let $\gamma \in \Gamma$, then the following is always true, $rank(G(J_\gamma)) \leq |J_\gamma|$. $rank(G(J_\gamma)) = |J_\gamma|$ follows as a special case so lets assume that $rank(G(J_\gamma)) < |J_\gamma|$. Choose a set $J \subset J_\gamma$ such that $rank(G(J)) = |J| = rank(G(J_\gamma))$. From Lemma 3 and Theorem 2 $rank(H(\backslash J)) = n-k$ and $rank(H(\backslash (J \cup \{0\}))) < n-k$. This means that h_0 is independent of h_i, $i \in \backslash J$. If we add indices to J to get J_γ we will remove indices from $\backslash J$. If h_0 is independent of all columns h_i, $i \in \backslash J$ then it must still be independent when we have removed some of the columns.

5 Conclusions

We have constructed a secret sharing scheme using linear block codes. We show how to choose the columns of a generator matrix so that we get a desired access structure. We have also given an algorithm that use the access structure Γ and the negative access structure $\overline{\Gamma}$ to construct the generator matrix of the code. The size of this generator matrix is only $\sim N \times N$ for N users. Finally we show how to retrieve the secret given the shares of a set of participants that should be able to retrieve the secret.

References

[1] F.J. MacWilliams, N. J. A. Sloane, *The Theory of ErrorCorrecting Codes*, North-Holland, First Edition 1977

[2] A. Shamir, *How to Share a Secret*, Comm. ACM, Vol.22, pp612, Nov 1979

[3] G. R. Blakely, *Safeguarding Cryptographic Keys*, Proc. AFIPS 1979 Natl. Computer Conference, New York, Vol.48, pp 313-317, June 1979

[4] G. Simmons Ed, *Contemporary Cryptology*, IEEE Press, 1992

[5] G. Simmons, W. Jackson, K. Martin, *The Geometry of Shared Secret Schemes*, Bulletin of the Institute of Combinatorics and its Application (ICA) to appear 1991

[6] R. J. McEliece, D. V. Sarwate, *On Sharing Secrets and Reed-Solomon Codes*, Comm. ACM, Vol.24, pp 583-584, Sep 1981

[7] E.D Karnin, J.W. Green, M.E. Hellman, *On Secret Sharing Systems*, IEEE Trans. Information Th., Vol. IT-29, No 1, pp. 35-41, Jan 1983

[8] K.M. Martin, *Discrete Structures in the Theory of Secret Sharing*, Doct. Thesis, Royal Holloway and Bedford New College, University of London, 1991

Session 3

SIGNATURES AND HASHING ALGORITHMS

Chair: Yvo Desmedt
(University of Wisconsin-Milwaukee, US)

HAVAL — A One-Way Hashing Algorithm with Variable Length of Output (Extended Abstract)

Yuliang Zheng *, Josef Pieprzyk ** and Jennifer Seberry ***

Department of Computer Science
University of Wollongong, Wollongong, NSW 2522, Australia
E-mail: {yuliang, josef, jennie}@cs.uow.edu.au

Abstract. A one-way hashing algorithm is a deterministic algorithm that compresses an arbitrary long message into a value of specified length. The output value represents the fingerprint or digest of the message. A cryptographically useful property of a one-way hashing algorithm is that it is infeasible to find two distinct messages that have the same fingerprint. This paper proposes a one-way hashing algorithm called HAVAL. HAVAL compresses a message of arbitrary length into a fingerprint of 128, 160, 192, 224 or 256 bits. In addition, HAVAL has a parameter that controls the number of passes a message block (of 1024 bits) is processed. A message block can be processed in 3, 4 or 5 passes. By combining output length with pass, we can provide fifteen (15) choices for practical applications where different levels of security are required. The algorithm is very efficient and particularly suited for 32-bit computers which predominate the current workstation market. Experiments show that HAVAL is 60% faster than MD5 when 3 passes are required, 15% faster than MD5 when 4 passes are required, and as fast as MD5 when full 5 passes are required. It is conjectured that finding two collision messages requires the order of $2^{n/2}$ operations, where n is the number of bits in a fingerprint.

1 Introduction

A one-way hashing algorithm is a deterministic algorithm that compresses an arbitrarily long message into a value of specified length. The output value represents the fingerprint or digest of the input message. A very useful property of a one-way hashing algorithm is that it is *collision intractable*, i.e., it is computationally infeasible to find a pair of messages that have the same fingerprint. One-way hashing algorithms are widely used in information authentication, in particular, in digital signature, and have received extensive research

* Supported in part by the Australian Research Council under the reference number A49232172.

** Supported in part by the Australian Research Council under the reference number A49131885.

*** Supported in part by the Australian Research Council under the reference numbers A49130102, A9030136, A49131885 and A49232172.

since the invention of public key cryptography by Diffie and Hellman [DH76] and by Merkle [Mer78]. Theoretical results on one-way hashing algorithms were obtained by Damgård [Dam87, Dam90]. Results on a weaker version of one-way hashing algorithms, universal one-way hashing algorithms, can be found in [NY89, ZMI91, Rom90].

Recently much progress has been made in the design of practical one-way hashing algorithms which are suited for efficient implementation by software. Notable work includes the MD family which consists of three algorithms MD2, MD4 and MD5 [Kal92, Riv92a, Riv92b], the federal information processing standard for secure hash (SHS) proposed by the National Institute of Standards and Technology (NIST) of the United States [NIS92], and Schnorr's hashing algorithm FFT-Hash based on fast Fourier transformations [Sch92, Vau92]. All these algorithms output fingerprints of fixed length. In particular, fingerprints of FFT-Hash and the algorithms in the MD family are of 128 bits, while fingerprints of SHS are of 160 bits which is designed primarily for NIST's proposed digital signature standard DSS [NIS91].

Despite the progress, little work has been done in the design of one-way hashing algorithms that can output fingerprints of variable length (except the proposal of Snefru, whose two-pass version has been found to be insecure [Mer90]). Such an algorithm would be more flexible and hence more suited for various applications where variable length fingerprints are required. The aim of this research is to design a one-way hashing algorithm that can output fingerprints of 128, 160, 192, 224 or 256 bits. These different lengths for fingerprints provide practical applications with a broad spectrum of choices. The algorithm, which we call HAVAL, uses some of the principles behind the design of the MD family. In addition, HAVAL makes an elegant use of Boolean functions recently discovered by Seberry and Zhang [SZ92]. These functions have nice properties which include

1. they are 0-1 balanced,
2. they are highly non-linear,
3. they satisfy the Strict Avalanche Criterion (SAC),
4. they can not be transformed into one another by applying linear transformation to the input coordinates and
5. they are not mutually correlated via linear functions or via bias in output.

In addition, the number of passes each 1024-bit block of an input message is processed can be 3, 4 or 5. This adds one more dimension of flexibility to the algorithm. Combination of the two variable parameters, pass and output length, provides practical applications with fifteen different levels of security.

When compared with MD2, MD4, SHS and FFT-Hash, MD5 is considered much superior in terms of speed and security. In particular, MD5 is about 15% faster than SHS (See for example the note posted on the `sci.crypt` news group by Kevin McCurley, 5 September 1992), although the latter is very likely to become a standard. Our preliminary experiments show that HAVAL is at least 60% faster than MD5 when 3 passes are required, at least 15% faster than MD5 when 4 passes are required, and about as fast as MD5 when full 5 passes are required.

Detailed specifications of HAVAL are presented in Section 2. Section 3 discusses rationale behind the design of HAVAL. This is followed by a discussion about security issues of HAVAL in Section 4. Extensions of HAVAL in several directions are discussed in Section 5. Finally, Section 6 presents some concluding remarks.

2 Specifications of HAVAL

We begin with a general description of the algorithm. Detailed specifications of all parts of the algorithm follows.

First we introduce a few notations and conventions. We consider, unless otherwise specified, strings (or sequences) on $GF(2)$. Throughout the paper, a single bit from $GF(2)$ will be denoted by a lower case letter, while a string of bits on $GF(2)$ will be denoted by a upper case letter. A *byte* is a string of 8 bits, a *word* is a string of 4 bytes (32 bits) and a *block* is the concatenation of 32 words (1024 bits). We assume that the most significant bit of a byte appears at the left end of the byte. Similarly we assume that the most significant byte of a word comes at the left end of the word, and the most significant word of a block appears at the left end of the block. Note that a binary string $X = x_{n-1}x_{n-2}\cdots x_0$ can be viewed as an integer whose value is $I_X = x_{n-1}2^{n-1}+x_{n-2}2^{n-2}+\cdots+x_0 2^0$. Conversely an integer I can also be viewed as a binary string $X_I = x_{n-1}x_{n-2}\cdots x_0$ with $I = x_{n-1}2^{n-1}+x_{n-2}2^{n-2}+\cdots+x_0 2^0$.

The modulo 2 multiplication and modulo 2 addition of $x_1, x_2 \in GF(2)$ are denoted by x_1x_2 and $x_1 \oplus x_2$ respectively. The bit-wise modulo 2 addition operation of two binary strings S_1 and S_2 of the same length is denoted by $S_1 \oplus S_2$, and the bit-wise modulo 2 multiplication of the two strings S_1 and S_2 is denoted by $S_1 \bullet S_2$. Note that $\bullet$ has precedence over $\oplus$ in computation. Another notation $\boxplus$ is also used in the specifications. Assume that $S_1 = W_{1,n-1}W_{1,n-2}\cdots W_{1,0}$ and $S_2 = W_{2,n-1}W_{2,n-2}\cdots W_{2,0}$, where each $W_{i,j}$ is a 32-bit word, the word-wise integer addition modulo 2^{32} of the two strings is denoted by $S_1 \boxplus S_2$, i.e., $S_1 \boxplus S_2 = (W_{1,n-1}+W_{2,n-1} \bmod 2^{32})(W_{1,n-2}+W_{2,n-2} \bmod 2^{32})\cdots(W_{1,0}+W_{2,0} \bmod 2^{32})$. Note that in the definition of $\boxplus$ we have viewed each $W_{i,j}$ as an integer in $[0, 2^{32}-1]$.

Given a message M to be compressed, HAVAL pads (extends) M first. The length of (i.e., the number of bits in) the message after padding is a multiple of 1024, and padding is always applied even when the length of M is already a multiple of 1024. The last block of the padded message contains the number of bits in the unpadded message, the required number of bits in the fingerprint and the number of passes each message block is processed. It also indicates the version number of HAVAL. The current version number is 1.

Now suppose that the padded message is $B_{n-1}B_{n-2}\cdots B_0$, where each B_i is a 1024-bit block. HAVAL starts from the block B_0 and a 8-word (256-bit) constant string $D_0 = D_{0,7}D_{0,6}\cdots D_{0,0}$, which is taken from the fraction part of

$\pi = 3.1415...$, and processes the message $B_{n-1}B_{n-2}\cdots B_0$ in a block-by-block way. More precisely, it compresses the message by repeatedly calculating

$$D_{i+1} = H(D_i, B_i)$$

where i ranges from 0 to $n-1$ and H is called the basic compressing part of HAVAL. See Section 2.4 for the actual values of the 8 constant 32-bit words $D_{0,7}, D_{0,6}, \cdots, D_{0,0}$.

Finally, HAVAL adjusts, if necessary, the last 256-bit string D_n into a string of the length specified in the last block B_{n-1}, and outputs the adjusted string as the fingerprint of the message M. In summary, HAVAL processes a message M in the following three steps:

1. Pad the message M so that its length becomes a multiple of 1024. The last (or the most significant) block of the padded message indicates the length of the original (unpadded) message M, the required length of the fingerprint of M, the number of passes each block is processed and the version number of HAVAL.
2. Calculate repeatedly $D_{i+1} = H(D_i, B_i)$ for i from 0 to $n-1$, where D_0 is a 8-word (256-bit) constant string and n is the total number of blocks in the padded message.
3. Adjust the 256-bit value D_n obtained in the above calculation according to the fingerprint length specified in the last block B_{n-1}, and output the adjusted value as the fingerprint of the message M.

The second and the third steps are graphically dipicted in Figure 1. Detailed descriptions of the three steps are provided in the following sections.

2.1 Padding

The purpose of padding is two-fold: to make the length of a message be a multiple of 1024 and to let the message indicate the length of the original message, the required number of bits in the fingerprint, the number of passes and the version number of HAVAL. HAVAL uses a 64-bit field MSGLEN to specify the length of an unpadded message. Thus messages of up to $(2^{64}-1)$ bits are accepted, which is long enough for practical applications. HAVAL also uses a 10-bit field FPTLEN to specify the required number of bits in a fingerprint. In addition HAVAL uses a 3-bit field PASS to specify the number of passes each message block is processed, and another 3-bit field VERSION to indicate the version number of HAVAL. The number of bits in a fingerprint can be 128, 160, 192, 224 and 256, while the number of passes can be 3, 4 and 5. The current version number of HAVAL is 1.

HAVAL pads a message by appending a single bit 1 next to the most significant bit of the message, followed by zero or more bit 0s until the length of the (new) message is 944 modulo 1024. Then, HAVAL appends to the message the 3-bit field VERSION, followed by the 3-bit field PASS, the 10-bit field FPTLEN and the 64-bit field MSGLEN.

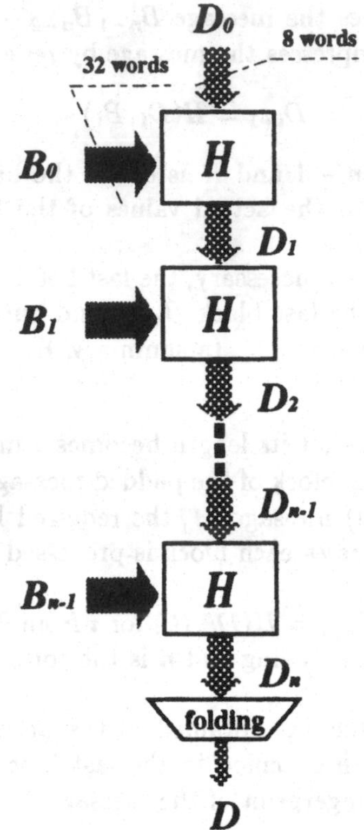

Fig. 1. Hashing a Message by HAVAL

2.2 The Basic Compressing Algorithm *H*

The basic compressing algorithm H processes a block in 3, 4 or 5 passes, which is specified by the 3-bit field PASS in the last block. Denote by H_1, H_2, H_3, H_4 and H_5 the five passes. Now suppose that the input to H is (D_{in}, B), here D_{in} is a 8-word string and B is a 32-word (1024-bit) block. Let D_{out} denote the 8-word output of H on input (D_{in}, B). Then processing of H can be described in th following way. (See also Figure 2.)

$$
\begin{aligned}
E_0 &= D_{in};\\
E_1 &= H_1(E_0, B);\\
E_2 &= H_2(E_1, B);\\
E_3 &= H_3(E_2, B);\\
E_4 &= H_4(E_3, B);\ \text{(if PASS=4, 5)}\\
E_5 &= H_5(E_4, B);\ \text{(if PASS=5)}
\end{aligned}
$$

$$D_{out} = \begin{cases} E_3 \boxplus E_0 \text{ if PASS=3} \\ E_4 \boxplus E_0 \text{ if PASS=4} \\ E_5 \boxplus E_0 \text{ if PASS=5} \end{cases}$$

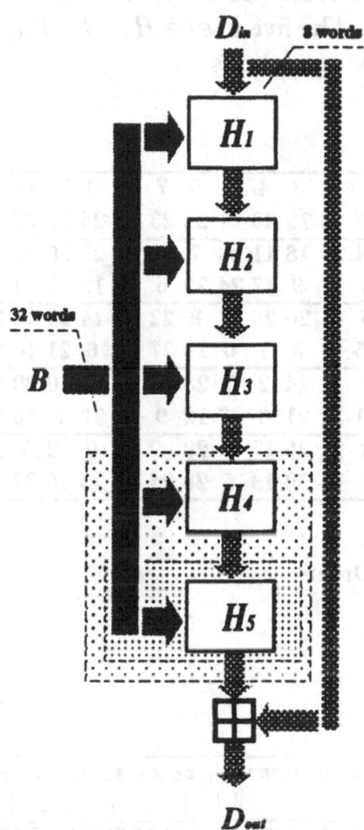

Fig. 2. The Basic Compressing Algorithm H

Each of the five passes H_1, H_2, H_3, H_4 and H_5 has 32 rounds of operations and each round processes a different word from B. The orders in which the words in B are processed differ from pass to pass. In addition, each pass employs a different Boolean function to perform bit-wise operations on words. The five functions employed by H_1, H_2, H_3, H_4 and H_5 are:

$$f_1(x_6, x_5, x_4, x_3, x_2, x_1, x_0) = x_1x_4 \oplus x_2x_5 \oplus x_3x_6 \oplus x_0x_1 \oplus x_0$$

$$f_2(x_6, x_5, x_4, x_3, x_2, x_1, x_0) = x_1x_2x_3 \oplus x_2x_4x_5 \oplus x_1x_2 \oplus x_1x_4 \oplus x_2x_6 \oplus x_3x_5 \oplus x_4x_5 \oplus x_0x_2 \oplus x_0$$

$$f_3(x_6, x_5, x_4, x_3, x_2, x_1, x_0) = x_1x_2x_3 \oplus x_1x_4 \oplus x_2x_5 \oplus x_3x_6 \oplus x_0x_3 \oplus x_0$$

$$f_4(x_6, x_5, x_4, x_3, x_2, x_1, x_0) = x_1x_2x_3 \oplus x_2x_4x_5 \oplus x_3x_4x_6 \oplus$$

$$x_1x_4 \oplus x_2x_6 \oplus x_3x_4 \oplus x_3x_5 \oplus$$
$$x_3x_6 \oplus x_4x_5 \oplus x_4x_6 \oplus x_0x_4 \oplus x_0$$
$$f_5(x_6, x_5, x_4, x_3, x_2, x_1, x_0) = x_1x_4 \oplus x_2x_5 \oplus x_3x_6 \oplus x_0x_1x_2x_3 \oplus x_0x_5 \oplus x_0$$

These five Boolean functions have very nice properties when their coordinates are permuted. This will be stated in Section 3 together with rationale behind the design of the functions. The five passes H_1, H_2, H_3, H_4 and H_5 are specified in more detail in the following sections.

Original (H_1)																
	0	1	2	3	4	5	6	7	8	9	10	11	12	13	14	15
	16	17	18	19	20	21	22	23	24	25	26	27	28	29	30	31
ord_2 (H_2)	5	14	26	18	11	28	7	16	0	23	20	22	1	10	4	8
	30	3	21	9	17	24	29	6	19	12	15	13	2	25	31	27
ord_3 (H_3)	19	9	4	20	28	17	8	22	29	14	25	12	24	30	16	26
	31	15	7	3	1	0	18	27	13	6	21	10	23	11	5	2
ord_4 (H_4)	24	4	0	14	2	7	28	23	26	6	30	20	18	25	19	3
	22	11	31	21	8	27	12	9	1	29	5	15	17	10	16	13
ord_5 (H_5)	27	3	21	26	17	11	20	29	19	0	12	7	13	8	31	10
	5	9	14	30	18	6	28	24	2	23	16	22	4	1	25	15

Table 1. Word Processing Orders

permutations	x_6 ↓	x_5 ↓	x_4 ↓	x_3 ↓	x_2 ↓	x_1 ↓	x_0 ↓
$\phi_{3,1}$	x_1	x_0	x_3	x_5	x_6	x_2	x_4
$\phi_{3,2}$	x_4	x_2	x_1	x_0	x_5	x_3	x_6
$\phi_{3,3}$	x_6	x_1	x_2	x_3	x_4	x_5	x_0
$\phi_{4,1}$	x_2	x_6	x_1	x_4	x_5	x_3	x_0
$\phi_{4,2}$	x_3	x_5	x_2	x_0	x_1	x_6	x_4
$\phi_{4,3}$	x_1	x_4	x_3	x_6	x_0	x_2	x_5
$\phi_{4,4}$	x_6	x_4	x_0	x_5	x_2	x_1	x_3
$\phi_{5,1}$	x_3	x_4	x_1	x_0	x_5	x_2	x_6
$\phi_{5,2}$	x_6	x_2	x_1	x_0	x_3	x_4	x_5
$\phi_{5,3}$	x_2	x_6	x_0	x_4	x_3	x_1	x_5
$\phi_{5,4}$	x_1	x_5	x_3	x_2	x_0	x_4	x_6
$\phi_{5,5}$	x_2	x_5	x_0	x_6	x_4	x_3	x_1

Table 2. Permutations on Coordinates

Pass 1 Assume that the input to H_1 is (E_0, B), where E_0 consists of 8 words $E_{0,7}, E_{0,6}, \cdots, E_{0,0}$ and B of 32 words $W_{31}, W_{30}, \cdots, W_0$. H_1 processes the block B in a word-by-word way and transforms the input into a 8-word output $E_1 = E_{1,7}E_{1,6}\cdots E_{1,0}$. Denote by $\overrightarrow{\mathrm{ROT}}(X, s)$ the s position rotate right operation on a word X and by $f \circ g$ the composition of two functions f and g (g is evaluated first). Then H_1 can be described in the following way. (See also Figure 3.)

1. Let $T_{0,i} = E_{0,i}$, $0 \leq i \leqq 7$.
2. Repeat the following steps for i from 0 to 31:

$$P = \begin{cases} F_1 \circ \phi_{3,1}(T_{i,6}, T_{i,5}, T_{i,4}, T_{i,3}, T_{i,2}, T_{i,1}, T_{i,0}) \text{ if PASS=3} \\ F_1 \circ \phi_{4,1}(T_{i,6}, T_{i,5}, T_{i,4}, T_{i,3}, T_{i,2}, T_{i,1}, T_{i,0}) \text{ if PASS=4} \\ F_1 \circ \phi_{5,1}(T_{i,6}, T_{i,5}, T_{i,4}, T_{i,3}, T_{i,2}, T_{i,1}, T_{i,0}) \text{ if PASS=5} \end{cases}$$

$$R = \overrightarrow{\mathrm{ROT}}(P, 7) \boxplus \overrightarrow{\mathrm{ROT}}(T_{i,7}, 11) \boxplus W_i;$$

$$T_{i+1,7} = T_{i,6};\ T_{i+1,6} = T_{i,5};\ T_{i+1,5} = T_{i,4};\ T_{i+1,4} = T_{i,3};$$

$$T_{i+1,3} = T_{i,2};\ T_{i+1,2} = T_{i,1};\ T_{i+1,1} = T_{i,0};\ T_{i+1,0} = R.$$

3. Let $E_{1,i} = T_{32,i}$, $0 \leqq i \leqq 7$, and output $E_1 = E_{1,7}E_{1,6}\cdots E_{1,0}$.

Note that the input to the i-th round $(T_{i,6}, T_{i,5}, T_{i,4}, T_{i,3}, T_{i,2}, T_{i,1}, T_{i,0})$ is permuted according to $\phi_{3,1}$ (when PASS=3), $\phi_{4,1}$ (when PASS=4) or $\phi_{5,1}$ (when PASS=5) before being passed to F_1. Here $\phi_{3,1}$, $\phi_{4,1}$ and $\phi_{5,1}$ are permutations on coordinates specified in Table 2, where permutations employed by the other four passes are also specified. F_1 performs bit-wise operations on its input words according to the Boolean function f_1 specified in Section 2.2.

$$F_1(X_6, X_5, X_4, X_3, X_2, X_1, X_0) =$$
$$X_1 \bullet X_4 \oplus X_2 \bullet X_5 \oplus X_3 \bullet X_6 \oplus X_0 \bullet X_1 \oplus X_0$$

The result of F_1 is rotated and added (modulo 2^{32}) to the rotated version of $T_{i,7}$. The i-th word W_i in B is also added to the rotated version of $T_{i,7}$. The sum is used to substitute (the old) $T_{i,7}$. After the substitution, the 8 words $T_{i,7}, T_{i,6}, \cdots, T_{i,0}$ are shifted with $T_{i,7}$ being replaced by $T_{i,6}$, $T_{i,6}$ by $T_{i,5}$, ..., $T_{i,1}$ by $T_{i,0}$, and $T_{i,0}$ by $T_{i,7}$. These words are then used as input to the $(i+1)$-th round. Finally, T_{32} is output as a result.

Pass 2 Assume that the input to H_2 is (E_1, B). H_2 processes the words in B according to the word processing order ord_2 specified in Table 1. It employs in its computation 32 constant words $K_{2,31}, K_{2,30}, \cdots, K_{2,0}$, all of which are taken from the fraction part of π. The actual values of these constant words are defined in Section 2.4. H_2 processes the words as follows (see also Figure 4):

1. Let $T_{0,i} = E_{1,i}$, $0 \leqq i \leqq 7$.
2. Repeat the following steps for i from 0 to 31:

$$P = \begin{cases} F_2 \circ \phi_{3,2}(T_{i,6}, T_{i,5}, T_{i,4}, T_{i,3}, T_{i,2}, T_{i,1}, T_{i,0}) \text{ if PASS=3} \\ F_2 \circ \phi_{4,2}(T_{i,6}, T_{i,5}, T_{i,4}, T_{i,3}, T_{i,2}, T_{i,1}, T_{i,0}) \text{ if PASS=4} \\ F_2 \circ \phi_{5,2}(T_{i,6}, T_{i,5}, T_{i,4}, T_{i,3}, T_{i,2}, T_{i,1}, T_{i,0}) \text{ if PASS=5} \end{cases}$$

$$R = \overrightarrow{\mathrm{ROT}}(P, 7) \boxplus \overrightarrow{\mathrm{ROT}}(T_{i,7}, 11) \boxplus W_{\mathrm{ord}_2(i)} \boxplus K_{2,i};$$

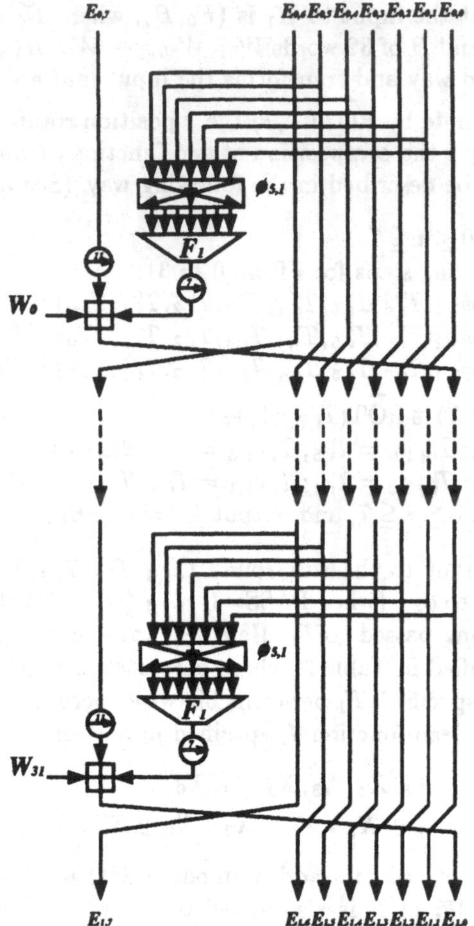

Fig. 3. Pass 1

$$T_{i+1,7} = T_{i,6};\ T_{i+1,6} = T_{i,5};\ T_{i+1,5} = T_{i,4};\ T_{i+1,4} = T_{i,3};$$
$$T_{i+1,3} = T_{i,2};\ T_{i+1,2} = T_{i,1};\ T_{i+1,1} = T_{i,0};\ T_{i+1,0} = R.$$

3. Let $E_{2,i} = T_{32,i}$, $0 \leqq i \leqq 7$, and output $E_2 = E_{2,7}E_{2,6} \cdots E_{2,0}$.

Similar to H_1, $(T_{i,6}, T_{i,5}, T_{i,4}, T_{i,3}, T_{i,2}, T_{i,1}, T_{i,0})$ is permuted according to $\phi_{3,2}$, $\phi_{4,2}$ or $\phi_{5,2}$ before being passed to F_2, where $\phi_{3,2}$, $\phi_{4,2}$ and $\phi_{5,2}$ are specified in Table 2. F_2 performs bit-wise operations on its 7 input words according to the Boolean function f_2:

$$F_2(X_6, X_5, X_4, X_3, X_2, X_1, X_0) =$$
$$X_1 \bullet X_2 \bullet X_3 \oplus X_2 \bullet X_4 \bullet X_5 \oplus$$
$$X_1 \bullet X_2 \oplus X_1 \bullet X_4 \oplus X_2 \bullet X_6 \oplus X_3 \bullet X_5 \oplus X_4 \bullet X_5 \oplus X_0 \bullet X_2 \oplus X_0$$

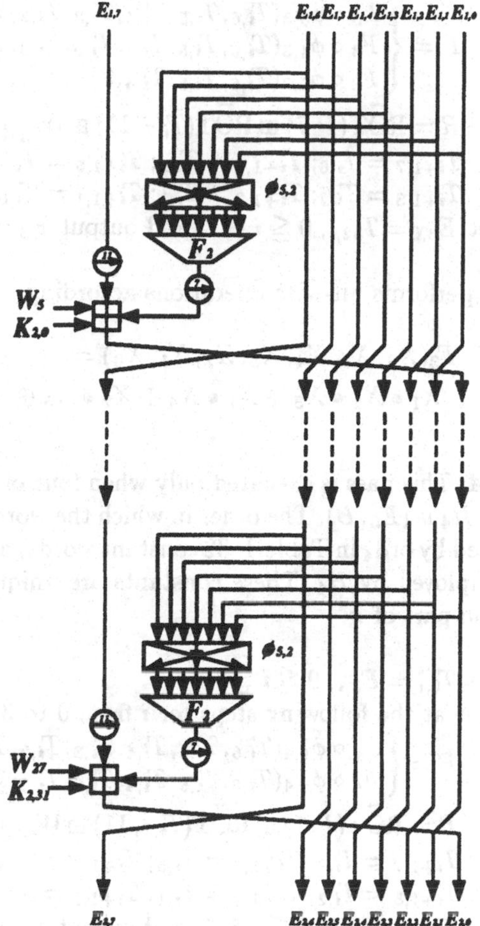

Fig. 4. Pass 2

The output value of F_2 is rotated and added to the rotated version of $T_{i,7}$. The i-th word $W_{\mathrm{ord}_2(i)}$ is also added to the rotated version of $T_{i,7}$. In addition, a constant $K_{2,i}$ which is unique to i is added to the rotated version of $T_{i,7}$. As in H_1, the 8 words are shifted before proceeding to the next round of operations. The output of H_2 is the result of the last round.

Pass 3 The input to H_3 is (E_2, B). H_3 processes the words in the block B according to the word processing order for ord_3 specified in Table 1. H_3 also employs 32 constant words $K_{3,31}, K_{3,30}, \cdots, K_{3,0}$, all of which are taken from the fraction part of π.

1. Let $T_{0,i} = E_{2,i}$, $0 \leqq i \leqq 7$.
2. Repeat the following steps for i from 0 to 31:

$$P = \begin{cases} F_3 \circ \phi_{3,3}(T_{i,6}, T_{i,5}, T_{i,4}, T_{i,3}, T_{i,2}, T_{i,1}, T_{i,0}) \text{ if PASS=3} \\ F_3 \circ \phi_{4,3}(T_{i,6}, T_{i,5}, T_{i,4}, T_{i,3}, T_{i,2}, T_{i,1}, T_{i,0}) \text{ if PASS=4} \\ F_3 \circ \phi_{5,3}(T_{i,6}, T_{i,5}, T_{i,4}, T_{i,3}, T_{i,2}, T_{i,1}, T_{i,0}) \text{ if PASS=5} \end{cases}$$

$R = \overrightarrow{\text{ROT}}(P, 7) \boxplus \overrightarrow{\text{ROT}}(T_{i,7}, 11) \boxplus W_{\text{ord}_3(i)} \boxplus K_{3,i}$;
$T_{i+1,7} = T_{i,6}$; $T_{i+1,6} = T_{i,5}$; $T_{i+1,5} = T_{i,4}$; $T_{i+1,4} = T_{i,3}$;
$T_{i+1,3} = T_{i,2}$; $T_{i+1,2} = T_{i,1}$; $T_{i+1,1} = T_{i,0}$; $T_{i+1,0} = R$.

3. Let $E_{3,i} = T_{32,i}$, $0 \leqq i \leqq 7$, and output $E_3 = E_{3,7}E_{3,6} \cdots E_{3,0}$.

F_3 performs bit-wise operations according to the Boolean function f_3 :

$$F_3(X_6, X_5, X_4, X_3, X_2, X_1, X_0) =$$
$$X_1 \bullet X_2 \bullet X_3 \oplus X_1 \bullet X_4 \oplus X_2 \bullet X_5 \oplus X_3 \bullet X_6 \oplus X_0 \bullet X_3 \oplus X_0$$

Pass 4 This pass is executed only when four or five passes are required. The input to H_4 is (E_3, B). The order in which the words in the block B are processed is specified by ord_4 in Table 1. 32 constant words, denoted by $K_{4,31}, K_{4,30}, \cdots, K_{4,0}$, are employed by H_4. These constants are unique to H_4 and all taken from the fraction part of π.

1. Let $T_{0,i} = E_{3,i}$, $0 \leqq i \leqq 7$.
2. Repeat the following steps for i from 0 to 31:

$$P = \begin{cases} F_4 \circ \phi_{4,4}(T_{i,6}, T_{i,5}, T_{i,4}, T_{i,3}, T_{i,2}, T_{i,1}, T_{i,0}) \text{ if PASS=4} \\ F_4 \circ \phi_{5,4}(T_{i,6}, T_{i,5}, T_{i,4}, T_{i,3}, T_{i,2}, T_{i,1}, T_{i,0}) \text{ if PASS=5} \end{cases}$$

$R = \overrightarrow{\text{ROT}}(P, 7) \boxplus \overrightarrow{\text{ROT}}(T_{i,7}, 11) \boxplus W_{\text{ord}_4(i)} \boxplus K_{4,i}$;
$T_{i+1,7} = T_{i,6}$; $T_{i+1,6} = T_{i,5}$; $T_{i+1,5} = T_{i,4}$; $T_{i+1,4} = T_{i,3}$;
$T_{i+1,3} = T_{i,2}$; $T_{i+1,2} = T_{i,1}$; $T_{i+1,1} = T_{i,0}$; $T_{i+1,0} = R$.

3. Let $E_{4,i} = T_{32,i}$, $0 \leqq i \leqq 7$, and output $E_4 = E_{4,7}E_{4,6} \cdots E_{4,0}$.

F_4 performs bit-wise operations on its input words according to the Boolean function f_4:

$$F_4(X_6, X_5, X_4, X_3, X_2, X_1, X_0) =$$
$$X_1 \bullet X_2 \bullet X_3 \oplus X_2 \bullet X_4 \bullet X_5 \oplus X_3 \bullet X_4 \bullet X_6 \oplus$$
$$X_1 \bullet X_4 \oplus X_2 \bullet X_6 \oplus X_3 \bullet X_4 \oplus X_3 \bullet X_5 \oplus$$
$$X_3 \bullet X_6 \oplus X_4 \bullet X_5 \oplus X_4 \bullet X_6 \oplus X_0 \bullet X_4 \oplus X_0$$

Pass 5 This pass is executed only when five passes are required. The input to H_5 is (E_4, B). The order in which the words in the block B are processed is specified by ord_5 in Table 1. The 32 constant words employed by H_5 are denoted by $K_{5,31}, K_{5,30}, \cdots, K_{5,0}$.

1. Let $T_{0,i} = E_{4,i}$, $0 \leqq i \leqq 7$.
2. Repeat the following steps for i from 0 to 31:

$P = F_5 \circ \phi_{5,5}(T_{i,6}, T_{i,5}, T_{i,4}, T_{i,3}, T_{i,2}, T_{i,1}, T_{i,0});$
$R = \overrightarrow{\mathrm{ROT}}(P, 7) \boxplus \overrightarrow{\mathrm{ROT}}(T_{i,7}, 11) \boxplus W_{\mathrm{ord}_5(i)} \boxplus K_{5,i};$
$T_{i+1,7} = T_{i,6};\ T_{i+1,6} = T_{i,5};\ T_{i+1,5} = T_{i,4};\ T_{i+1,4} = T_{i,3};$
$T_{i+1,3} = T_{i,2};\ T_{i+1,2} = T_{i,1};\ T_{i+1,1} = T_{i,0};\ T_{i+1,0} = R.$

3. Let $E_{5,i} = T_{32,i}$, $0 \leq i \leq 7$, and output $E_5 = E_{5,7}E_{5,6}\cdots E_{5,0}$.

F_5 performs bit-wise operations on its input words according to the Boolean function f_5:

$$F_5(X_6, X_5, X_4, X_3, X_2, X_1, X_0) =$$
$$X_1 \bullet X_4 \oplus X_2 \bullet X_5 \oplus X_3 \bullet X_6 \oplus X_0 \bullet X_1 \bullet X_2 \bullet X_3 \oplus X_0 \bullet X_5 \oplus X_0$$

2.3 Tailoring the Last Output of *H*

Recall that the last string $D_n = D_{n,7}D_{n,6}\cdots D_{n,0}$ output by H is of 256 bits. D_n is used directly as the fingerprint of M if a 256-bit fingerprint is required. Otherwise, D_n is tailored into a string of specified length. We discuss the four cases that need adjustment to D_n. These four cases are (1) Case-1 when 128-bit fingerprints are required, (2) Case-2 when 160-bit fingerprints are required, (3) Case-3 when 192-bit fingerprints are required and (4) Case-4 when 224-bit fingerprints are required. In the following discussions, we will use a superscript to indicate the length of a string. For instance, if X is a t-bit string, we use $X^{[t]}$ to indicate explicitly the length of X.

Case-1 (128-bit fingerprint): We divide $D_{n,7}$, $D_{n,6}$, $D_{n,5}$ and $D_{n,4}$ in the following way

$$D_{n,7} = X_{7,3}^{[8]}X_{7,2}^{[8]}X_{7,1}^{[8]}X_{7,0}^{[8]},$$
$$D_{n,6} = X_{6,3}^{[8]}X_{6,2}^{[8]}X_{6,1}^{[8]}X_{6,0}^{[8]},$$
$$D_{n,5} = X_{5,3}^{[8]}X_{5,2}^{[8]}X_{5,1}^{[8]}X_{5,0}^{[8]},$$
$$D_{n,4} = X_{4,3}^{[8]}X_{4,2}^{[8]}X_{4,1}^{[8]}X_{4,0}^{[8]}.$$

The 128-bit fingerprint is $Y_3Y_2Y_1Y_0$, where

$$Y_3 = D_{n,3} \boxplus (X_{7,3}^{[8]}X_{6,2}^{[8]}X_{5,1}^{[8]}X_{4,0}^{[8]}),$$
$$Y_2 = D_{n,2} \boxplus (X_{7,2}^{[8]}X_{6,1}^{[8]}X_{5,0}^{[8]}X_{4,3}^{[8]}),$$
$$Y_1 = D_{n,1} \boxplus (X_{7,1}^{[8]}X_{6,0}^{[8]}X_{5,3}^{[8]}X_{4,2}^{[8]}),$$
$$Y_0 = D_{n,0} \boxplus (X_{7,0}^{[8]}X_{6,3}^{[8]}X_{5,2}^{[8]}X_{4,1}^{[8]}).$$

Case-2 (160-bit fingerprint): We divide $D_{n,7}$, $D_{n,6}$ and $D_{n,5}$ in the following way

$$D_{n,7} = X_{7,4}^{[7]}X_{7,3}^{[6]}X_{7,2}^{[7]}X_{7,1}^{[6]}X_{7,0}^{[6]},$$
$$D_{n,6} = X_{6,4}^{[7]}X_{6,3}^{[6]}X_{6,2}^{[7]}X_{6,1}^{[6]}X_{6,0}^{[6]},$$
$$D_{n,5} = X_{5,4}^{[7]}X_{5,3}^{[6]}X_{5,2}^{[7]}X_{5,1}^{[6]}X_{5,0}^{[6]}.$$

Then the 160-bit fingerprint $Y_4Y_3Y_2Y_1Y_0$ is obtained by computing

$$Y_4 = D_{n,4} \boxplus (X^{[7]}_{7,4}X^{[6]}_{6,3}X^{[7]}_{5,2}),$$
$$Y_3 = D_{n,3} \boxplus (X^{[6]}_{7,3}X^{[7]}_{6,2}X^{[6]}_{5,1}),$$
$$Y_2 = D_{n,2} \boxplus (X^{[7]}_{7,2}X^{[6]}_{6,1}X^{[6]}_{5,0}),$$
$$Y_1 = D_{n,1} \boxplus (X^{[6]}_{7,1}X^{[6]}_{6,0}X^{[7]}_{5,4}),$$
$$Y_0 = D_{n,0} \boxplus (X^{[6]}_{7,0}X^{[7]}_{6,4}X^{[6]}_{5,3}).$$

Case-3 (192-bit fingerprint): Divide $D_{n,7}$ and $D_{n,6}$ into

$$D_{n,7} = X^{[6]}_{7,5}X^{[5]}_{7,4}X^{[5]}_{7,3}X^{[6]}_{7,2}X^{[5]}_{7,1}X^{[5]}_{7,0},$$
$$D_{n,6} = X^{[6]}_{6,5}X^{[5]}_{6,4}X^{[5]}_{6,3}X^{[6]}_{6,2}X^{[5]}_{6,1}X^{[5]}_{6,0}.$$

Let

$$Y_5 = D_{n,5} \boxplus (X^{[6]}_{7,5}X^{[5]}_{6,4}),$$
$$Y_4 = D_{n,4} \boxplus (X^{[5]}_{7,4}X^{[5]}_{6,3}),$$
$$Y_3 = D_{n,3} \boxplus (X^{[5]}_{7,3}X^{[6]}_{6,2}),$$
$$Y_2 = D_{n,2} \boxplus (X^{[6]}_{7,2}X^{[5]}_{6,1}),$$
$$Y_1 = D_{n,1} \boxplus (X^{[5]}_{7,1}X^{[5]}_{6,0}),$$
$$Y_0 = D_{n,0} \boxplus (X^{[5]}_{7,0}X^{[6]}_{6,5}).$$

Output $Y_5Y_4Y_3Y_2Y_1Y_0$ as the fingerprint.

Case-4 (224-bit fingerprint): We divide $D_{n,7}$ into

$$D_{n,7} = X^{[5]}_{7,6}X^{[5]}_{7,5}X^{[4]}_{7,4}X^{[5]}_{7,3}X^{[4]}_{7,2}X^{[5]}_{7,1}X^{[4]}_{7,0}.$$

The 224-bit fingerprint is $Y_6Y_5Y_4Y_3Y_2Y_1Y_0$, where

$$Y_6 = D_{n,6} \boxplus X^{[4]}_{7,0},$$
$$Y_5 = D_{n,5} \boxplus X^{[5]}_{7,1},$$
$$Y_4 = D_{n,4} \boxplus X^{[4]}_{7,2},$$
$$Y_3 = D_{n,3} \boxplus X^{[5]}_{7,3},$$
$$Y_2 = D_{n,2} \boxplus X^{[4]}_{7,4},$$
$$Y_1 = D_{n,1} \boxplus X^{[5]}_{7,5},$$
$$Y_0 = D_{n,0} \boxplus X^{[5]}_{7,6}.$$

2.4 The Constants from π

HAVAL uses totally 136 constant 32-bit words. Among them, 8 words are used as initial values $D_{0,7}$, $D_{0,6}$, $\cdots$, $D_{0,0}$, 32 words are employed by Pass 2 as $K_{2,31}$, $K_{2,30}$, $\cdots$, and $K_{2,0}$, 32 words by Pass 3 as $K_{3,31}$, $K_{3,30}$, $\cdots$, and $K_{3,0}$, 32 words by Pass 4 as $K_{4,31}$, $K_{4,30}$, $\cdots$, and $K_{4,0}$, and the remaining 32 words by Pass 5 as $K_{5,31}$, $K_{5,30}$, $\cdots$, and $K_{5,0}$. The first 8 constant words correspond to the first 256 bits of the fraction part of π. The 32 constant words used in Pass 2 correspond to the next 1024 bits of the fraction part of π, which is followed by the 32 constant words used by Pass 3, the 32 constant words used by Pass 4 and the 32 constant words used by Pass 5. The 136 constant words are listed in the following in hexadecimal form. They appear in the following order:

1. $D_{0,0}$, $D_{0,1}$, $\cdots$, $D_{0,7}$,
2. $K_{2,0}$, $K_{2,1}$, $\cdots$, $K_{2,31}$,
3. $K_{3,0}$, $K_{3,1}$, $\cdots$, $K_{3,31}$,
4. $K_{4,0}$, $K_{4,1}$, $\cdots$, $K_{4,31}$,
5. $K_{5,0}$, $K_{5,1}$, $\cdots$, $K_{5,31}$.

```
243F6A88 85A308D3 13198A2E 03707344 A4093822 299F31D0 082EFA98 EC4E6C89

452821E6 38D01377 BE5466CF 34E90C6C C0AC29B7 C97C50DD 3F84D5B5 B5470917
9216D5D9 8979FB1B D1310BA6 98DFB5AC 2FFD72DB D01ADFB7 B8E1AFED 6A267E96
BA7C9045 F12C7F99 24A19947 B3916CF7 0801F2E2 858EFC16 636920D8 71574E69
A458FEA3 F4933D7E 0D95748F 728EB658 718BCD58 82154AEE 7B54A41D C25A59B5

9C30D539 2AF26013 C5D1B023 286085F0 CA417918 B8DB38EF 8E79DCB0 603A180E
6C9E0E8B B01E8A3E D71577C1 BD314B27 78AF2FDA 55605C60 E65525F3 AA55AB94
57489862 63E81440 55CA396A 2AAB10B6 B4CC5C34 1141E8CE A15486AF 7C72E993
B3EE1411 636FBC2A 2BA9C55D 741831F6 CE5C3E16 9B87931E AFD6BA33 6C24CF5C

7A325381 28958677 3B8F4898 6B4BB9AF C4BFE81B 66282193 61D809CC FB21A991
487CAC60 5DEC8032 EF845D5D E98575B1 DC262302 EB651B88 23893E81 D396ACC5
0F6D6FF3 83F44239 2E0B4482 A4842004 69C8F04A 9E1F9B5E 21C66842 F6E96C9A
670C9C61 ABD388F0 6A51A0D2 D8542F68 960FA728 AB5133A3 6EEF0B6C 137A3BE4

BA3BF050 7EFB2A98 A1F1651D 39AF0176 66CA593E 82430E88 8CEE8619 456F9FB4
7D84A5C3 3B8B5EBE E06F75D8 85C12073 401A449F 56C16AA6 4ED3AA62 363F7706
1BFEDF72 429B023D 37D0D724 D00A1248 DB0FEAD3 49F1C09B 075372C9 80991B7B
25D479D8 F6E8DEF7 E3FE501A B6794C3B 976CE0BD 04C006BA C1A94FB6 409F60C4
```

We generated these constant words by Maple (Version 5 on a SPARCstation) with the following program:

```
printlevel := -1;
Digits := 2000;
pifrac := evalf(Pi) - 3;
```

```
K := 2 ^ 32;
for i from 1 by 1 while i <= 136 do
nextword := trunc(pifrac * K);
lprint(convert(nextword,hex));
pifrac:= frac(pifrac * K);
od;
```

3 The Design Rationale

3.1 Designing the Boolean Functions

The five boolean functions f_1, f_2, f_3, f_4 and f_5 used by H_1, H_2, H_3, H_4 and H_5 are of central importance to the hashing algorithm. We first introduce a few definitions before going into their design details.

Denote by V_n the the vector space of n-tuples of elements from $GF(2)$, where n is a positive integer. A Boolean function is a function from V_n to $GF(2)$. Note that a Boolean function f from V_n to $GF(n)$ can be "reduced" to a unique polynomial in n coordinate variables $x_n, x_{n-1}, \ldots, x_1$. In the following discussions, we will identify the function f with its unique polynomial $f(x_n, x_{n-1}, \ldots, x_1)$. The sequence of the function f is defined as the concatenation of the 2^n output bits of $f(x_n, x_{n-1}, \ldots, x_1)$ when $x_n, x_{n-1}, \ldots, x_1$ vary from $0, 0, \cdots, 0$ to $1, 1, \cdots, 1$. The function f is called a linear function if f has the form of $f(x_n, x_{n-1}, \ldots, x_1) = a_n x_n \oplus a_{n-1} x_{n-1} \oplus \cdots \oplus a_1 x_1 \oplus a_0$, where $a_i \in GF(2)$.

We say that a function f from V_n to $GF(2)$ is 0-1 *balanced* if the number of 1 bits and the number of 0 bits in the sequence of f are the same, both being 2^{n-1}. Let g be another function from V_n to $GF(2)$. The *distance* between f and g is the number of positions in the sequences of f and g at which the two functions have different values. The *non-linearity* of the function f is defined as the *minimum* distance between f and *all* linear functions from V_n to $GF(2)$. When $n = 2k$ for some $k > 1$, the maximum non-linearity a function from V_n to $GF(2)$ can attain is $2^{2k-1} - 2^{k-1}$. Such a functions is called a *bent* function [Rot76]. We say that f satisfies the *Strict Avalanche Criterion (SAC)* if for every $1 \leqq i \leqq n$, complementing x_i results in the output of f being complemented 50% of the time over all possible input vectors.

Two functions f and g are linearly *equivalent* (in structure) if f can be transformed into g via linear transformation of coordinates and complementation of functions, i.e., there is a non-singular $n \times n$ matrix A on $GF(2)$ as well as a vector $B \in V_n$ such that $f(xA \oplus B) = g(x)$ or $f(xA \oplus B) \oplus 1 = g(x)$, where $x = (x_n, x_{n-1}, \ldots, x_1)$. Otherwise we say that f and g are linearly *inequivalent*. A set of functions is said linearly inequivalent if all pairs of functions from the set are linearly inequivalent.

f and g are *mutually output-uncorrelated* if f, g and $f \oplus g$ are all *0-1 balanced non-linear* functions. A set of functions is mutually output-uncorrelated if all pairs of functions in the set are mutually output-uncorrelated. The set is said

perfectly output-uncorrelated if any non-zero linear combination of the functions in the set results in a 0-1 balanced non-linear function.

Linear equivalence and output-correlation can be used to examine from two different angles the structural similarity among functions. Our goal is to design five Boolean functions in seven variables so that each of the functions has the following properties P1, P2 and P3.

P1 Being 0-1 balanced.
P2 Having a high non-linearity.
P3 Satisfying the Strict Avalanche Criterion (SAC).

In addition, as a set of functions, they have the following properties P4 and P5.

P4 Being linearly inequivalent in structure.
P5 Being mutually output-uncorrelated.

These properties are considered as desirable ones for a cryptographic primitive such as a one-way hashing algorithm. P1 ensures that a function outputs a 0 bit and a 1 bit with the same probability 0.5 when the input to the function is picked randomly and uniformly over all possible vectors. P2 is desirable since a linear function would render a cryptographic algorithm easily breakable. P3 brings good avalanche effect to a cryptographic algorithm. P4 ensures that functions employed by a cryptographic algorithm bears no resemblance in structure (with respect to linear transformation of coordinates and complementation of functions.) Finally, P5 ensures that the sequences of the functions are not mutually correlated either via linear functions or via the bias in output bits.

In [SZ92], Seberry and Zhang presented a novel method for constructing Boolean functions that have the properties P1, P2 and P3. In particular, they showed that given a bent function from V_{2k} to $GF(2)$, where $k \geq 1$, one can obtain a Boolean function from V_{2k+1} to $GF(2)$ that has the properties P1, P2 and P3 and a non-linearity of $2^{2k} - 2^k$. Here is their construction method. Let $g(x_{2k}, x_{2k-1}, \ldots, x_1)$ be a bent function, and let $\ell(x_{2k}, x_{2k-1}, \ldots, x_1)$ be an arbitrary non-constant linear function. Let

$$h(x_{2k}, x_{2k-1}, \ldots, x_1) = g(x_{2k}, x_{2k-1}, \ldots, x_1) \oplus \ell(x_{2k}, x_{2k-1}, \ldots, x_1).$$

Note that $h(x_{2k}, x_{2k-1}, \ldots, x_1)$ is also a bent function. Also note that a bent function is ***not*** 0-1 balanced. Now assume that both the function sequence of $g(x_{2k}, x_{2k-1}, \ldots, x_1)$ and that of $h(x_{2k}, x_{2k-1}, \ldots, x_1)$ have more 1s (or 0s) than 0s (or 1s). Then the following function from V_{2k+1} to $GF(2)$

$$\begin{aligned} & f(x_{2k}, x_{2k-1}, \ldots, x_1, x_0) \\ & = (x_0 \oplus 1) g(x_{2k}, x_{2k-1}, \ldots, x_1) \oplus x_0 (h(x_{2k}, x_{2k-1}, \ldots, x_1) \oplus 1) \\ & = g(x_{2k}, x_{2k-1}, \ldots, x_1) \oplus x_0 \ell(x_{2k}, x_{2k-1}, \ldots, x_1) \oplus x_0 \end{aligned}$$

has properties P1, P2 and P3.

The five Boolean functions f_1, f_2, f_3, f_4 and f_5 employed by H_1, H_2, H_3, H_4 and H_5 are constructed from the following bent functions g_1, g_2, g_3 and g_4.

$$g_1(x_6,x_5,x_4,x_3,x_2,x_1) = x_1x_4 \oplus x_2x_5 \oplus x_3x_6$$
$$g_2(x_6,x_5,x_4,x_3,x_2,x_1) = x_1x_2x_3 \oplus x_2x_4x_5 \oplus x_1x_2 \oplus x_1x_4 \oplus x_2x_6 \oplus x_3x_5 \oplus x_4x_5$$
$$g_3(x_6,x_5,x_4,x_3,x_2,x_1) = x_1x_2x_3 \oplus x_1x_4 \oplus x_2x_5 \oplus x_3x_6$$
$$g_4(x_6,x_5,x_4,x_3,x_2,x_1) = x_1x_2x_3 \oplus x_2x_4x_5 \oplus x_3x_4x_6 \oplus x_1x_4 \oplus x_2x_6 \oplus x_3x_4 \oplus x_3x_5 \oplus x_3x_6 \oplus x_4x_5 \oplus x_4x_6$$

These four bent functions were discovered by Rothaus in his pioneering work [Rot76]. In the same paper, Rothaus also proved that these are the only bent functions from V_6 to $GF(2)$ which are linearly inequivalent in structure. Let

$$\ell_1(x_6,x_5,x_4,x_3,x_2,x_1) = x_1,$$
$$\ell_2(x_6,x_5,x_4,x_3,x_2,x_1) = x_2,$$
$$\ell_3(x_6,x_5,x_4,x_3,x_2,x_1) = x_3,$$
$$\ell_4(x_6,x_5,x_4,x_3,x_2,x_1) = x_4.$$

By applying Seberry and Zhang's method, we obtain the first four functions f_1, f_2, f_3 and f_4 as follows:

$$\begin{aligned} f_i(x_6,x_5,x_4,x_3,x_2,x_1,x_0) &= g_i(x_6,x_5,x_4,x_3,x_2,x_1) \oplus x_0\ell_i(x_6,x_5,x_4,x_3,x_2,x_1) \oplus x_0 \\ &= g_i(x_6,x_5,x_4,x_3,x_2,x_1) \oplus x_0x_i \oplus x_0 \end{aligned}$$

where $i = 1,2,3,4$. The fifth function, which also has the properties P1, P2 and P3, is obtained in the following way. Let

$$h_5(x_6,x_5,x_4,x_3,x_2,x_1) = g_1(x_6,x_5,x_4,x_3,x_2,x_1) \oplus x_1x_2x_3 \oplus x_6.$$

Then

$$\begin{aligned} &f_5(x_6,x_5,x_4,x_3,x_2,x_1,x_0) \\ &= (1 \oplus x_0)g_1(x_6,x_5,x_4,x_3,x_2,x_1) \oplus x_0(1 \oplus h_5(x_6,x_5,x_4,x_3,x_2,x_1)) \\ &= g_1(x_6,x_5,x_4,x_3,x_2,x_1) \oplus x_0x_1x_2x_3 \oplus x_0x_5 \oplus x_0 \end{aligned}$$

These functions have a non-linearity of $2^6 - 2^3 = 56$, which is in fact the *maximum* non-linearity of functions from V_7 to $GF(2)$ [SZ92].

Now we show that these functions are linearly inequivalent in structure. We call the product of several coordinate variables a *term*. The *degree* of a term is the number of coordinate variables in it. The degree of a Boolean function is the maximum degree among all terms of the function. Thus f_1 has five terms x_1x_4, x_2x_5, x_3x_6, x_0x_1 and x_0. The first four terms are of degree 2, the last term is of degree 1, and hence the degree of f_1 is 2. Consider the case when a linear transformation of coordinates is applied to a Boolean function f and a new Boolean function g is obtained. Each term of f generates one or more new terms. However no terms that have higher degrees than that of the original one can be created. Therefore, all the terms in g which have the highest degree are derived from terms in f which have the same degree. This implies that linear transformation of coordinates does not change the degree of a function.

The degrees of the five functions f_1, f_2, f_3, f_4 and f_5 are 2, 3, 3, 3 and 4 respectively. From the above discussions, we know that f_1 and f_5 are linearly inequivalent. In addition, neither f_1 nor f_5 can be transformed into any of the other three functions f_2, f_3 and f_4 by linear transformation of coordinates. The other direction is also true. Now consider f_2, f_3 and f_4. Note that f_2 has two degree-3 terms $x_1x_2x_3$ and $x_2x_4x_5$, f_3 has one degree-3 term $x_1x_2x_3$, and f_4 has three degree-3 terms $x_1x_2x_3$, $x_2x_4x_5$ and $x_3x_4x_6$. It was shown in [Rot76] that the above three sets of degree-3 terms can not be transformed into one another by linear transformation of coordinates. From this it follows that the three functions f_2, f_3 and f_4 are linearly inequivalent. In summary f_1, f_2, f_3, f_4 and f_5 are linearly inequivalent, and hence they have the property P4.

By now we have seen that the five functions f_1, f_2, f_3, f_4 and f_5 satisfy properties P1, P2, P3 and P4. Verification shows that these five functions do *not* have the property P5. By permuting the coordinates of the functions f_1, f_2 and f_3 according to $\phi_{3,1}$, $\phi_{3,2}$ and $\phi_{3,3}$ shown in Table 2, we obtain three functions $f_1 \circ \phi_{3,1}$, $f_2 \circ \phi_{3,2}$ and $f_3 \circ \phi_{3,3}$ that are mutually output-uncorrelated (i.e., satisfying the property P5). In fact these three functions are perfectly output-uncorrelated. As permuting coordinates does not affect the functions with respect to properties P1, P2, P3 and P4, we know that the three permuted functions $f_1 \circ \phi_{3,1}$, $f_2 \circ \phi_{3,2}$ and $f_3 \circ \phi_{3,3}$ which are used in the 3-pass case satisfy all the five properties P1, P2, P3, P4 and P5. All non-zero linear combinations of the three functions have the maximum non-linearity of 56.

Similarly, by permuting the coordinates of the functions f_1, f_2, f_3 and f_4 according to $\phi_{4,1}$, $\phi_{4,2}$, $\phi_{4,3}$ and $\phi_{4,4}$ shown in Table 2, we obtain four functions $f_1 \circ \phi_{4,1}$, $f_2 \circ \phi_{4,2}$, $f_3 \circ \phi_{4,3}$ and $f_3 \circ \phi_{4,3}$ that are perfectly output-uncorrelated and hence satisfy the property P5. Among the non-zero linear combinations of $f_1 \circ \phi_{4,1}$, $f_2 \circ \phi_{4,2}$, $f_3 \circ \phi_{4,3}$ and $f_4 \circ \phi_{4,4}$, ten achieve the maximum non-linearity of 56 and the remaining 5 achieve 48.

Permuting the coordinates of the functions f_1, f_2, f_3, f_4 and f_5 according to $\phi_{5,1}$, $\phi_{5,2}$, $\phi_{5,4}$, $\phi_{5,3}$ and $\phi_{5,5}$ shown in Table 2 yields five functions $f_1 \circ \phi_{5,1}$, $f_2 \circ \phi_{5,2}$, $f_3 \circ \phi_{5,3}$, $f_3 \circ \phi_{5,4}$ and $f_3 \circ \phi_{5,5}$ that are mutually output-uncorrelated and hence satisfy the property P5. Although the permutations do not yield perfectly output-uncorrelated functions, all the non-zero combinations are either 0-1 balanced or very close to 0-1 balanced. Eight of the combinations have the maximum non-linearity of 56, four have 52, fifteen have 48, three have 44 and one has 32.

The permutations shown in Table 2 are obtained by random sampling. We have also found many other alternative permutations. The permutations shown in Table 2 are chosen since they bring the highest average non-linearity to the linear combinations of the functions.

To compare with MD4, MD5 and SHS, we have listed the Boolean functions used by these algorithms in Table 3. The main design criterion for these functions is as follows [Riv92a, Riv92b]: ***if the input to a function is the result of flipping independent unbiased coins, then the output of the function should behave in the same way as the result of flipping an independent unbiased coin as well.*** This is equivalent to say that the functions are all 0-1 balanced, i.e., they satisfy the property P1, one of our five design criteria. Note that one of the functions, $x \oplus y \oplus z$, is linear. The other degree-2 functions can be transformed into one another by linear transformation on coordinates. In particular, $xy \oplus xz \oplus yz$, $xz \oplus yz \oplus y$, and $y \oplus z \oplus xz \oplus 1$ can all be transformed into $xy \oplus xz \oplus z$ by

$$\begin{bmatrix} x \to x \oplus y \oplus 1 \\ y \to y \\ z \to z \end{bmatrix}, \begin{bmatrix} x \to y \\ y \to z \\ z \to x \end{bmatrix}, \begin{bmatrix} x \to y \oplus z \\ y \to x \oplus z \oplus 1 \\ z \to x \end{bmatrix}$$

respectively. In addition, it is easy to check that correlations among the output sequences of the function are very poor.

	MD4	MD5	SHS
1	$xy \oplus xz \oplus z$	$xy \oplus xz \oplus z$	$xy \oplus xz \oplus z$
2	$xy \oplus xz \oplus yz$	$xz \oplus yz \oplus y$	$x \oplus y \oplus z$
3	$x \oplus y \oplus z$	$x \oplus y \oplus z$	$xy \oplus xz \oplus yz$
4		$y \oplus z \oplus xz \oplus 1$	$x \oplus y \oplus z$

Table 3. Boolean Functions Used by MD4, MD5 and SHS

3.2 Other Design Issues

At the i-th round of Pass 1, $T_{i,7}$ is updated essentially by adding to it the output of F_1 and the i-th word W_i. This can be viewed as the folding technique used in ordinary hashing (see Page 512, [Knu73]). Rotation is employed to destroy the symmetry of addition modulo 2^{32} operation. This technique is also used in the processing of Passes 2, 3, 4 and 5. Inversion of the basic compressing algorithm H is made computationally infeasible by the addition of its 8-word input to the last pass' output.

Processing of the five passes is made more distinct by allowing them to perform re-ordering operations upon the words. The word processing orders are selected in such a way that no word is processed by the same round at different passes and that the orders are as un-related as possible. In addition, constant words unique to each round are used in the later four passes. These constant words have been defined as consecutive bits in the fraction part of π to avoid possible allegation that a trap-door would have been planted in them.

In addition, different permutations on coordinates of f_1, f_2, f_3, f_4 and f_5 are employed according to the number of passes required. This makes the hashing algorithm behave more differently when the number of passes changes.

4 Security of HAVAL

Two messages are said to collide with each other with respect to a one-way hashing algorithm if they are compressed to the same fingerprint. For HAVAL, there are two possibilities for a pair of messages to collide: the number of passes the messages are processed can be the same or differ. Ideally, given a one-way hashing algorithm, we would like to prove formally that it is computationally infeasible to find a collision pair for the hashing algorithm. Like many other hashing algorithms such as the MD family, SHS and FFT-hash, however, HAVAL could not be formally proved to be secure. Recently, Berson has proposed an attack to a single pass of MD5 [Ber92]. His method applies to a single pass of HAVAL as well. However, it seems that the attack can not be extended to two or more passes.

It is conjectured that the best way to find a collision pair is by using the *birthday attack*. In such an attack, an attacker prepares two sets of $2^{n/2}$ distinct messages, and calculates their fingerprints. Here n denotes the number of bits in a fingerprint, and it can be 128, 160, 192, 224, 256. Also note that the number of passes the two sets of messages are compressed may differ. The attacker can check (by, for instance, sorting) if there is any collision pair of messages, one is from the first set and the other from the second set. The attacker will succeed with a probability about 0.5. However, such an attack requires the order of $2^{n/2}$ operations, which is impractical even for $n = 128$. It is also conjectured that given a fingerprint, it requires the order of 2^n operations to obtain a message that is mapped to the fingerprint.

5 Extensions and Future Work

The algorithm can be extended in several directions. Firstly, we note that the number of passes can be increased by adding more functions into the function set $\{f_1, f_2, f_3, f_4, f_5\}$.

It is well known that for any $k \geqq 4$, there are at least k linearly inequivalent bent functions from V_{2k} to $GF(2)$. Thus by using the same approach as described in Section 3.1, we can design, at least in theory, four or more functions from V_{2k+1} to $GF(2)$ that have the properties P1, P2, P3, P4 and P5. In this way, we can design one-way hashing algorithms that compress an arbitrarily long message into a fingerprint of $32(2k + 2)$ or less bits, where $k \geqq 4$.

We also note that although HAVAL is designed primarily for 32-bit machines, hashing algorithms suited to more advanced platforms such as 64-bit machines can be obtained by modifying the definition of a word.

The efficiency of the algorithm can be improved if we can find simpler replacements for the five functions. It is a future research subject to search for other approaches that might lead to simpler functions having the five properties.

6 Conclusions

We have proposed a new one-way hashing algorithm HAVAL that can compress an arbitrarily long message into a fingerprint of 128, 160, 192, 224 or 256 bits. To meet the needs of various practical applications, HAVAL also has provides the flexibility to change the number of passes message blocks are processed. A great deal of attention has been paid to the design of the five Boolean functions used by the algorithm. We expect that it requires the order of $2^{n/2}$ operations to find a pair of collision messages, where n is the length of a fingerprint. We also expect that the algorithm would be widely used in practical applications where fingerprints of variable length are required.

Acknowledgements

The authors are grateful to Xian-Mo Zhang for his invaluable contribution to this project. This work would be impossible without his insight in the construction of cryptographically useful Boolean funcitons. We also would like to thank Tor Nordhagen for his help in testing and programming, and Bart Preneel for bringing [Mer90] to our attention.

References

[Ber92] Thomas A. Berson. Differential cryptanalysis mod 2^{32} with applications to MD5. In *Advances in Cryptology - Proceedings of EuroCrypt'92*, Lecture Notes in Computer Science. Springer-Verlag, 1992. (to appear).

[Dam87] I. Damgård. Collision free hash functions and public key signature schemes. In *Advances in Cryptology - Proceedings of EuroCrypt'87*, Lecture Notes in Computer Science. Springer-Verlag, 1987.

[Dam90] I. Damgård. A design principle for hash functions. In G. Brassard, editor, *Advances in Cryptology - Proceedings of Crypto'89*, Lecture Notes in Computer Science, Vol. 435, pages 416–427. Springer-Verlag, 1990.

[DH76] W. Diffie and M. Hellman. New directions in cryptography. *IEEE Transactions on Information Theory*, IT-22(6):472–492, 1976.

[Kal92] B. Kaliski. The MD2 message digest algorithm, April 1992. Request for Comments (RFC) 1319.

[Knu73] Donald E. Knuth. *The Art of Computer Programming, Sorting and Searching*, volume 3. Addison-Wesley, 1973.

[Mer78] R. Merkle. Secure communication over insecure channels. *Communications of the ACM*, 21:294–299, 1978.

[Mer90] R. C. Merkle. A fast software one-way hash function. *Journal of Cryptology*, 3(1):43–58, 1990.

[NIS91] NIST. A proposed federal information processing standard for digital signature standard (DSS), August 1991.

[NIS92] NIST. A proposed federal information processing standard for secure hash (SHS), January 1992.

[NY89] M. Naor and M. Yung. Universal one-way hash functions and their cryptographic applications. In *Proceedings of the 21-st ACM Symposium on Theory of Computing*, pages 33–43, 1989.

[Riv92a] R. Rivest. The MD4 message digest algorithm, April 1992. Request for Comments (RFC) 1320. (Also presented at Crypto'90, 1990).

[Riv92b] R. Rivest. The MD5 message digest algorithm, April 1992. Request for Comments (RFC) 1321.

[Rom90] J. Rompel. One-way functions are necessary and sufficient for secure signatures. In *Proceedings of the 22-nd ACM Symposium on Theory of Computing*, pages 387–394, 1990.

[Rot76] O. S. Rothaus. On "bent" functions. *Journal of Combinatorial Theory (A)*, 20:300–305, 1976.

[Sch92] C. P. Schnorr. FFT-Hash II, efficient cryptographic hashing, April 1992. Presented at EuroCrypt'92.

[SZ92] J. Seberry and X.-M. Zhang. Highly nonlinear 0-1 balanced boolean functions satisfying strict avalanche criterion, 1992. AusCrypt'92, Gold Coast.

[Vau92] S. Vaudenay. FFT-Hash-II is not yet collision-free. In *Rump Session, Crypto'92*, 1992.

[ZMI91] Y. Zheng, T. Matsumoto, and H. Imai. Structural properties of one-way hash functions. In A. J. Menezes and S. A. Vanstone, editors, *Advances in Cryptology - Proceedings of Crypto'90*, Lecture Notes in Computer Science, Vol. 537, pages 303–311. Springer-Verlag, 1991.

On the Power of Memory in the Design of Collision Resistant Hash Functions

Bart Preneel*, René Govaerts, and Joos Vandewalle

Katholieke Universiteit Leuven, Laboratorium ESAT-COSIC,
Kardinaal Mercierlaan 94, B-3001 Heverlee, Belgium

Abstract. Collision resistant hash functions are an important basic tool for cryptographic applications such as digital signature schemes and integrity protection based on "fingerprinting". This paper proposes a new efficient class of hash functions based on a block cipher that allows for a tradeoff between security and speed. The principles behind the scheme can be used to optimize similar proposals.

1 Introduction

Although more theoretical definitions of collision resistant hash functions are available [Dam89], we will be satisfied with a more practical definition as given in [Mer89].

Definition 1 *A function $h()$ is a* **collision resistant hash function** *if:*

- *The description of $h()$ is* publicly known *and does not require any secret information for its operation (extension of Kerckhoff's principle).*
- *The argument X can be of* arbitrary length *and the result has a fixed length of h (with $h \geq 112-128$ bits in order to avoid the birthday or square-root attack* [QD89, Yuv79]*).*
- *Given $h()$ and X, the computation of $h(X)$ must be* "easy".
- *The hash function must be* one-way *in the sense that given a Y in the image of h, it is "hard" to find a message X such that $h(X) = Y$, and given X and $h(X)$, it is "hard" to find a message $X' \neq X$ such that $h(X') = h(X)$.*
- *If is* "hard" *to find two distinct arguments that hash to the same result (the collision resistant property).*

Here "easy" and "hard" can to be substituted by adequate definitions. In this paper, "hard" will mean that it requires at least 2^S encryptions, with S the security level of the hash function. Note that under certain circumstances one-wayness is implied by the collision resistant property [Dam89].

Two arguments can be indicated to construct a hash function based on a block cipher. The first argument is the minimization of the design and implementation effort. The major advantage however is that the trust in an existing

* NFWO aspirant navorser, sponsored by the National Fund for Scientific Research (Belgium).

block cipher can be transferred to a hash function. It is important to note that for the time being significantly more research has been spent on the design of secure block ciphers compared to the effort to design hash functions. It is also not obvious at all that the limited number of design principles for encryption algorithms are also valid for hash functions. The main disadvantage of this approach is that dedicated hash functions are likely to be more efficient. Moreover some block ciphers show certain weaknesses that are only relevant if they are used in a hash function. One also has to take into account that in some countries export restrictions apply to encryption algorithms but not to hash functions.

For a hash function based on a block cipher, the following notations have to be fixed. The block length i.e., the size of plaintext and ciphertext in bits will be denoted with n. The argument of the hash function is padded with an unambiguous padding rule such that it is a multiple t of the block size. The hash function can subsequently be described as follows:

$$H_i = f(X_i, H_{i-1}) \qquad i = 1, 2, \ldots t\,.$$

Here f is the round function, H_0 is equal to the initial value (IV), that should be specified together with the scheme, and H_t is the hashcode. Finally the **rate** R of a hash function based on a block cipher is defined as the number of encryptions to process a block of n bits.

A large number of hash functions based on a block cipher have been proposed [MPW91, Pre93]. For most schemes the size of the hashcode is equal to the block length. However, the block ciphers that have been proposed in literature comprising DES [Fi46] have only a block length of $n = 64$ bits, which implies that for a collision resistant hash function one needs that the size of the hashcode $h = 2n$. An additional problem is that the key length of DES is only 56 bits. Existing proposals of this type are MDC-2 and MDC-4 [MS88], with rate 2 and 4 respectively, three schemes by R. Merkle [Mer89] with rate 18.3, 5.8, and 3.62. More efficient schemes have been proposed with a rate close to 1 [BPS90, PBG89, QG89], but currently all these proposals have been broken [Cop92, Pre93]. The constructions by R. Merkle are based on a collision resistant function (the "meta-method" [Dam89, Mer89]), and for these schemes it is possible to write down a proof based on a black box model of DES. On the other hand, if one wants to use the scheme with DES, it still has to be modified to take into account properties like the weak keys and the complementation property, and it should be checked for vulnerability to specific attacks (e.g., differential attacks). Two new schemes based on a block cipher with a double length key have been proposed recently [LM92]. The security level of all these schemes is 64 bits, 56 bits or even smaller.

In this paper a new class of schemes will be proposed, that allow for a tradeoff between security level, rate, and size of the hashcode. Their rate lies between 4 and 8, but they are *faster* than most other schemes because the key remains fixed during the evaluation of the hash function. This also implies that they can be applied *under more general circumstances.*

2 Design Principles

The basic principle is that the key remains fixed during the hashing process. This has the following advantages:

performance: in general, the key scheduling is significantly slower than the encryption operation. A first argument to support this is that the key scheduling can be designed as a very complex software oriented process to discourage exhaustive attacks. Here software oriented means that the variables are updated sequentially, which reduces the advantages of a parallel hardware implementation. Even when the key schedule is simple, it can be harder to optimize its implementation. E.g., for highly optimized DES software, an encryption with key change will between 2.5 and 4.5 times slower. Moreover, encryption hardware is in general not designed to allow fast modification of the key, as a key change can cause loss of pipelining, resulting in a serious speed penalty.

security: an attacker has no control at all over the key. Hence attacks based on weak keys can be eliminated completely in the design stage.

generality: the hash function can be based on any one-way function with small dimensions (e.g., 64 bit input and output).

collision resistant MAC: if the keys are kept secret, the scheme gives a construction for a MAC for which it is hard to produce collisions even for someone who knows the secret key. This is not possible with the widespread schemes for a MAC. An application where this property might be useful has been discussed in [MW88].

The other design principles of the scheme are:

atomic operation: the one-way function that is used will be encryption of the argument X with the key K followed by the addition modulo 2 of X to the ciphertext: $E(K,X) \oplus X$. It has shown to be very useful for single length hash functions (e.g., [MMO85]) and was also used by R. Merkle and in MDC-2 and MDC-4.

parallel operation : the one-way function will be used more than once, but it will be possible to evaluate the function in parallel; this opens the possibility of a fast parallel hardware implementation. It is also clear that a scheme in which several instances are used in a serial way is much harder to analyze.

tradeoff between memory, rate, and security level : the rate of the system can be decreased at the cost of a decreasing security level; it will also be possible to decrease the rate by increasing the size of the hashcode. This could also be formulated in a negative way, namely that the security level will be smaller than what one would expect based on the size of the hashcode. Observe that this property is also present in a limited way in MDC-2, MDC-4 [MS88], and in the schemes of R. Merkle.

the basic function is not collision resistant: the construction is not based on a collision resistant function, because it can be shown that this would not yield an acceptable performance (the efficiency will decrease with a factor

4 or more). This means that producing collisions for different values of the chaining variable is easy. As it should be hard to produce collisions for the data input, this input will be protected more strongly.

3 Description of the New Scheme

One iteration step consists of k parallel instances of the one-way function, each parameterized with a fixed and different key. In the following, these instances will be called 'blocks'. These k keys should be tested for specific weaknesses: in the case of DES it is recommended that all 16 round keys are different, and that no key is the complement of any other one. The total number of input bits is equal to kn. These inputs are selected from the x bits of the data X_i and from the h bits of the previous value of the chaining variable H_{i-1}. Every bit of X_i will be selected α times, and every bit of H_{i-1} will be selected only once (this can be generalized), which implies the following basic relation:

$$\alpha \cdot x + h = k \cdot n . \tag{1}$$

The rate R of this scheme is given by the expression:

$$R = \frac{n \cdot k}{x} . \tag{2}$$

As it does not make sense to enter the same bit twice to a single block, it will always be the case that $2 \leq \alpha \leq k$.

The output of the functions has also size kn. This will be reduced to a size of h by selecting only h/k bits from every output. Subsequently a simple mixing operation is executed, comparable to the exchange of left and right halves in MDC-2. The goal of this operation is to avoid that the hashing operation consists of k independent chains. If h is a multiple of k^2, this mixing operation can be described as follows. The selected output block of every function (consisting of h/k bits) is divided into k parts, and part j of block i is denoted with H^{ij} $(1 \leq i, j \leq k)$. Then $H^{ji}_{\text{out}} \longleftarrow H^{ij}_{\text{in}}$. Figure 1 depicts one iteration for the case $k = 4$ and $\alpha = 4$. It will become clear that the complexity of the scheme does not depend on α, but on the difference between k and α. Therefore, the parameter ϕ is defined as $k - \alpha$.

The next step in the design is the decision on how the data bits are to be distributed over the different blocks if $\phi > 0$. The construction is obtained by considering the following attack. Let $\mathcal{S}$ be a subset of the blocks. For a given value of H_{i-1}, fix the data input of the blocks in $\mathcal{S}$, which means that the output of these blocks will be the same for both input values. Subsequently, match the remaining outputs with a birthday attack. In order to maximize the effort for this attack, it is required that after fixing the input of the blocks in $\mathcal{S}$, the number of bits that can still be freely chosen by an attacker is minimal (this will be explained in more detail in Sect. 4). This number will be denoted with $A(\mathcal{S})$.

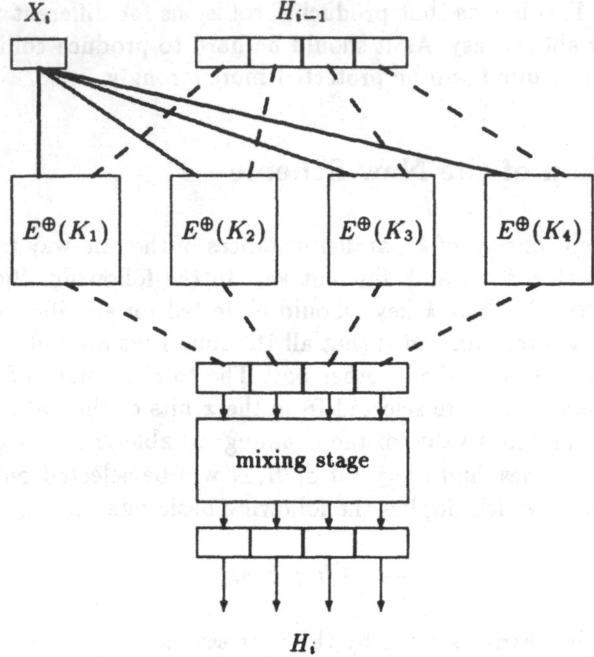

Fig. 1. One iteration of the new hash function proposal.

Theorem 1 *If the data is divided into $\binom{k}{\phi}$ parts, and every part goes to a different combination of $\binom{k}{\phi}$ blocks, an optimal construction is obtained. Let A_s be defined as $\max_{|S|=s} A(S)$, then this construction results in the following expression for A_s:*

$$A_s = \frac{\binom{k-s}{k-\phi}}{\binom{k}{\phi}} \cdot x = \frac{\binom{\phi}{s}}{\binom{k}{s}} \cdot x \qquad 1 \leq s \leq \phi \tag{3a}$$

$$= 0 \qquad \textit{else} . \tag{3b}$$

This construction is optimal in the sense that for any other construction there will be an s (with $1 \leq s \leq k$) such that $A'_s \geq A_s$. If equality holds for all values of s both constructions are equivalent.

In order to clarify the scheme, a detailed description will be given for the case $n = 64$, $k = 4$, $\phi = 2$, $h = 148$, and hence $x = 54$. In this case Theorem 1 states that in order to optimize the security level, X_i has to be split into 6 parts of 9 bits each. The first 9-bit part of X_i goes to blocks 1 and 2, the second 9-bit part to encryption block 1 and 3, etc., and the sixth part goes to block 3 and 4. The 64-bit input of a single block cipher consists of $148/4 = 37$ bits of h and 3 parts of 9 bits each. The rate of this scheme is 4.7 (but with a fixed key), and from our evaluation it follows that the security level is about 55 bits. In the next sections it will be explained how the security level can be determined.

4 Attacks on the Scheme

A security proof for the scheme can not be given for the time being. This disadvantage is shared with all other schemes (including MDC-2 and MDC-4); the only exceptions are the schemes by R. Merkle [Mer89]. The main difference with the other schemes is that the system is parameterized, and that the security level depends on the size of hashcode h.

In the following, the number of operations to produce a preimage and a collision for the hash function will be studied by considering a number of attacks that are faster than a straightforward exhaustive or birthday attack. Such attacks are possible as not all output bits depend in a strong way on the inputs of a single iteration step. Indeed, the data only enter α blocks, and hence if $\alpha < k$, the output of $\phi = k - \alpha$ blocks does not depend on these input bits. The diffusion of the H_{i-1} is limited to one block. Note that this property is shared with MDC-2. This limited dependency is solved by increasing the size of the hashcode. The required number of bits for the hashcode is estimated from studying a set of attacks that exploit the structure of the scheme. The generality of the proposed attacks should form an argument for the security. However, it is for the time being not possible to prove that there does not exist any more sophisticated attack. The advantage of the scheme is that the security level can always be increased at the cost of an increased memory and decreased efficiency. By construction the scheme is not vulnerable to attacks based on weak keys or based on the complementation property.

Before discussing the collision attacks in detail, expressions are required for the number of operations to produce a collision under certain constraints. Assume one has a random function with B output bits and A input bits that can be chosen arbitrarily. The function might have C inputs bits that can not be chosen freely; these input bits will be called parameters. If a collision is to be produced for this function for a certain value of a parameter, i.e., two inputs that result in the same output bits, two cases have to be distinguished:

$A > B/2$: in this case producing a collision requires $2^{B/2}$ function evaluations.

$A < B/2$: in this case, the number of expected collisions after a single trial is equal to $p = (2^A)^2/2^B$. This process will be repeated for several values of the parameter (it is assumed that C is sufficiently large). The expected number of trials is given by $1/p$ and the effort for a single trial is 2^A function evaluations. Hence the expected number of function evaluations is equal to $2^B/2^A$, which is always larger than $2^{B/2}$.

For the evaluation of the scheme one has to determine an expression for the number of operations to produce a 2^c-fold collision. It can be shown that for large values of c this number is given by 2^{B+c} (if A is sufficiently large). For smaller values of c expressions have been derived in [Pre93].

Four types of birthday attacks that exploit the structure of the scheme will be discussed. They are only valid if $\phi > 0$. All these attacks yield a ***piece-wise linear relation*** between memory h and security level S. Because of (1) and (2), there will be a hyperbolic relation between the rate R and the security level S.

Attack A: fix the input bits to the first s blocks and match the output of the remaining $k-s$ blocks with a birthday attack. The number of output bits of these blocks is denoted with B_s. It is clear that $B_s = \frac{k-s}{k} \cdot h$. The binary logarithm of number of operations for this attack is given by the following expression:

$$\frac{B_s}{2} + 1 + \log_2(k-s) \quad \text{if } A_s \geq \frac{B_s}{2}$$
$$B_s - A_s + 1 + \log_2(k-s) \quad \text{if } A_s < \frac{B_s}{2}.$$

Attack B: a more effective attack consists of generating a 2^c-fold collision for the first s blocks. In the next step, one has $A_s + c$ degrees of freedom available to match the remaining $k-s$ blocks. This attack has already two parameters: s and c. A problem that should be considered is the following: for large values of c, an attacker needs about $h\frac{s}{k} + c$ degrees of freedom to produce such a multiple collision. In most cases, there are not that many data bits that enter the first block(s). However, one can assume that an attacker can also introduce variations in the previous iteration steps. If we would have designed a collision resistant function, this assumption would not have been valid. In that case any attack has to produce a collision for a single iteration step.

Attack C: under certain circumstances, attack B can be optimized by exploiting the block structure: first a 2^{c_1}-fold collision is produced for the output of the first block (possibly using variations in previous rounds), subsequently a 2^{c_2}-fold collision is produced for the output of the second block. This continues until a 2^{c_s}-fold collision is produced for the output of block s. Finally the last s blocks are matched with a birthday attack with $A_s + c_s$ degrees of freedom. The attack is optimized with respected to the parameter c_s; the choice of c_s fixes the other c_i's as follows: in order to produce a 2^{c_s}-fold collision for block s, $h/k + c_s$ trials are necessary (assuming c_s is not too small). There are only a_s $(= A_{s-1} - A_s)$ bits available at the input of block s, which means that a $2^{c_{s-1}}$-fold collision will be required at the output of block $s-1$, with $c_{s-1} = h/k + c_s - a_s$. This procedure is repeated until the first block is reached. It is assumed that there are sufficient degrees of freedom are available through variations in the previous iterations.

Attack D: This attack is different from attacks B and C because it makes a more explicit use of interaction with the previous iteration step. It is based on the observation that if h is significantly smaller than the value that is being evaluated, it should be easy to produce collisions or even multiple collisions for H_{i-1}. Therefore it should also be easy to produce multiple collisions for the first s' blocks of H_{i-1}, that contain $h' = s'/h$ bits. The data bits that enter the first s' block are now fixed, and this implies that the output of these blocks will be identical. From block $s'+1$ on, the attack continues with a type C attack.

The next step consists in determining the number of operations to produce a multiple collision for the first s' blocks of H_{i-1}. This can be done by

determining the optimal attack on a reduced scheme. It might be that again attack D is optimal and then the evaluation program works in a recursive mode.
The problem that remains to be solved is how to evaluate the number of operations to produce a multiple collision for the first s' blocks of H_{i-1}. This number is estimated by calculating the number of operations for a collision S'; in general S' will be smaller than $h'/2$. Subsequently, the expression for the number of operations for a multiple collision is used where the size of the block is replaced by the effective block length $2S'$. The number S' can be found by comparing the efficiency of attacks A, B, C, and D for a scheme with the same configuration as the original one (this means the number of data bits has not been modified), but with h replaced by h' (at the output). If attack D is optimal, the program works in a recursive mode.

The non-linear behavior of the number of operations for the birthday attacks and the recursive nature of the attack D forced us to evaluate the security with a computer program. However, the results can be verified by using approximations for the expressions.

The applicability of differential attacks to hash functions based on block ciphers has been studied in [Pre93]. The main reasons why differential attacks on this scheme will not work if DES is used as the underlying block cipher is that there has not been found a good characteristic with an even number of rounds. Differential attacks on this scheme are harder because the attacker has no complete control over the plaintext, and because the keys can be selected in a special way to minimize the probability of iterative characteristics (which is not possible for MDC-2). On the other hand, the characteristic has only to hold for the subset of the output bits that has been selected. This means that in the last round the characteristic must not be completely satisfied. However, it should be noted that the success probability will decrease very fast if the characteristic is not satisfied in earlier rounds. Like in the case of MDC-2, every data bit is used at least twice, which implies that a characteristic with a high probability is required for an attack faster than exhaustive search. Once a detailed scheme has been fixed, more work can be done on selecting the keys in such a way that differential attacks are less efficient.

When looking for a preimage, one can also exploit the limited dependencies between H_i and H_{i-1}. Going one step backwards in the chain (for a fixed X_i) requires only $2^{h/k+\log_2(k)}$ operations. Subsequently one will try to obtain a match for the chaining variable with a meet in the middle attack. This implies that one can find a preimage with less than 2^h operations. However, for the time being no attack has been identified that is faster than the attacks that look for a collision.

5 A Detailed Study of the Security Level

In this section the security level of schemes with ϕ between 0 and 4 will be discussed. In order to keep the schemes practical, the value of k will be limited to 6. If $\phi = 0$ it can be shown that for practical values of the security level the schemes with small values of k are more efficient. For all other schemes, larger values of k would imply that there are too many small parts (recall that the number of parts is equal to $\binom{k}{\phi}$), which would increase the overhead of bit manipulations to an unacceptable level. Finally note that the restriction $\alpha \geq 2$ is equivalent to $k \geq \phi + 2$.

5.1 The Case $\phi = 0$

In this case $\alpha = k$, which means that every data bit goes to every block. There are no problems in determining an optimal configuration of the scheme. As indicated above, none of above attacks applies, which means that the security level is equal to its upper bound $S = h/2$. It can be shown that under certain circumstances the security level is 1 bit lower.

The expression for the rate of this scheme reduces to $R = k/(1 - \frac{2S}{kn})$. It can be seen that it becomes advantageous to increase k by one if the security level is given by $S = n \cdot \frac{k(k+1)}{2k+1}$. This means that $k = 4$ will be the optimal solution for a security level between 54.9 and 71.1 bits. A graphical representation of the relation between R and S is given in Fig. 2.

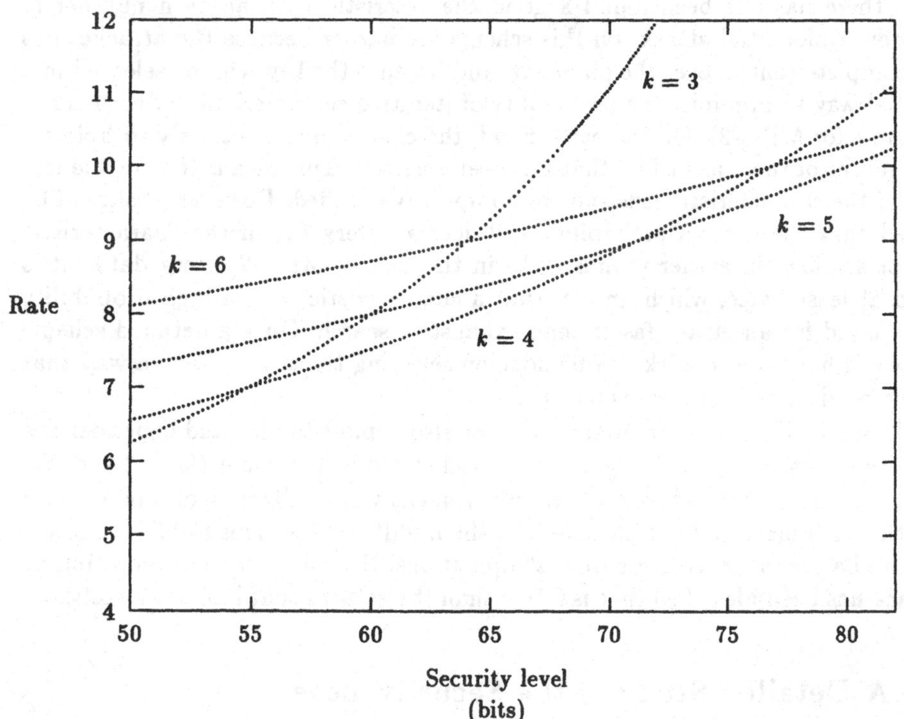

Fig. 2. Relation between the rate R and the security level S for $\phi = 0$ and k between 3 and 6. The optimal value of k increases with S.

5.2 The Case $\phi = 1$

In this case the data input is split into k parts, and every part goes to $k-1$ blocks. Computer calculation has shown that the most efficient attack is attack D with $s = s' = 1$. The number of operations can be approximated by the following expressions:

1. number of operations to produce a collision for blocks 2 to k:

$$\frac{k-1}{k}h - \frac{x}{k} + 1 + \log_2(k-1) - c\,.$$

2. number of operations to produce a 2^c-fold collision for the first block of H_{i-1}:

$$\frac{k-1}{k}\frac{h}{k} + 2(1 + \log_2(k-1)) + c\,.$$

The approximation lies in the fact that the logarithm of the number of operations to produce a 2^c-fold collision is not a linear function of c for small values of c. The total number of operations should be minimized with respect to c, which yields the following expression for the security level:

$$S = \frac{h}{2}\left[\frac{k^2-1}{k^2} + \frac{1}{k(k-1)}\right] - n\,\frac{2}{k-1} + \frac{5}{2} + \frac{3}{2}\log_2(k-1)\,.$$

For k between 3 and 5, the relation between h and S is indicated in Table 1. For $k \geq 6$, the resulting expression is larger than $h/2$ for all values of h, which means that a simple birthday attack is more efficient. The theoretical results agree very well with the result obtained from computer calculations. The program shows that $k = 4$ is the best choice for S between 51 and 72 bits.

k	Security level S (bits)
3	$\frac{19}{36}h - 12.0$
4	$\frac{49}{96}h - 5.8$
5	$\frac{101}{200}h - 2.5$

Table 1. Relation between h and S for $\phi = 1$ and $k = 3$, 4, and 5.

5.3 The Case $\phi = 2$

In this case the data input is split into $k(k-1)/2$ parts, and every part goes to $k-2$ blocks. Computer calculation has shown that the most efficient attack is attack D with $s = 2$ and $s' = 1$. The number of operations can be approximated by the following expressions:

1. number of operations to produce a collision for blocks 3 to k:

$$h\left[\frac{k-2}{k}+\frac{2}{k(k-1)(k-2)}\right]-n\,\frac{2}{(k-1)(k-2)}+1+\log_2(k-2)-c\,.$$

2. number of operations to produce a 2^c-fold collision for block 2:

$$\frac{h}{k}+c\,.$$

3. number of operations to produce a 2^c-fold collision for the first block of H_{i-1}:

$$h\left[\frac{1}{k}+\frac{k-2}{k^2}+\frac{2}{k(k-1)}\right]-n\,\frac{2}{k-1}+2(1+\log_2(k-3))+c\,.$$

This number of operations should be minimized with respect to c. For smaller values of h, the third term is negligible, while for larger values of h, the second term is negligible. For k between 4 and 6, the relation between h and S is indicated in Table 2. The program shows that $k=4$ is the best choice for S smaller than 69 bits.

k	Security level S (bits)			
4	$h \geq 132$	$\frac{27}{48}h-28.0$	$h \leq 132$	$\frac{10}{24}h-8.7$
5	$h \geq 122$	$\frac{79}{150}h-16.5$	$h \leq 122$	$\frac{5}{12}h-3.0$
6	$h \geq 110$	$\frac{37}{72}h-10.5$	$h \leq 110$	$\frac{17}{40}h-0.7$

Table 2. Relation between h and S for $\phi=2$ and $k=4$, 5, and 6.

5.4 The Case $\phi = 3$

In this case the data input is split into $\binom{k}{3}$ parts, and every part goes to $k-3$ blocks. Computer calculation has shown that the most efficient attack is attack D with $s=3$ and $s'=1$. The number of operations can be approximated by the following expressions:

1. number of operations to produce a collision for blocks 4 to k:

$$h\left[\frac{k-3}{k}+\frac{6}{k(k-1)(k-2)(k-3)}\right]-n\,\frac{6}{(k-1)(k-2)(k-3)}+1+\log_2(k-3)-c\,.$$

2. number of operations to produce a 2^c-fold collision for block 3:

$$\frac{h}{k}+c\,.$$

3. number of operations to produce a $2^{c'}$-fold collision for block 2 (with $c' = h/k + c - a_3$):

$$h\left[\frac{2}{k}+\frac{6}{k(k-1)(k-2)}\right]-n\,\frac{6}{(k-1)(k-2)}+2(1+\log_2(k-3))+c\,.$$

4. number of operations to produce a $2^{c''}$-fold collision for the first block of H_{i-1} (with $c'' = h/k + c' - a_2$). It can be shown that this number is significantly smaller than $\frac{h}{k} + c$.

This number of operations should be minimized with respect to c. The second term is always smaller than the third term. This results in the following expression for the security level:

$$S=\frac{h}{2}\left[\frac{k-1}{k}+\frac{6}{k(k-1)(k-3)}\right]-n\,\frac{3}{(k-1)(k-3)}+\frac{3}{2}+\frac{1}{2}\log_2(k-3)\,.$$

For k equal to 5 and 6, the relation between h and S is indicated in Table 3. The program shows that $k = 5$ is the best choice for S smaller than 82 bits.

k	Security level S (bits)
5	$\frac{19}{40}h - 22.0$
6	$\frac{9}{20}h - 10.5$

Table 3. Relation between h and S for $\phi = 2$ and $k = 4$, 5, and 6.

5.5 The Case $\phi = 4$

The only case that has been studied is $k = 6$. Here it is not possible to derive simple analytic expressions that are sufficiently accurate. This is because the result obtained by the different attacks lie very closely together; depending on the value of h, the best attack is attack D with $s = 4$ and $s' = 2$ or 4. Moreover, the optimal values of c are very small, which means that the system is non-linear, and the attacks are strongly dependent on the use of recursion. An upper bound on S can be easily obtained using method A:

$$S \leq \frac{11}{30}h - 10.8\,.$$

A least squares fitting the computer evaluation yields a correlation coefficient of 0.99958 with the following expression:

$$S = 0.3756h - 14.974\,.$$

This can be approximated very well by $S = 3/8\,h - 15$.

6 Extensions

Before the scheme can be applied in practice, one has to consider the following limitation: h and x are not continuous variables, but integers that have to satisfy certain constraints. The scheme can be extended by using every bit of H_i more than once, but this complicates the evaluation. Another extension is to design a collision resistant round function based on this scheme: this simplifies the evaluation but decreases the efficiency. Finally it is explained how the basic principles can be applied to other hash functions based on block ciphers.

The study of the previous scheme assumed that h and x are continuous variables. However, in practice they will have to be integers that satisfy certain constraints:

1. x has to be an integer multiple of $\binom{k}{\phi}$. Therefore define $x' = x/\binom{k}{\phi}$.
2. $nk - h$ has to be an integer multiple of $k - \phi$.
3. h has to be an integer multiple of k. Therefore define $h' = h/k$.

Note that in order to perform the mixing stage on the output of the k blocks, one needs in fact the requirement that h is an integer multiple of k^2. However, this mixing stage is not critical in the security analysis of the scheme. The following algorithm steps through all values of x and h that satisfy the constraints, for which $h > h_0$. First the starting values are generated:

$$x'_1 = \left\lceil \frac{n - \lceil \frac{h_0}{k} \rceil}{\binom{k-1}{\phi}} \right\rceil \tag{4}$$

$$x_1 = \binom{k}{\phi} x'_1 \tag{5}$$

$$h_1 = k\left(n - \binom{k-1}{\phi} x'_1\right) . \tag{6}$$

Here $\lceil x \rceil$ denotes the smallest integer greater than or equal to x. The next values are calculated as follows:

$$x_{i+1} = x_i + \binom{k}{\phi} \tag{7}$$

$$h_{i+1} = h_i + k\binom{k-1}{\phi} . \tag{8}$$

In the overview of the results it will be graphically indicated which schemes satisfy the requirements. It is of course always possible to think of schemes for which the parts differ 1 or 2 bits in size just to match the constraints. This will affect the security level compared to the ideal situation, but the decrease will certainly be only marginal. These schemes are certainly less elegant, but as long as the asymmetry in the bit manipulations has no negative influence on the performance, this is not so important.

If the schemes are extended by using every bit of H_{i-1} more than once, the following elements will have to be considered. First, the allocation of the bits of

H_{i-1} to the different blocks will have to be made in a similar way as for the data bits. However, some additional work has to be done because both allocations should be as independent as possible, i.e., the bits of H_{i-1} and X_i will have to occur in as many combinations as possible. The study of attacks on this scheme is more complicated, especially for type D attacks.

Another issue is how to extend this method to construct a collision resistant function. In this case the attacks to be considered would be simpler, because the interaction with the previous iteration steps can be neglected. This is not completely true however, because an attacker could exploit the fact that the function is not complete, by producing collisions for part of the output blocks. However, if the program is adapted to evaluate this type of schemes, it becomes clear that they will never be very efficient: the best scheme with a security level of 56 bits under the constraint $k \leq 16$ has a rate of 20 (which is in case of software a little worse than MDC-4). It is a scheme with $\alpha = 3$ or $\phi = 12$. The size of the hashcode would be 272 bits, and every iteration processes 48 bits. The scheme is however very impractical because X_i has to be split into 455 blocks of 2 or 3 bits.

The basic principles used here can also be applied to other hash functions based on block ciphers where the key is modified in every iteration step. As an example it is indicated how MDC-2 could be extended in two ways to obtain a security level larger than 54 bits. The 2 parallel DES operations will be replaced by 3 parallel DES operations ($k = 3$).

- A trivial way would be $\alpha = 3$: every data bit is used 3 times as plaintext. The size of the hashcode would be 192 bits, and the effective security level is equal to 81 bits. The rate of this scheme is equal to 3 (comprising a key change).
- A second scheme can be obtained by selecting $\alpha = 2$: the data input of 96 bits is divided into 3 32-bit parts, that are allocated in the optimal way to the 3 blocks. The rate of this scheme is equal to 2 (the same as MDC-2), but the security level is slightly larger than 60 bits.

Of course it is possible to extend this for $k > 3$, which will result in faster schemes that require more memory.

A disadvantage of all these new schemes is that the decreased rate has to be paid for by increasing the memory. The additional 64 to 80 bits are no problem for the chaining variables (certainly not when the computations are performed in software), but the increased size of the hashcode might cause problems. This is not the case for digital signatures, as most signature schemes sign messages between 256 and 512 bits long. Exceptions to this rule are the scheme by Schnorr [Sch89] and DSA, the draft standard proposed by NIST [Fi91], where the size of the argument is 160 bits. If the hash function is used for fingerprinting computer files, an increased storage can pose a more serious problem. However, it can be solved by compressing the result to a number of bits equal to twice the security level S. This can be done with a (slow) hash function with $\phi = 0$, security level S, and size of the hashcode $2S$.

7 Conclusion

A new class of hash functions based on block ciphers has been proposed, that allows for a tradeoff between security, memory, and speed. The parameters for some efficient schemes for a given value of ϕ and a given security level S are indicated in Table 4. For $k = 4$ and $k = 5$, the relation between the rate R and the security level S with parameter ϕ, with $0 \leq \phi \leq k-2$ can be found in Fig. 3 and 4. The solutions that take into account the discrete character are marked with a ◇. The underlying principles can also be used to improve other proposals for hash functions based on block ciphers.

ϕ	k	security level $\simeq$ 54 bits				security level $\simeq$ 64 bits			
		R	h	S	$h(k-\phi)/\phi$	R	h	S	$h(k-\phi)/\phi$
0	4	6.92	108	54	108	8.00	128	64	128
1	4	5.82	124	58	93	6.40	136	64	102
2	4	4.74	148	55	74	6.10	172	69	86
3	5	4.00	160	53	64	4.57	180	62	72
4	6					4.27	204	62	68

Table 4. Overview of best results for small values of ϕ and k. The quantity $h(k-\phi)/\phi$ indicates how many output bits depend on a single input bit.

References

[BPS90] L. Brown, J. Pieprzyk, and J. Seberry, "LOKI - a cryptographic primitive for authentication and secrecy applications," *Advances in Cryptology, Proc. Auscrypt'90, LNCS 453*, J. Seberry and J. Pieprzyk, Eds., Springer-Verlag, 1990, pp. 229–236.

[Cop92] D. Coppersmith, "Two broken hash functions," *IBM T.J. Watson Center, Yorktown Heights, N.Y., 10598, Research Report RC 18397*, October 6, 1992.

[Dam89] I.B. Damgård, "A design principle for hash functions," *Advances in Cryptology, Proc. Crypto'89, LNCS 435*, G. Brassard, Ed., Springer-Verlag, 1990, pp. 416–427.

[Fi46] *"Data Encryption Standard,"* Federal Information Processing Standard (FIPS), Publication 46, National Bureau of Standards, U.S. Department of Commerce, Washington D.C., January 1977.

[Fi91] *"Digital Signature Standard,"* Federal Information Processing Standard (FIPS), Draft, National Institute of Standards and Technology, US Department of Commerce, Washington D.C., August 1991.

[GCC88] M. Girault, R. Cohen, and M. Campana, "A generalized birthday attack," *Advances in Cryptology, Proc. Eurocrypt'88, LNCS 330*, C.G. Günther, Ed., Springer-Verlag, 1988, pp. 129–156.

[LM92] X. Lai and J.L. Massey "Hash functions based on block ciphers," *Advances in Cryptology, Proc. Eurocrypt'92, LNCS*, R.A. Rueppel, Ed., Springer-Verlag, to appear.

[MMO85] S.M. Matyas, C.H. Meyer, and J. Oseas, "Generating strong one-way functions with cryptographic algorithm," *IBM Techn. Disclosure Bull.*, Vol. 27, No. 10A, 1985, pp. 5658–5659.

[Mer89] R. Merkle, "One way hash functions and DES," *Advances in Cryptology, Proc. Crypto'89, LNCS 435*, G. Brassard, Ed., Springer-Verlag, 1990, pp. 428–446.

[MS88] C.H. Meyer and M. Schilling, "Secure program load with manipulation detection code," *Proc. SECURICOM 1988*, pp. 111–130.

[MW88] C. Mitchell and M. Walker, "Solutions to the multidestination secure electronic mail problem," *Computers & Security*, Vol. 7, 1988, pp. 483–488.

[MPW91] C.J. Mitchell, F. Piper, and P. Wild, "Digital signatures," in *"Contemporary cryptology: the science of information integrity,"* G.J. Simmons, Ed., IEEE Press, 1991, pp. 325–378.

[PBG89] B. Preneel, A. Bosselaers, R. Govaerts, and J. Vandewalle, "Collision free hash functions based on blockcipher algorithms," *Proc. 1989 International Carnahan Conference on Security Technology*, pp. 203-210.

[Pre93] B. Preneel, "Analysis and design of cryptographic hash functions," *Doctoral Dissertation*, Katholieke Universiteit Leuven, 1993.

[QD89] J.-J. Quisquater and J.-P. Delescaille, "How easy is collision search ? Application to DES," *Advances in Cryptology, Proc. Eurocrypt'89, LNCS 434*, J.-J. Quisquater and J. Vandewalle, Eds., Springer-Verlag, 1990, pp. 429–434.

[QG89] J.-J. Quisquater and M. Girault, "2n-bit hash-functions using n-bit symmetric block cipher algorithms," *Advances in Cryptology, Proc. Eurocrypt'89, LNCS 434*, J.-J. Quisquater and J. Vandewalle, Eds., Springer-Verlag, 1990, pp. 102–109.

[Sch89] C.P. Schnorr, "Efficient identification and signatures for smart cards," *Advances in Cryptology, Proc. Crypto'89, LNCS 435*, G. Brassard, Ed., Springer-Verlag, 1990, pp. 239–252.

[Yuv79] G. Yuval, "How to swindle Rabin," *Cryptologia*, Vol. 3, 1979, pp. 187–189.

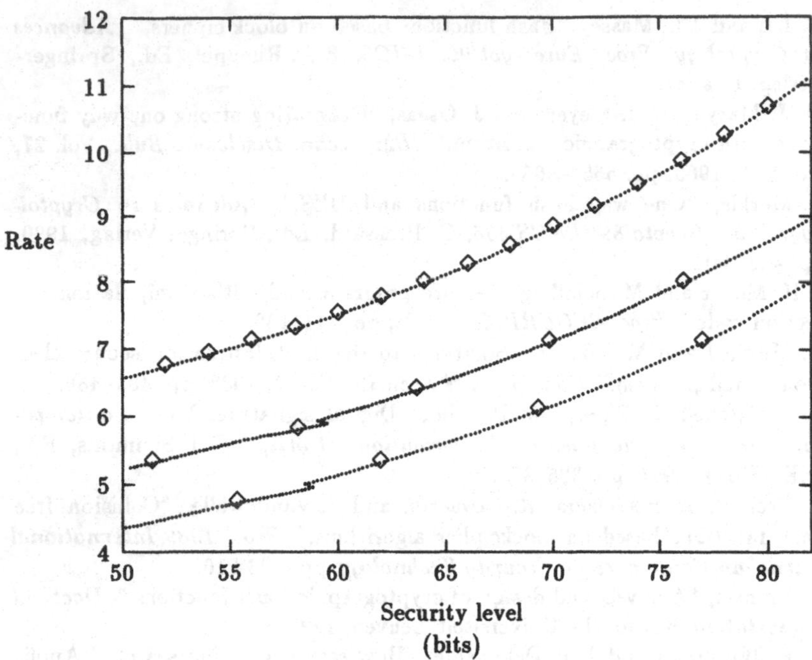

Fig. 3. Relation between the rate R and the security level S for $k = 4$ with parameter $\phi = 0$, 1, and 2.

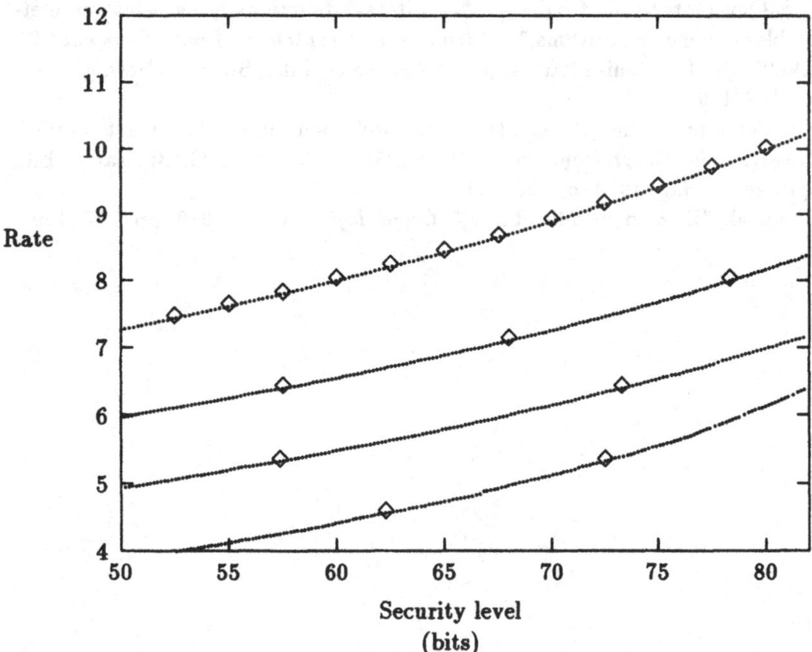

Fig. 4. Relation between the rate R and the security level S for $k = 5$ with parameter $\phi = 0$, 1, 2, and 3.

A Practical Digital Multisignature Scheme Based on Discrete Logarithms (Extended Abstract)

Thomas Hardjono[1] * and Yuliang Zheng[2] **

[1] ATR Communications Research Laboratories
2-2 Hikaridai, Seika-Cho, Soraku-gun, Kyoto 619-02, Japan
[2] Department of Computer Science, University of Wollongong, Australia

Abstract. This paper proposes a practical digital multisignature scheme based on the C^{*}_{sig} cryptosystem derived from the C_{sig} cryptosystem of Zheng and Seberry (1993). The simple scheme consists of three phases. In the first phase the issuer of the document prepares the document, the list of prospective signatories and a pad on which signatories are to write their signatures. In the second phase each signatory verifies the document, signs it and forwards it to the next signatory. In the third phase a trusted verifier or notary decides on the validity of the signatures. The scheme prevents cheating by dishonest signatories from going undetected. The scheme is practical and offers at least the same security level afforded by its underlying cryptosystem against external attacks. Internal attacks in the form of forging or cheating by a dishonest issuer or by one or more of the signatories (alone or by collaboration) requires the solving of instances of the discrete logarithm problem.

Keywords: Data Encryption, Information Security. Digital Signatures, Digital Multisignatures.

1 Introduction

The problem of providing digital signatures and digital multisignatures for electronic documents has been addressed by a number of researchers over the past years [1, 2, 3, 4, 5]. The aim of electronically signing an electronic document is in principle the same as handwritten signatures on paper documents. A useful analogy is that of the signing of a two-party contract. The party issuing the contract on paper requires the second party involved in the contract to sign it in the

* A substantial part of this work was completed when the author was at the Centre for Computer Security Research, Australia.

** Supported in part by the Australian Research Council under the reference number A49232172.

presence of a witness who is acceptable to both parties. The witness has the task of verifying that the signatures of both parties are authentic. This verification can take the form of a signature from the witness, which will be used to solve disputes that may occur in the future.

When this analogy is carried over into the world of computers a number of problems immediately occur. First and foremost is the problem of the signature representation to be attached to the document file. If signatures were to be represented electronically simply a unique string of bits to be appended to a document file, then both the document file and the signatures are vulnerable to modifications which may disadvantage the document issuer, the signatory or even the witness. Hence a method is required to ensure that any changes done to the signed document be detectable, the instant at which the validity of the document comes into question.

A number of sophisticated digital signature schemes have emerged in the past two decades. The strength of these schemes rely on their underlying cryptosystem which in turn must be resistant against various types of attacks. Chief among these attacks is the chosen ciphertext attack in which the attacker has access to the deciphering algorithm of a cryptosystem. In order to cryptanalyze the object ciphertext he or she can query the deciphering algorithm any number of times using any ciphertext (except the object ciphertext) as input to the deciphering algorithm. In this way he or she indirectly gains some knowledge about the object ciphertext that he or she wishes to cryptanalyze.

In this paper we present a practical digital multisignature scheme based on a modified version of the C_{sig} cryptosystem of [6]. The C_{sig} is one of a family of cryptosystems proposed in [6] which grew out of the earlier work in [7]. The family of cryptosystems represents an important step in public-key cryptography since they are practical and secure against *adaptively* chosen ciphertext attacks in a provable manner. In this paper we do not formally prove the security of the digital multisignature scheme since the constructs of the original C_{sig} are still intact. Hence the reader is directed to [6] for a formal proof of the security of the cryptosystem.

The outline of this paper is as follows. Section 2 presents the background notation for C_{sig}. Section 3 presents the digital multisignature scheme, while Section 4 continues with a brief discussion on its security. This paper is closed with a conclusion in Section 5

2 Notation

The notation used in this paper is taken from the original work in [6], and is presented briefly in this section. The cryptosystems of [6] are reminiscent of the Diffie-Hellman cryptosystem [8] and El Gamal cryptosystem [9] in their use of a n-bit prime p (public) and a generator g (public) of the multiplicative group $GF(p)^*$ of the finite field $GF(p)$. Here n is a security parameter which is greater that 512 bits, while the prime p must be chosen such that $p-1$ has a large prime factor [10]. In this paper the alphabet $\Sigma = \{0,1\}$ will be employed, and

$|x|$ denotes length of a string x over Σ. Concatenation of string are denoted using the "$||$" symbol and the bit-wise XOR operations of two strings is symbolized using "$\oplus$". The notation $x_{[i \cdots j]}$ $(i \leq j)$ is used to indicate the substring obtained by taking the bits of string x from the i-th bit (x_i) to the j-th bit (x_j). It is important to see that there is a natural one-to-one correspondence from strings in Σ^n to elements in the finite field $GF(2^n)$, and from from strings in Σ^n to integers in $[0, 2^n - 1]$.

Other notations are as follows. The action of choosing an element x randomly and uniformly from set S is denoted by $x \in_R S$. G is a cryptographically strong pseudo-random string generator based on the difficulty of computing discrete logarithms in finite fields [11, 12, 13]. G stretches an n-bit input string into an output string whose length can be an arbitrary polynomial in n. This generator produces $O(\log n)$ bits output at each exponentiation. The function h is a one-way hash function compressing input strings into n-bit output strings. All messages to be encrypted are chosen from the set $\Sigma^{P(n)}$, where $P(n)$ is an arbitrary polynomial with $P(n) \geq n$ and where padding can be used for messages of length less than n bits. The polynomial $\ell = \ell(n)$ specifies the length of tags. In the remainder of this paper all exponentiations are assumed to be done over the underlying groups. The reader is directed to [6] for a comprehensive discussion on the constructs of the family of cryptosystems.

The cryptosystem employed in this paper is a modified version of the C_{sig} cryptosystem of [6]. To distinguish it from C_{sig} we will refer to it as $C^{\star}_{sig}$ The algorithm is shown in the following, where Bob having his secret-public key pair (x_B, y_B) wants to send a $P(n)$-bit message m to Alice whose key pair is (x_A, y_A). Here Alice's (Bob's) secret key x_A (x_B) is element chosen randomly from $[1, p-1]$ and $y_A = g^{x_A}$ ($y_B = g^{x_B}$). Bob enciphers using $E^{\star}_{sig}$ (Algorithm 1) while Alice deciphers using $D^{\star}_{sig}$ (Algorithm 2).

Algorithm 1 $E^{\star}_{sig}(x_B, y_A, p, g, m)$

1. $k \in_R [1, p-1]$ s.t. $\gcd(k, p-1) = 1$.
2. $r = y_A^{x_B + k}$.
3. $z = G(r)_{[1 \cdots P(n)]}$.
4. $c_1 = g^{x_B}$.
5. $c_2 = g^k$.
6. $t = (h(m) - x_B r)/k \bmod (p-1)$.
7. $c_3 = {c_2}^t$.
8. $c_4 = z \oplus m$.
9. output (c_1, c_2, c_3, c_4).

end

Algorithm 2 $D^{\star}_{sig}(x_A, p, g, c_1, c_2, c_3, c_4)$

1. $r' = (c_1 c_2)^{x_A}$.
2. $z' = G(r')_{[1 \cdots P(n)]}$.
3. $m' = z' \oplus c_4$.
4. if $g^{h(m')} = c_1^{r'} c_3$ then
   output (m')
   else
   output $(\emptyset)$.

end

To distinguish between the problem of message tampering during transit and the problem of a dishonest issuer or signatory we will employ the terms *integrity* and *authenticity*. Corresponding to this is the difference between a *message* and a *document*, the later being contained in the former. A receiver of a message must himself of herself verify the integrity of the message, namely the fact that the document and signatures so far accumulated in the message has arrived from the sender unmodified. He or she must also verify the authenticity of the document, namely the fact that it is the same as the original document produced by the issuer of the document. These two terms will become clear in the next section, in which authenticity refers to the state of the document as produced by the issuer and integrity refers to the safety of the message containing the document with the signatures so far collected during their transit from sender to a receiver.

3 A Digital Multisignature Scheme

The scheme based on the $C^{\star}_{sig}$ cryptosystem follows the traditional approach of digital signatures where an *issuer* S_I of a *document* D requires a number of *signatories* $S_1, S_2, \ldots, S_u$ to electronically sign the document, which is to be verified by a *trusted verifier* S_V at the end of the signing process. More specifically, the issuer requires the signature of u number of signatories, each assumed to hold a secret key x_i and a public key $y_i = g^{x_i}$ $(i = 1, \ldots, u)$. The issuer's secret and public key are similarly denoted as x_I and y_I respectively. In addition x_V and y_V are the secret and public keys of the verifier.

The digital multisignature scheme consists of three phases. In the first phase the issuer prepares the document and the "list" of the identities of the prospective signatories of the document. The issuer then submits the document-list pair to the verifier who selects randomly and uniformly a value x_d (secret) associated with the document-list. The verifier also calculates $y_d = g^{x_d}$ and broadcasts y_d. The value x_d can be looked upon as a unique document identifier, and can in fact be derived from the secret random number and the true document or contract number. The tag associated with document-list mirrors the case where a document issuer must choose a trusted witness or notary that is allowed to view the document before it is passed around all the signatories who in turn require such an action be taken by the issuer before they enter into the signing process.

On receiving a tag for the document-list from the verifier the issuer creates an authenticator incorporating the document-list and the tag. The issuer then sends a message to the first signatory containing the document, pad and other parameters necessary for the authentication process by the signatory.

In the second phase the first signatory S_1 tries to establish the authenticity of the document-list pair and the integrity of the message that carried the document-list and pad. If the document has not been tampered with during its transit and the document is authentic (not substituted illegally by the issuer) then the signatory S_1 signs the pad and forwards the document-list pair and pad to signatory S_2. This verification, signing and forwarding process is repeated by each subsequent signatory S_i ($i = 2, \ldots, u$), with the last signatory S_u forwarding the document-list and pad to the verifier. The second phase is then completed.

In the third phase the trusted verifier receives the message and tries to establish its integrity. The verifier then checks the authenticity of the document-list and checks whether all signatories have signed. Before starting the verification process the verifier broadcasts x_d. Using this value all signatories and the general public can check whether $g^{x_d} = y_d$.

In the following the three phases will be presented more precisely. The document D is assumed to contain a list of prospective signatories and the verifier, similar to the way that multi-party agreements on paper documents have a list of the names of signatories and a witness to the signing of the document. Throughout this paper it is assumed that an electronic document contains a unique identifier and a timestamp for duration of time in which the signing process must be completed. This is to prevent replay of messages by an active attacker.

Phase 1 In the first phase the issuer prepares the document D and the list PAD, and registers to the verifier $M = (D \parallel PAD)$ through $V_{register}$ (Algorithm 3). The verifier then chooses x_d, broadcasts $y_d = g^{x_d}$ and returns the tag $g^{h(M \oplus x_d)}$ for M. The issuer then signs the pad and sends $(C_{I_1}, C_{I_2}, C_{I_3}, c_1)$ to the first signatory S_1. The actions of the issuer is shown in E_I (Algorithm 4).

Algorithm 3 $V_{register}(M, p, g)$

1. $x_d \in_R [1, p-1]$ such that $\gcd(k_I, p-1) = 1$.
2. $y_d = g^{x_d}$.
3. $\alpha = h(M \oplus x_d)$.
4. $\beta = g^{\alpha}$.
5. broadcast y_d.
6. output (β).

end

Algorithm 4 $E_I(x_I, y_1, \ldots, y_u, y_V, p, g, D)$

1. $k_I \in_R [1, p-1]$ such that $\gcd(k_I, p-1) = 1$.
2. $C_{I_1} = g^{x_I}$.
3. $C_{I_2} = g^{k_I}$.
4. $PAD = (y_I^{h(D)})^{x_I+k_I} \oplus (y_1^{h(D)})^{x_I+k_I} \oplus \cdots \oplus (y_u^{h(D)})^{x_I+k_I} \oplus (y_V^{h(D)})^{x_I+k_I}$.
5. $M = D \,||\, PAD$.
6. $R = V_{register}(M)$.
7. $A = (h(MR) - x_I y_d)/k_I \bmod (p-1)$.
8. $T = (C_{I_2})^A$.
9. $pad_1 = ((C_{I_1} C_{I_2})^{x_I})^{h(D)}$.
10. $m_1 = M \,||\, R \,||\, T \,||\, pad_1$.
11. $r_1 = y_1^{x_I+k_I}$.
12. $z_1 = G(r_1)_{[1 \cdots P(n)]}$.
13. $t_1 = (h(m_1) - x_I r_1)/k_I \bmod (p-1)$.
14. $C_{I_3} = (C_{I_2})^{t_1}$.
15. $c_1 = z_1 \oplus m_1$.
16. output $(C_{I_1}, C_{I_2}, C_{I_3}, c_1)$.

end

The important components of the message of m_1 are T and pad_1. T ensures the authenticity of M and thus D, while pad_1 represents the signature of the issuer S_I who is also involved in the event. All three M, R and T remain unchanged throughout the remainder of the signing process. Each signatory and the verifier uses T to verify that no malicious signatory has modified M. In contrast, t_1 is used only by the next signatory to verify the integrity of m_1 containing the signed pad. Note that Steps 11 to 16 is the basic encryption steps of the $C^{\star}_{sig}$ cryptosystem in order to provide the secure and authentic transfer of m_1.

<u>**Phase 2**</u> The second phase is started when the first signatory S_1 verifies the integrity and authenticity of the unsigned document using D_{S_1} (Algorithm 5). If all is well the signatory S_1 keeps a copy of the values $(C_{I_1}, C_{I_2}, C_{I_3}, c_1)$ which he or she received and verified.

Algorithm 5 $D_{S_1}(x_1, y_d, p, g, C_{I_1}, C_{I_2}, C_{I_3}, c_1)$

1. $r_1' = (C_{I_1}C_{I_2})^{x_1}$.
2. $z_1' = G(r_1')_{[1\cdots P(n)]}$.
3. $m_1' = z_1' \oplus c_1$.
4. if $g^{h(m_1')} = ({C_{I_1}}^{r_1'})\,C_{I_3}$ then
   output (m_1') and continue to Step 5
   else
   output $(\emptyset)$ and stop.
5. Separate m_1' into M, R, T and pad_1.
6. if $g^{h(MR)} = ({C_{I_1}}^{y_d})\,T$ then
   output (M)
   else
   output $(\emptyset)$ and stop.

end

Next the signatory S_1 signs the pad using E_{S_1} (Algorithm 6) and forwards the message containing the document and pad to signatory S_2. Signatory S_1 also keeps a copy of the message $(C_{I_1}, C_{I_2}, C_{I_3}, c_{2_1}, c_{2_2}, c_{2_3}, c_{2_4})$ that he or she sent to S_2. Note that the output of E_{S_1} has more components compared to the output of E_I. This has no effect on the scheme as a whole since the components $(C_{I_1}, C_{I_2}, C_{I_3})$ important for the verification of the authenticity of the document is the same as in E_I. Each subsequent signatories after S_1 will receive the same format of the message as that output by E_{S_1}.

Algorithm 6 $E_{S_1}(x_1, y_2, p, g, C_{I_1}, C_{I_2}, C_{I_3}, D, M, R, T, pad_1)$

1. $k_1 \in_R [1, p-1]$ such that $\gcd(k_1, p-1) = 1$.
2. $c_{2_1} = g^{x_1}$.
3. $c_{2_2} = g^{k_1}$.
4. $pad_2 = pad_1 \oplus ((C_{I_1}C_{I_2})^{x_1})^{h(D)}$.
5. $m_2 = M \,||\, R \,||\, T \,||\, pad_2$.
6. $r_2 = y_2^{x_1+k_1}$.
7. $z_2 = G(r_2)_{[1\cdots P(n)]}$.
8. $t_2 = (h(m_2) - x_1 r_2)/k_1 \bmod (p-1)$.
9. $c_{2_3} = (c_{2_2})^{t_2}$.
10. $c_{2_4} = z_2 \oplus m_2$.
11. output $(C_{I_1}, C_{I_2}, C_{I_3}, c_{2_1}, c_{2_2}, c_{2_3}, c_{2_4})$.

end

In general the signatory S_i $(1 \le i \le u)$ employs E_{S_i} (Algorithm 7) to sign the pad associated with the document, and to prepare the document and the pad for the next signatory S_{i+1}. Note that for E_{S_i} signatory S_0 is identical to the issuer S_I and that signatory S_{u+1} is equivalent the verifier.

Algorithm 7 $E_{S_i}(x_i, y_{i+1}, p, g, C_{I_1}, C_{I_2}, C_{I_3}, D, M, R, T, pad_i)$

1. $k_i \in_R [1, p-1]$ such that $\gcd(k_i, p-1) = 1$.
2. $c_{(i+1)_1} = g^{x_i}$.
3. $c_{(i+1)_2} = g^{k_i}$.
4. $pad_{i+1} = pad_i \oplus ((C_{I_1} C_{I_2})^{x_i})^{h(D)}$.
5. $m_{i+1} = M \,||\, R \,||\, T \,||\, pad_{i+1}$.
6. $r_{i+1} = y_{i+1}^{x_i + k_i}$.
7. $z_{i+1} = G(r_{i+1})_{[1 \cdots P(n)]}$.
8. $t_{i+1} = (h(m_{i+1}) - x_i r_{i+1})/k_i \bmod (p-1)$.
9. $c_{(i+1)_3} = (c_{(i+1)_2})^{t_{i+1}}$.
10. $c_{(i+1)_4} = z_{i+1} \oplus m_{i+1}$.
11. output $(C_{I_1}, C_{I_2}, C_{I_3}, c_{(i+1)_1}, c_{(i+1)_2}, c_{(i+1)_3}, c_{(i+1)_4})$.

end

The signatory S_i uses D_{S_i} (Algorithm 8) to verify the integrity and authenticity of the document and pad from signatory S_{i-1}. Phase 2 is ended when the last signatory S_u employs D_{S_i} to verify the message he or she received from signatory S_{u-1} and employs E_{S_i} to sign the document and prepare the enclosing message for the verifier. Note that each signatory S_i must keep a copy of the message $(C_{I_1}, C_{I_2}, C_{I_3}, c_{i_1}, c_{i_2}, c_{i_3}, c_{i_4})$ received from signatory S_{i-1} and a copy of the message $(C_{I_1}, C_{I_2}, C_{I_3}, c_{(i+1)_1}, c_{(i+1)_2}, c_{(i+1)_3}, c_{(i+1)_4})$ sent to signatory S_{i+1}.

Algorithm 8 $D_{S_i}(x_i, y_d, p, g, C_{I_1}, C_{I_2}, C_{I_3}, c_{i_1}, c_{i_2}, c_{i_3}, c_{i_4})$

1. $r'_i = (c_{i_1} c_{i_2})^{x_i}$.
2. $z'_i = G(r'_i)_{[1 \cdots P(n)]}$.
3. $m'_i = z'_i \oplus c_{i_4}$.
4. if $g^{h(m'_i)} = (c_{i_1}{}^{r'_i})\, c_{i_3}$ then
   output (m'_i) and continue to Step 5
   else
   output $(\emptyset)$ and stop.
5. Separate m'_i into M, R, T and pad_i.
6. if $g^{h(MR)} = (C_{I_1}{}^{y_d})\, T$ then
   output (M)
   else
   output $(\emptyset)$ and stop.

end

<u>**Phase 3**</u> In Phase 3 the verifier receives the document and pad from the last signatory S_u and verifies their authenticity and integrity. These tasks are done using D_V (Algorithm 9) which is similar in form to D_{S_i}. However, D_V contains the additional steps of checking that the $pad_{u+1} = pad_V$ received from signatory S_u contains the signature of all signatories S_i $(i = 1, \ldots, u)$. Furthermore, after broadcasting x_d, which is used by all participants to check y_d, the verifier must see that M is still authentic using x_d. Each signatory can in fact check whether $R = g^{h(M \oplus x_d)}$.

Algorithm 9 $D_V(x_V, x_d, y_d, p, g, C_{I_1}, C_{I_2}, C_{I_3}, c_{V_1}, c_{V_2}, c_{V_3}, c_{V_4})$

1. $r'_V = (c_{V_1} c_{V_2})^{x_V}$.
2. $z'_V = G(r'_V)_{[1 \cdots P(n)]}$.
3. $m'_V = z'_V \oplus c_{V_3}$.
4. if $g^{h(m'_V)} = (c_{V_1}{}^{r'_V})\, c_{V_3}$ then
   output (m'_V) and continue to Step 5
   else
   output $(\emptyset)$ and stop.
5. Separate m'_V into M, R, T and pad_V.
6. if $g^{h(MR)} = (C_{I_1}{}^{y_d})\, T$ then
   output (M) and continue to Step 7
   else
   output $(\emptyset)$ and stop.
7. Separate M into D and PAD.
8. $pad_{check} = pad_V \oplus ((C_{I_1} C_{I_2})^{x_V})^{h(D)}$.
9. if $pad_{check} = PAD$
   output message "All Signed"
   else
   output $(\emptyset)$ and stop.

end

4 Security

The third cryptosystem of [6] (instead of the first or second cryptosystems) was deemed suitable for the foundation of the multisignature scheme since it already featured a basic authentication capability in the sense of El Gamal [9]. In [6] it was shown that both the first and the second cryptosystems are semantically secure against adaptively chosen ciphertext attacks under the assumption that the space induced by the two cryptosystems were sole-samplable (see [6]). However, C_{sig} was not able to be proven as being semantically secure against chosen plaintext attacks due to the difficulty in measuring the leak of information m from $t = (h(m) - xr)/k \bmod (p-1)$ in the ciphertext. Thus C_{sig} is semantically secure against adaptively chosen ciphertext attacks only with the assumption that it is semantically secure against chosen plaintext attacks.

Our proposed $C^{\star}_{sig}$ algorithm improves C_{sig} by reducing this possible leak. Previously a receiver of a message would already know the values of m and r at the successful decipherment of the ciphertext message. Using these two values he or she could work towards obtaining x and k from $t = (h(m) - xr)/k \bmod (p-1)$. The $C^{\star}_{sig}$ makes harder this effort of obtaining x and k from t at the expense of an apparent ease in substituting m by using the known r, $c_1 (= g^x)$ and $c_2 (= g^k)$. Note, however, that in our application of the $C^{\star}_{sig}$ algorithm the registration value $R = g^{h(M \oplus x_d)}$ guards against this trivial substitution. In the case of the typical one-to-one communication between two trusting parties $C^{\star}_{sig}$ represents a more secure approach against external attacks.

From the point of view of a dishonest signatory (excluding the issuer and the verifier) the security of the digital multisignature scheme lies in the strength of

T (Algorithm 4). Unlike the case of the native form of C_{sig} and $C^{\star}_{sig}$ where the attacker's aim is to derive and modify M, in this case each signatory already has M on receipt and correct decipherment of the message. Thus, his or her aim would be to cheat the remainder of signatories into signing a false document D' (note that a cheating signatory cannot do more than some random modifications to the PAD). In order to do this he or she must forge T, which requires knowing x_I, k_I and x_d. Deriving x_I from C_{I_1}, k_I from C_{I_2} and x_d from y_d at the very least constitutes the solving of instances of the discrete logarithm problem.

One possible scenario is when the first and the last signatories (S_1 and S_u) collaborate to cheat the rest of the signatories. This can be done if they can obtain x_I and k_I. The first signatory then calculates a false D' and makes M'. He or she then calculates A' using a false x'_I and k'_I, and in turn generates T' using A' (Algorithm 4). Hence signatories $S_2, \ldots, S_{u-1}$ will be signing a forged document M', while the last signatory can restore T before signing the pad and delivering them to the verifier. Note, however, that if D is modified into D', signatory S_i ($2 \leq i \leq u-1$) will be signing using $((C_{I_1} C_{I_2})^{x_i})^{h(D')}$. Thus the test of $pad_{check} = PAD$ by the verifier will fail at the first instant (Algorithm 9). The value PAD cannot be intelligently modified by anyone (except the issuer) since breaking it at least constitutes the solving discrete logarithm problems. Random modifications of PAD will also result in the failure of the $pad_{check} = PAD$ test by the verifier. Each signatory will be able to verify the M on which they based their signatures in the third phase when x_d is broadcasted. If the issuer cheats (alone or by collaboration) by creating $M' = D' \mid\mid PAD'$, then it will be detected by the verifier when he or she computes $R' = g^{h(M' \oplus x_d)}$ in the third phase.

Though unlikely to occur in real world situations where all signatories to a contract are aware of the contract, one possible weakness of the scheme is that a dishonest issuer can discredit some parties by mentioning their identities in D, yet purposely fail to include them in the computation of PAD. The issuer can get away with this only by the collaboration of all actual signatories $S_1, \ldots, S_u$. For if even one of these signatories is honest, he or she may deliver the document to be signed to one or more of the discredited parties. They who would then unknowingly add their signatures to the pad, thus frustrating the efforts of the dishonest issuer. This problem can be avoided by allowing the verifier to generate the PAD within $V_{register}$ using x_d and a random k_d. However, a better solution to this problem remains for future investigation, where the verifier should be able to know precisely the identities of those who have actually signed the pad.

5 Conclusion

In this paper we have presented a practical digital multisignature scheme based on the $C^{\star}_{sig}$ cryptosystem derived from the C_{sig} cryptosystem of [6]. The scheme, which consists of three phases, allows for the detection of a dishonest issuer of the document and of dishonest signatories. Any dishonesty by a signatory is detected in the first place by the next signatory during the second phase, and in the second place by the verifier and all signatories in the third phase. Any detected dishonesty will result in the verifier annulling any effects of the contract specified in the document. The scheme is practical and offers at the very least the same security level afforded by the C_{sig} cryptosystem of [6] against external attacks. Internal attacks in the form of forging or cheating by a dishonest issuer or by one or more of the signatories (alone or by collaboration) requires the solving of instances of the discrete logarithm problem.

Acknowledgements

We thank Prof. J. Seberry for her continual support and interest in this work, and for valuable comments.

References

1. R. L. Rivest, A. Shamir, and L. Adleman, "A method for obtaining digital signatures and public-key cryptosystems," *Communications of the ACM*, vol. 21, no. 2, pp. 120–128, 1978.
2. D. W. Davies, "Applying the RSA digital signature to electronic mail," *IEEE Computer*, vol. 16, pp. 55–62, February 1983.
3. A. Fiat and A. Shamir, "How to prove yourself: practical solutions of identification and signature problems," in *Advances in Cryptology - Proceedings of Crypto'86*, Lecture Notes in Computer Science, Vol. 263, pp. 186–194, Springer-Verlag, 1987.
4. T. Okamoto, "A digital multisignature scheme using bijective public-key cryptosystems," *ACM Transactions on Computer Systems*, vol. 6, no. 8, pp. 432–441, 1988.
5. K. Ohta and T. Okamoto, "A digital multi-signature scheme based on the fiat-shamir scheme," in *Advances in Cryptology - Proceedings of ASIACRYPT '91*, (Fujiyoshida, Japan), pp. 75–79, November 1991.
6. Y. Zheng and J. Seberry, "Immunizing public key cryptosystems against chosen ciphertext attacks," *IEEE Journal of Selected Areas in Communications*, 1993. (to appear).
7. Y. Zheng, T. Hardjono, and J. Seberry, "A practical non-malleable public key cryptosystem," Technical Report CS91/28, Computer Science Department, University College, University of New South Wales, Australia, 1991.
8. W. Diffie and M. E. Hellman, "New directions in cryptography," *IEEE Transactions on Information Theory*, vol. IT-22, no. 6, pp. 644–654, 1976.
9. T. ElGamal, "A public key cryptosystem and a signature scheme based on discrete logarithms," *IEEE Transactions on Information Theory*, vol. IT-31, no. 4, pp. 469–472, 1985.
10. B. A. LaMacchia and A. M. Odlyzko, "Computation of discrete logarithms in prime fields," *Designs, Codes and Cryptography*, vol. 1, pp. 47–62, 1991.
11. M. Blum and S. Micali, "How to generate cryptographically strong sequences of pseudo-random bits," *SIAM Journal on Computing*, vol. 13, no. 4, pp. 850–864, 1984.
12. D. L. Long and A. Wigderson, "The discrete logarithm hides O(log n) bits," *SIAM Journal on Computing*, vol. 17, no. 2, pp. 363–372, 1988.
13. R. Peralta, "Simultaneous security of bits in the discrete log," in *Advances in Cryptology - Proceedings of EuroCrypt'85*, Lecture Notes in Computer Science, Vol. 219 (F. Pichler, ed.), pp. 62–72, Springer-Verlag, 1985.

Group-Oriented Undeniable Signature Schemes without the Assistance of a Mutually Trusted Party

Lein Harn[1] and Shoubao Yang[2]

[1] Computer Science Telecommunications Program
University of Missouri - Kansas City
Kansas City, MO 64110 USA
Tel: (816)235-2367
E-mail: harn@cstp.umkc.edu

[2] Department of Computer Science
University of Science and Technology of China
Hefei, Anhui 230026 PRC

Abstract. In a group-oriented (t, n) undeniable signature scheme, where t is the threshold value and n is the total number of group members, each group, instead of each individual member within the group, publishes a single group public key. During an initial "commitment phase", at least t group members work together to sign a message. In a "verification phase", all signers work together to prove the validity of the signature to an outsider. There is only a single group public key required for an outsider to verify the group signature. In case some group members disavow the signature, a judge can resolve the dispute between the group and a verifier. The group-oriented (t, n) undeniable signature scheme has the following four properties: (1) the group signature is mutually generated by at least t group members; (2) the signature verification process is simplified because there is only one group public key required; (3) the signature can only be verified with the consent of all signers; (4) the signers hold the responsibility to the signed messages. More specifically, we propose two schemes where the threshold value t can be either n or 1. In other words, it requires either all group members or a single member to sign a group signature. Moreover, these two schemes do not require the assistance of a mutually trusted party. Each member selects his own secret and the group public key is determined by all group members.

1 Introduction

The group-oriented cryptography was first introduced by Desmedt [1] in 1987. In a group-oriented cryptosystem, each group, instead of each individual member within the group, publishes a single group public key. An outsider can use this group public key to send an encrypted message to this group. However, the received ciphertext can only be deciphered correctly by some subsets of the internal members of this group according to a predetermined policy.

All earlier solutions [1-3] to the group-oriented cryptosystem require a mutually trusted center to decide the group public key and generate corresponding secrets for all internal members. Most recently, Desmedt and Frankel [4] proposed a group-oriented digital signature scheme based on the RSA [5] scheme. In their scheme, a mutually trusted center is still required.

As pointed out by Ingemarsson and Simmons [6], however, in most applications there does not exist any trusted party within a group. This situation becomes more common in some commercial and/or international applications. The group-oriented signature schemes presented in this paper are developed in accordance with this situation. In our system, each member within the group selects his own secret key and all members mutually decide the group public key. Unfortunately, the RSA scheme cannot be used in our algorithm. This is due to the fact that, if the modulus n is universal and each member needs to know the factoring of n in order to decide his secret key, there will be no secret at all among all internal members.

The undeniable signature scheme was first proposed by D. Chaum [7] in 1989. An undeniable signature, like a digital signature, is a random number, which is generated based on a message and a secret key known only to the signer. Unlike a digital signature, on the other hand, an undeniable signature can only be verified with the cooperation of the signer. However, the signer still holds the responsibility to the signed message. There are two phases in an undeniable signature scheme: the "commitment phase" and the "verification phase". In the "commitment phase", a commitment signature will be produced by the signer, which cannot be verified by outsiders themselves. In the "verification phase", the outsiders will be able to verify the validity of the signature only with the cooperation of the signer. A zero-knowledge verification procedure presented by D. Chaum in [8] can also be used for signature verification. Chaum pointed out that the undeniable signature scheme is preferable to digital signature for many upcoming applications.

In this paper, we incorporate these two concepts, the group-oriented cryptography and the undeniable signature, to propose "group-oriented undeniable signature schemes without the assistance of a mutually trusted party". Generally speaking, in a group-oriented (t, n) undeniable signature scheme, where t is the threshold value and n is the total number of group members, each group, instead of each individual member within the group, publishes a single group public key. During an initial "commitment phase", at least t group members work together to sign a message. In a "verification phase", all signers work together to prove the validity of the signature to an outsider. More specifically, we propose two schemes where the threshold value t can be either n or 1. In other words, it requires either all group members or a single member to sign a group signature. Moreover, these two schemes do not require the assistance of a mutually trusted party. As an example, (n, n) scheme can be illustrated as the following statement:

A company is funded by n international investors. These investors do not trust each other and further, there does not exist any third party who can be trusted by all investors. Suppose all the released products of the company must be signed mutually by all investors. However, these digital signatures cannot be verified as authentic by anyone using the corresponding company public key(s). Only paying customers who really purchased the products are able to verify the validity of the signatures with the cooperation of the signers. The investors may come up with the following two arrangements:

Arrangement 1. Let each investor select his own secret key and the company publish a list of public keys collected from all n investors. Using their secret keys, all investors take turns to sign a product. Later, a paying customer needs to execute a verification process to each signer to authenticate the undeniable signature by using one of the corresponding public keys. The problem with arrangement 1 is that the number of listed public keys of the company and the number of individual undeniable signatures is linear in the number of investors.

Arrangement 2. Let all investors agree on a secret key and each investor keep a copy of this secret key. The company publishes a single public key. All released products are signed by using the same secret key. The paying customer needs to execute a verification process to verify the undeniable signature by using the single group public key. The problem with arrangement 2, however, is that in case of a dispute in the future, it is impossible to identify who should take the responsibility upon a signed product since all investors have the same secret key.

The group-oriented undeniable signature schemes we present in this paper preserve the merits of these two arrangements without any trade-off. More generally, the group-oriented (t, n) undeniable signature scheme without the assistance of a mutually trusted party has the following properties:

1. only a single group public key is required which is mutually generated by all group members;
2. any t or more than t group members can represent the group to sign messages;
3. only paying customers can verify the validity of the undeniable signature with the cooperation of all signers;
4. in case of a dispute in the future, the signers still hold the responsibility to the signed messages.

It is interesting to note that a very similar definition of a group-oriented $(1, n)$ signature scheme has been proposed by D. Chaum and E. van Heyst in Eurocrypt '91 [9]. The major differences lie in properties (1) and (2) as stated above. In their definition, in addition to the requirement that the identity of a signer is anonymous to outsiders, it is also guaranteed that the outsiders cannot discover that two or more messages are signed by the same member. Thus, in their scheme, the number of listed public keys of the group is not limited to one. In a very recent paper [10] written by current authors, two group-oriented (n, n) digital signature schemes have also been proposed. In that paper, any outsider is able to verify the group signature independently.

Since our scheme is based on the first undeniable signature scheme proposed by D. Chaum [7], we will review this scheme in the next section. In Section 3 and Section 4, we propose our group-oriented (n, n) and $(1, n)$ undeniable signature schemes respectively.

2 Review of the Chaum's Undeniable Signature Scheme

We will now review the first undeniable signature scheme proposed by D. Chaum based on the discrete logarithm.

We begin with a large prime p and a primitive element α of $GF(p)$, which are publicly known. In order to provide adequate security, Pohlig and Hellman [11] indicate that p should be selected such that $p-1$ contains at least one large prime factor. They recommend choosing $p = 2p' + 1$, where p' is also a large prime.

In this cryptosystem, each user selects a random exponent x from $GF(p)$ as his private key. Suppose A randomly selects a number x_A from $[0, p-1]$ with $gcd(x_A, p-1) = 1$. Then A computes

$$y_A = \alpha^{x_A} \; mod \; p$$

as the public key from A. Assuming A wants to sign a message m, where $0 \leq m \leq p-1$, we have the following two processing phases:

Commitment Phase:

A computes $z = m^{x_A} \; mod \; p$ as his undeniable signature of message m and sends z , together with m, to a verifier.

Verification Phase:

A challenge-response protocol is applied between the verifier and the signer A to establish the signature validity.

Step 1: The verifier randomly selects two integers a and b, from $GF(p)$, and computes
$w = z^a y_A^b \; mod \; p$.
w is sent to A.

Step 2: A computes
$R = w^{x_A^{-1}} \; mod \; p$,
where x_A^{-1} is the multiplicative inverse of $x_A \; mod \; p-1$. R is sent back to the verifier.

Step 3: The verifier checks whether
$R = m^a \alpha^b \; mod \; p$.
If it does hold this equality, the signature has been verified.

Chaum also proposed a disavowal protocol. If the receiver receives a forged commitment, the signer can prove that the commitment is a forged one during the verification phase.

3 Group-Oriented (n, n) Undeniable Signature Scheme without the Assistance of a Mutually Trusted Party

The group-oriented (n, n) undeniable signature scheme without the assistance of a mutually trusted party consists of three phases: the group public key generation phase, group undeniable signature generation phase, and group undeniable signature verification phase. During the key generation phase, all group members are connected in a ring structure and mutually determine the group public key.

Group Public Key Generation Phase

This phase is almost the same as in [10]. Assuming there are n members in a group, each member selects a random exponent x from $GF(p)$ as his private key. Suppose member u_i randomly selects a number x_i from $[0, p-1]$ with $gcd(x_i, p-1) = 1$, (i.e., x_i^{-1} $mod\ p-1$ does exist). Then the group public key y is determined mutually by all members as

$$y = \alpha^{x_1 x_2 \ldots x_n}\ mod\ p$$

This group public key is mutually generated by all members without revealing each member's secret key. As stated previously, all members must be connected in a ring in any order to generate the group public key. Specifically, we assume the members are connected in the following order: $u1, u2, \ldots, u_n$. Since each member within this ring needs to work in the same way to mutually generate the group public key, we select member u_i as an example to illustrate the procedure:

Step 1: Member u_i uses his secret key x_i to compute $y_{i,1}$ as
$y_{i,1} = \alpha^{x_i}\ mod\ p$,
and transmits this value to his successor u_{i+1}.

Step $t(2 \leq t < n)$: u_i receives $y_{i-1,t-1}$ from his predecessor u_{i-1} and uses his secret key x_i to compute
$y_{i,t} = (y_{i-1,t-1})^{x_i}\ mod\ p$.
The result $y_{i,t}$ is transmitted to the successor from u_i which is u_{i+1}.

Step n: The group public key y can be obtained by u_i as
$y = (y_{i-1,n-1})^{x_i}\ mod\ p$.

After n steps, each member can obtain the same group public key y individually. In the event all members are required to release the computed group public key simultaneously after the n-th step, this group public key can be verified by all group members.

Group Undeniable Signature Generation Phase

This scheme requires the group undeniable signature to be mutually generated by all group members. Only after all group members have signed their individual undeniable signatures, the group undeniable signature can then be obtained and an outsider will be able to verify it with the cooperation of all group members. We propose two protocols to generate the group undeniable signature. The first one is a so-called blind multisignature scheme in which most of the signers, except the first one, cannot verify the content of a message before they sign it. The second one enables every signer to validate the group undeniable signature before releasing it to the public. The first protocol is given as below:

Blind Multisignature Scheme:

This scheme allows all group members to sign a message in any order. For the sake of convenience, we assume the members are connected sequentially in the following order: $u_1, u_2, \ldots, u_n$, and sign a message successively on a progressive basis. u_1 prepares the message m, which will be signed by all members, and signs his individual signature with his secret key x_1 as

$$z_1 = m^{x_1} \; mod \; p.$$

z_1 is sent to the successor from u_1 which is u_2. u_2 signs z_1 with his secret key x_2 and gets z_2, where

$$z_2 = m^{x_1 x_2} \; mod \; p.$$

Just by repeating the same procedure until all members have signed their signatures, the group undeniable signature can be obtained as

$$z = z_n = m^{x_1 x_2 \ldots x_n} \; mod \; p.$$

Discussion:

In this scheme, since each member uses his secret key as an exponent and releases the computed exponential result as a partial group signature, the security of these secret keys is the same as solving the discrete logarithm problem. Except the first signer who prepares the message and knows the message, all the rest of signers have no ability to verify the content of the message before they sign it with their secret keys. Therefore, this scheme only provides a blind signature for most of the signers. Under the assumption that the members do not trust each other, this scheme can only be used in some very limited applications.

Verifiable Multisignature Scheme:

This scheme allows each group member to verify the validity of the group undeniable signature before releasing it to the public. However, without the cooperation of all the group members, the outsiders cannot verify the signature by themselves.

The verifiable multisignature scheme is based on the observation derived from the blind multisignature scheme. If we let the entire group execute the blind multisignature protocol n times, where n is the number of group members, to sign the same message m that they all agree to sign, and each member take turns to be the first signer once, all these n results should be identical if all members follow the protocol honestly. Therefore, the group undeniable signature can be verified by all members before being released to the public.

A more efficient way to implement the above concept is to connect all group members in a ring in any order and execute the procedure described in the group public key generation phase by using message m instead of the public primitive element α. Thus, the group undeniable signature z can be obtained simultaneously by each member after the n-th step as

$$z = (z_{n-1,n-1})^{x_i} \ mod \ p.$$

Discussion:

After the n-th step, each member can obtain the same group signature z individually. In the event all members are required to release the computed group signature simultaneously after the n-th step, this group signature can be verified internally by all members before being released to the public. Therefore, it is a verifiable signature scheme only for the internal members.

Group Undeniable Signature Verification

With the consent of all group members, a verifier starts the verification procedure given as below:

Step 1: The verifier randomly selects two integers a and b, from $GF(p)$, and computes
$w = z^a y^b \; mod \; p,$
where y is the single group public key and z is the group undeniable signature. w is sent to the group.

Step 2: All group members need to be connected sequentially in any order (Note that this verification order can be different from the group signature generation order). The first member, say u_1, computes
$R_1 = w^{x_1^{-1}} \; mod \; p,$
where x_1^{-1} is the multiplicative inverse of $x_1 \; mod \; p-1$. R_1 is sent to the successor from u_1, say u_2. With the secret key x_2, u_2 computes
$R_2 = R_1^{x_2^{-1}} \; mod \; p.$
Repeat the same process until all members have been covered, and a result R is obtained and sent to the verifier as
$R = w^{x_1^{-1} x_2^{-1} \dots x_n^{-1}} \; mod \; p.$

Step 3: The verifier receives R and checks whether
$R = m^a \alpha^b \; mod \; p.$
If it does hold this equality, the group undeniable signature has been verified.

Discussion:

Some features of these proposed schemes should be pointed out here:

1. Each group member is required to contribute a partial signature in order to produce the final group signature. Any partial signature cannot be verified successfully because the group public key is determined by all members.
2. A single authenticated group public key is only required for the verification purpose. Thus, it simplifies the public key authentication problem significantly.
3. The group undeniable signature cannot be verified by any outsider without the consent of all group members. Only the paying customers are able to verify the validity of the signature with the consent of all group members.

4 Group-Oriented $(1, n)$ Undeniable Signature Scheme without the Assistance of a Mutually Trusted Party

The group-oriented $(1, n)$ undeniable signature scheme allows any group member to represent the group to sign signature with his own secret key, that means the signer will take the full responsibility to the signed messages, and the verifier can only verify the signature with the consent of the single signer. The basic idea of our proposed scheme is given as below:

Just like the previous scheme, during the initial phase, all group members mutually generate a single group public key. Then, each member selects his own secret key and has his corresponding public key signed (certified) by all group members. This public key certificate can be computed off-line by using the group-oriented (n, n) signature scheme developed in [10]. All group member work together to generate the public key certificate that can be verified by any outsider independently. When a group member is required to represent the group to sign a message, he generates an undeniable signature with his own secret key. During the verification phase, the signer passes his public key along with the public key certificate to the verifier. The verifier first checks the validity of the signer's public key through the public key certificate by using the single group public key, and then verifies the signature with the signer. Since each member has different public key certificate, the signer holds the responsibility to the signed message.

5 Conclusion

The first two group-oriented undeniable signature schemes (n, n) and $(1, n)$, have been proposed in this paper. The same approach can also be applied to develop a zero-knowledge verification process. The question of how to design more general group-oriented (t, n) undeniable signature scheme is still an open research problem.

References

[1] Y. Desmedt.: Society and group-oriented cryptography: a new concept. Advances in Cryptology, Proc. of Crypto '87, pp. 120-127, August 16-20,1988.

[2] Y. Frankel.: A practical protocol for large group-oriented networks. In Advances in Cryptology, Proc. of Eurocrypt '89, pp. 56-62, April 10-13,1989.

[3] Y. Desmedt and Y. Frankel.:Threshold cryptosystem. In Advances in Cryptology, Proc. of Crypto '89, pp. 307-315, August 20-24, 1989.

[4] Y. Desmedt and Y. Frankel.: Shared generation of authenticators. In Advances in Cryptology, Proc. of Crypto '91, August 11-15, 1991.

[5] R. L. Rivest, A. Shamir, and L. Adelman.: A method for obtaining digital signatures and public-key cryptosystem. In Commun. of ACM, Vol. 21, No.2, pp. 120-126, Feb. 1978.

[6] I. Ingemarsson, and G. J. Simmons.: A protocol to set up shared secret schemes without the assistance of a mutually trusted party. In Advances in Cryptology, Proc. of Eurocrypt '90, May 21-24, 1990.

[7] D. Chaum and H. van Antwerpen.: Undeniable signature. In Advances in Cryptology, Proc. of Crypto '89, pp. 212-216, August 20-24, 1989.

[8] D. Chaum.: Zero-knowledge undeniable signatures. In Advances in Cryptology, Proc. of Eurocrypt '90, May 21-24, 1990.

[9] D. Chaum and E. van Heyst.: Group signature. In Advances in Cryptology, Proc. of Eurocrypt '91, pp. 257-265, April 8-11, 1991.

[10] L. Harn and S. Yang.: Group-oriented (n, n) digital signature schemes without the assistance of a mutually trusted party. Submitted to The Journal of Computer Security.

[11] S. Pohlig, and M. Hellman.: An improved algorithm for computing logarithms over GF(p) and its cryptographic significance. In IEEE Trans. Inform. Theory, Vol. IT-24, pp. 106-110, 1978.

Session 4

THEORY OF S-BOXES

Chair: Tsutomu Matsumoto
(Yokohama National University, Japan)

Highly Nonlinear 0-1 Balanced Boolean Functions Satisfying Strict Avalanche Criterion (Extended Abstract)

Jennifer Seberry * and Xian-Mo Zhang **

Department of Computer Science
The University of Wollongong
Wollongong, NSW 2522, AUSTRALIA

Abstract. Nonlinearity, 0-1 balancedness and strict avalanche criterion (SAC) are important criteria for cryptographic functions. Bent functions have maximum nonlinearity and satisfy SAC however they are not 0-1 balanced and hence cannot be directly used in many cryptosystems where 0-1 balancedness is needed. In this paper we construct

(i) 0-1 balanced boolean functions on V_{2k+1} ($k \geq 1$) having nonlinearity $2^{2k} - 2^k$ and satisfying SAC,

(ii) 0-1 balanced boolean functions on V_{2k} ($k \geq 2$) having nonlinearity $2^{2k-1} - 2^k$ and satisfying SAC.

We demonstrate that the above nonlinearities are very high not only for the 0-1 balanced functions satisfying SAC but also for all 0-1 balanced functions.

1 Basic Definitions

Let V_n be the vector space of n tuples of elements from $GF(2)$. Let α, $\beta \in V_n$. Write $\alpha = (a_1 \cdots a_n)$, $\beta = (b_1 \cdots b_n)$, where $a_i, b_i \in GF(2)$. Write $\langle \alpha, \beta \rangle = \sum_{j=1}^{n} a_j b_j$ for the scalar product of α and β. We write $\alpha = (a_1 \cdots a_n) < \beta = (b_1 \cdots b_n)$ if there exists k, $1 \leq k \leq n$, such that $a_1 = b_1, \ldots, a_{k-1} = b_{k-1}$ and $a_k = 0$, $b_k = 1$. Hence we can order all vectors in V_n by the relation $<$

$$\alpha_0 < \alpha_1 < \cdots < \alpha_{2^n-1},$$

where

$$\alpha_0 = (0 \cdots 00), \ldots, \alpha_{2^{n-1}-1} = (01 \cdots 1),$$

$$\alpha_{2^{n-1}} = (10 \cdots 0), \ldots, \alpha_{2^n-1} = (11 \cdots 1).$$

* Supported in part by the Australian Research Council under the reference numbers A49130102, A9030136, A49131885 and A49232172.

** Supported in part by the Australian Research Council under the reference number A49130102.

Definition 1. Let $f(x)$ be a function from V_n to $GF(2)$ (simply, a function on V_n). We call the (1 -1)-sequence $\eta_f = ((-1)^{f(\alpha_0)}\ (-1)^{f(\alpha_1)}\ \ldots\ (-1)^{f(\alpha_{2^n-1})})$ the *sequence of* $f(x)$. $f(x)$ is called the *function* of η_f. The (0, 1)-sequence $(f(\alpha_0)\ f(\alpha_1)\ \ldots\ f(\alpha_{2^n-1}))$ is called the *truth table* of $f(x)$. In particular, if the truth table of $f(x)$ has 2^{n-1} zeros (ones) $f(x)$ is called *0-1 balanced.*

Let $\xi = (a_1 \cdots a_{2^n})$ and $\eta = (b_1 \cdots b_{2^n})$ be (1 , -1)-sequences of length 2^n. The operation $*$ between ξ and η, denoted by $\xi * \eta$, is the sequence $(a_1b_1 \cdots a_{2^n}b_{2^n})$. Obviously if ξ and η are the sequences of functions $f(x)$ and $g(x)$ on V_n respectively then $\xi * \eta$ is the sequence of $f(x) + g(x)$.

Definition 2. We call the function $h(x) = a_1x_1 + \cdots + a_nx_n + c$, $a_j, c \in GF(2)$, an *affine function*, in particular, $h(x)$ will be called a *linear function* if the constant $c = 0$. The sequence of an affine function (a linear function) will be called an *affine sequence* (a *linear sequence*).

Definition 3. Let f and g be functions on V_n. $d(f, g) = \sum_{f(x)\neq g(x)} 1$ is called the *Hamming distance* between f and g. Let $\varphi_1, \ldots, \varphi_{2^n}, \varphi_{2^n+1}, \ldots, \varphi_{2^{n+1}}$ be all affine functions on V_n. $N_f = min_{i=1,\ldots,2^{n+1}} d(f, \varphi_i)$ is called the *nonlinearity* of $f(x)$.

The nonlinearity is a crucial criterion for a good cryptographic design. It prevents the cryptosystems from being attacked by a set of linear equations. The concept of nonlinearity was introduced by Pieprzyk and Finkelstein [16].

Definition 4. Let $f(x)$ be a function on V_n. If $f(x) + f(x + \alpha)$ is 0-1 balanced for every $\alpha \in V_n$, $W(\alpha) = 1$, where $W(\alpha)$ denotes the number of nonzero coordinates of α (*Hamming weight*) of α, we say that $f(x)$ satisfies the *strict avalanche criterion* (*SAC*).

We can give an equivalent description of SAC: let f be a function on V_n. If if we change any single input the probability that the output changes is $\frac{1}{2}$ (see [2]). The strict avalanche criterion was originally defined in [20], [21], later it has been generalized in many ways [2], [3], [6], [10], [13], [18]. The SAC is relevant to the completeness and the avalanche effect. The 0-1 balancedness, the nonlinearity and the avalanche criterion are important criteria for cryptographic functions [1], [3], [4], [13].

Definition 5. A (1, -1)-matrix H of order h will be called an *Hadamard matrix* if $HH^T = hI_h$.

If h is the order of an Hadamard matrix then h is 1, 2 or divisible by 4 [19]. A special kind of Hadamard matrix, defined as follows will be relevant:

Definition 6. The *Sylvester-Hadamard matrix* (or *Walsh-Hadamard matrix*) of order 2^n, denoted by H_n, is generated by the recursive relation

$$H_n = \begin{bmatrix} H_{n-1} & H_{n-1} \\ H_{n-1} & -H_{n-1} \end{bmatrix}, \quad n = 1, 2, \ldots, \quad H_0 = 1.$$

Definition 7. Let $f(x)$ be a function from V_n to $GF(2)$. If

$$2^{-\frac{n}{2}} \sum_{x \in V_n} (-1)^{f(x)+\langle \beta, x \rangle} = \pm 1,$$

for every $\beta \in V_n$. We call $f(x)$ a *bent function* on V_n.

¿From Definition 7, bent functions on V_n only exist for even n. Bent functions were first introduced and studied by Rothaus [17]. Further properties, constructions and equivalence bounds for bent functions can be found in [1], [7], [9], [15], [22]. Kumar, Scholtz and Welch [8] defined and studied the bent functions from Z_q^n to Z_q. Bent functions are useful for digital communications, coding theory and cryptography [2], [4], [9], [11], [12], [13], [14], [15]. Bent functions on V_n (n is even) not only attain the upper bound of nonlinearity, $2^{n-1} - 2^{\frac{1}{2}n-1}$, but also satisfy SAC. However 0-1 balancedness is often required in cryptosystems and bent functions are not 0-1 balanced since the Hamming weight of bent functions on V_n is $2^{n-1} \pm 2^{\frac{1}{2}n-1}$ [17]. In this paper we construct 0-1 balanced functions with high nonlinearity satisfying high-order SAC from bent functions.

Notation 8. Let X be an indeterminant. We give X a binary subscript that is $X_{i_1 \cdots i_n}$ where $i_1, \ldots, i_n \in GF(2)$. For any sequence of constants $i_1, \ldots, i_p$ from $GF(2)$ define a function $D_{i_1 \cdots i_p}$ from V_p to $GF(2)$ by

$$D_{i_1 \cdots i_p}(y_1, \ldots, y_p) = (y_1 + \bar{i_1}) \cdots (y_p + \bar{i_p})$$

where $\bar{i} = 1 + i$ is the complement of i modulo 2.

2 The Properties of Balancedness and Nonlinearity

Lemma 9. *Let $\xi_{i_1 \cdots i_p}$ be the sequence of a function $f_{i_1 \cdots i_p}(x_1, \cdots, x_q)$ from V_q to $GF(2)$. Write $\xi = (\xi_{0 \cdots 00}\ \xi_{0 \cdots 01} \cdots \xi_{1 \cdots 11})$ for the concatenation of $\xi_{0 \cdots 00}$, $\xi_{0 \cdots 01}$, $\cdots$, $\xi_{1 \cdots 11}$. Then ξ is the sequence of the function from V_{q+p} to $GF(2)$ given by*

$$f(y_1, \ldots, y_p, x_1, \ldots, x_q) = \sum_{(i_1 \cdots i_p) \in V_p} D_{i_1 \cdots i_p}(y_1, \ldots, y_p) f_{i_1 \ldots i_p}(x_1, \cdots, x_q).$$

Proof. It is obvious that:

$$D_{i_1 \cdots i_p}(y_1, \ldots, y_p) = \begin{cases} 1 \text{ if } (y_1 \cdots y_p) = (i_1 \cdots i_p), \\ 0 \text{ otherwise.} \end{cases}$$

Hence, by exhaustive choice,

$$f(i_1, \ldots, i_p, x_1, \ldots, x_q) = D_{i_1 \cdots i_p}(i_1, \ldots, i_p) f_{i_1 \ldots i_p}(x_1, \cdots, x_q) = f_{i_1 \ldots i_p}(x_1, \cdots, x_q).$$

By the definition of sequence of functions (Definition 1) the lemma is true. □

Lemma 10. *Write* $H_n = \begin{bmatrix} l_0 \\ l_1 \\ \vdots \\ l_{2^n-1} \end{bmatrix}$ *where* l_i *is a row of* H_n. *Then* l_i *is the sequence of* $h_i(x) = \langle \alpha_i, x\rangle$ *where* α_i *is defined before Definition 1.*

Proof. By induction on n. Let $n = 1$. Since $H_1 = \begin{bmatrix} + & + \\ + & - \end{bmatrix}$, $l_0 = (+\ +)$, the sequence of $\langle 0, x\rangle$ and $l_1 = (+\ -)$, the sequence of $\langle 1, x\rangle$ where $x \in V_1$, + and − stand for 1 and −1 respectively. Suppose the lemma is true for $n = 1, 2, \ldots, k-1$.

Since $H_k = H_1 \times H_{k-1}$, where $\times$ is the Kronecker product, each row of H_n can be expressed as $\delta \times l$ where $\delta = (+\ +)$ or $(+\ -)$, and l is a row of H_{n-1}. By the assumption l is the sequence of a function, say $h(x) = \langle \alpha, x\rangle$, where $\alpha, x \in V_{k-1}$. Thus $\delta \times l$ is the sequence of $\langle \beta, y\rangle$ where $y \in V_k$, $\beta = (0\ \alpha)$ or $(1\ \alpha)$ according as $l = (+\ +)$ or $(+\ -)$. Thus the lemma is true for $n = k$. □

¿From Lemma 10 all the rows of H_n comprise all the sequences of linear functions on V_n and hence all the rows of $\pm H_n$ comprise all the sequences of affine functions on V_n.

Lemma 11. *Let f and g be functions on V_n whose sequences are η_f and η_g respectively. Then* $d(f,g) = 2^{n-1} - \frac{1}{2}\langle \eta_f, \eta_g\rangle$.

Proof. $\langle \eta_f, \eta_g\rangle = \sum_{f(x)=g(x)} 1 - \sum_{f(x)\neq g(x)} 1 = 2^n - 2\sum_{f(x)\neq g(x)} 1 = 2^n - 2d(f,g)$. This proves the lemma. □

Let $H_n = (h_{ij})$ and $L_i = (h_{i1} \cdots h_{i2^n})$ i.e. the i-th row of H_n. Write $L_{i+2^n} = -L_i$, $i = 1, \ldots, 2^n$. Since L_i, $i = 1, \ldots, 2^n$, is a linear sequence $L_1, \ldots, L_{2^n}$, $L_{2^n+1}, \ldots, L_{2^{n+1}}$ comprise all affine sequences. Let f be a function on V_n whose sequence is η_f and φ_i be the function of L_i.

Write $\eta_f = (a_1 \cdots a_{2^n})$. Since $\langle \eta_f, L_i\rangle = \sum_{j=1}^{2^n} a_j h_{ij}$

$$\langle \eta_f, L_i\rangle^2 = 2^n + 2\sum_{j<t} a_j a_t h_{ij} h_{it}. \tag{1}$$

and

$$\sum_{i=1}^{2^n} \langle \eta_f, L_i\rangle^2 = 2^{2n} + 2\sum_{i=1}^{2^n}\sum_{j<t} a_j a_t h_{ij} h_{it} = 2^{2n} + 2\sum_{j<t} a_j a_t \sum_{i=1}^{2^n} h_{ij} h_{it}.$$

Since H_n is an Hadamard matrix $\sum_{i=1}^{2^n} h_{ij} h_{it} = 0$ for $j \neq t$ and hence

$$\sum_{i=1}^{2^n} \langle \eta_f, L_i\rangle^2 = 2^{2n}. \tag{2}$$

Thus there exists an integer, say i_0, such that $\langle \eta_f, L_{i_0}\rangle^2 = \langle \eta_f, L_{i_0+2^n}\rangle^2 \geq 2^n$ and hence $\langle \eta_f, L_{i_0}\rangle \geq 2^{\frac{1}{2}n}$ or $\langle \eta_f, L_{i_0+2^n}\rangle \geq 2^{\frac{1}{2}n}$. Without any loss of generality suppose $\langle \eta_f, L_{i_0}\rangle \geq 2^{\frac{1}{2}n}$. By Lemma 11 $d(f, \varphi_{i_0}) \leq 2^{n-1} - 2^{\frac{1}{2}n-1}$. This proves

Lemma 12. *$N_f \leq 2^{n-1} - 2^{\frac{1}{2}n-1}$ for any function on V_n.*

Lemma 13. *If both (1, -1)-sequences ξ and η of length 2t consist of an even number of ones and an even number of minus ones then $d(\alpha, \beta)$ is even.*

Proof. Write $\xi = (a_1 \cdots a_{2t})$ and $\eta = (b_1 \cdots b_{2t})$. Let n_1 denote the number of pairs (a_i, b_i) such that $a_i = +1$, $b_i = +1$; let n_2 denote the number of pairs (a_i, b_i) such that $a_i = +1$, $b_i = -1$; let n_3 denote the number of pairs (a_i, b_i) such that $a_i = -1$, $b_i = +1$; and let n_4 denote the number of pairs (a_i, b_i) such that $a_i = -1$, $b_i = -1$. Hence $n_1 + n_2$, $n_3 + n_4$, $n_1 + n_3$ and $n_2 + n_4$ are all even and hence $2n_1 + n_2 + n_3$ is even. Thus $n_2 + n_3 = d(\alpha, \beta)$ is even. □

The following result can be found in [5]

Lemma 14. *Let $f(x)$ be a function from V_n to $GF(2)$. $f(x)$ and ξ be the sequence of $f(x)$. Then the following four statements are equivalent*

(i) $f(x)$ is bent,
(ii) for any affine sequence of length 2^n, denoted by l, $\langle \xi, l \rangle = \pm 2^{\frac{1}{2}n}$,
(iii) $f(x) + f(x + \alpha)$ is 0-1 balanced for every nonzero $\alpha \in V_n$,
(iv) $f(x) + \langle \alpha, x \rangle$ contains $2^{n-1} \pm 2^{\frac{1}{2}n-1}$ zeros for every $\alpha \in V_n$.

Let L_j and φ, $j = 1, \ldots, 2^{n+1}$, be the same as in the proof of Lemma 12. If f is a bent function then $\langle \eta_f, L_i \rangle^2 = 2^n$ and hence $\langle \eta_f, L_i \rangle = 2^{\frac{1}{2}n}$ or $\langle \eta_f, L_{i+2^n} \rangle = 2^{\frac{1}{2}n}$ for each fixed i, $1 \leq i \leq 2^n$. By Lemma 11 $d(f, \varphi_i) = 2^{n-1} - 2^{\frac{1}{2}n-1}$ or $d(f, \varphi_{i+2^n}) = 2^{n-1} - 2^{\frac{1}{2}n-1}$ for each fixed i, $1 \leq i \leq 2^n$. Thus $N_f = 2^{n-1} - 2^{\frac{1}{2}n-1}$. In other words, bent functions attain the upper bound for nonlinearities given in Lemma 12. Conversely, if a function f on V_n attains the upper bound for nonlinearities, $2^{n-1} - 2^{\frac{1}{2}n-1}$, then $\langle \eta_f, L_i \rangle^2 = 2^n$ for $i = 1, \ldots, 2^{n+1}$ i.e. f is bent, otherwise $\langle \eta_f, L_i \rangle^2 = 2^n$ does not hold for some i, $1 \leq i \leq 2^{n+1}$. Note that $L_{i+2^n} = -L_i$. ¿From (2) there exist i_1 and i_2, $1 \leq i_1, i_2, \leq 2^n$, such that $\langle \eta_f, L_{i_1} \rangle^2 > 2^n$ and $\langle \eta_f, L_{i_2} \rangle^2 < 2^n$. Thus $\langle \eta_f, L_{i_1} \rangle > 2^{\frac{1}{2}n}$ or $\langle \eta_f, L_{i_1+2^n} \rangle > 2^{\frac{1}{2}n}$. Without any loss generality, suppose $\langle \eta_f, L_{i_1} \rangle > 2^{\frac{1}{2}n}$. By using Lemma 11 $d(f, \varphi_{i_1}) < 2^{n-1} - 2^{\frac{1}{2}n-1}$ and hence $N_f < 2^{n-1} - 2^{\frac{1}{2}n-1}$. This is a contradiction to the assumption that f attains the maximum nonlinearity $2^{n-1} - 2^{\frac{1}{2}n-1}$. Hence we have proved

Corollary 15. *A function on V_n attains the upper bound for nonlinearities, $2^{n-1} - 2^{\frac{1}{2}n-1}$, if and only if it is bent.*

¿From (1) we have

Corollary 16. *Let f be a function on V_n whose sequence is $\eta_f = (a_1 \cdots a_{2^n})$. Then f is bent if and only if $\sum_{j<t} a_j a_t h_{ij} h_{it} = 0$ for $i = 1, \ldots, 2^n$ where $(h_{ij}) = H_n$.*

¿From Corollary 15 0-1 balanced functions cannot attain the upper bound for nonlinearities $2^{n-1}-2^{\frac{1}{2}n-1}$. However we can construct a class of 0-1 balanced functions with high nonlinearity by using bent functions.

Corollary 17. *Let f be a 0-1 balanced function on V_n $(n \geq 3)$. Then $N_f \leq 2^{n-1}-2^{\frac{1}{2}n-1}-2$ if n is even number and $N_f \leq \lfloor\lfloor 2^{n-1}-2^{\frac{1}{2}n-1}\rfloor\rfloor$ if n is odd where $\lfloor\lfloor x\rfloor\rfloor$ denotes the maximum even number less than or equal to x.*

Proof. Note that f and each φ_i, where φ_i is the same as in Definition 3, have an even number of ones and an even of number of zeros. By Lemma 13 $d(f,\varphi_i)$ is even. By corollary 15 $d(f,g_i) < 2^{n-1}-2^{\frac{1}{2}n-1}$. This proves the corollary. □

Lemma 18. *Let $f_j(x_1,\ldots,x_{2k})$ be a bent function on V_{2k-2}, $j=1,2$. Set*

$$g=(u,x_1,\ldots,x_{2k})=(1+u)f_1(x)+uf_2(x).$$

Then $N_g \geq 2^{2k}-2^k$.

Proof. Write ξ_j for the sequence of f_j, $j=1,2$. By Lemma 9 $\gamma=(\xi_1\ \xi_2)$ is the sequence of g, of length 2^{2k+1}. Let L be the sequence of an affine function, say φ. By Lemma 10 L is a row of $\pm H_{2k+1}$. Since $H_{2k+1}=H_1\times H_{2k}$ and $H_1=\begin{bmatrix}1 & 1\\ 1 & -1\end{bmatrix}$, where $\times$ is the Kronecker product, L can be expressed as $L=(l'\ l')$ or $L=(l'\ -l')$, by Lemma 10, where l' is a row of $\pm H_{2k}$. Since both f and $f+h$ are bent, by (ii) of Lemma 14, $\langle\xi_j,l'\rangle=\pm 2^k$. $\langle\gamma,L\rangle=\langle\xi_1,l'\rangle\pm\langle\xi_2,l'\rangle$. Thus $|\langle\gamma,L\rangle|\leq 2^{k+1}$. By Lemma 11 $d(g,\varphi)\geq 2^{2k}-2^k$. Since φ is arbitrary $N_g\geq 2^{2k}-2^k$. □

Lemma 19. *Let $f_j(x_1,\ldots,x_{2k-2})$ be a bent function on V_{2k-2}, $j=1,2,3,4$. Set*

$$g(u,v,x_1,\ldots,x_{2k-2})=(1+u)(1+v)f_1(x)+(1+u)vf_2(x)+u(1+v)f_3(x)+uvf_4(x).$$

Then $N_g\geq 2^{2k-1}-2^k$.

Proof. Let ξ_j be the sequence of $f_j(x)$, $j=1,2,3,4$ and $\eta=(\xi_1\ \xi_2\ \xi_3\ \xi_4)$ be the sequence of g. Let L be an affine sequence of length 2^{2k} whose function is $h(z)$, an affine function. By Lemma 10 L is a row of $\pm H_{2k}$. Since $H_{2k}=H_2\times H_{2k-2}$ and L can be expressed as $L=l_2\times l_{k-2}$ where l_2 is a row of $\pm H_2$ and l_{2k-2} is a row of $\pm H_{2k-2}$. Since each ξ_i is bent, by (ii) of Lemma 14, $\langle\xi_i,l\rangle=\pm 2^{k-1}$. Note that $|\langle\eta,L\rangle|\leq\sum_{i=1}^{4}|\langle\xi_i,l\rangle|$ and hence $|\langle\eta,L\rangle|\leq 4\cdot 2^{k-1}$. By Lemma 11 $d(g,h)\geq 2^{2k-1}-2^k$. Since h is an arbitrary affine function $N_g\geq 2^{2k-1}-2^k$. □

Lemma 20. *$f(x_1,\ldots,x_n)+\psi(u_1,\ldots,u_t)$ is a 0-1 balanced function on V_{n+t} if f is a 0-1 balanced function on V_n or ψ is a 0-1 balanced function on V_t.*

Proof. Set $g(x_1,\ldots,x_n,u_1,\ldots,u_t)=f(x_1,\ldots,x_n)+\psi(u_1,\ldots,u_t)$. Without any loss of generality, suppose f is a 0-1 balanced function on V_n. Note that for every fixed $(u_1^0\cdots u_t^0)\in V_t$, $g(x_1,\ldots,x_n,u_1^0,\ldots,u_t^0)=f(x_1,\ldots,x_n)+\psi(u_1^0,\ldots,u_t^0)$ is a 0-1 balanced function on V_n thus $g(x_1,\ldots,x_n,u_1,\ldots,u_t)$ is a 0-1 balanced function on V_{n+t}. □

3 Construction

3.1 On V_{2k+1}

Let $k \geq 1$ and $f(x_1, \ldots, x_{2k})$ be a bent function on V_{2k}. Write $x = (x_1 \cdots x_{2k})$. Let $h(x)$ be a non-constant affine function on V_{2k}. Note that $f(x) + h(x)$ is also bent (see Property 2, p95, [8]) and hence $f + h$ assumes the value zero $2^{2k-1} \pm 2^{k-1}$ times and assumes the value one $2^{2k-1} \mp 2^{k-1}$ times.

Without any loss of generality we suppose $f(x)$ assumes the value zero $2^{2k-1} + 2^{k-1}$ times (if $f(x)$ assumes the value zero $2^{2k-1} - 2^{k-1}$ times, the bent function $f(x)+1$ assumes the value zero $2^{2k-1}+2^{k-1}$ times and hence we can replace $f(x)$ by $f(x) + 1$). Also we suppose $f(x) + h(x)$ assumes the value zero $2^{2k-1} - 2^{k-1}$ times (if $f(x)+h(x)$ assumes the value zero $2^{2k-1}+2^{k-1}$ times, the bent function $f(x) + h(x) + 1$ assumes the value zero $2^{2k-1} - 2^{k-1}$ times so we can replace $f(x) + h(x)$ by $f(x) + h(x) + 1$). Set

$$g(u, x_1, \ldots, x_{2k}) = f(x_1, \ldots, x_{2k}) + uh(x_1, \ldots, x_{2k}). \tag{3}$$

Lemma 21. *$g(u, x_1, \ldots, x_{2k})$ defined by (3) is a 0-1 balanced function on V_{2k+1}.*

Proof. Note that $g(0, x_1, \ldots, x_{2k}) = f(x_1, \ldots, x_{2k})$ assumes the value zero $2^{2k-1} + 2^{k-1}$ times and $g(1, x_1, \ldots, x_{2k}) = f(x_1, \ldots, x_{2k}) + h(x_1, \ldots, x_{2k})$ assumes the value zero $2^{2k-1} - 2^{k-1}$ times. Thus $g(u, x_1, \ldots, x_{2k})$ assumes the value zero 2^k times (one 2^k times). □

Lemma 22. *$N_g \geq 2^{2k} - 2^k$ where g is defined by (3).*

Proof. $g = f + uh = (1 + u)f + u(f + h)$. Note that both f and $f + h$ are bent functions on V_{2k}. By Lemma 18 $N_g \geq 2^{2k} - 2^k$. □

Lemma 23. *$g(u, x_1, \ldots, x_{2k})$ defined by (3) satisfies the strict avalanche criterion.*

Proof. Let $\gamma = (b\ a_1 \cdots a_{2k})$ with $W(\gamma) = 1$. Write $\alpha = (a_1 \cdots a_{2k})$, $z = (u\ x_1 \ldots x_{2k})$ and $x = (x_1 \ldots x_{2k})$. $g(z + \gamma) = f(x + \alpha) + (u + b)h(x + \alpha)$ and hence $g(z) + g(z + \gamma) = f(x) + f(x + \alpha) + u(h(x) + h(x + \alpha)) + bh(x + \alpha)$.

Case 1: $b = 0$ and hence $W(\alpha) = 1$. $g(z) + g(z + \gamma) = f(x) + f(x + \alpha) + u(h(x) + h(x + \alpha))$. Since h is a non-constant affine function $h(x) + h(x + \alpha) = c$ where c is a constant. Thus $g(z) + g(z + \gamma) = f(x) + f(x + \alpha) + cu$.
By (iii) of Lemma 14 $f(x) + f(x + \alpha)$ is a 0-1 balanced function on V_{2k} and hence by Lemma 20 $g(z) + g(z + \gamma)$ is a 0-1 balanced function on V_{2k+1}.

Case 2: $b = 1$ and hence $W(\alpha) = 0$ i.e. $\alpha = 0$. $g(z) + g(z + \gamma) = h(x)$. Since $h(x)$ is a non-constant affine function on V_{2k} $h(x)$ is a 0-1 balanced and hence by Lemma 20 $g(z) + g(z + \alpha)$ is a 0-1 balanced function on V_{2k+1}. □

Summarizing Lemmas 21, 22, 23 we have

Theorem 24. *For $k \geq 1$, $g(u, x_1, \ldots, x_{2k})$ defined by (3) is a 0-1 balanced function on V_{2k+1} having $N_g \geq 2^{2k} - 2^k$ and satisfying the strict avalanche criterion.*

3.2 On V_{2k}

Let $k \geq 2$ and $f(x_1, \ldots, x_{2k-2})$ be bent function on V_{2k-2}. Write $x = (x_1 \cdots x_{2k-2})$. Let $h_j(x)$, $j = 1, 2, 3$, be three non-constant affine functions on V_{2k-2} such that $h_i(x) + h_j(x)$ is non-constant for any $i \neq j$. Such $h_1(x)$, $h_2(x)$, $h_3(x)$ exist for $k \geq 2$. Note that each $f(x) + h_j(x)$ is also bent (see Property 2, p95, [8]) and hence $f + h_j$ assumes the value zero $2^{2k-3} \pm 2^{k-2}$ times and assumes the value one $2^{2k-3} \mp 2^{k-2}$ times.

Without any loss of generality we suppose both $f(x)$ and $f(x)+h_1(x)$ assume the value zero $2^{2k-3}+2^{k-2}$ times and both $f(x)+h_2(x)$ and $f(x)+h_3(x)$ assume the value zero $2^{2k-3}-2^{k-2}$ times. This assumption is reasonable because $f(x)+h_j(x)$ assumes the value zero $2^{2k-3} - 2^{k-2}$ times if and only if $f(x) + h_j(x) + 1$ assumes the value zero $2^{2k-3} + 2^{k-2}$ times and $h_j(x) + 1$ is also a non-constant affine function thus we can choose one of $f(x) + h_j(x)$ and $f(x) + h_j(x) + 1$ so that the assumption is satisfied. Set

$$g(u, v, x_1, \ldots, x_{2k-2}) = \\ f(x) + vh_1(x) + uh_2(x) + uv(h_1(x) + h_2(x) + h_3(x)). \tag{4}$$

Lemma 25. *$g(u, v, x_1, \ldots, x_{2k-2})$ defined by (4) is a 0-1 balanced function on V_{2k}.*

Proof. Note that $g(0, 0, x_1, \ldots, x_{2k-2}) = f(x)$, $g(0, 1, x_1, \ldots, x_{2k-2}) = f(x) + h_1(x)$, $g(1, 0, x_1, \ldots, x_{2k-2}) = f(x) + h_2(x)$, $g(1, 1, x_1, \ldots, x_{2k-2}) = f(x) + h_1(x) + h_2(x) + (h_1(x) + h_2(x) + h_3(x)) = f(x) + h_3(x)$. By the assumption the first two functions assume the value zero $2^{2k-2} + 2^{k-1}$ times in total and the second two functions assume the value zero $2^{2k-2} - 2^{k-1}$ times in total. Hence $g(u, v, x_1, \ldots, x_{2k-2})$ assumes the value zero 2^{2k-1} times in total and thus it is a 0-1 balanced function on V_{2k}. □

Lemma 26. *$N_g \geq 2^{2k-1} - 2^k$ where g is defined by (4).*

Proof. Note that $g = f(x) + vh_1(x) + uh_2(x) + uv(h_1(x) + h_2(x) + h_3(x)) = (1 + u)(1 + v)f(x) + (1 + u)v(f(x) + h_1(x)) + u(1 + v)(f(x) + h_2(x)) + uv(f(x) + h_3(x))$. By Lemma 19 $N_g \geq 2^{2k-1} - 2^k$. □

Lemma 27. *$g(u, v, x_1, \ldots, x_{2k-2})$ defined by (4) satisfies the strict avalanche criterion.*

Proof. Let $\gamma = (b\ c\ a_1 \cdots a_{2k-2})$ with $W(\gamma) = 1$. Write $\alpha = (a_1 \cdots a_{2k-2})$, $z = (u\ v\ x_1 \ldots x_{2k-2})$ and $x = (x_1 \ldots x_{2k-2})$.

Note that $g(z + \gamma) = f(x + \alpha) + (v + c)h_1(x + \alpha) + (u + b)h_2(x + \alpha) + (u + b)(v + c)(h_1(x + \alpha) + h_2(x + \alpha) + h_3(x + \alpha))$.

Case 1: $b = 1$ and hence $c = 0$, $W(\alpha) = 0$ i.e. $\alpha = 0$. $g(z) + g(z + \gamma) = h_2(x) + v(h_1(x) + h_2(x) + h_3(x))$ will be $h_2(x)$ when $v = 0$ and $h_1(x) + h_3(x)$ when $v = 1$. Both $h_2(x)$ and $h_1(x) + h_3(x)$ are non-constant affine functions on V_{2k-2} and hence $g(z) + g(z + \gamma)$ is 0-1 balanced on V_{2k}.

Case 2: $c = 1$ and hence $b = 0$, $W(\alpha) = 0$ i.e. $\alpha = 0$. The proof is similar to Case 1.

Case 3: $W(\alpha) \neq 0$ and hence $b = c = 0$. Since h_j is an affine function we can write $h_j(x) + h_j(x + \alpha) = a_j$ where a_j is a constant. Hence $g(z) + g(z + \gamma) = f(x) + f(x+\alpha) + va_1 + ua_2 + uv(a_1 + a_2 + a_3)$. By (iii) of Lemma 14 $f(x) + f(x+\alpha)$ is a 0-1 balanced function on V_{2k-2} and hence by Lemma 20 $g(z) + g(z + \gamma)$ is a 0-1 balanced function on V_{2k}. This proves that $g(u, v, x_1, \ldots, x_{2k-2})$ satisfies the strict avalanche criterion. □

Summarizing Lemmas 25, 26, 27 we have

Theorem 28. *For $k \geq 2$, $g(u, v, x_1, \ldots, x_{2k-2})$ defined by (4) is a 0-1 balanced function on V_{2k} having $N_g \geq 2^{2k-2} - 2^k$ and satisfying the strict avalanche criterion.*

4 Remarks

We note that the nonlinearities of 0-1 balanced functions satisfying SAC in Theorems 24 and 28 are the same as those for ordinary 0-1 balanced functions (see [13]). Next we give two examples of the theorems.

Example 1. In Theorem 24 let $k = 2$. Consider V_5. As we know, $f(x_1, x_2, x_3, x_4) = x_1x_2 + x_3x_4$ is a bent function in V_4. Choose the non-constant affine function $h(x_1, x_2, x_3, x_4) = 1 + x_1 + x_2 + x_3 + x_4$. Note f assumes the value zero $2^{4-1} + 2^{2-1} = 10$ times and $f + h$ assumes the value zero $2^{4-1} - 2^{2-1} = 6$ times. Hence we set $g(u, x_1, x_2, x_3, x_4) = f(x_1, x_2, x_3, x_4) + uh(x_1, x_2, x_3, x_4) = x_1x_2 + x_3x_4 + u(1 + x_1 + x_2 + x_3 + x_4)$. By Theorem 24 $g(u, x_1, x_2, x_3, x_4)$ is a 0-1 balanced function with $N_g \geq 2^4 - 2^2 = 12$, satisfying the strict avalanche criterion. On the other hand, by Corollary 17 the bound for nonlinearly 0-1 balanced functions on V_5 is $\lfloor\lfloor 2^4 - 2^{2-\frac{1}{2}} \rfloor\rfloor = \lfloor\lfloor 13.1818\cdots \rfloor\rfloor = 12$ where $\lfloor\lfloor x \rfloor\rfloor$ denotes the maximum even number no larger than x. This means that $N_g = 12$ attains the upper bound for nonlinearly 0-1 balanced functions on V_5.

Example 2. In Theorem 28 let $k = 3$. Consider V_6. Choose $f(x_1, x_2, x_3, x_4) = x_1x_2 + x_3x_4$, a bent function in V_4. Also choose non-constant affine functions $h_1(x_1, x_2, x_3, x_4) = x_1$, $h_2(x_1, x_2, x_3, x_4) = 1 + x_2$, $h_3(x_1, x_2, x_3, x_4) = 1 + x_3$. Note both f and $f + h_1$ assume the value zero $2^{4-1} + 2^{2-1} = 10$ times and both $f + h_3$ and $f + h_4$ assume the value zero $2^{4-1} - 2^{2-1} = 6$ times. Hence we set $g(u, v, x_1, x_2, x_3, x_4) = f + vh_1 + uh_2 + uv(h_1 + h_2 + h_3)$. By Theorem 28 $g(u, v, x_1, x_2, x_3, x_4)$ is a 0-1 balanced function with $N_g \geq 2^5 - 2^3 = 24$, satisfying the strict avalanche criterion. On the other hand, by Corollary 17 the upper bound for nonlinearly 0-1 balanced functions on V_6 is $2^5 - 2^2 - 2 = 26$. This means that $N_g = 24$ is very high.

Recently Zheng, Pieprzyk and Seberry [23] constructed a very efficient one way hashing algorithm using boolean functions constructed by the method given in Theorem 24. These functions have further cryptographically useful properties.

References

1. C. M. Adams and S. E. Tavares. Generating and counting binary bent sequences. *IEEE Transactions on Information Theory*, IT-36 No. 5:1170–1173, 1990.
2. C. M. Adams and S. E. Tavares. The use of bent sequences to achieve higher-order strict avalanche criterion. to appear, 1990.
3. M. H. Dawson and S. E. Tavares. An expanded set of S-box design criteria based on information theory and its relation to differential-like attacks. In *Advances in Cryptology-EUROCRYPT'91*, volume 547, Lecture Notes in Computer Science, pages 352–367. Springer-Verlag, 1991.
4. John Detombe and Stafford Tavares. Constructing large cryptographically strong S-boxes. Presented in AUSCRYPT'92, 1992.
5. J. F. Dillon. A survey of bent functions. *NSA Mathematical Meeting*, pages 191–215, 1972.
6. R. Forre. The strict avalanche criterion: Special properties of boolean functions and extended definition. In *Advances in Cryptology: Crypto'88 Proceedings*, volume 403, Lecture Notes in Computer Science, pages 450–468. Springer-Verlag, New York, 1989.
7. P. V. Kumar and R. A. Scholtz. Bounds on the linear span of bent sequences. *IEEE Transactions on Information Theory*, IT-29 No. 6:854–862, 1983.
8. P. V. Kumar, R. A. Scholtz, and L. R. Welch. Generalized bent functions and their properties. *Journal of Combinatorial Theory*, Ser. A, 40:90–107, 1985.
9. A. Lempel and M. Cohn. Maximal families of bent sequences. *IEEE Transactions on Information Theory*, IT-28 No. 6:865–868, 1982.
10. S Lloyd. Couting functions satisfying a higher order strict avalanche criterion. In *Advances in Cryptology-EUROCRYPT'89*, volume 434, Lecture Notes in Computer Science, pages 64–74. Springer-Verlag, New York, 1990.
11. V. V. Losev. Decoding of sequences of bent functions by means of a fast Hadamard transform. *Radiotechnika i elektronika*, 7:1479–1492, 1987.
12. F. J. MacWilliams and N. J. A. Sloane. *The Theory of Error-Correcting Codes.* New York: North-Holland, 1977.
13. Willi Meier and Othmar Staffelbach. Nonlinearity criteria for cryptographic functions. In *Advances in Cryptology-EUROCRYPT'89*, volume 434, Lecture Notes in Computer Science, pages 549–562. Springer-Verlag, 1990.
14. Kaisa Nyberg. Perfect nonlinear S-boxes. In *Advances in Cryptology-EUROCRYPT'91*, volume 547, Lecture Notes in Computer Science, pages 378–386. Springer-Verlag, 1991.
15. J. D. Olsen, R. A. Scholtz, and L. R. Welch. Bent-function sequences. *IEEE Transactions on Information Theory*, IT-28 No. 6:858–864, 1982.
16. J. Pieprzyk and G. Finkelstein. Towards effective nonlinear cryptosystem design. *IEE Proceedings (Part E)*, 135:325–335, 1988.
17. O. S. Rothaus. On bent functions. *Journal of Combinatorial Theory*, Ser. A, 20:300–305, 1976.
18. S. E. Tavares, M. Sivabalan, and L. E. Peppard. On the designs of SP networks from an information theoretic point of view. In *Advances in Cryptology: Crypto'92 Proceedings*, 1992.
19. W. D. Wallis, A. Penfold Street, and J. Seberry Wallis. *Combinatorics: Room Squares, sum-free sets, Hadamard Matrices*, volume 292 of Lecture Notes in Mathematics. Springer-Verlag, Berlin- Heidelberg- New York, 1972.

20. A. F. Webster. *Plaintext/Ciphertext Bit Dependencies in Cryptographic System.* Master's Thesis, Department of Electrical Engineering, Queen's University, 1985.
21. A. F. Webster and S. E. Tavares. On the designs of S-boxes. In *Advances in Cryptology: Crypto'85 Proceedings*, volume 219, Lecture Notes in Computer Science, pages 523–534. Springer-Verlag, New York, 1986.
22. R. Yarlagadda and J. E. Hershey. Analysis and synthesis of bent sequences. *IEE Proceeding (Part E)*, 136:112–123, 1989.
23. Yuliang Zheng, Josef Pieprzyk, and Jennifer Seberry. Haval — one-way hashing algorithm with variable length of output. Presented in AUSCRYPT'92, 1992.

Linear Nonequivalence versus Nonlinearity*

Chris Charnes and Josef Pieprzyk

Department of Computer Science
University of Wollongong
PO Box 1144
Wollongong, NSW 2500, AUSTRALIA
e-mail: charnes@cs.uow.edu.au
e-mail: josef@cs.uow.edu.au

Abstract. The choice of a collection of cryptographically strong Boolean functions is crucial in designing a strong hashing algorithm. The paper shows that it is possible to obtain five linearly nonequivalent functions with five Boolean variables which are cryptographically strong and easy to implement. They can be readily used to design hashing algorithms (of the MD5 structure).

1 Introduction

In [27] Shannon formulated general principles for designing "strong" cryptographic algorithms from cryptographically "weak" components. Feistel [13] used this philosophy to design the LUCIFER encryption algorithm. Another example of the same approach is the well known DES algorithm [18].

The DES algorithm along with its components has been tested and examined very carefully since its description was first published. The main purpose of the investigations has been to reveal the details of the design criteria used. Now we know a collection of design properties which were probably used while designing the DES. Some of them relate to the structure of the system ([7],[8]), the other provide justification for S-boxes selection.

S-boxes are the core of the DES and the strength of it mainly rests on the proper selection of permutations which create the S-boxes. The list of desirable properties of S-boxes includes: nonlinearity ([2], [17], [19], [21], [23], [24]), completeness ([15]), 0-1 balance, and the strict avalanche criterion (SAC) together with its higher orders ([2], [14], [25], [29]).

Invention of a new cryptographic attack on iterated cryptosystems by Biham and Shamir ([5],[6]), prompted a new criterion - the flat XOR profile to be added to the list. Constructions of S-boxes with uniform XOR profiles can be found in [20] (also see [1]).

It is well-known that linearity in S-boxes weakens an encryption algorithm. Chaum and Evertse [10] used the so called linear structures present in S-boxes to

* Support for this project was provided in part by the Australian Research Council under the reference number A49131885

cryptanalyse the DES with a reduced number of rounds. Later a partial linearity has been analysed in [3] and [22]. The notion of linear equivalence of Boolean functions has been mentioned in several papers (see [11] [12]), but there was no systematic treatment of it in the open literature.

In this work we are going to examine the relation between nonlinearity and linear nonequivalence.

2 Linear Equivalence

Let $\mathcal{S}$ be the set of all Boolean functions from $GF(2^n)$ to $GF(2)$. The n-dimensional affine group $GA(n)$, acts on $\mathcal{S}$ as follows. (The elements of $GA(n)$ are ordered pairs (A, v), where A is an $n \times n$ non-singular matrix with entries from $GF(2)$ and v is a n-ary binary vector.) Let $\hat{A} = (A, v)$ belong to $GA(n)$ and let f belong to S, then $\hat{A} \cdot f$ is a Boolean function defined as $\hat{A} \cdot f(x) = f(Ax + v)$ for all $x \in GF(2^n)$. The action of $GA(n)$ induces a decomposition of S into disjoint sets; these are the ***orbits*** of the group $GA(n)$ acting on S.

The design of a cryptographic system imposes two important criteria which must be satisfied by the Boolean functions which comprise such systems. Namely, that these functions are ***non-linear*** and ***linearly inequivalent***. These requirements can be rephrased in terms of the action of $GA(n)$ on the space S.

Definition 1. Two Boolean functions $f, g : GF(2^n) \rightarrow GF(2)$ are linearly equivalent if there is an affine transformation $\hat{A} = (A, v)$ such that

$$f(x) = g(Ax + v) \tag{1}$$

for all $x \in GF(2^n)$.

Thus the classes of linearly equivalent functions are precisely the orbits of the affine group $GA(n)$. Clearly, knowing the decomposition of S into orbits under the action of $GA(n)$ would be very useful in the design of a cryptographic system. In general the orbit problem for $GA(n)$ is unsolved, but we would like to point out some results which offer a partial solution.

Proposition 2. *Given $f, g : GF(2^n) \longrightarrow GF(2)$. If $g(x) = f(Ax + v)$ for all $x \in GF(2^n)$ and some $(A, n) \in GA(n)$, then*

$$\mathcal{N}_f = \mathcal{N}_g.$$

This result will be established in the next section. The following is an immediate consequence.

Corollary 3. *If $f, g : GF(2^n) \longrightarrow GF(2)$ are Boolean functions and $\mathcal{N}_f \neq \mathcal{N}_g$, then f and g belong to different orbits of $GA(n)$.*

However if two Boolean functions f and g have the same nonlinearity, they may or may not belong to different orbits. Other invariants may help to resolve such ambiguities. We will discuss some of these next.

3 More Orbit Invariants

In this section we introduce some other orbit invariants of the affine group. These require familiarity with the binary Reed-Muller codes, and we refer the reader to [16] for a full account of these. Here we just recall those facts needed to state our results.

A Boolean function f from $GF(2^n)$ to $GF(2)$, can be represented as a binary string $[f]$, which has length 2^n. This string is the truth table of f. We can thus think of f in two ways: as a Boolean polynomial and as a binary string $[f]$. As a Boolean polynomial, f has a unique representation as a combination of exclusive ors (XORs) and of terms which consist of the products of the variables of f. This is commonly called the algebraic normal form of f. The highest degree, r, of a product term in the algebraic normal form is the *degree* of f. If f has degree r and has n-variables $x_1, \ldots, x_n$, then the corresponding binary string $[f]$, is a code word of the r-th order binary Reed-Muller code $\mathcal{R}(r, n)$.

It can be readily checked that the degree of the algebraic normal form of a Boolean function f, is preserved when the variables $x_1, \ldots, x_n$ of f are transformed by the elements of $GA(n)$. ($GA(n)$ is a subgroup of the automorphism group of $\mathcal{R}(r, n)$; see [16].) Thus the degree of a Boolean function is an orbit invariant of $GA(n)$.

Another orbit invariant of a Boolean function f is the Hamming weight of the corresponding binary string, i.e. the number of non-zero coordinates of $[f]$. This can been seen as follows. If f and g are linearly equivalent then $f(x) = g(Ax+v)$ for all $x \in GF(2^n)$, where $(A, v) \in GA(n)$. Now (A, v) is a one-to-one mapping of $GF(2^n)$ onto $GF(2^n)$, so as x varies through $GF(2^n)$ so does $Ax + v$. These two facts suffice to establish the validity of the remark.

The set of affine Boolean functions from $GF(2^n)$ to $GF(2)$ correspond to the code words of $\mathcal{R}(1, n)$. Moreover, this code is a sub-code of $\mathcal{R}(r, n)$ for every $r > 1$. So to any n-variable Boolean function f we can associate the coset $f+\mathcal{R}(1, n)$ — the coset containing f. The vectors of $f+\mathcal{R}(1, n)$ are the sums of the binary string $[f]$ with all the code words belonging to $\mathcal{R}(1, n)$. The weight distributor of $f+\mathcal{R}(1, n)$ is defined to be the the set $\{w_i\}$, where w_i is the total number of vectors in $f + \mathcal{R}(1, n)$ whose Hamming weight equals i.

The following theorem is proved in [16].

Theorem 4. *If f and g are Boolean functions from $GF(2^n)$ to $GF(2)$ which are linearly equivalent then the cosets of $\mathcal{R}(1, n)$ containing f and g have the same weight distribution.*

The above theorem states that the weight distributor of the coset $f+\mathcal{R}(1, n)$ is an orbit invariant of $GA(n)$. Furthermore, the Hadamard transform provides a method for calculating this invariant for a given Boolean function f as we shall explain.

Let F be a real valued function obtained from the binary string $[f]$ of a Boolean function f by replacing each occurrence of 1 by -1 and 0 by 1. Thus $F(u) = (-1)^{f(u)}$ for $u \in GF(2^n)$. The Hadamard transform $\hat{F}$, of a real vector F is defined in Chapter 14 of [16], where the next result is proved.

Theorem 5. *The weight distribution of that coset of* $\mathcal{R}(1,n)$ *which contains* f *is*

$$\frac{1}{2}\{2^n \pm \hat{F}(u)\} \quad for\ u \in GF(2^n).$$

Thus the Hadamard transform of the real function F which corresponds to f, gives a formula for calculation of the weight distributor of the coset $f + \mathcal{R}(1,n)$.

Equivalently we can think of the weight distributor of the coset $f + \mathcal{R}(1,n)$, as the set of Hamming distances from the Boolean function f to the set of all affine Boolean functions i.e. $\mathcal{R}(1,n)$. Now, $\mathcal{N}_f$, the non-linearity of f was defined in [23] to be the minimum of these distances, thus we obtain the following result.

Corollary 6. *The nonlinearity* $\mathcal{N}_f$ *of a Boolean function* f *is given by the following expression*

$$\mathcal{N}_f = \frac{1}{2}Min\{2^n \pm \hat{F}(u)\} \quad for\ u \in GF(2^n).$$

This formulation of nonlinearity, and the above theorems establish Proposition 2.1, i.e. the invariance of $\mathcal{N}_f$ under transformations belonging to $GA(n)$.

In enumerating the orbits of $GA(n)$ on $\mathcal{S}$, it is sometimes useful to ignore the subspace of $\mathcal{S}$ comprising the affine functions. Since $\mathcal{R}(1,n)$ is invariant under the group $GA(n)$, we can define an action of $GA(n)$ on the set of cosets of $\mathcal{R}(1,n)$ as follows.

Definition 7. Two Boolean functions $f, g : GF(2^n) \to GF(2)$ are linearly equivalent if there is an affine transformation $\hat{A} = (A, v)$ such that

$$f(x) = g(Ax + v) + a(x) \tag{2}$$

for all $x \in GF(2^n)$, and some affine function $a(x)$.

This definition in independent of the affine function $a(x)$. Thus there is a well defined action of the group $GA(n)$ on the space of cosets of $\mathcal{R}(1,n)$. Evidently if two Boolean functions belong to an orbit of $GA(n)$ in the sense of Definition 2.1, then they are also in an orbit of cosets. However the converse need not hold. The orbits of the induced action of $GA(5)$ on the cosets of $\mathcal{R}(1,5)$ were determined by Berlekamp and Welch in [4], we make use of their result in section 5.

4 The Number of Orbits

In Section 3 we have defined an action of the group of affine transformations $GA(n)$ on the set $\mathcal{S}$ of 2^{2^n} Boolean functions. This action induces on $\mathcal{S}$ a permutational representation of $GA(n)$ of degree 2^{2^n}. (For these notions, see [9].) So Burnside's theorem presents us with a possibility of counting the number of orbits of $GA(n)$ on $\mathcal{S}$. In other words, the number of linearly inequivalent n-variable Boolean functions.

Theorem 8. *The number of orbits t of a finite permutation group G acting on a finite set is the following*

$$t = \frac{1}{|G|}\sum_{a\in G} F(a)$$

where $|G|$ denotes the order of G.

This theorem is proved in Chapter 10 of [9]. This result, reduces the counting of the number of orbits of $GA(n)$ to the problem of determining the set of fixed-points $\{F(a)\}$. In an equivalent formulation of Burnside's theorem, it suffices to determine $\{F(a)\}$ for a set of representatives of the conjugacy classes in the group. The set of fixed-points can also be abtained using character theory ([9].) With this approach, the permutation character is decomposed as a sum of the irreducible complex characters of the group, where each character occurs with the appropriate multiplicity. The value of the permutation character is the number of fixed points.

Slepian in [28], has used these results to determine the orbits of n-variable Boolean functions under the hyper-octahedral group, i.e. permuting and changing the signs of the variables in all combinations.

5 Five Boolean Functions

In this section we apply the principles enunciated previously to the problem of finding five Boolean functions of five variables which

- are balanced (their weights are 16),
- satisfy SAC (SAC=40),
- have high nonlinearity (12 or 10),
- are pairwise linearly nonequivalent,
- are far away from each other (in terms of the Hamming distance),
- can be expressed by short algebraic normal forms (ANFs).

It is intended that these functions are used in the design of a five round hashing algorithm, where each round is encoded by one of the functions. We prove that our functions are optimal in the sense that their algebraic normal forms involve the least number of operations. Thus a hashing algorithm designed around these functions will have a favorable rate of operation.

Table 1. Boolean Functions

Function	ANF	Nonlinearity
$f_1(a,b,c,d,e)$	$ae \oplus cb \oplus cd \oplus bd$	12
$f_2(a,b,c,d,e)$	$ade \oplus ab \oplus bd \oplus ce \oplus a$	12
$f_3(a,b,c,d,e)$	$abcd \oplus cde \oplus ac \oplus bd \oplus a$	12
$f_4(a,b,c,d,e)$	$acde \oplus ac \oplus de \oplus b$	10
$f_5(a,b,c,d,e)$	$bcde \oplus abc \oplus de \oplus a \oplus d \oplus e$	10

The Boolean functions in Table 1 are $0-1$ balanced and satisfy the strict avalanche criterion; their SAC parameter is 40 (measuring the changes in one output bit). They are also mutually linearly nonequivalent. This is seen as follows. Since $\{f_1, f_2, f_3\}$ have nonlinearity 12, and $\{f_4, f_5\}$ have nonlinearity 10 we only have to check (by Proposition 2.2), the nonequivalence of the functions contained within each set. (The highest nonlinearity for a five variable function is 12, but a nonlinearity of 10 is acceptable.) Noting the degrees of the functions within each set shows that the functions are nonequivalent.

Table 2 gives the mutual Hamming distances of the corresponding binary strings of the functions. The functions can be scheduled so that the Hamming distance of two contigous functions is either: 14, 16, or 18. So the linear combination of two such functions is close to being $0-1$ balanced.

Table 2.

Function pair	Distance	Function pair	Distance
(f_1, f_2)	14	(f_2, f_4)	16
(f_1, f_3)	14	(f_2, f_5)	12
(f_1, f_4)	12	(f_3, f_4)	16
(f_1, f_5)	18	(f_3, f_5)	20
(f_2, f_3)	14	(f_4, f_5)	16

We have exhibited five Boolean functions which satisfy the cryptographic criteria listed at the beginning of this section. Next we shall explain how these functions were found and prove that they are optimal.

The functions were found by a computer search of five variable Boolean functions whose ANF contained at most four terms of degree greater or equal to two. We made use of the representation of the Boolean functions as codewords of $\mathcal{R}(r,5)$. The r-th order Reed-Muller code $\mathcal{R}(r,5)$ has a canonical basis whose vectors are labelled by the product terms of the algebraic normal form. Therefore a given binary string can be expressed as a linear combination of the canonical basis, which is the ANF of the function.

To prove that $\{f_1, \ldots, f_5\}$ are optimal we use of the result obtained by Berlekamp and Welch in [4], which provides us with the weight distributions of the cosets of $\mathcal{R}(1,5)$. Table I given in [4] contains the information we need.

Since we are only interested in $0-1$ balanced functions, we can ignore the odd cosets given in that table. In view of the remarks made in section 3, we can read off the nonlinearity of a coset representative from the weight distribution given in Table I. We look for those cosets whose minimal weight is 10 and which have a non-zero entry in the column labelled by weight 16. Since only these coset representatives can be made $0-1$ balanced by the addition of a suitable element of $\mathcal{R}(1,5)$.

These considerations eliminate all but 11 cosets, i.e. those whose coset representatives are: $2345 \oplus 23 \oplus 45$; $2345 \oplus 12 \oplus 34$; $2345 \oplus 123 \oplus 45$; $2345 \oplus 123 \oplus 12 \oplus 34$; $2345 \oplus 123 \oplus 14 \oplus 35$; $2345 \oplus 123 \oplus 24 \oplus 35$; $2345 \oplus 123 \oplus 145 \oplus 24 \oplus 45$; $2345 \oplus 123 \oplus 145 \oplus 35 \oplus 24$; $123 \oplus 14 \oplus 25$; $123 \oplus 145 \oplus 23 \oplus 24 \oplus 35$; and $12 \oplus 34$. (This notation of Berlekamp and Welch translates to ours by substituting 1 for a, 2 for b, etc.)

Counting of the number of basic operations involved in the algebraic normal forms of the functions given in Table 1, gives the following tally: f_1 – 7 operations; f_2 – 9 operations; f_3 – 11 operations; f_4 – 8 operations; and f_5 – 11 operations.

Now we examine the 11 coset representatives of Table I of [4] more closely. We find for example, that $123 \oplus 14 \oplus 25 \oplus 3$, and $2345 \oplus 12 \oplus 34 \oplus 5$, are $0-1$ balanced but their SAC parameters are 44 and 42 repectively. It is conceivable that the SAC parameters of these two functions could be changed to 40 after a transformation of their variables by some element of $GA(5)$. (SAC is only invariant under the subgroup of $GA(5)$ comprised of the permutations of the variables.) This was checked by computer, and turns out not to be the case. So these functions are rejected on this basis and we are left with 9 coset representatives. The remaining cosets give Boolean functions which either have the same number of basic operations as the functions given in Table 1, or more.

Thus we have established the following theorem.

Theorem 9. *The five Boolean functions $\{f_1, \ldots, f_5\}$ given in Table 1, are optimal.*

6 Conclusions

We have considered the linear nonequivalence property of Boolean functions. It turns out that finding a collection of linearly nonequivalent cryptographically strong Boolean functions is a difficult problem. For example the well-known hashing algorithm MD5 (see [26]) applies a collection of 4 Boolean functions of which one is linear and the rest are linearly equivalent. All the functions are defined with three Boolean variables. It can be easily proved that it is impossible to do better than the designers of MD5 if the choice is restricted to functions of three Boolean variables.

In this paper we have shown that for Boolean functions of five variables it is possible to find five linearly nonequivalent functions which are cryptographically strong. They can be used to design a faster version of the HAVAL hashing algorithm (see [30]).

ACKNOWLEDGMENT

The authors wish to thank Jennifer Seberry for her comments and corrections, Yuliang Zheng for his critical discussions, and the members of the Crypto group for their constant support and help.

References

1. C. Adams. On immunity against Biham and Shamir's differential cryptanalysis. *Information Processing Letters*, 41:77–80, 1992.
2. C. Adams and S. Tavares. The structured design of cryptographically good S-boxes. *Journal of Cryptology*, 3:27–41, 1990.
3. M. Beale and M.F. Monaghan. Encryption using random boolean functions. Cryptography and Coding (H. Beker and F. Piper Eds), 1989.
4. E.R. Berlekamp and L.R. Welch. Weight distribution of the cosets of the (32, 6) Reed-Muller code. *IEEE Transactions on Information Theory*, IT-18(1):203–207, 1972.
5. E. Biham and A. Shamir. Differential cryptanalysis of DES-like cryptosystems. *Journal of Cryptology*, 4(1):3–72, 1991.
6. E. Biham and A. Shamir. Differential cryptanalysis of Snefru, Khafre, REDOC-II, LOKI and Lucifer. *Proceedings of CRYPTO'91, Lecture Notes in Computer Science, Advances in Cryptology*, 576:156–171, 1992.
7. L. Brown and J. Seberry. On the design of permutation P in DES type cryptosystems. *Proceedings of EUROCRYPT'89, Lecture Notes in Computer Science, Advances in Cryptology*, 434, 1989.
8. L. Brown and J. Seberry. Key scheduling in DES type cryptosystems. *Advances in Cryptology - AUSCRYPT'90, Lecture Notes in Computer Science*, 453:176–183, 1990.
9. W. Burnside. *Theory of Groups of Finite Order*. Dover Publications, second edition, New York, 1955.
10. D. Chaum and J.H. Evertse. Cryptanalysis of DES with a reduced number of rounds. *Proceedings of CRYPTO'85, Lecture Notes in Computer Science, Advances in Cryptology*, 218:192–211, 1986.
11. J. Detombe and S. Tavares. Constructing large cryptographically strong S-boxes. Preprint, Department of Electrical Engineering, Queen's University at Kingston, April 1992.
12. J. Detombe and S. Tavares. Constructing near-bent boolean functions of five variables. Technical Report, Department of Electrical Engineering, Queen's University at Kingston, April 1992.
13. H. Feistel. Cryptography and computer privacy. *Scientific American*, 228(5):15–23, 1973.
14. R. Forre. The strict avalanche criterion: spectral properties of boolean functions and an extended definition. *Proceedings of CRYPTO'88, Lecture Notes in Computer Science, Advances in Cryptology*, 403:450–468, 1989.

15. J. Kam and G. Davida. Structured design of substitution-permutation networks. *IEEE Transactions on Computers*, C-28:747–753, 1979.
16. F.J. MacWilliams and N.J.A. Sloane. *The theory of error-correcting codes*. North-Holland, Amsterdam, 1977.
17. W. Meier and O. Staffelbach. Nonlinearity criteria for cryptographic functions. *Proceedings of EUROCRYPT'89, Lecture Notes in Computer Science, Advances in Cryptology*, 434:549–562, 1989.
18. NBS. Data Encryption Standard DES. FIPS PUB46, US National Bureau of Standards,Washington, DC, January 1977.
19. K. Nyberg. Constructions of bent functions and difference sets. In *Advances in Cryptology - EUROCRYPT'90, Lecture Notes in Computer Science, Vol.473*, pages 151–160. Springer Verlag, May 1990.
20. K. Nyberg. Perfect nonlinear S-boxes. In *Advances in Cryptology - EUROCRYPT'91, Lecture Notes in Computer Science, Vol.547*, pages 378–386. Springer Verlag, 1991.
21. K. Nyberg. On the construction of highly nonlinear permutations. In *Extended Abstracts - Eurocrypt'92*, pages 89–94, May 1992.
22. L. O'Connor. An analysis of product ciphers based on the properties of boolean functions. PhD thesis, the University of Waterloo, 1992. Waterloo, Ontario, Canada.
23. J. Pieprzyk and G. Finkelstein. Towards effective nonlinear cryptosystem design. *IEE Proceedings-E, Computers and Digital Techniques*, 135(6):325–335, November 1988.
24. J.P. Pieprzyk. On bent permutations. In *Proceedings of the International Conference on Finite Fields, Coding Theory, and Advances in Communications and Computing, Las Vegas*, August 1991.
25. B. Preneel, W. Van Leewijck, L. Van Linden, R. Govaerts, and J. Vandewalle. Propagation characteristics of boolean functions. In *Advances in Cryptology - EUROCRYPT'90, Lecture Notes in Computer Science, Vol.473*, pages 161–173. Springer Verlag, May 1990.
26. R. Rivest. The MD5 message digest algorithm. Request for Comments, RFC 1321, 1992.
27. C. Shannon. Communication theory of secrecy systems. *Bell Systems Technical Journal*, 28:656–715, 1949.
28. D. Slepian. On the number of symmetry types of boolean functions of n variables. *Canadian Journal of Mathematics*, 5:185–193, 1953.
29. A.F. Webster and S.E. Tavares. On the design of S-boxes. In *Lecture Notes in Computer Science, Advances in Cryptology, Proceedings of Crypto'85*, pages 523–534. Springer-Verlag, 1985.
30. Y. Zheng, J. Pieprzyk, and J. Seberry. HAVAL - A one-way hashing algorithm with variable length of output. Abstracts of AUSCRYPT'92, Gold Coast, Australia, December 1992.

Constructing Large Cryptographically Strong S-boxes[1]

John Detombe and Stafford Tavares
Department of Electrical Engineering
Queen's University at Kingston
Kingston, Ontario, Canada. K7L 3N6

Phone: (613) 545-2925
FAX: (613) 545-6615
email: tavares@ee.queensu.ca

Abstract - While there is evidence that large substitution boxes (S-boxes) have better cryptographic properties than small S-boxes, they are much harder to design. The difficulty arises from the relative scarcity of suitable boolean functions as the size of the S-box increases. We describe the construction of cryptographically strong 5x5 S-boxes using near-bent boolean functions of five variables. These functions, where the number of variables is odd, possess highly desirable cryptographic properties and can be generated easily and systematically. Moreover, the S-boxes they compose are shown to satisfy all the important design criteria. Further, we feel that it is possible to generalize near-bent functions to any odd number of variables, thereby making construction of yet larger S-boxes feasible.

1. Introduction

Substitution boxes (S-boxes) provide a block cipher with the *confusion* property described by Shannon in his classic paper [1]. They form the only nonlinear component of the cipher. Consequently for substitution-permutation (DES-like) ciphers to be cryptographically strong, so must its S-boxes. An S-box is *strong* if various criteria are satisfied. Further, research indicates that large S-boxes are stronger than small S-boxes [2, 3, 4, 5]. Since the Data Encryption Standard (DES) was introduced in 1977 [6], a great deal of research has been conducted into the formulation of these criteria and the efficient construction of small S-boxes (i.e., 4x4) that meet them. However, for an $m x n$ S-box, when m becomes large, i.e., $m \geq 5$, the previously known construction techniques used for small S-boxes are no longer efficient.

2. Background

An S-box is a transformation of m input bits to n output bits where m is not necessarily equal to n (see Figure 1). Each output bit corresponds to the output of a boolean function of

1 This work was partially supported by a grant from the Natural Sciences and Engineering Research Council of Canada

the m input bits such that

$$y_i = f_i(x_1, x_2, ..., x_m), \quad 1 \leq i \leq n.$$

Alternatively, we may write $f : Z_2^m \rightarrow Z_2$.

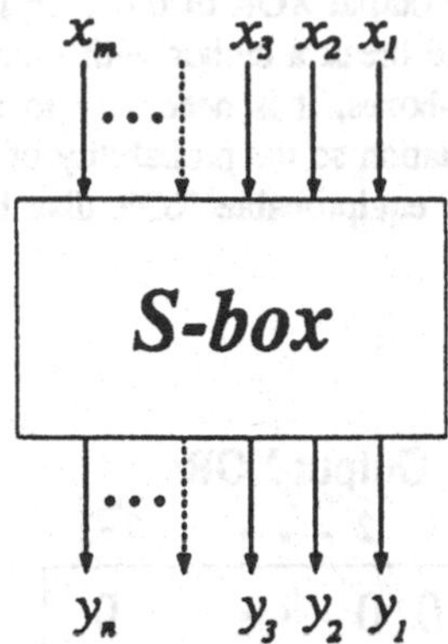

Figure 1. An *mxn* S-box

If $m = n$, the S-box is square and can be bijective (each input maps to a unique output). This is the case with the 4x4 DES sub S-boxes where $m=n=4$. These S-boxes have been intensively studied and have withstood numerous attempts at the identification of cryptographic weaknesses. However, regardless of the cryptographic strength of DES, as technology advances, the computing power required to exhaustively break DES with its 56-bit key becomes increasingly available. Moreover, the success of Biham and Shamir with their differential attack [7, 8] of DES demonstrates that the need for a replacement for the Data Encryption Standard is rapidly growing.

Since the introduction of DES and the subsequent controversy regarding its design criteria, much research has focussed on what makes a good block cipher. As a result several criteria have evolved that indicate the cryptographic strength of a cryptosystem and its component parts. Certain criteria apply to individual boolean functions while others apply to entire S-boxes. The former include nonlinearity [9, 10], completeness [11], Strict Avalanche Criterion (SAC) [12], higher-order SAC [13, 14, 15], and 0-1 balancedness; while the latter include avalanche [16], the Bit Independence Criterion (BIC) [12, 14] and the input/output XOR table distribution [7, 17].

Ideally, each boolean function should behave independently of the others that make up an S-box. This prevents a cryptanalyst from predicting how any function will behave based upon the knowledge of any other function. Two criteria in particular provide the S-box designer with an indication of how the functions perform together. These are the BIC and the input/output XOR table distribution. The former states that for any S-box input change, the output changes should be pairwise independent for any pair of boolean functions and is usually measured in the form of a correlation. The latter criterion looks at the distribution of the S-box's XOR table entries.

The pairs XOR distribution table is a 2^mx2^n matrix where rows represent all possible input changes, or XORs, and columns represent all possible output changes, or XORs.

The table entries are the number of times a particular output XOR occurs for a given input XOR. Differential cryptanalysis uses imbalances in the occurrences of output XORs to obtain information about the key. Intuitively, it would seem that for an S-box or even an entire cryptosystem to resist the differential attack, the XOR distribution should appear as equiprobable as possible. However, recent work has shown that this is not necessarily the case [18, 19]. Since all output XORs appear equiprobably, there is a probability of $1/2^n$ that for any given input XOR (aside from 0) an output XOR of 0 can be found. Differential cryptanalysis can make use of this information to break a cipher with much less work than an exhaustive search. Therefore, in designing S-boxes, it is necessary to avoid large variations in output XORs but yet maintain enough variation so the probability of specific output XORs occurring is minimized. Figure 2 shows the equiprobable XOR distribution of a bijective S-box that meets these requirements.

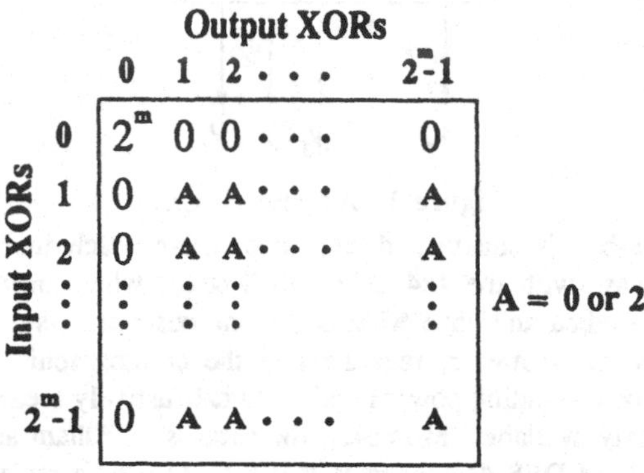

Figure 2 An equiprobable XOR distribution table for a bijective S-box.

An important tool emerging in S-box design is information theory [20, 21]. The amount of mutual information between the inputs and outputs $I(X;Y)$ of an S-box gives an indication of the strength of the S-box. As $I(X;Y)$ increases, so does the information about the input that can be obtained from the output (or vice versa). An entire framework for S-box design based on information theory was presented in [21]. Included in this framework are several criteria that are useful in determining the strength of an S-box both statically (inputs and outputs fixed) and dynamically (inputs and outputs changing).

3 Construction strategy

Since, generally, all the nonlinearity within a block cipher is derived from its substitution boxes, care must be exercised in their construction. Further, as large S-boxes are thought to be inherently stronger that small boxes, the ability to design and construct large, cryptographically strong S-boxes would be useful. However, large S-boxes are difficult to design because the search space for suitable boolean functions increases as 2^{2^m}. As a result, the percentage of cryptographically strong boolean functions within that search space decreases exponentially as m increases [2, 9]. In [22], a construction technique is described that easily and quickly generates a class of cryptographically strong boolean functions of five variables that we call *near-bent*. In this paper, we will use these functions to construct 5x5 S-boxes that possess very good cryptographic properties.

4 Near-Bent Boolean Functions

The desirable cryptographic properties of bent functions are well known [10, 23, 24]. They are maximally nonlinear and satisfy all orders of SAC[2]. However since bent functions are not 0-1 balanced, they cannot be used for constructing bijective S-boxes. In [25], Pieprzyk introduced the notion of bent permutations. In that paper it was shown that boolean functions of the form

$$x_1x_2 \oplus \ldots \oplus x_{m-1}x_m \oplus x_mx_1\ ,\ m \text{ odd}$$

are maximally nonlinear and are 0–1 balanced. Our investigation of the cryptographic properties of boolean functions of this form revealed some remarkable similarities to bent functions. Besides being maximally nonlinear, they also satisfy all orders of SAC except for one particular input change which causes the output to change with probability 1.0. Further, we have found that boolean functions of odd m that possess these properties are not restricted to the form shown above. An analysis of the algebraic normal form of these types of functions has produced an entire class of boolean functions that we will call *near-bent*.

Definition 1

A boolean function $f(x_1, x_2, \ldots, x_m)$, m odd, is near-bent if:

a.) it is maximally nonlinear,
b.) its algebraic normal form has nonlinear order 2 (quadratic) , and
c.) every input variable appears in at least one product term (nondegenerative).

In [26], Nyberg described the class of quadratic functions which are maximally nonlinear. These functions coincide with our near-bent functions. Further, Nyberg's results are a generalization of Pieprzyk's paper [25].

Example 1:

The following boolean functions, all of five variables, are near-bent:

a.) $x_2 \oplus x_1x_2 \oplus x_1x_3 \oplus x_4x_5$,
b.) $x_1x_2 \oplus x_1x_3 \oplus x_2x_3 \oplus x_4x_5$, and
c.) $x_3 \oplus x_1x_2 \oplus x_1x_3 \oplus x_1x_4 \oplus x_2x_3 \oplus x_2x_5$.

All three near–bent functions depicted in the above example are from different *equivalence classes*. We define an equivalence class on boolean functions as the set of functions derived from an algebraic normal form (anf) that has been reduced by removal of all linear terms. We call this the *generator* of the class. All functions within an equivalence class have the same nonlinearity and the same nonlinear order (*ord*). Functions are derived from a generator by the addition of linear terms and/or by permuting the input variables. For functions of five variables, we have exhaustively identified a total of 19 equivalence classes that produce near-bent functions. Table 1 lists the generator functions for these equivalence classes. Note

[2] Here we use Adams' definition [23] of higher order SAC (also known as propagation criterion [15]) rather than Forré's [13].

that no combination of linear terms added to or any permutation of the input variables of one generator function will result in a function derived from a different generator function.

Function	Distribution of Input Variables	Generator Functions
f_1	21111	$x_1x_2 \oplus x_1x_3 \oplus x_4x_5$
f_2	22211	$x_1x_2 \oplus x_1x_3 \oplus x_2x_3 \oplus x_4x_5$
f_3	22211	$x_1x_2 \oplus x_1x_3 \oplus x_2x_4 \oplus x_3x_5$
f_4	22222	$x_1x_2 \oplus x_2x_3 \oplus x_3x_4 \oplus x_4x_5 \oplus x_1x_5$
f_5	32111	$x_1x_2 \oplus x_1x_3 \oplus x_1x_4 \oplus x_2x_5$
f_6	32221	$x_1x_2 \oplus x_1x_3 \oplus x_1x_4 \oplus x_2x_3 \oplus x_4x_5$
f_7	32221	$x_1x_2 \oplus x_1x_3 \oplus x_1x_5 \oplus x_2x_4 \oplus x_3x_4$
f_8	33211	$x_1x_2 \oplus x_1x_3 \oplus x_1x_4 \oplus x_2x_3 \oplus x_2x_5$
f_9	33222	$x_1x_2 \oplus x_1x_3 \oplus x_1x_4 \oplus x_2x_4 \oplus x_2x_5 \oplus x_3x_5$
f_{10}	33321	$x_1x_2 \oplus x_1x_3 \oplus x_1x_4 \oplus x_2x_3 \oplus x_2x_4 \oplus x_3x_5$
f_{11}	33332	$x_1x_2 \oplus x_1x_3 \oplus x_1x_4 \oplus x_2x_3 \oplus x_2x_4 \oplus x_3x_5 \oplus x_4x_5$
f_{12}	42211	$x_1x_2 \oplus x_1x_3 \oplus x_1x_4 \oplus x_1x_5 \oplus x_2x_3$
f_{13}	42222	$x_1x_2 \oplus x_1x_3 \oplus x_1x_4 \oplus x_1x_5 \oplus x_2x_3 \oplus x_4x_5$
f_{14}	43221	$x_1x_2 \oplus x_1x_3 \oplus x_1x_4 \oplus x_1x_5 \oplus x_2x_3 \oplus x_2x_4$
f_{15}	43322	$x_1x_2 \oplus x_1x_3 \oplus x_1x_4 \oplus x_1x_5 \oplus x_2x_3 \oplus x_2x_4 \oplus x_3x_5$
f_{16}	43331	$x_1x_2 \oplus x_1x_3 \oplus x_1x_4 \oplus x_1x_5 \oplus x_2x_3 \oplus x_2x_4 \oplus x_3x_4$
f_{17}	44332	$x_1x_2 \oplus x_1x_3 \oplus x_1x_4 \oplus x_1x_5 \oplus x_2x_3 \oplus x_2x_4 \oplus x_2x_5 \oplus x_3x_4$
f_{18}	44433	$x_1x_2 \oplus x_1x_3 \oplus x_1x_4 \oplus x_1x_5 \oplus x_2x_3 \oplus x_2x_4 \oplus x_2x_5 \oplus x_3x_4 \oplus x_3x_5$
f_{19}	44444	$x_1x_2 \oplus x_1x_3 \oplus x_1x_4 \oplus x_1x_5 \oplus x_2x_3 \oplus x_2x_4 \oplus x_2x_5 \oplus x_3x_4 \oplus x_3x_5 \oplus x_4x_5$

Table 1 Generator Functions for the equivalence classes of the near-bent functions of 5 variables

In order to ensure that the boolean functions making up an S-box are not linearly related, we will specify that no two or more functions composing the S-boxes we construct be from the same equivalence class.

Preneel et al [15] present results which characterize higher order SAC properties of quadratic functions. Tables 2 and 3 show how balanced functions derived from each of the 19 equivalence classes respond to all possible input bit changes (all orders of SAC). The columns give the probability that each function's output bit will change. The row headings indicate the input bits changed

	f1	f2	f3	f4	f5	f6	f7	f8	f9	f10
x1	0.50	0.50	0.50	0.50	0.50	0.50	0.50	0.50	0.50	0.50
x2	0.50	0.50	0.50	0.50	0.50	0.50	0.50	0.50	0.50	0.50
x3	0.50	0.50	0.50	0.50	0.50	0.50	0.50	0.50	0.50	0.50
x4	0.50	0.50	0.50	0.50	0.50	0.50	0.50	0.50	0.50	0.50
x5	0.50	0.50	0.50	0.50	0.50	0.50	0.50	0.50	0.50	0.50
x1x2	0.50	0.50	0.50	0.50	0.50	0.50	0.50	0.50	0.50	0.50
x1x3	0.50	0.50	0.50	0.50	0.50	0.50	0.50	0.50	0.50	0.50
x2x3	1.00	0.50	0.50	0.50	0.50	0.50	1.00	0.50	0.50	0.50
x1x4	0.50	0.50	0.50	0.50	0.50	0.50	0.50	0.50	0.50	0.50
x2x4	0.50	0.50	0.50	0.50	0.50	0.50	0.50	0.50	0.50	0.50
x3x4	0.50	0.50	0.50	0.50	1.00	0.50	0.50	0.50	0.50	0.50
x1x5	0.50	0.50	0.50	0.50	0.50	0.50	0.50	0.50	0.50	0.50
x2x5	0.50	0.50	0.50	0.50	0.50	0.50	0.50	0.50	0.50	0.50
x3x5	0.50	0.50	0.50	0.50	0.50	0.50	0.50	0.50	0.50	0.50
x4x5	0.50	0.50	0.50	0.50	0.50	0.50	0.50	0.50	0.50	0.50
x1x2x3	0.50	1.00	0.50	0.50	0.50	0.50	0.50	0.50	0.50	0.50
x1x2x4	0.50	0.50	0.50	0.50	0.50	0.50	0.50	0.50	0.50	1.00
x1x3x4	0.50	0.50	0.50	0.50	0.50	0.50	0.50	0.50	0.50	0.50
x2x3x4	0.50	0.50	0.50	0.50	0.50	0.50	0.50	0.50	0.50	0.50
x1x2x5	0.50	0.50	0.50	0.50	0.50	0.50	0.50	0.50	0.50	0.50
x1x3x5	0.50	0.50	0.50	0.50	0.50	0.50	0.50	0.50	0.50	0.50
x2x3x5	0.50	0.50	0.50	0.50	0.50	0.50	0.50	0.50	0.50	0.50
x1x4x5	0.50	0.50	1.00	0.50	0.50	0.50	0.50	0.50	0.50	0.50
x2x4x5	0.50	0.50	0.50	0.50	0.50	0.50	0.50	0.50	0.50	0.50
x3x4x5	0.50	0.50	0.50	0.50	0.50	0.50	0.50	1.00	0.50	0.50
x1x2x3x4	0.50	0.50	0.50	0.50	0.50	0.50	0.50	0.50	0.50	0.50
x1x2x3x5	0.50	0.50	0.50	0.50	0.50	1.00	0.50	0.50	1.00	0.50
x1x2x4x5	0.50	0.50	0.50	0.50	0.50	0.50	0.50	0.50	0.50	0.50
x1x3x4x5	0.50	0.50	0.50	0.50	0.50	0.50	0.50	0.50	0.50	0.50
x2x3x4x5	0.50	0.50	0.50	0.50	0.50	0.50	0.50	0.50	0.50	0.50
x1x2x3x4x5	0.50	0.50	0.50	1.00	0.50	0.50	0.50	0.50	0.50	0.50

Table 2 Output Bit change probabilities for functions derived from generator functions f_1 to f_{10} in Table 1

	f11	f12	f13	f14	f15	f16	f17	f18	f19
x1	0.50	0.50	0.50	0.50	0.50	0.50	0.50	0.50	0.50
x2	0.50	0.50	0.50	0.50	0.50	0.50	0.50	0.50	0.50
x3	0.50	0.50	0.50	0.50	0.50	0.50	0.50	0.50	0.50
x4	0.50	0.50	0.50	0.50	0.50	0.50	0.50	0.50	0.50
x5	0.50	0.50	0.50	0.50	0.50	0.50	0.50	0.50	0.50
x1x2	0.50	0.50	0.50	0.50	0.50	0.50	0.50	0.50	0.50
x1x3	0.50	0.50	0.50	0.50	0.50	0.50	0.50	0.50	0.50
x2x3	0.50	0.50	0.50	0.50	0.50	0.50	0.50	0.50	0.50
x1x4	0.50	0.50	0.50	0.50	0.50	0.50	0.50	0.50	0.50
x2x4	0.50	0.50	0.50	0.50	0.50	0.50	0.50	0.50	0.50
x3x4	1.00	0.50	0.50	1.00	0.50	0.50	0.50	0.50	0.50
x1x5	0.50	0.50	0.50	0.50	0.50	0.50	0.50	0.50	0.50
x2x5	0.50	0.50	0.50	0.50	0.50	0.50	0.50	0.50	0.50

Table 3 Output Bit change probabilities for functions derived from generator functions f_{11} to f_{19} in Table 1 (Continued . . .)

x3x5	0.50	0.50	0.50	0.50	0.50	0.50	0.50	0.50	0.50
x4x5	0.50	**1.00**	0.50	0.50	0.50	0.50	0.50	**1.00**	0.50
x1x2x3	0.50	0.50	0.50	0.50	**1.00**	0.50	0.50	0.50	0.50
x1x2x4	0.50	0.50	0.50	0.50	0.50	0.50	0.50	0.50	0.50
x1x3x4	0.50	0.50	0.50	0.50	0.50	0.50	0.50	0.50	0.50
x2x3x4	0.50	0.50	0.50	0.50	0.50	0.50	0.50	0.50	0.50
x1x2x5	0.50	0.50	0.50	0.50	0.50	0.50	**1.00**	0.50	0.50
x1x3x5	0.50	0.50	0.50	0.50	0.50	0.50	0.50	0.50	0.50
x2x3x5	0.50	0.50	0.50	0.50	0.50	0.50	0.50	0.50	0.50
x1x4x5	0.50	0.50	0.50	0.50	0.50	0.50	0.50	0.50	0.50
x2x4x5	0.50	0.50	0.50	0.50	0.50	0.50	0.50	0.50	0.50
x3x4x5	0.50	0.50	0.50	0.50	0.50	0.50	0.50	0.50	0.50
x1x2x3x4	0.50	0.50	0.50	0.50	0.50	0.50	0.50	0.50	0.50
x1x2x3x5	0.50	0.50	0.50	0.50	0.50	0.50	0.50	0.50	0.50
x1x2x4x5	0.50	0.50	0.50	0.50	0.50	0.50	0.50	0.50	0.50
x1x3x4x5	0.50	0.50	0.50	0.50	0.50	0.50	0.50	0.50	0.50
x2x3x4x5	0.50	0.50	0.50	0.50	0.50	**1.00**	0.50	0.50	0.50
x1x2x3x4x5	0.50	0.50	**1.00**	0.50	0.50	0.50	0.50	0.50	**1.00**

Table 3 Output Bit change probabilities for functions derived from generator functions f_{11} to f_{19} in Table 1

Not all generator functions produce 0–1 balanced functions directly. It is sometimes necessary to add linear terms (as in Examples 1a and 1c) to obtain balanced functions. Preneel observed in [27] that the addition of appropriate linear terms to a boolean function with at least one zero in its Walsh spectrum will render that function balanced. Recall that the addition of linear terms to a function will not affect its nonlinearity or SAC properties. Motivated by these observations, we used the locations of the Walsh spectrum zeroes to provide us with the information we need to determine exactly what linear term combinations will balance the function.

Recall that if $f(\mathbf{x})$, $\mathbf{x} \in Z_2^m$ is any real valued function, the Walsh transform of $f(\mathbf{x})$ is defined as:

$$F(\mathbf{w}) = \sum_{\mathbf{x} \in Z_2^m} f(\mathbf{x}) \cdot (-1)^{\mathbf{x} \cdot \mathbf{w}}$$

where $\mathbf{w} \in Z_2^m$ and $\mathbf{x} \cdot \mathbf{w}$ denotes the dot-product of $\mathbf{x}$ and $\mathbf{w}$, which is, in turn, defined as

$$\mathbf{x} \cdot \mathbf{w} = x_{m-1} w_{m-1} \oplus \ldots \oplus x_0 w_0$$

The following proofs result from [28].

Lemma 1

If $\sum_{\mathbf{x} \in \mathbf{X}} \hat{f}(\mathbf{x}) \hat{g}(\mathbf{x}) = 0$, where $\hat{f}(x) = (-1)^{f(x)}$, then $\sum_{\mathbf{x} \in \mathbf{X}} f(\mathbf{x}) \oplus g(\mathbf{x}) = \frac{1}{2}|\mathbf{X}|$.

Proof

$$\text{Let } \sum_{\mathbf{x}\in\mathbf{X}} \hat{f}(\mathbf{x})\hat{g}(\mathbf{x}) = 0$$

$$\Rightarrow \#\left\{\mathbf{x}|\hat{f}(\mathbf{x}) = \hat{g}(\mathbf{x})\right\} = \#\left\{\mathbf{x}|\hat{f}(\mathbf{x}) \neq \hat{g}(\mathbf{x})\right\} = \frac{1}{2}|\mathbf{X}|$$

$$\Rightarrow \#\{\mathbf{x}|f(\mathbf{x}) = g(\mathbf{x})\} = \#\{\mathbf{x}|f(\mathbf{x}) \neq g(\mathbf{x})\} = \frac{1}{2}|\mathbf{X}|$$

$$\Rightarrow \sum_{\mathbf{x}\in\mathbf{X}} f(\mathbf{x}) \oplus g(\mathbf{x}) = \frac{1}{2}|\mathbf{X}|.$$

□

Theorem 1:

Let $f(\mathbf{x}) : Z_2^m \to Z_2$ be a Boolean function where $\hat{F}(\mathbf{w})$ is the Walsh transform of $\hat{f}(\mathbf{x})$. If

$$\hat{F}(\mathbf{w}) = 0 \text{ for } \mathbf{w} = w_{m-1}w_{m-2}\ldots w_0$$

then

$$f(\mathbf{x}) \oplus w_{m-1}x_{m-1} \oplus w_{m-2}x_{m-2} \oplus \ldots \oplus w_0x_0$$

is 0–1 balanced.

Proof

Let $g(\mathbf{x}) = \mathbf{x}\cdot\mathbf{w} = w_{m-1}x_{m-1} \oplus \ldots \oplus w_0x_0$.

If $\hat{F}(\mathbf{w}) = 0$

$$\Rightarrow \sum_{\mathbf{x}\in Z_2^m} \hat{f}(\mathbf{x})(-1)^{\mathbf{x}\cdot\mathbf{w}} = 0$$

$$\Rightarrow \sum_{\mathbf{x}\in Z_2^m} \hat{f}(\mathbf{x})(-1)^{g(\mathbf{x})} = \sum_{\mathbf{x}\in Z_2^m} \hat{f}(\mathbf{x})\hat{g}(\mathbf{x}) = 0$$

and finally by Lemma 3.1

$$\Rightarrow \sum_{\mathbf{x}\in Z_2^m} f(\mathbf{x}) \oplus g(\mathbf{x}) = \frac{1}{2}|2^m| = 2^{m-1}$$

□

Example 2:

Take f_1 from Table 1:

$$f_1(x) = x_1x_2 \oplus x_1x_3 \oplus x_4x_5$$

Now

$$\hat{F}_1(00010) = 0$$

which indicates that

$$f(x) = x_2 \oplus x_1x_2 \oplus x_1x_3 \oplus x_4x_5$$

is 0–1 balanced.

5 Constructing S-boxes

The 19 equivalence classes described in §4 provide an abundant source of near-bent boolean functions of five variables. If five of these functions, each from a different equivalence class, are selected such that the resulting S-box is bijective (invertible), the S-box is guaranteed to satisfy all the individual criteria (nonlinearity, SAC, etc). However, to ensure the S-box as a whole is cryptographically strong, further restrictions must be placed on the function selection. For example, while the XOR distribution should not be completely flat, neither should it be widely varied. Here, we shall show it is possible to design an S-box where a suitable trade-off exists in its XOR distribution that we feel sufficiently resists the differential attack.

5.1 Input/Output XOR Distribution

Due to the symmetrical nature of the exclusive-or operation, the entries of an XOR table are always non-negative even-valued integers. Since the entries in each of the 2^m rows sum to 2^m and there are 2^n possible output XORs, it follows that the entries of an equiprobable XOR table (for non-zero input XORs) must be $2^m/2^n$. However, for a square S-box $m = n$ and $2^m/2^n = 1$. Since no entry for a square S-box can be 1, the most equiprobable its XOR table can be is for the entries in each row to consist of 2^{m-1} zeroes and 2^{m-1} twos. For purposes of this paper, since we are constructing square, bijective S-boxes, we will refer to XOR tables of this type as equiprobable (see Figure 2).

Clearly, when the input XOR is 0, the output XOR is always 0 and there are no occurrences of any other output XOR. With an invertible or bijective S-box, the inverse is also true. This means in order to obtain output XORs of 0, the input XOR is always 0 and there are no occurrences of any other input XOR. This also means that bijective S-boxes are not susceptible to the weakness described earlier in the paper since there is no probability of an input XOR of 0 occurring (unless, of course one chooses an output XOR of 0). Our aim then is to design a bijective S-box with an equiprobable XOR distribution.

It is computationally inefficient to randomly construct an S-box that has equiprobable XOR table entries. However, in [17], a theorem was presented and proved which will be used here. The theorem states that if all the boolean functions that make up an S-box are bent and all non-zero linear combinations of these functions are also bent functions, the S-box will have equiprobable XOR table entries. Clearly we cannot use this theorem directly for our purpose because we are dealing with near-bent functions. However, a modified form of this theorem is presented and proved here that ensures the construction of S-boxes with near-bent functions which have an equiprobable XOR table. Before stating and proving our theorem, we restate two lemmas presented and proved in [17].

Consider a binary mapping of m input variables to n output variables, m not necessarily equal to n. This mapping may be represented in the form of a binary matrix 2^m rows long by n columns wide.

Lemma 2:

A 2^mxn binary matrix where $m \geq n$ will have 2^{m-n} occurrences of each row if and only if all nonzero linear combinations (modulo 2) of the columns have Hamming weight 2^{m-1}.

Lemma 3:

A 2^mxn binary matrix where $m < n$ will have a single occurrence of each row if there exist m columns such that all nonzero linear combinations (modulo 2) of these m columns have Hamming weight 2^{m-1}.

Let $\mathfrak{F}$ be the set of near-bent boolean functions of five variables (i.e., $\mathfrak{F}$ is all the boolean functions that can possibly be derived from the 19 equivalence classes defined in [22]).

Given a boolean function $F \in \mathfrak{F}$, let $\mathfrak{X}$ be the set of input XORs

$$\mathfrak{X} = \{\Delta x_1, \Delta x_2, \ldots, \Delta x_{2^m-1}\}$$

of F where the output changes with probability 0.5. Let $\Delta\hat{x}$ be the single input XOR of F where the output changes with probability 1.0.

We will use the following notation from [17]:

a.) Let ϕ_i represent the outputs of a boolean function $F_i \in \mathfrak{F}$, $1 \leq i \leq n$, listed columnwise that correspond to all possible inputs arranged in lexicographic order.

b.) Let $v_i = \phi_i^\alpha \oplus \phi_i^\beta$, where the jth bit in v_i is the sum modulo 2 of bits ϕ_{ij}^α and ϕ_{ij}^β, represent the avalanche vector of ϕ_i .

Recall that avalanche vectors v are defined [12] as:

$$v = F(X) \oplus F(X \oplus X_j),\ 1 \leq j \leq m$$

where X and X_j differ only in bit j.

For example, if $j = m$ (MSB) then, ϕ_i^α represents the top half of ϕ_i while ϕ_i^β represents the bottom half.

Theorem 2:

An mxn S-box S composed of near-bent boolean functions and represented as a $2^m xn$ binary matrix with columns ϕ_i will have equiprobable output XORs if:

a.) all nonzero linear combinations (modulo 2) of the ϕ_i result in near-bent functions, and

b.) all $\Delta\hat{x}$ are different, $\Delta\hat{x}_i \neq \Delta\hat{x}_j$, $1 \leq i, j \leq n$, $i \neq j$.

Proof:

Let $\mathcal{V}$ represent the n avalanche vectors v_i of the boolean functions that compose S for a particular input XOR $\in \mathfrak{X}$. $\mathcal{V}$ is in the form of a $2^{m-1}xn$ binary matrix.

We will use the notation $\Sigma a_i y_i$ to represent a linear combination of boolean functions, $y_i \in \mathfrak{F}$, $a_i \in \{0,1\}$, not all $a_i = 0$.

Given an S-box S composed of n boolean functions $\in \mathfrak{F}$, there are n input XORs $\notin \mathfrak{X}$, such that for the remaining $2^m - n - 1$ input XORs, the following is true:

$$\begin{aligned}(\Sigma a_i y_i) \in \mathfrak{F} &\Rightarrow \mathrm{wt}\left[(\Sigma a_i y_i)^\alpha \oplus (\Sigma a_i y_i)^\beta\right] = 2^{m-2} \\ &\Rightarrow \mathrm{wt}\left[(\Sigma a_i y_i^\alpha) \oplus \left(\Sigma a_i y_i^\beta\right)\right] = 2^{m-2} \\ &\Rightarrow \mathrm{wt}\left[\Sigma a_i \left(y_i^\alpha \oplus y_i^\beta\right)\right] = 2^{m-2} \\ &\Rightarrow \mathrm{wt}[\Sigma a_i v_i] = 2^{m-2}\end{aligned}$$

Recalling that $\mathcal{V}$ is a $2^{m-1}xn$ binary matrix; since $m-1<n$, then by Lemma 2, $\mathcal{V}$ contains a single occurrence of each row.

For the remaining n input XORs $\notin \mathfrak{X}$, the weight of one column in $\mathcal{V}$ will be 2^{m-1} (i.e., will be all ones). If this column is ignored, $\mathcal{V}$ can be considered to be a $2^{m-1}x(n-1)$ binary matrix. In this case, $m-1=n-1$ and $\mathcal{V}$ contains a single occurrence of each row by Lemma 1. The XOR distribution is equiprobable. □

In [14] Adams and Tavares prove that a necessary and sufficient condition for an S-box S to be bijective is that any linear combination of the columns of S has Hamming weight 2^{m-1}. If we ensure the near-bent functions we are working with are 0–1 balanced, then this condition is entirely satisfied by Theorem 1. We have therefore obtained our goal of designing an invertible S-box where its XOR table entries are all zeroes and twos.

Example 3:

A 5x5 S-box composed of near-bent boolean functions chosen according to Theorem 1 (in hex notation) is shown below.

1C F 12 1E D 0 14 6 11 17 13 A 8 10 1D 1A 1F 4 1 5 B E 2 18 15 1B 7 16 9 19 C 3

The hex values represent the S-box outputs when the inputs are listed in lexicographic order $(0,1,\ldots,2^n-1)$. Its corresponding XOR distribution table follows:

		OUTPUT XORS																													
		0	1	2	3	4	5	6	7	8	9	A	B	C	D	E	F	10	11	12	13	14	15	16	17	18	19	1A	1B	1C	1D	1E	1F
	0	32	0	0	0	0	0	0	0	0	0	0	0	0	0	0	0	0	0	0	0	0	0	0	0	0	0	0	0	0	0	0	0
	1	0	0	0	0	2	2	2	2	0	0	0	0	2	2	2	2	2	2	2	2	0	0	0	0	2	2	2	2	0	0	0	0
	2	0	2	2	0	0	2	2	0	0	2	2	0	0	2	2	0	0	2	2	0	0	2	2	0	0	2	2	0	0	2	2	0
	3	0	0	2	2	2	2	0	0	0	0	2	2	2	2	0	0	0	0	2	2	2	2	0	0	0	0	2	2	2	2	0	0
	4	0	0	2	2	0	0	2	2	0	0	2	2	0	0	2	2	2	2	0	0	2	2	0	0	2	2	0	0	2	2	0	0
	5	0	2	2	0	2	0	0	2	0	2	2	0	2	0	0	2	0	2	2	0	2	0	0	2	0	2	2	0	2	0	0	2
	6	0	0	0	0	0	0	0	0	2	2	2	2	2	2	2	2	0	0	0	0	0	0	0	0	2	2	2	2	2	2	2	2
	7	0	0	2	2	0	0	2	2	0	0	2	2	0	0	2	2	0	0	2	2	0	0	2	2	0	0	2	2	0	0	2	2
	8	0	2	2	0	0	2	2	0	0	2	2	0	0	2	2	0	2	0	0	2	2	0	0	2	2	0	0	2	2	0	0	2
	9	0	2	2	0	2	0	0	2	2	0	0	2	0	2	2	0	0	2	2	0	2	0	0	2	2	0	0	2	0	2	2	0
I	A	0	2	0	2	0	2	0	2	0	2	0	2	0	2	0	2	2	0	2	0	2	0	2	0	2	0	2	0	2	0	2	0
N	B	0	0	2	2	2	2	0	0	2	2	0	0	0	0	2	2	2	2	0	0	0	0	2	2	0	0	2	2	2	2	0	0
P	C	0	0	0	0	2	2	2	2	0	0	0	0	2	2	2	2	0	0	0	0	2	2	2	2	0	0	0	0	2	2	2	2
U	D	0	0	2	2	0	0	2	2	2	2	0	0	2	2	0	0	2	2	0	0	2	2	0	0	0	0	2	2	0	0	2	2
T	E	0	2	0	2	0	2	0	2	2	0	2	0	2	0	2	0	0	2	0	2	0	2	0	2	2	0	2	0	2	0	2	0
	F	0	0	2	2	0	0	2	2	2	2	0	0	2	2	0	0	0	0	2	2	0	0	2	2	2	2	0	0	2	2	0	0
X	10	0	2	0	2	2	0	2	0	0	2	0	2	2	0	2	0	0	2	0	2	2	0	2	0	0	2	0	2	2	0	2	0
O	11	0	0	2	2	2	2	0	0	0	0	2	2	2	2	0	0	2	2	0	0	0	0	2	2	2	2	0	0	0	0	2	2
R	12	0	2	0	2	2	0	2	0	2	0	2	0	0	2	0	2	0	2	0	2	2	0	2	0	2	0	2	0	0	2	0	2
S	13	0	2	2	0	2	0	0	2	2	0	0	2	0	2	2	0	2	0	0	2	0	2	2	0	0	2	2	0	2	0	0	2
	14	0	2	0	2	2	0	2	0	0	2	0	2	2	0	2	0	2	0	2	0	0	2	0	2	2	0	2	0	0	2	0	2
	15	0	0	0	0	2	2	2	2	2	2	2	2	0	0	0	0	2	2	2	2	0	0	0	0	0	0	0	0	2	2	2	2
	16	0	2	2	0	0	2	2	0	2	0	0	2	2	0	0	2	2	0	0	2	2	0	0	2	0	2	2	0	0	2	2	0
	17	0	2	0	2	2	0	2	0	2	0	2	0	0	2	0	2	2	0	2	0	0	2	0	2	0	2	0	2	2	0	2	0
	18	0	0	2	2	2	2	0	0	2	2	0	0	0	0	2	2	0	0	2	2	2	2	0	0	2	2	0	0	0	0	2	2
	19	0	0	0	0	2	2	2	2	2	2	2	2	0	0	0	0	0	0	0	0	2	2	2	2	2	2	2	2	0	0	0	0
	1A	0	2	0	2	0	2	0	2	2	0	2	0	2	0	2	0	2	0	2	0	2	0	2	0	0	2	0	2	0	2	0	2
	1B	0	0	0	0	0	0	0	0	2	2	2	2	2	2	2	2	2	2	2	2	2	2	2	2	0	0	0	0	0	0	0	0
	1C	0	0	0	0	0	0	0	0	0	0	0	0	0	0	0	0	2	2	2	2	2	2	2	2	2	2	2	2	2	2	2	2
	1D	0	2	2	0	0	2	2	0	2	0	0	2	2	0	0	2	0	2	2	0	0	2	2	0	2	0	0	2	2	0	0	2
	1E	0	2	2	0	2	0	0	2	0	2	2	0	2	0	0	2	2	0	0	2	0	2	2	0	2	0	0	2	0	2	2	0
	1F	0	2	0	2	0	2	0	2	0	2	0	2	0	2	0	2	0	2	0	2	0	2	0	2	0	2	0	2	0	2	0	2

Corollary:

If an *mxn* S-box S is composed of n boolean functions $F_i \in \mathfrak{I}$ where any two or more F_i have a common $\Delta\hat{x}_i$, then S will not have an equiprobable XOR distribution table.

Proof:

Say $\Delta\hat{x}_i$ is common to both F_j and F_k in S. Then for the input XOR $\Delta\hat{x}_i$, $\mathcal{V}$ will have two columns representing $\mathbf{v}_j$ and $\mathbf{v}_k$ that have Hamming weight 2^{m-1} (i.e., consist of all ones). In this case, if these two columns are ignored, $\mathcal{V}$ may be considered a 2^{m-1}x$(n-2)$ binary matrix. Now by Lemma 1, there will be $2^{(m-1)-(n-2)}=2$ occurrences of each row. The XOR distribution is no longer equiprobable. □

5.2 Output Bit Independence Criterion (BIC)

BIC is measured by determining the bitwise correlation between pairs of avalanche vectors for all possible input changes. In [14], it was shown that if the pairwise sum of any two boolean functions that compose an S-box results in a SAC or near-SAC fulfilling function, then that S-box will come close to satisfying BIC. Say S-box S is composed of boolean functions $F_i \in \mathfrak{F}$. Further, these F_i are selected according to Theorem 1. It clearly follows that the pairwise sum of any two of the F_i in S will result in another function $F_j \in \mathfrak{F}$. We know that all functions in $\mathfrak{F}$ satisfy all orders of SAC except for a single input change $\notin \mathfrak{X}$. It follows that, aside for that one input change, the avalanche vectors of the two functions that make up F_j must have a correlation of zero. For the one input change $\notin \mathfrak{X}$, the correlation will be -1.0. Consequently, for an S-box constructed with near-bent functions according to Theorem 1, all possible pairs of avalanche vectors will be bitwise uncorrelated except for one particular input change per pair.

Table 4 illustrates the BIC values for the S-box shown in Example 3. The * entries in the table result from one of the two vectors under test having the same value (1) at every coordinate. This results in a divide-by zero error in the correlation algorithm.

INPUT CHANGES	f1f2	f1f3	f1f4	f1f5	f2f3	f2f4	f2f5	f3f4	f3f5	f4f5
	0.00	0.00	0.00	0.00	0.00	0.00	0.00	0.00	-1.00	0.00
	-1.00	0.00	0.00	0.00	0.00	0.00	0.00	0.00	0.00	0.00
1 bit	0.00	0.00	0.00	0.00	-1.00	0.00	0.00	0.00	0.00	0.00
	0.00	0.00	0.00	0.00	0.00	0.00	-1.00	0.00	0.00	0.00
	0.00	0.00	0.00	0.00	0.00	0.00	0.00	0.00	0.00	0.00
	0.00	0.00	*	0.00	0.00	*	0.00	*	0.00	*
	*	0.00	0.00	0.00	*	*	*	0.00	0.00	0.00
	0.00	0.00	0.00	0.00	0.00	0.00	0.00	0.00	0.00	0.00
	0.00	0.00	0.00	0.00	0.00	0.00	0.00	0.00	0.00	0.00
	0.00	0.00	0.00	-1.00	0.00	0.00	0.00	0.00	0.00	0.00
2 bits	0.00	0.00	0.00	0.00	0.00	0.00	0.00	0.00	0.00	0.00
	0.00	*	0.00	0.00	*	0.00	0.00	*	*	0.00
	0.00	0.00	0.00	0.00	0.00	0.00	0.00	0.00	0.00	0.00
	0.00	0.00	-1.00	0.00	0.00	0.00	0.00	0.00	0.00	0.00
	0.00	0.00	0.00	0.00	0.00	-1.00	0.00	0.00	0.00	0.00
	0.00	-1.00	0.00	0.00	0.00	0.00	0.00	0.00	0.00	0.00
	0.00	0.00	0.00	0.00	0.00	0.00	0.00	0.00	0.00	0.00
	0.00	0.00	0.00	0.00	0.00	0.00	0.00	0.00	0.00	0.00
	0.00	0.00	0.00	0.00	0.00	0.00	0.00	0.00	0.00	0.00

Table 4 BIC values for the S-box shown in Example 3. (Continued . . .)

	0.00	0.00	0.00	0.00	0.00	0.00	0.00	0.00	0.00	0.00
3 bits	0.00	0.00	0.00	0.00	0.00	0.00	0.00	0.00	0.00	0.00
	0.00	0.00	0.00	0.00	0.00	0.00	0.00	0.00	0.00	0.00
	0.00	0.00	0.00	0.00	0.00	0.00	0.00	0.00	0.00	0.00
	0.00	0.00	0.00	0.00	0.00	0.00	0.00	0.00	0.00	0.00
	0.00	0.00	0.00	0.00	0.00	0.00	0.00	-1.00	0.00	0.00
	0.00	0.00	0.00	0.00	0.00	0.00	0.00	0.00	0.00	0.00
	0.00	0.00	0.00	0.00	0.00	0.00	0.00	0.00	0.00	-1.00
4 bits	0.00	0.00	0.00	*	0.00	0.00	*	0.00	*	*
	0.00	0.00	0.00	0.00	0.00	0.00	0.00	0.00	0.00	0.00
	0.00	0.00	0.00	0.00	0.00	0.00	0.00	0.00	0.00	0.00
5 bits	*	*	*	*	0.00	0.00	0.00	0.00	0.00	0.00

Table 4 BIC values for the S-box shown in Example 3.

The columns in the Table represent all the possible pairwise combinations of the Boolean functions that compose the S-box. The rows represent all the possible input changes of one to five bits. The table entries represent the correlation between the assorted pairs of Boolean functions for all the given input changes. A correlation of 0.00 means the function pair is uncorrelated for a particular input bit change. An entry of -1.00 means for that particular input bit change, the Boolean function pair is negatively correlated (i.e., every time the output of function A changes from 0 to 1, then the output of function B changes from 1 to 0 and vice versa).

5.3 S-box Construction Efficiency

With the restrictions placed upon how boolean functions can be combined to create good S-boxes, one may wonder at the rate at which they can be constructed and how many actually exist. We wrote an S-box generating program in C++ and, on a SPARC workstation, constructed 64 different 5x5 S-boxes possessing equiprobable XOR tables within a short period of time. Further, from the state of the program following its run, we determined that a great many more S-boxes of this type exist. Here is a list of eight of the 64 S-boxes generated:

```
14 4 11 1E 8 6 1A B 16 13 1F 5 2 19 1C 18 15 D 0 7 C A E 17 10 1D 9 1B 1 12 F 3
10 3 7 A 18 14 D 1F 16 2 19 13 17 1C 1A F 11 B 1 5 9 C 1B 0 15 8 1D 1E 4 6 E 12
4 3 9 10 14 C D B 1 13 2 E 18 15 F 1C 6 8 1A A 11 0 19 16 5 1E 17 12 1B 1F 1D 7
C 10 2 1 F D 8 15 4 13 1F 17 3 A 11 7 9 18 5 B 1B 14 1E E 0 1A 19 1C 16 12 6 1D
1C B 7 F 1B 12 9 1F 14 8 1A 19 17 15 10 D 18 2 1 4 E A 1E 5 11 0 1D 13 3 C 6 16
10 2 7 B 18 15 C 1F 16 3 19 12 17 1D 1B F 11 A 1 4 9 D 1A 0 14 8 1C 1E 5 6 E 13
C 11 2 1 F D 8 14 4 12 1E 16 3 A 10 7 9 19 5 B 1A 15 1F E 0 1B 18 1D 17 13 6 1C
1C A 6 E 1B 12 8 1F 14 9 1A 19 17 15 10 C 18 3 0 5 F B 1E 4 11 1 1D 13 2 D 7 16
```

6 S-box Information Theoretic Strength

It has already been demonstrated that near-bent functions possess good cryptographic properties [22]. We also know, as shown in this paper, that S-boxes constructed with these functions according to Theorem 1 have equiprobable input/output XOR tables and largely satisfy the BIC. Obviously, these S-boxes are good candidates for use in a cryptosystem. However, information theoretic properties have not yet been considered. In [21], several information theoretic criteria pertaining to S-boxes were introduced. These criteria can be used as bases for tests to determine the information theoretic strength of an S-box. Three criteria are particularly useful for this purpose:

a.) Static input/output independence,
b.) Dynamic input/output independence, and
c.) Dynamic output/output independence

The Static I/O independence test of order k determines the uncertainty in the output variables given knowledge of k ($1 \leq k < m$) input variables. The Dynamic I/O independence test of order k is similar except that it looks at how knowledge of *changes* in k ($1 \leq k \leq m$) input variables affects uncertainty in *changes* in the output variables. Note that extensions to both of these tests which look at all possible combinations of one or more of the output bits are possible. The Dynamic output/output independence test of order k ($1 \leq k < n$) determines the uncertainty in changes of $n - k$ unknown output bits given knowledge of changes in the k remaining output bits and a particular pattern of input bit changes. We constructed several 5x5 S-boxes and subjected them to the above tests. Typical results are given in Table 5. For comparison purposes, we also subjected several 4x4 S-boxes taken from [3] to the same tests (see Table 6). Note that these 4x4 S-boxes were constructed for optimum information theoretic properties.

k	Static I/O	Dynamic I/O	Dynamic O/O
1	0.026941	0.001667	0.064516
2	0.080354	0.004812	0.129032
3	0.206982	0.010971	0.258065
4	0.500000	0.023590	0.516129
5	1.000000	0.062500	NA

Table 5 Average information theoretic strengths of our 5x5 S-boxes (the numbers indicate information leakage, therefore low values are desired).

k	Static I/O	Dynamic I/O	Dynamic O/O
1	0.014951	0.000925	0.266697
2	0.090819	0.012493	0.580043
3	0.371094	0.047714	0.784552
4	1.000000	0.158920	NA

Table 6 Average information theoretic strengths of 4x4 S-boxes chosen to optimize their information theoretic properties.

These results appear to give further credence to the claim that as S-boxes increase in size, their cryptographic properties strengthen.

7 Inverse S-boxes

Dawson observed in [3] that since cryptanalysis can work both from plaintext-to-ciphertext and from ciphertext-to-plaintext, it is important that inverse S-boxes be cryptographically strong. Nyberg [26] gives results showing that it is possible to construct an S-box such that the functions that compose both an S-box and its inverse are maximally nonlinear.

Independently, we constructed S-boxes according to Theorem 2 and found that their inverse functions were always maximally nonlinear (but not necessarily quadratic). Several cryptographic strength tests were performed upon the inverse S-boxes. Our test results show that the cryptographic strength of the inverse S-boxes generally matched, and in a few cases exceeded, the forward S-box strength.

8 Conclusions

With a fast computer, one can construct cryptographically strong 4x4 S-boxes with relative ease. However, there has been no efficient, systematic way to find suitable boolean functions of five or more variables. For odd *m*, we define a class of maximally nonlinear boolean functions called near-bent functions. With the techniques described in this paper, we are able to generate near-bent boolean functions of five variables quickly. Further we show how to combine near-bent boolean functions into S-boxes that have excellent cryptographic properties. These S-boxes satisfy all the important design criteria and have good resistance to differential cryptanalysis. We have also demonstrated that in an information theoretic sense, larger S-boxes are better than smaller ones. This gives further evidence to the hypothesis that large S-boxes are cryptographically stronger than small S-boxes. We are continuing our research in this area with the intent to adapt the techniques described here toward the construction of yet larger (7x7) S-boxes possessing similar properties.

Bibliography

[1] C. Shannon, "Communication theory of secrecy systems," *Bell Systems Technical Journal*, vol. 28, pp. 656–715, 1949.

[2] J. Gordon and H. Retkin, "Are big S–boxes best?," in *Lecture Notes in Computer Science: Proc of the Workshop on Cryptography*, pp. 257–262, Springer-Verlag, 1982.

[3] M. Dawson, "A unified framework for substitution box design based on information theory," Master's thesis, Queen's University, 1991.

[4] L. O'Connor, "Affinity and degeneracy in boolean functions with applications to cryptography," *submitted for publication*, September, 1991.

[5] J. Detombe, "An efficient design methodology for large substitution boxes," Master's thesis, Queen's University at Kingston, Ontario, Canada, August 1992.

[6] National Bureau of Standards (U.S.), "Data Encryption Standard (DES)," tech. rep., Federal Information Processing Standards, 1977. Publication 46.

[7] E. Biham and A. Shamir, "Differential cryptanalysis of DES-like cryptosystems," *Journal of Cryptology*, vol. 4, no. 1, pp. 3–72, 1991.

[8] E. Biham and A. Shamir, "Differential cryptalanysis of the full 16-round DES," in *Proceedings of CRYPTO 92, to appear*, August 1992.

[9] J. Pieprzyk and G. Finkelstein, "Towards effective nonlinear cryptosystem design," *IEE Proceedings, Part E: Computers and Digital Techniques*, vol. 135, pp. 325–335, 1988.

[10] W. Meier and O. Staffelbach, "Nonlinearity criteria for cryptographic functions," in *Advances in Cryptology: Proc of EUROCRYPT '89*, pp. 549–562, Springer-Verlag, 1990.

[11] J. Kam and G. Davida, "Structured design of substitution-permutaton networks," *IEEE Transactions on Computers*, vol. C-28, pp. 747–753, 1979.

[12] A. Webster and S. Tavares, "On the design of S-boxes," in *Advances in Cryptology: Proc of CRYPTO '85*, pp. 523–534, Springer-Verlag, 1986.

[13] R. Forré, "The strict avalanche criterion: spectral properties of boolean functions and an extended definition," in *Advances in Cryptology: Proc of CRYPTO '88*, pp. 450–468, Springer-Verlag, 1989.

[14] C. Adams and S. Tavares, "The structured design of cryptographically good S-boxes," *Journal of Cryptology*, vol. 3, no. 1, pp. 27–41, 1990.

[15] B. Preneel, W. Van Leewijck, L. Van Linden, R. Govaerts, and J. Vandewalle, "Propagation characteristics of boolean functions," in *Advances in Cryptology: Proc of EUROCRYPT '90*, pp. 161–173, Springer-Verlag, 1991.

[16] H. Feistel, "Cryptography and computer privacy," *Scientific American*, vol. 228, no. 5, pp. 15–23, 1973.

[17] C. M. Adams, "On immunity against Biham and Shamir's "differential cryptanalysis"," *Information Processing Letters*, vol. 41, pp. 77–80, 1992.

[18] L. Brown, M. Kwan, J. Pieprzyk, and J. Seberry, "Improving resistance to differential cryptanalysis and the redesign of LOKI," in *Asiacrypt '91 Abstracts*, (Fujiyoshida, Japan), pp. 25–30, November 1991.

[19] E. Biham, *Differential Cryptanalysis of Iterated Cryptosystems*. PhD thesis, The Weizmann Institute of Science, Rehovot, Israel, 1992.

[20] R. Forré, "Methods and instruments for designing S-boxes," *Journal of Cryptology*, vol. 2, no. 3, pp. 115–130, 1990.

[21] M. Dawson and S. Tavares, "An expanded set of S-box design criteria based on information theory and its relation to differential—like attacks," in *Advances in Cryptology: Proc of EUROCRYPT '91*, pp. 352–367, Springer-Verlag, 1991.

[22] J. Detombe and S. Tavares, "Constructing near-bent boolean functions of five variables," tech. rep., Department of Electrical Engineering, Queen's University, Kingston, Ontario, April, 1992.

[23] C. Adams, *A Formal and Practical Design Procedure for Substitution Permutation Network Cryptosystems*. PhD thesis, Queen's University, 1990.

[24] C. M. Adams and S. E. Tavares, "The use of bent sequences to achieve higher-order avalanche criterion in S-box design," Tech. Rep. TR 90–013, Department of Electrical Engineering, Queen's University, May 1990.

[25] J. Pieprzyk, "On bent permutations," in *Proc of International Conference on Finite Fields, Coding Theory, and Advances in Communications and Computing*, (University of Nevada, L.V.), 1991.

[26] K. Nyberg, "On the construction of highly nonlinear permutations," in *Proceedings of Eurocrypt '92, to appear*, May 1992.

[27] B. Preneel, R. Govaerts, and J. Vandewalle, "Boolean functions satisfying higher order propagation criteria," in *Advances in Cryptology: Proc of EUROCRYPT '91*, pp. 141–152, Springer-Verlag, 1991.

[28] H. Meijer. Private Communication, 27 August 1992.

Session 5

CRYPTANALYSIS

Chair: Peter Landrock
(Aarhus University, Denmark)

Nonasymptotic Estimates of Information Protection Efficiency for the Wire-tap Channel Concept

Valery Korzhik[1] and Victor Yakovlev[2]

LEIC, Moika 61, St.Petersburg, 191065, Russia

1 Introduction

Wyner [6] introduced the concept of the wire-tap channel. He gave asymptotic conditions to enable messages to be transmitted reliably over the main channel and simultaneously to ensure that the transmission of information over the wire-tap channel approached zero. However for some practical applications we need to use randomized codes with finite block length and therefore it is very important to find criteria for the information protection efficiency when there are different sources. (Korzhik and Yakovlev [4])

In this paper we suggest two criteria for information protection in the wire-tap channel. The first concerns the capacity of the code noising channel for sources with a given redundancy. The second concerns the probability that the real transmitted message assumes a value in the list, of given size, which was formed during the optimal list decoding.

Similarly to ordinary error correcting codes we introduce the notion of the coding noise gain (CNG) for *code noising* (the name we have given Wyner's approach) and give formulas to calculate it.

We illustrate our theorems with a number of different examples, these indicate that it is possible to provide adequate information protection over wire-tap channels with the help of very simple codes while obtaining coding noise gains, in some cases of more than $30dB$.

2 Estimation of the Information Transmission Protection with a Redundant Source

Consider a linear binary block (n, n')-code G with a $(n, n'-k)$-subcode V. The random coding procedure for *code noising* the k-bit binary message is as follows. The standard array G/V is used and each of the cosets is compared with one of the 2^k variants of the message m_i, $i = 1, 2, \ldots, 2^k$. When a source generates some particular message m_i, the binary block is chosen with equal probability from the words of the coset that had been compared with m_i and transmitted over the channel. To describe the set of all received words at the output of the wire-tap channel we make a standard array V^n/G, where V^n is the space of all binary tuples of length n. Then the first coset G_1 (ie the code G itself) we represent as the standard array G/V. The remaining cosets we denote as G_i,

$i = 2,3,\ldots,2^{n-n'}$. Then each word $\bar{z}_{ij\tau}$ from the standard array V^n/G has three indices. The first index $i = 1,2,\ldots,2^{n-n'}$ shows the number of cosets G_i in V^n/G, the second index $j = 1,2,\ldots,2^k$ is the number of the cosets in G_i/V and the third one $\tau = 1,2,\ldots,2^{n'-k}$ gives the ordinal number of the word in V^n/G for given i and j. Let us denote the subcosets G_i/V of the standard array by V_{ij}. It is easy to see that placing all V_{ij} in one column under V_{11} we obtain the standard array V^n/V.

The *code noising channel* comprises a random coder with a pair of codes (G,V) and a wire-tap channel. Thus the input to the random coder is the input to the code noising channel s^k and the output of the wire-tap channel is the output z^n of the code noising channel.

The capacity C_{cn} of such a code noising channel per binary input symbol when the wire-tap channel is a BSC without memory and with symbol error probability p is determined by the following theorem.

Theorem 1.

$$C_{cn} = 1 - \frac{1}{k}\sum_{j=1}^{2^{n-n'}} P(G_j)\log P(G_j) + \frac{1}{k}\sum_{s=1}^{2^{n-(n'-k)}} P(V_s)\log P(V_s) \qquad (1)$$

where

$$P(G_j) = \sum_{i=0}^{n} N_{ij}p^i(1-p)^{n-i};\ P(V_s) = \sum_{i=0}^{n} M_{is}p^i(1-p)^{n-i}\ , \qquad (2)$$

and N_{ij}, M_{is} are the number of words in the cosets G_j and V_s respectively with Hamming weights i (spectra V^n/G and V^n/V).

Proof. It is well known that the capacity of the discrete channel without memory is

$$C^* = \max_{P(S^k)} I(S^k, Z^n)\ , \qquad (3)$$

where $I(S^k, Z^n)$ is the mutual information between its input S^k and its output Z^n and $P(S^k)$ is the probability distribution at the channel input.

For channels which are ***strongly symmetric at the input*** the optimal distribution in (3) is uniform [1, Gallager]. We recall that such a channel, with transition probability matrix $(P(\bar{z} \mid \bar{s})$ satisfies the following two conditions:

(i) all the rows of the matrix are permutations of each other,
(ii) the set of all possible $\bar{z}$ can be divided into disjoint subsets $\bar{z}_1$, $\bar{z}_2, \cdots$, for which the transition probability matrix $(P_i(\bar{z} \mid \bar{s})$ corresponding to $\bar{z}_i$ has rows and columns formed by transpositions of the same number set.

Denote by x^n the output words of the random coder. Since the random vectors $\bar{s}$, $\bar{x}$ and $\bar{z}$ are connected by a simple Markov chain we have

$$P(\bar{z}_{ij\tau} \mid \bar{s}_d) = \sum_{\ell=1}^{K'} P(\bar{z}_{ij\tau} \mid \bar{z}_{1d\ell}) \cdot P(\bar{z}_{1d\ell} \mid \bar{s}_d) = \frac{1}{K'}\sum_{\ell=1}^{K'} P(\bar{z}_{ij\tau} \mid \bar{z}_{1d\ell})\ , \qquad (4)$$

where $K' = 2^{n'-k}$.

Since we have a memoryless BSC we obtain

$$P(\bar{z}_{ij\tau} \mid \bar{z}_{1d\ell}) = P(\bar{z}_{ij\tau} \oplus \bar{z}_{1d\ell}) = P(\bar{e}) \tag{5}$$

where $\oplus$ is addition modulo two and $\bar{e}$ is the error pattern.

Suppose $\bar{z}_{ij\tau}$ is given, then for any d and ℓ, using the group properties of V^n/G we have the vector $\bar{e} \in G_i$.

As d is fixed and ℓ runs through all the values $1, 2, \cdots, K'$, each error pattern *be* belongs to some third coset V_{ib}, where the number b is determined by j and d. Thus the expression (4) can be written as

$$P(\bar{z}_{ij\tau} \mid \bar{s}_d) = \frac{1}{K'} \sum_{b=1}^{K'} P(\bar{e} \in V_{ib}) = P(V_{i\psi(j,d)}) \tag{6}$$

and hence

$$P(\bar{z}_{ij\tau} \mid \bar{s}_d) = P(\bar{z}_{ijm} \mid \bar{s}_d) \tag{7}$$

for any (τ, m).

Taking into account the above relations we can represent the transition probability matrix of the code noising channel in the form

$$P_{cd} = [P_1, P_2, \cdots, P_{N'}] \tag{8}$$

where $N' = 2^{n-n'}$, $K = 2^k$ and

$$P_i = \begin{bmatrix} P(V_{i\psi(1,1)}) & P(V_{i\psi(2,1)}) & \cdots & P(V_{i\psi(K,1)}) \\ P(V_{i\psi(1,2)}) & P(V_{i\psi(2,2)}) & \cdots & P(V_{i\psi(K,2)}) \\ \cdots & \cdots & \cdots & \cdots \\ P(V_{i\psi(1,K)}) & P(V_{i\psi(2,K)}) & \cdots & P(V_{i\psi(K,K)}) \end{bmatrix} \tag{9}$$

(We keep only one column from each coset G_i/V due to (7).)

Given j, if d runs through all values from 1 to K the function $b = \psi(j,d)$ gives exactly K different values and moreover all $V_{ib} \in G_i$. Similarly, for given d, if j runs through all values from 1 to K the function $\psi(j,d)$ gives exactly the same K values. Therefore we see that each row and column of the matrix (9) contains the same set of values and hence all the rows and columns of P_i are permutations of the first row and the first column respectively. This proves condition (ii) holds.

This row symmetry of the matrix P_i follows from the fact that all the rows of P are formed by transpositions of the first row and hence condition (i) also holds. Hence the code noising channel is strongly symmetric at the input.

To find the capacity of the code noising channel we now suppose that $P(\bar{s}^k) = \frac{1}{2^k}$ for any s^k.

By definition the mutual information is given by

$$I(S^k, Z^n) = H(Z^k) - H(Z^n \mid S^k) \ , \tag{10}$$

where $H(Z^k)$ and $H(Z^n \mid S^k)$ are the entropy and conditional entropy respectively.

Using known results from information theory [1, Gallager] we obtain

$$\begin{aligned} H(\bar{z}^k) &= -\sum_{\bar{z}} P(z^n) \log P(\bar{z}) \\ &= \sum_{\bar{z}} \sum_{\bar{s}} P(\bar{z} \mid \bar{s}) P(\bar{z}) \log \sum_{\bar{s}} P(\bar{z} \mid \bar{s}) P(\bar{s}) \\ &= -\frac{1}{K} \sum_{\bar{z}} \sum_{\bar{s}} P(\bar{z} \mid \bar{s}) \log \frac{1}{K} \sum_{\bar{s}} P(\bar{z} \mid \bar{s}) \ . \end{aligned} \tag{11}$$

The relation (11) can be reduced using the standard arrays V^n/G and G_i/V to the following

$$\begin{aligned} H(\bar{z}^n) &= -\frac{1}{K} \sum_{i=1}^{N'} \sum_{j=1}^{K} \sum_{\tau=1}^{K'} \sum_{d=1}^{K} P(\bar{z}_{ij\tau} \mid \bar{s}_d) \log \frac{1}{K} \sum_{d=1}^{K} P(\bar{z}_{ij\tau} \mid \bar{s}_d) \\ &= -\frac{1}{K} \sum_{i=1}^{N'} \sum_{j=1}^{K} \sum_{d=1}^{K} P(V_{i\psi(j,d)}) \log \frac{1}{KK'} \sum_{j=1}^{K} P(V_{i\psi(j,d)}) \ . \end{aligned} \tag{12}$$

□

As was mentioned previously, given j and with d running through all values from 1 to k, the function $\psi(j,d)$ takes all possible values from the set $\{1, 2, \cdots, K\}$. From (12) we obtain

$$H(\bar{z}^n) = -\sum_{i=1}^{N'} \sum_{j=1}^{K} P(V_{ij}) \log \frac{1}{KK'} \sum_{j=1}^{K} P(V_{ij}) \ . \tag{13}$$

Using

$$\sum_{j=1}^{K} P(V_{ij}) = P(G_i)$$

we obtain

$$\begin{aligned} H(\bar{z}^n) &= -\log KK' \sum_{i=1}^{N'} P(G_i) - \sum_{i=1}^{N'} P(G_i) \log P(G_i) \\ &= n' - \sum_{i=1}^{N'} P(G_i) \log P(G_i) \ . \end{aligned} \tag{14}$$

Now consider the second term in (10)

$$H(\bar{z} \mid \bar{s}) = -\sum_{\bar{z}} \sum_{\bar{s}} P(\bar{z} \mid \bar{s}) P(\bar{s}) \log P(\bar{z} \mid \bar{s}) \ . \tag{15}$$

Using transformations similar to those for deriving (14) and the fact that $P(\bar{s}^k) = \frac{1}{2^k}$ we obtain from (15)

$$
\begin{aligned}
H(\bar{z}^n \mid \bar{s}^k) &= -\frac{1}{K}\sum_{j=1}^{K}\sum_{i=1}^{N'}\sum_{d=1}^{K}\sum_{\tau=1}^{K'} P(\bar{z}_{ij\tau} \mid \bar{s}_d)\log P(\bar{z}_{ij\tau} \mid \bar{s}_d) \\
&= -\frac{1}{K}\sum_{d=1}^{K}\sum_{i=1}^{N'}\sum_{j=1}^{K} P(V_{i\psi(j,d)})\log\frac{1}{K'}P(V_{i\psi(j,d)}) \ .
\end{aligned} \tag{16}
$$

Because for given d, with j running through all values from 1 to K the function $\psi(j,d)$ takes all possible values from the set $\{1,2,\ldots,K\}$, we obtain from (16)

$$
\begin{aligned}
H(\bar{z}^n \mid \bar{s}^k) &= -\sum_{i=1}^{N'}\sum_{j=1}^{K} P(V_{ij})\log\frac{1}{K'}P(V_{ij}) \\
&= \sum_{i=1}^{N'}\sum_{j=1}^{K} P(V_{ij})\log K' - \sum_{i=1}^{N'}\sum_{j=1}^{K} P(V_{ij})\log P(V_{ij}) \\
&= n' - k - \sum_{s=1}^{2^{n-(n'-k)}} P(V_s)\log P(V_s) \ .
\end{aligned} \tag{17}
$$

Putting (14) and (17) into (10) we have

$$
\max_{P(\bar{s}^k)} I(S^k, Z^k) = k - \sum_{j=1}^{2^{n-n'}} P(G_j)\log P(G_j) + \sum_{s=1}^{2^{n-(n'-k)}} P(V_s)\log P(V_s) \ . \tag{18}
$$

If the expression (11) is divided by k we obtain (1) and the theorem is proved. □

Consider the case where the main channel is noiseless. Here we can take $G = V^n$ and the code V to have parameters $(n, n-k)$. Now the random coder coincides with the random code book described in Wyner [6].

Thus expression (1) yields

$$
\tilde{C}_{cn} = 1 + \frac{1}{k}\sum_{j=1}^{2^k} P(V_j)\log P(V_j) \ . \tag{19}
$$

(A similar result has been derived in (Korzhik and Yakovlev [4].)

To find the spectra of the cosets for an arbitrary linear code V is a well known NPC-problem and hence we can calculate $\tilde{C}_{cn}$ only for some classes of codes: Hamming, BCH $(2^m-1, 2^m-1-2m)$; Golay; Reed-Muller (16,5), (32,6); selfdual $(2m, m)$, $m = 4, 8, 16, 24$; simplex $(2^m-1, m)$, $m \leq 5$ and some others.

In table 1 the values of $\tilde{C}_{cn}$ are given as functions of the symbol error probability p_w in the wire-tap channel when the main channel is noiseless.

In the last column of table 1 the data rates $R_d = \frac{k'}{n}$, where $k' = n - k$ are given.

Table 1. The capacities of the code noising channel for noiseless main channel

Codes \ p_w	10^{-3}	10^{-2}	5×10^{-2}	10^{-1}	2×10^{-1}	R_d
Hamming(31,26)	0.94	0.605	0.083	0.0041	6.2×10^{-7}	0.16
BCH(31,21)	0.965	0.751	0.235	0.092	9.7×10^{-5}	0.32
Golay(24,12)	0.977	0.838	0.458	0.156	6.3×10^{-3}	0.5
Reed-Muller(16,5)	0.986	0.901	0.698	0.422	0.12	0.687
Without code noising	≈ 1	0.91	0.77	0.53	0.28	1

The last row shows the values for the ordinary *BSC* capacity with symbol error probability equal to p_w calculated using the formula

$$C_{BSC} = 1 + p_w \log p_w + (1-p_w)\log(1-p_w)\ . \qquad (20)$$

We can see that the difference between $\tilde{C}_{cn}$ and C_{BSC} is significant especially for large values of p_w.

In table 2 we give the values of the secrecy capacity C_s of p_w which has been considered in Wyner [6] and given by

$$C_s = -p_w \log p_w - (1-p_w)\log(1-p_w)\ . \qquad (21)$$

By definition for such values, in the asymptotic case (when the block length approaches ∞), we can, when $R_d < C_s$ obtain $I(S^u, Z^u) = 0$.

Table 2. The secrecy capacity

p_w	5×10^{-3}	10^{-2}	5×10^{-2}	0.11	0.2	0.3
C_s	0.045	0.081	0.256	0.5	0.72	0.88

Comparison of table 2 and the last column of table 1 shows that the codes chosen in table 1 are far from asymptotic results.

So for $p_w = 2\times10^{-1}$ we can obtain $I(\bar{s}^u, \bar{z}^u) = 0$ for $R_d < 0.72$ while we have only $I(\bar{s}^u, \bar{z}^u) \le 6.2\times10^{-7}$ for $R_d < 0.16$ if we chose the Hamming (31, 26) code. (Later we will show that these codes are actually better that they first appear.)

In table 3 we give the values of C_{cn} for p_w calculated using (1) for the pair of codes $G = (32,26); V = (32,21)$.

Comparison of tables 1 and 3 shows that coding by a pair of codes decreases a few of the capacities but it allows us to improve the main channel. (We confirm this fact later.)

The natural criterion to estimate the coding efficiency is the so called coding gain. We introduce a similar notion, the *code noising gain* (CNG).

Table 3. The capacity C_{cn} for a noisy main channel

p_w	10^{-3}	10^{-2}	5×10^{-2}	10^{-1}	2×10^{-1}
C_{cn}	0.9996	0.961	0.466	0.087	3×10^{-4}

We consider the situation where the capacity C_r of the wire-tap channel is given, the probabilities p_{wcn} and p_{cn} are known from (19) and (20) and modulation/demodulation assumes binary signalling over a continuous channel with demodulation function $p = f(h^2)$ where h^2 is the signal to noise ratio. Then for given $f(h^2)$ we can calculate h^2_{wcn} and h^2_{cn} which let us show that if some sources of messages have entropy $H > C_r$ then it is impossible to receive any information after code noising for any decoding.

Define the code noising gain (CNG) from using code noising as follows:

$$CNG = 10 \log \frac{h^2_{cn}}{h^2_{wcn}} \qquad (dB) \; . \tag{22}$$

(Note that the factor R (code rate) is absent from (22) as would be expected for error correcting codes because we believe that the channel rate is kept constant for code noising and consequently the source rate is decreased).

Table 4 gives in the numerators the values of the CNG in dB which are calculated by (22) for some codes and C_r. For the maximum information protection the CNG is significant. The values of CNG for unlimited block lengths with the same code rate for corresponding codes are obtained from Wyner's results [6] and are shown in the denominator. We observe the numerator and denominator are sufficiently close that there is no significant reason to search for longer codes.

For a noisy main channel additional conditions can be imposed on the CNG to ensure the correct decoding probability over the main channel.

Table 4. The code noising gains

$C_r \setminus V$	(15,11)	(24,12)	(31,21)	(31,26)	(63,57)
10^{-4}	$\frac{33.9}{40.9}$	$\frac{32.2}{38.8}$	$\frac{35.2}{41.2}$	$\frac{37.3}{42.7}$	$\frac{39.7}{44.0}$
10^{-3}	$\frac{25.4}{31.3}$	$\frac{23.7}{28.6}$	$\frac{26.4}{31.2}$	$\frac{28.3}{32.8}$	$\frac{30.0}{33.5}$
10^{-2}	$\frac{17.2}{21.4}$	$\frac{15.7}{18.5}$	$\frac{17.8}{21.0}$	$\frac{19.6}{22.9}$	$\frac{21.3}{23.5}$
10^{-1}	$\frac{9.6}{11.0}$	$\frac{7.9}{9.8}$	$\frac{9.7}{10.9}$	$\frac{11.1}{13.0}$	$\frac{12.5}{12.7}$

3 Estimation of the Information Transmission Protection of the Key Exchange

Consider some cryptosystem which uses code noising to protect the key exchange before ciphertext is transmitted. Given interception of sufficient ciphertext symbols the wiretapper can recover the plaintext without obtaining the key by total search. Therefore the information theoretic criterion does not adapt directly to the wire-tap channel to be used for a key transmission. We now suggest other criteria to ensure that the key transmission protection is ensured by enabling the probability for the real transmitted message to assume a value in the list of given size, which was formed during the optimal list decoding. Let us denote this value by $P_{a\ell}(G)$ which is the probability that the message is *available* in the list of size L.

If we consider the case of code noising with one code V (where the main channel is noiseless) we can easily show, using the Sullivan inequality [5], that the optimal likelihood decoding algorithm for $L = 1$ is to form the message m_j when the word $\bar{z} \in V_{ij}$ is received over the wire-tap channel. Let us call this decoding procedure the *removal of code noising.*

The *optimal list L-size decoding* and the above probability $P_{a\ell}$ are described by the the following theorem.

Theorem 2. *The optimal list L-size decoding algorithm after code noising and transmission of messages over BSC consists of two stages. First, the preliminary list* $\bar{z} \oplus \bar{z}_{ij_s\tau}$, $s = 1, 2, \ldots, L$ *is formed where* $\bar{z} \in G_i$ *is the received word,* j_s, *and an arbitrary index,* τ, *are chosen to satisfy the following condition*

$$P(V_{ij_1}) \geq P(V_{ij_2}) \geq \ldots \geq P(V_{ij_s}) \geq P(V_{ij})$$

for any $j \notin (j_1, j_2, \ldots, j_s)$, *where* $P(V_{ij}) = \sum_{t=0}^{n} A_{tij} p_w^t (1-p)^{n-t}$ *where* A_{tij} *is the number of words in the coset* V_{ij} *with Hamming weight* t *(spectrum* V_{ij}*). Second, the final list is formed by processing the preliminary list and removing the code noising. The probability of the actual message availability in this list is*

$$P_{a\ell}(L) = \sum_{i=1}^{2^{n-n'}} \sum_{s=1}^{L} P(V_{ij_s}) \tag{23}$$

Proof. Consider expression (6) obtained in the proof of theorem 1. For a maximum likelihood coding list decoding rule we should to place into the list those $s_{d1}, s_{d2}, \cdots, s_{dL}$ which give the maximum value of $P(V_{i\psi(j,d)})$. Let us suppose that we found these for every $i = 1, 2, \ldots, N'$ and rearranged them into order of decreasing probability $P(V_{ij})$ denoted by $\tilde{V}_{i1}, \tilde{V}_{i2}, \ldots, \tilde{V}_{iL}$. Then choosing in each of these subcosets any word $\tilde{z}_j \in \tilde{V}_{ij}$, $j = 1, 2, \ldots, L$, we obtain the following rule for optimal L-size list decoding of the transmitted words

$$\bar{z}_j = \bar{z}_{ij\tau} \oplus \tilde{z}_j, j = 1, 2, \ldots, L \ . \tag{24}$$

However, due to the group properties of V^n/G, the words $\tilde{x}_j \in G_1 = G$ and thus the L-size list for the messages can be obtained of $\tilde{x}_j, j = 1, 2, \ldots, L$ by the procedure of removal of code noising. This proves the first part of the theorem. To derive (23) note that the actual transmitted message is available in the list if and only if the error pattern belongs to the set $\bigcup_{i=1}^{2^{n-n'}} \bigcup_{j=1}^{L} \tilde{V}_{ij}$. We obtain (23) from this fact by taking into account that all $\tilde{V}_{ij}, i = 1, 2, \ldots, 2^{n-n'}, j = 1, 2, \ldots, L$ are disjoint. □

For the particular case $L = 1$ the optimal list decoding algorithm is similar to the maximum likelihood decoding of Hugget [2]. Then the probability of correct decoding is

$$P_{cd} = \sum_{i=1}^{2^{n-n'}} P(\tilde{V}_{i1}) . \tag{25}$$

The problem of calculating $P_{al}(L)$ or P_{cd} by (23) or (25) is more complicated then the calculation of C_{cn}, because it requires a knowledge of the spectra of the subcosets V_{ij}.

Consider the particular case, when G is the (31,26) Hamming code and V is the (31,21) BCH code. The coset weight distributions of these codes are shown in the table 5. Using this table it is easy to show, that for $p \leq 0.45$ and $L \leq 16$ the following expressions hold for this pair of codes.

$$P_{al}(L) = P(V) + (L-1)P(V^{(3)}) + 31[P(V^{(1)}) + (L-1)P(V^{(2)})] \ , \tag{26}$$

$$P_{cd} = P(V) + 31P(V^{(1)}) \ . \tag{27}$$

Where $P(V), P(V^{(1)}), P(V^{(2)}), P(V^{(3)})$ are calculated using the spectra given in the table 5.

In table 6 P_{cd} is shown for such pairs of codes as a function of the symbol error probability p_m of the main channel.

We can see, that if the main channel has good quality the use of the code noising by a pair of codes can even improve the information transmission over the main channel and simultaneously give a very low capacity for the wire-tap channel if it has poor quality (see table 3). Note that the criterion of correct probability decoding is more useful for the main channel than the capacity because we need simple decoding for the main channel.

Unfortunately the above pair of codes is unique and it is unlikely we will be able to find a "computable "pair for the real size of key "K". "It is proposed that the following method for code noising of large sizes K is used.

Let us divide $\ell - bit$ blocks into $M = \frac{l}{k}$ subblocks with lengths k and using code noising by a pair of codes (G, V) for each short subblock. It can be shown that this approach is equivalent to code noising by a pair codes $(\tilde{G}, \tilde{V})$, where $\tilde{G}$ and $\tilde{V}$ are the direct sums of M codes G codes and V respectively. Using

Table 5. The coset weight distributions for the Hamming $G = (31, 26)$ and BCH $V = (31, 21)$ codes

Weights	Types of cosets for code G and their cardinalities		Types of cosets for code V and their cardinalities				Weights
	G,I	$G^{(1)}$,31	V,I	$V^{(1)}$,31	$V^{(2)}$,465	$V^{(3)}$,527	
0	1	0	1	0	0	0	31
1	0	1	0	1	0	0	30
2	0	15	0	0	1	0	29
3	155	140	0	0	4	5	28
4	1085	980	0	30	26	35	27
5	5208	5313	186	156	171	162	26
6	22568	23023	806	751	736	702	25
7	82615	82160	2635	2610	2546	2580	24
8	247845	246480	7905	7530	7674	7740	23
9	628680	630045	18910	19285	19736	19670	22
10	1383096	1386099	41602	43780	43329	43270	21
11	2648919	2645916	85560	83382	82634	82689	20
12	4414865	4409860	142600	137100	137844	137815	19
13	5440560	6445565	195300	200800	201427	201460	18
14	8280720	8287155	251100	295515	258888	259020	17
15	9398115	9391680	301971	293556	293556	293424	16

Table 6. The probability of correct decoding after optimal decoding for a pair of codes $G = (31, 26)$ and $V = (31, 21)$

p_w	10^{-4}	10^{-3}	5×10^{-2}	10^{-2}	10^{-1}	0.15	0.2
P_{cd}	0.999995	0.9995	0.9894	0.9616	0.18	0.06	0.036

computer calculations we obtained for above the pair of codes $G = (31, 26), V = (31, 21), \ell = 130, M = 26$ and $P_w = 0.091$ the values of $P_{a\ell}(L)$ which are shown in table 7.

Table 7. The values $P_{a\ell}$ as a function of the list size L

L	10^{8}	10^{10}	10^{12}	10^{14}
$P_{a\ell}$	3×10^{-12}	7×10^{-11}	1.4×10^{-9}	2.2×10^{-8}

Simultaneously for the main channel with symbol error probability $P_m = 10^{-3}$ we found that the probability of correct transmission of a 130-bit key is about 0.998. These results show, that is is possible to provide high protection for a large size key transmission in the presence of a wiretapper and with sufficient reliable transmission of the the key over the main channel.

4 Conclusions

In this paper we have been considered the construction of code noising for the general case of the noising main channel and for codes with finite block length. It can be proved that such a construction is equivalent to the combined use of ordinary error correction codes and codes noising for a noiseless main channel. Therefore the complexity of code noising in the general case is at most that for ordinary error correcting codes.

Simultaneously we have shown that real codes with a moderate block length give significant information protection and the losses in energy in the asymptotic case are not so very large. We emphasized that for information transmission protection for key exchanges it is a more suitable criterion to consider the real transmitted message assuming a value in the list of given size which was found during the optimal list decoding than theoretical information criteria. The examples, which are given in the paper confirm that code noising has good prospects for practical applications.

Acknowledgement

The authors thank Professor Jennifer Seberry for useful discussion of this paper and support in its submission.

References

1. G. Gallager, Information Theory and Reliable Communication,
2. U. Hugget, Coding scheme for a wire-tap channel using regular codes, Discrete Math., 56 (1988) 191-206
3. V. Korzhik, Error control for the "unique" messages, Problemy Peredachi Informatcii, 1, 22 (1986) 26-31
4. V. Korzhik and V, Yakovlev, Nonasymptotic bounds for jamming in the wire-tap channel, Problemy Peredachi Informatcii, 4, 17 (1981) 11-17
5. D. D. Sullivan, A fundamental inequality between the probabilities of binary subgroups and cosets, IEEE Trans. on Inform. Theory, IT-13 (1967) 91-94
6. A. Wyner, The wire-tap channel, Bell Syst. Tech. Journal, 8, 54 (1975) 1355-1387

Cryptanalysis of LOKI 91

Lars Ramkilde Knudsen

Aarhus Universitet
Comp. Science Dept.
DK-8000 Århus C.
e-mail:ramlodi@daimi.aau.dk

Abstract. In this paper we examine the redesign of LOKI, LOKI 91 proposed in [5]. First it is shown that there is no characteristic with a probability high enough to do a successful differential attack on LOKI 91. Secondly we show that the size of the image of the F-function in LOKI 91 is $\frac{8}{13} \times 2^{32}$. Finally we introduce a chosen plaintext attack that reduces an exhaustive key search on LOKI 91 by almost a factor 4 using $2^{33}+2$ chosen plaintexts.

1 Introduction

In 1990 Brown *et al* [4] proposed a new encryption primitive, called LOKI, later renamed LOKI 89, as an alternative to the Data Encryption Standard (DES), with which it is interface compatible. Cryptanalysis showed weaknesses in LOKI 89 [2, 5, 8] and a redesign, LOKI 91 was proposed in [5]. The ciphers from the LOKI family are DES-like iterated block ciphers based on iterating a function, called the F-function, sixteen times. The block and key size is 64 bits. Each iteration is called a round. The input to each round is divided into two halves. The right half is fed into the F-function together with a 32 bit round key derived from the keyschedule algorithm. The output of the F-function is added (modulo 2) to the left half of the input and the two halves are interchanged except for the last round. The LOKI ciphers run 16 rounds. The plaintext is the input to the first round and the ciphertext is the output of the last round. The input to the F-function is the xor'ed value of a 32 bit input text and a 32 bit round key. The 32 bits are expanded to 48 bits and divided into blocks of 12 bits. The 12 bit blocks are the inputs to the 4 S-boxes in LOKI 91, each of which produces an 8 bit output. The 32 bits are permuted making the output of the F-function.

In section 2 we do differential cryptanalysis of LOKI 91 and show that there is no characteristic with a probability high enough to do a successful differential attack. Differential cryptanalysis was introduced by Biham and Shamir [1]. The underlying theory was later described by Lai and Massey [6]. For the remainder of this paper we expect the reader to be familiar with the basic concepts of differential cryptanalysis. Please consult the papers [1, 3, 6] for further details.

In section 3 we examine the size of the image of the F-function, the round function in LOKI 91. Because the key is added to the input text before the expansion in the F-function, the inputs to the 4 S-boxes are dependent. We show that this

has the effect that the size of the image of the F-function is $\frac{8}{13} \times 2^{32}$.
In section 4 we show a weakness in the keyschedule of LOKI 91, i.e. that for every key K there exists a key K^*, such that K and K^* have 14 common round keys. We exploit this weakness in a chosen plaintext attack that reduces an exhaustive key search by almost a factor 4 using $2^{33} + 2$ chosen plaintexts.

2 Differential cryptanalysis of LOKI 91

In [5] it is indicated that LOKI 91 is resistant against differential cryptanalysis, a chosen plaintext attack introduced in [1]. The first thing to do in differential cryptanalysis is to look for good characteristics or differentials. In [3] Biham and Shamir introduced an improved differential attack on DES. The attack shows how to extend an r-round characteristic to an $(r+1)$-round characteristic with unchanged probability by picking the chosen plaintexts more carefully. The cost is a more complex analysis. The improvement can be obtained in attacks on any DES-like iterated cipher. Thus the existence of a 13-round characteristic with a too high probability might enable a successful differential attack on LOKI 91. The probability of an r-round characteristic is found by multiplying the probabilities of r 1-round characteristics. As stated in [6] this way of calculating the probabilities for characteristics requires the cipher to be a Markov cipher. Since the round keys are dependent, LOKI 91 is not a Markov cipher, however tests for LOKI 89 show that the probabilities hold in practice at least for small characteristics [8]. Furthermore we have found no way of incorporating the key dependencies in the calculation of longer characteristics.

2.1 Characteristics for LOKI 91

The best one-round characteristic in LOKI 91 has probability 1 and comes from the fact that equal inputs always lead to equal outputs. A round with equal inputs is called a **zero round** (since the xor-sum of the inputs is zero).
The pairs xor table (see [1]) for LOKI 91 is a table with 2^{20} entries. Table 1 shows the most likely combinations for input/outputxors for one S-box isolated. Note that although inputxor 004_x leads to outputxor 01_x with probability $\frac{132}{4096}$ for one S-box it doesn't mean we can find a one round characteristic with this probability. Because the key is added to the input text before the E-expansion in LOKI 91 the inputs to two neighbouring S-boxes are dependent. In the above case a neighbouring S-box will have inputxor $4ij_x$, where $i, j \in \{0, ..., 15\}$.
The best one-round characteristic with a nonzero input difference has probability $\frac{52}{4096} \simeq 2^{-6.29}$. Therefore to find a 13-round characteristic with a probability high enough to enable a successful differential attack some of the 13 rounds must be zero rounds. The best characteristic for an attack on DES is based on a 2-round iterative characteristic [1], where every second round is a zero round. The best characteristic for an attack on LOKI 89 is based on a 3-round iterative characteristic [5, 8], where every third round is a zero round. We need a few definitions:

Input	Output	Prob. (n/4096)	Input	Output	Prob. (n/4096)
4	1	132	c	1	76
80	4	52	a0	e8	46
173	f7	46	185	90	46
37b	cd	48	3e0	24	48
42a	41	46	498	cf	56
49e	97	46	790	46	50
a20	0	46	a21	d7	48
c43	76	46	c76	f0	48
deb	c9	46	e7b	5f	48
ea6	5d	46	eec	ab	46
f33	e9	46			

Table 1. The most likely combinations from the pairs xor table.

Definition 1 *If the rounds no. $(i-1)$ and $(i+1)$ are zero-rounds, round no. (i) is of* **type A**.
If the rounds no. $(i-1)$ and $(i+2)$ are zero-rounds, rounds no. (i) and $(i+1)$ are a **pair of type B**.
If the rounds no. $(i-1)$ and $(i+3)$ are zero-rounds and the rounds no. (i), $(i+1)$ and $(i+2)$ are nonzero rounds, then rounds no. (i), $(i+1)$ and $(i+2)$ are a **triple of type C**.

A round of type A must have the following form $\phi \rightarrow 0_x$. The best probability of such a round for LOKI 91 is $\frac{122}{2^{20}} \simeq 2^{-13}$ [5].
A pair of rounds of type B must have the following forms,

$$\begin{array}{ll} \text{round no. } (i): & \phi \rightarrow \psi \\ \text{round no. } (i+1): & \psi \rightarrow \phi \end{array} \quad [9]$$

Lemma 1 *The highest probability for a pair of rounds of type B is* $(\frac{16}{2^{12}})^2 = 2^{-16}$.

Proof: Consider the case where ϕ and ψ differ in the input to only one S-box each, Si and Sj respectively. Because of the P-permutation (see [4]) it follows easily that the input/outputxor combinations for Si and Sj must have one of the following four forms:

Input	Output	Prob. (n/4096)
080_x	80_x	10
040_x	20_x	16
020_x	08_x	6
010_x	02_x	12

Table 2.

The highest probability for a pair of type B is therefore when the combination in both ϕ and ψ is $040_x \rightarrow 20_x$. It is exactly the situation that occurs for fixpoints [8]. From the pairs xor table it follows easily that if ϕ and ψ differ in the inputs to more than 2 S-boxes totally, we obtain probabilities smaller than 2^{-16}. □
A triple of rounds of type C must have the following forms,

$$\begin{array}{ll} \text{round } (i): & \phi \rightarrow \gamma \\ \text{round } (i+1): & \gamma \rightarrow \phi \oplus \psi \\ \text{round } (i+2): & \psi \rightarrow \gamma \end{array} \qquad [9]$$

Lemma 2 *The probability for a triple of rounds of type C is at most* 2^{-22}.

Proof: By a similar argument as for type B, assume that ϕ, γ and ψ differ in the inputs to only one S-box each. Obviously ϕ and ψ must differ in the inputs to the same S-box. It means that the combination in round $(i+1)$ must have one of the forms from Table 2. The combination in both round (i) and $(i+2)$ must have the following form: $0Y0_x \rightarrow Z$, where $Z \in \{2_x, 8_x, 20_x, 80_x\}$ and $Y \in \{0_x, \ldots\ldots, f_x\}$. The two highest probabilities for $0Y0_x \rightarrow Z$ are $\frac{34}{4096}$ and $\frac{28}{4096}$, therefore the probability of a triple of type C has probability at most

$$\frac{34 * 16 * 28}{2^{36}} < 2^{-22}.$$

Note that $\phi \neq \psi$ otherwise γ would have to differ in the inputs to at least two neighbouring S-boxes [5]. □
Now we can prove the following theorem.

Theorem 1 *There is no 13-round characteristic for LOKI91 with probability higher than* 2^{-63}.

Proof: The best one-round characteristic with a nonzero input difference has probability $\frac{52}{4096} \simeq 2^{-6.29}$. Because $(\frac{52}{4096})^n > 2^{-63} \Rightarrow n \leq 10$, we must have at least 3 rounds with equal inputs in the 13-round characteristic (13R). Since two consecutive zero-rounds force all rounds to be zero-rounds we can have at most 7 zero-rounds.
<u>7 zero-rounds:</u> Every second round is of type A, therefore

$$P(13R) \leq (2^{-13})^6 = 2^{-78}$$

<u>6 zero-rounds:</u> At least 3 rounds are of type A, the remaining 6 rounds can have probability at most $2^{-6.29}$, therefore

$$P(13R) \leq (2^{-13})^3 \times (2^{-6.29})^4 = 2^{-64.2}$$

<u>5 zero-rounds:</u> There can be at most one round of type A, since

$$(2^{-13})^n \times (2^{-6.29})^{8-n} > 2^{-63} \Rightarrow n \leq 1$$

There are two cases to consider

1. No rounds of type A, thereby 4 pairs of type B:

$$P(13R) \leq (2^{-16})^4 = 2^{-64}$$

2. One round of type A, thereby at least 2 pairs of type B:

$$P(13R) \leq 2^{-13} \times (2^{-16})^2 \times (2^{-6.29})^3 = 2^{-63.9}$$

<u>4 zero-rounds:</u> There can be no rounds of type A, since

$$(2^{-13})^n \times (2^{-6.29})^{9-n} > 2^{-63} \Rightarrow n < 1$$

There can be at most one pair of type B, since

$$(2^{-16})^n \times (2^{-6.29})^{9-2n} > 2^{-63} \Rightarrow n \leq 1$$

There are two cases to consider

1. No pairs of type B, thereby 3 triples of type C:

$$P(13R) \leq (2^{-22})^3 = 2^{-66}$$

2. One pair of type B, thereby at least one triple of type C:

$$P(13R) \leq 2^{-16} \times 2^{-22} \times (2^{-6.29})^4 = 2^{-63.2}$$

<u>3 zero-rounds:</u> All 10 nonzero rounds must be based on the best combination $080_x \rightarrow 4_x$, since the second best combination has probability $2^{-6.47}$ and $(2^{-6.29})^9 \times 2^{-6.47} < 2^{-63}$. However it is not possible to construct a 13-round characteristic based solely on the best combination. □

The above does not prove that LOKI 91 is resistant against differential attacks. As stated in [7] to prove this resistance for a DES-like cipher we need to find the best possible differentials. However this seems to be extremely difficult for LOKI 91 and we have done no work in that direction.

3 The F-function of LOKI 91

In the redesign of LOKI89 [5] one of the guidelines was

- to ensure that there is no way to make all S-boxes give 0 outputs, to increase the ciphers security when used in hashing modes.

The 4 S-boxes in LOKI 91 are equal. Each S-box takes a 12 bit input and produces an 8 bit output. Each output value occurs exactly 16 times. The inputs to one S-box that result in a 0 output are listed in Table 1. Because the key is added to the input text before the E-expansion, the input to one S-box is dependent on the inputs to neighbouring S-boxes. Let the input to one S-box be hij_x, then the input to one of the neighbouring S-boxes is jkl_x. From Table 1 we see that to get 0 output from both S-boxes we must have $h, j \in \{0, 5, a, c, f\}$ and $j, l \in \{3, 4, 9, e\}$ leaving no possible values for j. Therefore we cannot get 0

4	49	8e	d3	514	559	59e	5e3
a24	a69	aae	af3	c03	f34	f79	fbe

Table 3. Inputs yielding 0 output for one S-box (hex notation)

outputs from any two neighbouring S-boxes. Let the output from the F-function be $B = \{b_1, b_2, b_3, b_4\}$ where b_i represents the output byte from S-box i. Then $B = \{0, 0, *, *\}$, $B = \{*, 0, 0, *\}$, $B = \{*, *, 0, 0\}$ and $B = \{0, *, *, 0\}$, where '$*$' represents any byte value, are impossible values in the image of the F-function. We found a lot of other impossible values. Therefore we made an exhaustive search for the size of the image of the F-function in LOKI 91.

Theorem 2 *The F-function is not surjective, indeed the size of the image of F is about* $2^{31.3}$.

Note that once we found that $B = \{b_1, b_2, b_3, b_4\}$, where b_i represents the output byte from S-box i, is not in the image of F, then because the 4 S-boxes in LOKI 91 are equal we know that any rotation of the four bytes yields a value not in the image of F. The exact number of impossible values is $1,638,383,180 \simeq \frac{5}{13} \times 2^{32}$. It means that about 5 out of 13 values are never hit in the output of the F-function. In DES we do not have that the inputs to the S-boxes are dependent, because the key is added after the E-expansion. Therefore the size of the image of the F-function in DES is 2^{32}. We have not found any ways to exploit the above observation in an attack on LOKI 91. A consequence of the observation is that the left and right halves of a ciphertext reveals 0.7 bit of information about the inputs (before addition of the keys) to the second last round respectively the third last round of the encryption.

4 A chosen plaintext attack reducing key search

We begin by giving the notation used in this section.
<u>Notation:</u>

- $\overline{X}$ is the bitwise complement of X.
- $Rol_n(X)$ is bitwise rotation of X n positions to the left.
- $E_{16}(P, K)$ is a full 16 round encryption of P using K.
- $E_2(P, K')$ is a 2 round encryption of P using the 32 bit key K' in the first round and $Rol_{13}(K')$ in the second round.
- $Swap(X, Y)$ is the swapping of X and Y.
- $Swap(Z)$ is the swapping of the left and right halves of Z.
- $X \| Y$ is the concatenation of X and Y.

The attack we are to describe makes use of a property of the key schedule in LOKI 91. The key size is 64 bits. The key is divided into two 32 bit halves K_L, K_R and the 16 round keys $K(i)$, $i = 1, ..., 16$, are derived as follows:

1. $i := 1$
2. $K(i) = K_L\,;\; i = i + 1$
3. $K_L = Rol_{13}(K_L)$
4. $K(i) = K_L\,;\; i = i + 1$
5. $K_L = Rol_{12}(K_L)$
6. $Swap(K_L, K_R)$
7. go to 2.

The key schedule allows two different keys to have several round keys in common.

Theorem 3 *For every key K there exists a key K^*, such that K and K^* have 14 round keys in common.*

Proof: Let $K(1), \ldots\ldots, K(16)$ be the roundkeys for $K = K_L \| K_R$. Let $K^* = K_R \| Rol_{25}(K_L)$. Then $K(2+i) = K^*(i)$, $i = 1, .., 14$. □
If $K = K^*$ Theorem 3 is trivial, but this happens for only two keys, because $K = K^* \Rightarrow (K_L = K_R) \wedge (K_R = Rol_{25}(K_L)) \Rightarrow K_R = K_L = Rol_{25}(K_L) \Rightarrow K = 00....00 \vee K = 11....11$, since $gcd(25, 32) = 1$.

Corollary 1 *There exists 2^{36} pairs of keys, K and K^*, such that K and K^* have 16 round keys in common.*

Let $K = K_L \| K_R$, $K_L = hh...hh_x$ for some hex digit h and let K_R be any 32 bits. From Theorem 3 we have a key K^* such that K and K^* have 14 round keys in common and we have furthermore $K^*(15) = Rol_{100}(K_L) = K_L = K(1)$ and $K^*(16) = Rol_{113}(K_L) = Rol_{13}(K_L) = K(2)$, i.e. K and K^* have 16 round keys in common. □
Theorem 3 can be used in a chosen plaintext attack to reduce an exhaustive key search by almost a factor 2. It is well known that the complementation property[1] of DES can be used to reduce an exhaustive key search of DES by a factor 2 in an attack that needs two chosen plaintexts [1]. The complementation property holds also for LOKI 91. This property and Theorem 3 can be used to reduce an exhaustive key search by almost a factor 4 in a chosen plaintext attack that needs $2^{33} + 2$ plaintexts.
<u>Algorithm:</u>

1. Pick $P = P_L \| P_R$ at random. Get encryptions C, C^* for P, $\overline{P}$.
2. For all $a \in \{0, 1, ..., (2^{32} - 1)\}$:
   Let $P(a)$ be $E_2(P, a)$. More precisely $P(a) = P_L(a) \| P_R(a)$, where
   $$P_L(a) = F(P_R, a) \oplus P_L$$
   $$P_R(a) = F(P_L(a), Rol_{13}(a)) \oplus P_R.$$
3. Get encryptions $C(a)$, $C^*(a)$ for $P(a)$, $\overline{P(a)}$ for all a.
4. Let all keys be non discarded.

---

[1] Let C be the encryption of P using K, then $\overline{C}$ is the encryption of $\overline{P}$ using $\overline{K}$ as the key.

5. Exhaustive search for key:
   For every non discarded key $K = K_L || K_R$, do
   (a) Find $C' = E_{16}(P, K)$
   (b) then
      - if $C' = C$ return K and stop
      - if $C' = \overline{C^*}$ return $\overline{K}$ and stop
      - if $E_2(Swap(C'), Rol_{100}(K_L)) = C(K_L)$ return $(K_R || Rol_{25}(K_L))$ and stop
      - if $E_2(Swap(C'), Rol_{100}(K_L)) = \overline{C^*(K_L)}$ return $(\overline{K_R} || Rol_{25}(\overline{K_L}))$ and stop
   (c) Discard the four keys in (b).

Upon termination we have found either the secret key or a collision for LOKI 91, i.e. $K \neq K^*$, such that $E_{16}(P, K) = E_{16}(P, K^*)$. Note that in step 5, once we have encrypted P using key $K = K_L || K_R$ without success, we do not have to encrypt P using neither $\overline{K}$, $(K_R || Rol_{25}(K_L))$ nor $(\overline{K_R} || Rol_{25}(\overline{K_L}))$. If one of these three keys is the secret key, then the algorithm would have terminated before. At some points in the algorithm some of the four keys in 5(b) are equal, for example the all zero key will appear twice in the same iteration of step 5. Therefore we cannot find an enumeration of the keys in step 5, s.t. the total no. of iterations of step 5 is exactly one quarter of the size of the key space, i.e. 2^{62}. There exists however an enumeration, s.t. the no. of iterations of step 5 is about $2^{62} + 2^{48}$. It is given in Section 4.1 in every glory detail. Table 4 shows the estimates for space, time and number of chosen plaintexts for the attack, where one time unit is a full 16 round encryption and one space unit is 64 bits. The estimate for *Time* is the number of encryptions made in the analysis. In

Estimates for	Time	Space	Chosen plaintexts
	1.07×2^{62}	$2^{33} + 2$	$2^{33} + 2$

Table 4. Complexity of the chosen plaintext attack

every iteration of step 5 we do one full 16-round encryption in 5(a). For the two last tests in step 5(b) we do at most 2 rounds of encryption. For most iterations however, we need only to do one round of encryption, because we can test for equality of the right halves of $E_2(Swap(C'), Rol_{100}(K_L))$ and $C(K_L)$ (resp. $\overline{C^*(K_L)}$) already after one round of encryption of $Swap(C')$. If the tests fail we need not do the second round of encryption. Therefore for only about one out of 2^{31} iterations we need to do two rounds of encryption in 5(b). The total amount of time therefore is

$$(2^{62} + 2^{48}) \times \frac{17}{16} + (\frac{2^{62} + 2^{48}}{2^{31}}) \times \frac{1}{16} \simeq 1.07 \times 2^{62}.$$

Compared to this the time used in step 2 is negligible. The above attack is a weak attack. First of all, it is not very likely that we can get the encryptions for $2^{33}+2$ chosen plaintexts, furthermore an exhaustive search for 2^{62} keys is computationally infeasible. The LOKI cipher is meant as an alternative to DES, with which it is interface compatible. The so far best known attack on DES was introduced by Biham and Shamir in [3]. The attack is a chosen plaintext attack that needs 2^{47} chosen plaintexts. The time used in the analysis phase is 2^{37}. The time needed for the above attack on LOKI 91 is significantly higher than for Biham and Shamirs attack on DES, however the requirements for ever getting to the analysis phase, i.e. the number of encryptions of chosen plaintexts needed, are much higher for the attack on DES.
The steps 2, 3 and 5 can be carried out in parallel, for instance by letting $K_L = a$ in step 5, in that way we don't have to store the 2^{32} $C(a), C^*(a)$'s in step 3. It seems impossible however to obtain an enumeration that at the same time makes the total no. of iterations of step 5 be close to 2^{62} and enables a parallel run of the algorithm.

4.1 Enumeration of the keys in the chosen plaintext attack

We use the same notation as in the previous Sect. Let $\mathcal{A}$ be a function from $GF(2)^{64}$ to itself

$$\mathcal{A} : K_L \| K_R \rightarrow K_R \| Rol_{25}(K_L)$$

As stated above, once we have tried the key $K = K_L \| K_R$ in step 5 of the algorithm without success, we don't have to try the keys

$$\mathcal{A}(K),\ \overline{K},\ \mathcal{A}(\overline{K})$$

The idea is to use $\mathcal{A}$ to construct a set of keys about half the size of the key space and s.t.

- the biggest block of bits in every key consists of 1's.
- for every key K, $\mathcal{A}(K)$ is also in the set.

Then let the enumeration of the keys be every second key from the above constructed set of keys. Later in this Sect. we show that the enumeration obtained this way makes the total no. of keys tried in the attack be very close to 2^{62}.
Let $\mathcal{A}list(K)$ be the set of 64 keys $\{K, \mathcal{A}(K), \mathcal{A}^2(K), \ldots\ldots\ldots, \mathcal{A}^{63}(K)\}$. Note that $\mathcal{A}^{64}(K) = K$. Define for $K = K_L \| K_R$

$$\mathcal{M}_K = \cup_{p,q}\{\mathcal{A}list(Rol_p(K_L)\|Rol_q(K_R)) \cup \mathcal{A}list(Rol_p(K_R)\|Rol_q(K_L))\}$$

for $p = 0, 1, 2, 3$ and $q = 0, 8, 16, 24$.

Lemma 3 *For $K = K_L \| K_R$, $\mathcal{M}_K$ is the set of all keys of the forms:*

$$Rol_x(K_L)\|Rol_y(K_R)$$

$$Rol_x(K_R)\|Rol_y(K_L)$$

for all $x, y \in \{0, 1, \ldots\ldots, 31\}$

Proof: For fixed K there are $2 \times 32 \times 32 = 2^{11}$ keys of the above form. Since *Alist* produces 64 keys, the total no. of keys in $\mathcal{M}_K$ is $2 \times 16 \times 64 = 2^{11}$. Therefore it suffices to show that the pairs of rotations of the keys in $\mathcal{M}_K$ are distinct, i.e. that $Rol_a(K_L)||Rol_b(K_R)$ does not appear twice for any a, b. It is obvious that $Rol_a(K_L)||Rol_b(K_R)$ does not appear twice in one *Alist*. There are two cases to consider, $Rol_a(K_L)||Rol_b(K_R)$ appears in

1. $\underline{Alist(Rol_p(K_L)||Rol_q(K_R)) \text{ and } Alist(Rol_{p'}(K_L)||Rol_{q'}(K_R))}$
   $Rol_a(K_L)||Rol_b(K_R) = Rol_{p+25n}(K_L)||Rol_{q+25n}(K_R) \quad \wedge$
   $Rol_a(K_L)||Rol_b(K_R) = Rol_{p'+25n}(K_L)||Rol_{q'+25n}(K_R) \quad \Rightarrow$
   $p + 25n = p' + 25n \bmod 32 \wedge q + 25n = q' + 25n \bmod 32 \quad \Rightarrow$
   $p - p' = q - q' \bmod 32 \Rightarrow (p, q) = (p', q')$,
   since $p - p' \in \{-3, -2, -1, 0, 1, 2, 3\}$ and $q - q' \in \{0, 8, 16, 24\}$.

2. $\underline{Alist(Rol_p(K_L)||Rol_q(K_R)) \text{ and } Alist(Rol_{p'}(K_R)||Rol_{q'}(K_L))}$
   $Rol_a(K_L)||Rol_b(K_R) = Rol_{p+25n}(K_L)||Rol_{q+25n}(K_R) \quad \wedge$
   $Rol_a(K_L)||Rol_b(K_R) = Rol_{q'+25n}(K_L)||Rol_{p'+25+25n}(K_R) \quad \Rightarrow$
   $p + 25n = q' + 25n \bmod 32 \wedge q + 25n = p' + 25 + 25n \bmod 32 \quad \Rightarrow$
   $p + p' + 25 = q + q' \bmod 32$
   A contradiction, since $p+p'+25 \in \{25, 26, ..., 31\}$ and $q+q' \in \{0, 8, 16, 24\}$.

□

Let Ka and Kb be two 32-bit keyhalves, s.t. Ka and Kb are no rotations of each other, i.e. $Rol_x(Ka) \neq Kb$ for any x, $0 < x < 32$. For $K = Ka||Kb$, $\mathcal{M}_K$ is a set of distinct keys except in the cases where $Rol_x(Ka) = Ka$ and/or $Rol_y(Kb) = Kb$.

Lemma 4 *Let H be a 32-bit key and $Rol_n(H)$ any rotation to the left of H, where $0 < n < 32$. Then there are $2^{gcd(n,2)}$ possible values of H, such that $H = Rol_n(H)$.*

From Lemma 4 it follows for $K = K_L||K_R$, where K_L and K_R are no rotations of each other, that

Lemma 5 *There at most 2^{49} keys for which the elements in $\mathcal{M}_K$ are not distinct.*

Proof: Assume we have two equal keys K' and $K*$ from $\mathcal{M}_K$. Then

$$K' = Rol_a(K_L)||Rol_b(K_R), \; K* = Rol_c(K_L)||Rol_d(K_R)$$

Clearly from the proof of Lemma 1 $(a, b) \neq (c, d)$. Then

$$Rol_a(K_L) = Rol_c(K_L) \wedge Rol_b(K_R) = Rol_d(K_R) \Rightarrow$$
$$Rol_{a-c}(K_L) = K_L \wedge Rol_{b-d}(K_R) = K_R$$

If $a = c$ then there are 2^{32} possible values for K_L, but then there are at most 2^{16} possible values for K_R according to Lemma 4, since $(a, b) \neq (c, d)$. If $b = d$

then $a \neq b$ and we get a total no. of $2 \times 2^{32} \times 2^{16} = 2^{49}$ keys. □

The following algorithm makes a list of 32 bit strings, where no two strings are rotations of each other and where the biggest block of bits in every string consists of 1's.

ALGORITHM - No-rotations-of-keys (NRK)

For all positive $k \leq 32$, list all k-tuples $(a_1, a_2, ..., a_k)$, s.t.

1. $\sum_{i=1}^{k} a_i = 32$
2. $a_i \geq 1$ for $0 < i \leq k$
3. $\sum_{i=1}^{k} a_i \times 32^{k-i} \geq \sum_{i=1}^{k} a_{i+n \bmod (k+1)} \times 32^{k-i}$, for all $n \leq k$.

Method: For every k-tuple $(a_1, ..., a_k)$ output the 32-bit key, where the a_1 MSB are 1-bits, the next a_2 bits are 0-bits and so on.

Lemma 6 *No two keys in the output from (NRK) are rotations of each other.*

Proof: Because of the inequality in 3. above if $k > 1$, then k is even. Therefore for $k > 1$ the a_k LSB are 0-bits and furthermore $a_1 \geq a_i$ for $i \leq k$.

Let A and A' be two 32 bit keys from (NRK), s.t. $A = Rol_x(A')$ for some fixed x. Write A and A' as tuples $(a_1, ..., a_m)$ and $(a'_1, ..., a'_l)$ according to the method in (NRK). Clearly $l = m$ otherwise A cannot be a rotation of A'. Because $A = Rol_x(A')$ we have for some i

$$a_{1+n} = a'_{i+n \bmod (m+1)}, \quad 0 < n \leq m$$

Especially we have $a_1 = a'_i$ and $a'_1 = a_{m-i+2}$. Because of the inequality in 3. above we have

$$a'_i \leq a'_1 \ \wedge \ a_1 \geq a_{m-i+2}$$

Therefore $a'_i = a'_1 \Rightarrow a_1 = a'_1$. Now $a_1 = a_{m-i+2} \Rightarrow a_2 \geq a_{m-i+3}$. Similar as before

$$a_2 = a'_{i+1} \leq a'_2 = a_{m-i+3} \Rightarrow a'_2 = a_2$$

By induction we obtain $A = A'$ □

ALGORITHM - Enumeration (EN)

1. $i = 1$
2. Let K_L be the i'th output from (NRK).
3. For $j = 1$ to i do
   (a) Let K_R be the j'th output from (NRK)
   (b) For $K = K_L \| K_R$ output the first and then every second key from all *Alists* in $\mathcal{M}_K$
   (c) For $K = \overline{K_L} \| K_R$ do as in 3b
4. $i = i + 1$, goto 1.

We are left to check whether the set

$$KS = \cup_{Ki}\{Ki, \mathcal{A}(Ki), \overline{Ki}, \overline{\mathcal{A}(Ki)}\}$$

where the $Ki's$ are the keys output from (EN), contains the entire keyspace. Let $K^* = K_L^* || K_R^*$ be an arbitrary key. Rotate K_L^* and K_R^* such that the biggest blocks of bits (0's or 1's) are the MSB. Let $K(j) = K_L' || K_R'$ be that key.
If the MSB in both K_L' and K_R' are 1's then they are both output from (NRK). Then at some point $K(j)$ or $K(l) = K_R' || K_L'$, say $K(j)$, are the key considered in step 3(b) of (EN). Let $K(n)$, $0 < n \leq 2^{10}$ be all keys output in step 3(b) when $K = K(j)$. Then $L = \{K(n), \mathcal{A}(K(n))\}$, $0 < n \leq 2^{10}$ are all rotations of the key halves in $K(j)$ according to Lemma 3. Therefore $K^* \in L \in KS$.
If MSB in both K_L' and K_R' are 0's, then at some point either $\overline{K(j)}$ or $\overline{K(l)}$ are the key considered in step 3(b). Let $K(n)$ be as before, when $K = \overline{K(j)}$. Then $L = \{K(n), \mathcal{A}(K(n))\}$, $0 < n \leq 2^{10}$ are all rotations of the key halves in $K(j)$ according to Lemma 3. Therefore $\overline{K^*} \in L \in KS \Rightarrow K^* \in \overline{L} \in KS$.
If the MSB in K_L' and K_R' are 1's and 0's resp. or vice versa similar arguments hold for step 3(c).
We have implemented (NRK) on a SUN-Sparc workstation. The number of key halves output from (NRK) is $2^{26}+2068$. It means that the number of keys output from (NRK) in 2. and 3(a) above is about $2^{51} + 2^{37}$. Every second key from $\mathcal{M}_K$ gives 2^{10} keys for each K. The total number of keys in the enumeration therefore is about

$$(2^{51} + 2^{37}) \times 2 \times 2^{10} = 2^{62} + 2^{48}.$$

We have given an enumeration of the keys, s.t. the total no. of iterations of step 5 in the algorithm of the chosen plaintext attack is close to 2^{62}. The time used in the enumeration (EN) is negligible compared to the 1.07×2^{62} full 16 rounds of encryption of LOKI 91, since it runs only one time per every 2×2^{10} runs of step 5 in the algorithm of the chosen plaintext attack.

5 Conclusion and open problems

We have shown that we cannot find a characteristic for LOKI 91 good enough to do a successful differential attack on LOKI 91. Still it is not enough to conclude that LOKI 91 is secure against this kind of attack. To do that we need an efficient way of calculating the probabilities of differentials.
We have shown that the size of the image of the F-function in LOKI 91 is only $\frac{8}{13}$ of the size of the image of the F-function in DES. We have found no way of exploiting this fact in an attack on LOKI 91. Whether it represents a weakness of the algorithm is left as an open question.
Finally we introduced a chosen plaintext attack on LOKI 91 that reduces an exhaustive key search by almost a factor 4. The attack exploits a weakness in the key schedule of LOKI 91. It might also be possible to use this weakness, the common round key property, to find collisions for LOKI 91 when used as a hash function. This is left as an open question.

6 Acknowledgements

We wish to thank Prof. Ivan Damgård for valuable discussions on the enumeration of the keys in the chosen plaintext attack.

References

1. E. Biham, A. Shamir. *Differential Cryptanalysis of DES-like Cryptosystems.* Journal of Cryptology, Vol. 4 No. 1 1991.
2. E. Biham, A. Shamir. *Differential Cryptanalysis of Snefru, Khafre, REDOC-II, LOKI and Lucifer.* Extended abstract appears in Advances in Cryptology, proceedings of CRYPTO 91.
3. E. Biham, A. Shamir. *Differential Cryptanalysis of the full 16-round DES.* Technical Report # 708, Technion - Israel Institute of Technology.
4. L. Brown, J. Pieprzyk, J. Seberry. *LOKI - A Cryptographic Primitive for Authentication and Secrecy Applications.* Advances in Cryptology - AUSCRYPT '90. Springer Verlag, Lecture Notes 453, pp. 229-236, 1990.
5. L. Brown, M. Kwan, J. Pieprzyk, J. Seberry. *Improving Resistance to Differential Cryptanalysis and the Redesign of LOKI.* Abstracts from ASIACRYPT'91.
6. X. Lai, J. L. Massey, S. Murphy. *Markov Ciphers and Differential Cryptanalysis.* Advances in Cryptology - Eurocrypt '91. Lecture Notes in Computer Science 547, Springer Verlag.
7. K. Nyberg, L. Ramkilde Knudsen. *Provable Security Against Differential Cryptanalysis.* Presented at the rump session of CRYPTO'92. To appear in the proceedings of CRYPTO'92.
8. L. Ramkilde Knudsen. *Cryptanalysis of LOKI.* Abstracts from ASIACRYPT'91.
9. L. Ramkilde Knudsen. *Iterative Characteristics of DES and s^2-DES.* To appear in the proceedings from CRYPTO'92.

Cryptanalysis of Summation Generator

Ed Dawson

Information Security Research Centre, Queensland University of Technology, Gardens Point Campus, GPO Box 2434, Brisbane, Queensland 4001, Australia

Abstract. In this paper two known plaintext attacks on the summation generator will be described. The first attack is a new method of attack while the second method is due to Meier and Staffelbach. These two methods will be compared and contrasted.

1 Introduction

Many communications systems such as Fascimile machines require high speed algorithms to encrypt binary coded plaintext messages which may be several million bits long and display notable structure. The most commonly used cipher in this case is a stream cipher. In a stream cipher the plaintext is divided into 'characters' corresponding to one or more bits, and each character is encrypted separately by a time varying function. After encrypting each character the function changes state so that two occurences of the same plaintext character do not necessarily define the same ciphertext character. The simplest and most often used stream cipher for encrypting binary plaintext is where the bit at time interval t of a pseudorandom sequence expressed by, $z(t)$, is combined using modulo two addition with the plaintext bit, $p(t)$, at time interval t to produce the ciphertext bit at time interval t, denoted by $c(t)$. The sequence $z(t)$ is called the keystream for the stream cipher. The encryption process can be expressed as

$$c(t) = p(t) \bigoplus z(t) \tag{1}$$

where $\bigoplus$ denotes modulo two addition. The decryption process can be expressed as

$$p(t) = c(t) \bigoplus z(t) \tag{2}$$

A common method to form the keystream is to apply a nonlinear Boolean function to the output of several linear feedback shift registers (LFSR's). One such method is the summation generator from [2]. A description of the summation generator will be given in Section 2. In Sections 3 and 4 two known plaintext attack algorithms are described for the summation generator. The first algorithm is a new method for attacking the summation generator. The second algorithm is from [1]. It should be noted that the first algorithm was derived without knowledge of the second algorithm. Both algorithms rely on the same assumptions. A brief comparison of the advantages of attacking the summation generator using each algorithm will be given.

2 The Summation Generator

The summation generator from [2] is formed by combining two or more sequences using one memory bit. In this fashion suppose that there are k m-sequences as denoted by $a(i,t)$ for $i = 1, \ldots, k$ produced by shift registers whose lengths are pairwise relatively prime. The summation generator has output sequence $z(t)$ and memory bit sequence $s(t)$ defined by

$$z(t) = \sum_{i=1}^{k} a(i,t) \bigoplus s(t-1) \tag{3}$$

$$s(t) = \sum_{i=1}^{k}\sum_{j=1}^{k} a(i,t)a(j,t) \bigoplus \sum_{i=1}^{k} a(i,t)s(t-i) \bigoplus \sum_{i=1}^{k} a(i,t) \tag{4}$$

where the arithmetic is modulo two. The possible keys for the summation generator are the initial state vectors and the tap settings for the shift registers, and the selection of the initial value of the memory bit $s(-1)$.

Suppose that two LFSR's are used to form the summation generator where the two input sequences $a(t)$ and $b(t)$ are produced by LFSR(1) and LFSR(2) respectively. In this case the output sequence $z(t)$ of the summation generator is defined by the Boolean equations

$$z(t) = a(t) \bigoplus b(t) \bigoplus s(t-1) \tag{5}$$

$$s(t) = a(t)b(t) \bigoplus (a(t) \bigoplus b(t))s(t-1) \tag{6}$$

The formation of the sequence $z(t)$ as defined by (5) and (6) is indicated in Figure 1.

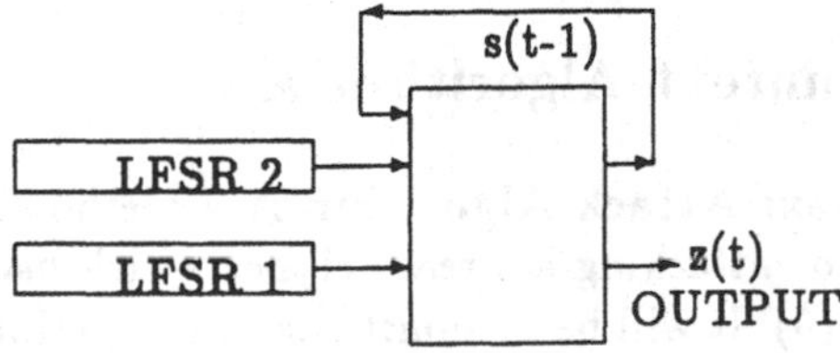

Figure 1
Summation Generator

Theorem 1. *The sequence* $z(t)$

1. *has a period whose length is ultimately* $(2^{L(1)} - 1)(2^{L(2)} - 1)$*;*
2. *has a linear complexity approximately equal to the period length;*
3. *is first order correlation immune.*

As demonstrated by this theorem the sequence $z(t)$, defined by (5) and (6), seems to satisfy the requirements of a pseudorandom sequence for use in a stream cipher secure from cryptanalytic attack. However, it will be demonstrated that such a cipher can be attacked based on the following three assumptions:

Assumption 1 The cryptanalyst knows a small number r of consecutive plaintext bits.

Assumption 2 The cryptanalyst knows the ciphertext sequence c(t).

Assumption 3 The cryptanalyst knows the type of generator being used; the lengths $L(1)$ and $L(2)$; and the tap settings for LFSR(1) and LFSR(2).

In many applications of stream ciphers it is a relatively easy task for an attacker to obtain a small number of consecutive plaintext bits $p(t)$ due to the redundancy of the language set being used. For example if a stream cipher is used to encrypt digital speech the cryptanalyst would only need to search for a silent section.

Under Assumption 2 the cryptanalyst is conducting passive wiretapping to obtain ciphertext bits $c(t)$. By substitution of the known r bits of plaintext together with the corresponding ciphertext bits into the equation $c(t) = p(t) \bigoplus z(t)$ the cryptanalyst can derive the corresponding r bits of the summation generator.

It should be noted that both the attack algorithms described in Sections 3 and 4 enable the cryptanalyst to derive the state vectors for the two shift registers for the time interval corresponding with the first of the r known consecutive plaintext-ciphertext pairs $p(t)$, $c(t)$. However, due to the deterministic nature of LFSR(1) and LFSR(2) once the cryptanalyst has derived state vectors at any time interval it is possible to derive all the preceeding and succeeding outputs of each shift register. This being the case it will be assumed without loss of generality that the cryptanalyst has available the first r plaintext-ciphertext pairs. This would imply, as shown above, that the cryptanalyst has available the first r bits of the summation generator $z(0), z(1), \ldots, z(r-1)$.

3 Known Plaintext Algorithm A

In **Known Plaintext Attack Algorithm A** a method is described, based on Assumptions 1-3, for attacking a stream cipher which uses a keystream $z(t)$ as defined by (5) and (6). It will be demonstrated by experimental results that this attack is possible provided the number of known consecutive plaintext-ciphertext pairs is approximately $L(1) + L(2) + 1$. The maximum number of seeds required to test in this attack is $2(2^{L(1)} - 1)$.

Known Plaintext Attack Algorithm A

Start of Algorithm

Step 1. Select $a(0)$ and $b(0)$. There are four possible such selections. Each of these selections defines a value of $s(-1)$ in (5) to match $z(0)$. By substitution of $a(0)$, $b(0)$ and $s(-1)$ into (6), $s(0)$ is found.

Step 2. There are two choices for each of $a(1), \ldots, a(L(1)-1)$. For each choice the corresponding values $b(1), \ldots, b(L(1)-1)$ and $s(1), \ldots, s(L(1)-1)$ are fixed by (5) and (6) respectively using the known values of $z(1), \ldots, z(L(1)-1)$.

Step 3. After Step 2 there are a total of $2(2^{L(1)}-1)$ possible seeds to select defined by $b(0)$ and $a(0), a(1), \ldots, a(L(1)-1)$. After a choice of $a(0), \ldots, a(L(1)-1)$ the terms $a(L(1)), \ldots, a(r-1)$ are derived from the recurrence relation for LFSR(1).

Step 4. Derive pairs $b(j)$,$s(j)$ for $j = L(1), \ldots, L(2)-1$ to match pairs $a(j)$, $z(j)$ by substitution into (5) and (6) respectively.

Step 5. Given $b(0), b(1), \ldots, b(L(2)-1)$ from Steps 1,2,4 the terms $b(L(2)), \ldots, b(r-1)$ are derived from the recurrence relation for LFSR(2).

Step 6. At this stage the seeds in LFSR(1) and LFSR(2) are tested against the known values $z(i)$ for $i = L(2), \ldots, r-1$. Let

$$u(i) = a(i) \bigoplus b(i) \bigoplus s(i-1)$$

where $s(i-1)$ for $i = L(2), \ldots, r-1$ is derived from (6) using the values of $a(i-1), b(i-1), s(i-2)$. Whenever a value of i is found such that $u(i)$ and $z(i)$ are not equal, it is necessary to return to either Steps 1 or 2 to select a new choice for one of $b(0)$ and $a(0), \ldots, a(L(1)-1)$. If $u(i) = z(i)$ for $i = L(2), \ldots, r-1$ then exit. The correct seed with a high probability is the current values $a(0), \ldots, a(L(1)-1), b(0), \ldots, b(L(2)-1), s(-1)$ provided a sufficiently large value of r has been used.

End of Algorithm

Provided the number of known plaintext bits r is slightly larger than $L(1)+L(2)+1$ there is a high probability that the above algorithm will find the correct seeds used to form the ciphertext. The reason why this is the critical value of r in the algorithm for successfully attacking the cipher is that firstly one requires $L(2)$ initial values of $z(j)$ in Steps 1-4 to fix current test seeds $a(0), \ldots, a(L(1)-1), b(0), \ldots, b(L(2)-1), s(-1)$ for Step 6. Secondly, in testing a false seed in Step 6 there is approximately a 50% chance that a given $u(i)$ will correspond to a given $z(i)$ for i selected from $L(2), \ldots, r-1$. Hence approximately one half of the seeds will define a value of $u(L(2))$ that is not equal to $z(L(2))$. For the remaining seeds approximately one half will define a value of $u(L(2)+1)$ that is not equal to $z(L(2)+1)$ and so forth. There is a total of $2(2^{L(1)}-1)$ seeds to test so that it will take approximately $L(1)+1$ trials to eliminate all the wrong seeds. In this fashion approximately $L(1)+1$ values of $z(i)$ are needed in Step 6. Hence provided r is slightly larger than $L(1)+L(2)+1$ the above algorithm will find the correct seed with a very high probabilty.

Known Plaintext Attack Algorithm A as described above was applied to several sequences formed by summation generators. Table 1 contains the results. The lengths $L(1)$ and $L(2)$ of the shift registers for each generator attacked are included in this table. The table contains the number of seeds that fail Step 6 of the algorithm after $L(2)+i$ bits of $z(t)$ are known for the various values of i. As demonstrated in Table 1 approximately one half of the seeds fail for each increase in i by one and in addition approximately all the false seeds fail provided $L(1)+L(2)+1$ bits of $z(t)$ are known. This experimental result is consistent with the theoretical result for r derived above.

If the tap settings of LRSR(1) and LFSR(2) are unknown it is only necessary to test for the tap settings for LRSR(1). One can determine the tap settings of LFSR(2) provided on determines $2L(2)$ terms of sequence $b(t)$ in Step 4 of the algorithm instead of $L(2)$ terms. Given $2L(2)$ consecutive terms of sequence $b(t)$ the tap settings can be determined by applying the Berlekamp-Massey Algorithm (see [2]).

$L(2)+i$	$L(1)=7$ $L(2)=9$	$L(1)=9$ $L(2)=11$	$L(1)=9$ $L(2)=13$	$L(1)=11$ $L(2)=19$	$L(1)=13$ $L(2)=19$
$L(2)+1$	126	509	509	2047	8191
$L(2)+2$	59	258	259	1023	4094
$L(2)+3$	34	128	117	510	2048
$L(2)+4$	18	66	64	265	1024
$L(2)+5$	9	22	37	112	519
$L(2)+6$	2	18	16	61	248
$L(2)+7$	2	9	11	37	134
$L(2)+8$	1	7	3	15	60
$L(2)+9$	0	2	4	11	34
$L(2)+10$	1	0		3	15
$L(2)+11$		0		4	6
$L(2)+12$		1		3	4
$L(2)+13$				1	3
$L(2)+14$				0	0
$L(2)+15$				1	1

Table 1
Number of Seeds Eliminated

4 Known Plaintext Attack Algorithm B

In [1] a second algorithm for attacking the summation generator is given. This shall be denoted as Known Plaintext Attack Algorithm B. Only a brief description of this technique will be given since a full description can be found in the above reference. The use of Algorithm B depends on the same assumptions as Algorithm A. It is assumed that the attacker has derived a portion of the keystream. The application of Algorithm B depends on the result that in a run of zeroes or ones in the keystream the memory bit sequence becomes highly predictable for a few terms. For example in (5) and (6):

$$\text{if } z(t) = z(t+1) = \ldots = z(t+j-1) = 0 \text{ and } s(t-1) = 1$$
$$\text{then } s(t) = s(t+1) = \ldots = s(t+j-1) = 1.$$

Similarly

$$\text{if } z(t) = z(t+1) = \ldots = z(t+j-1) = 1 \text{ and } s(t-1) = 0$$
$$\text{then } s(t) = s(t+1) = \ldots = s(t+j-1) = 0.$$

The cryptanalyst thus examines the known portion of the keystream for runs of consecutive zeroes or ones. It is suggested that the length of each such run be at least seven. Each set of runs enables the cryptanalyst to generate a system of linear equations. Provided sufficient sets of linear equations can be generated the cryptanalyst can solve these equations by back substitution for the initial state vectors for the two LFSR's used to form the keystream.

As claimed in [1], Algorithm B enables a cryptanalyst to attack a summation generator even in the case where both shift registers used are of considerable length. For example in [1] it is claimed that it may be possible to attack such a cipher where both shift registers are of length approximately 200 provided the number of known keystream bits is of the order of 50000. In the case of Algorithm A an attack on such a system is not feasible since the number of seeds required to test would be of the order of 2^{200}. From the above discussion Algorithm B has an advantage over applying Algorithm A in the case where both shift registers are large in length. However, the application of Algorithm B requires the cryptanalyst to know significantly more bits of the keystream in most cases than the application of Algorithm A. For example suppose that two LFSR's are used of length 23 and 25. Using Algorithm A the cryptanalyst would need to know only about 50 consecutive bits of keystream in order to conduct a successful attack. The number of seeds required to test in this case would be of the order of 2^{24} at the most. If the cryptanalyst uses Algorithm B instead to attack such a system it would require in most cases knowledge of at least 1000 keystream bits. Clearly, in this case, the summation generator is more susceptible to an attack using Algorithm A.

Firstly, it should be noted that in certain cases, by using methods from Algorithm B, it is possible to reduce the number of seeds required to attack the summation generator using Algorithm A. If the known keystream begins with a run of zero or one bits of length s, then as shown in [1] the memory bit after j terms can be predicted with a probabilty of at least $1-2^{-j}$. Hence, suppose that the known plaintext bits begin with a run j bits long where j is greater than six. By delaying the attack procedure in Algorithm A until after the first j terms of the keystream, the attacker needs only test at the most approximately $2^{L(1)}$ seeds in the case where the corresponding term of the memory bit sequence is determined with a high probability. Hence, in this case it is possible to reduce by a factor of one half the total number of seeds required to be tested in an attack on the summation generator.

Secondly, it should be noted that any sequence $z(t)$ which is defined as in (3) and (4) using two LFSR's, and one memory bit determined by a fixed Boolean function g can be attacked using Algorithm A. The derivation of the memory bit sequence $s(t)$ using Algorithm A only depends on the Boolean formula being fixed. However, it is possible to design a system which is secure from attack using Algorithm B by selecting an appropriate Boolean function g. For example let $z(t)$ be a sequence defined by

$$z(t) = a(t) \bigoplus b(t) \bigoplus s(t-1) \tag{7}$$

$$s(t) = a(t)s(t-1) \bigoplus s(t-1)b(t) \bigoplus b(t) \tag{8}$$

In this case the memory bit sequence $s(t)$ no longer becomes highly predictable in the case of a run of zeroes or ones in $z(t)$. For example if $z(t) = z(t+1) = 0$ and $s(t-1) = 0$ then $s(t)$ can be either zero or one. Similarly if $z(t) = z(t+1) = 0$ and $s(t-1) = 1$ then $s(t)$ can be either zero or one. In this fashion an attack on the cipher as defined by (7) and (8) using Algorithm B is not possible.

References

[1] W. Meier and O. Staffelbach: Correlation properties of combiners with memory in stream ciphers, Abstracts of EUROCRYPT 90, (1990), 188-196. i

[2] R.A. Rueppel: New Approaches to Stream Ciphers, PhD Thesis. Swiss Federal Institute of Technology, (1984).

[3] T. Siegenthaler: Decrypting a class of stream ciphers using ciphertext only, IEEE Transactions on Computing, V.C-34, (1985), 81-85.

Session 6

PROTOCOLS I

Chair: Rei Safavi-Naini
(University of Wollongong, Australia)

Secure Addition Sequence and Its Applications on the Server-Aided Secret Computation Protocols

Chi-Sung Laih and Sung-Ming Yen

Communication Laboratory,
Department of Electrical Engineering,
National Cheng Kung University,
Tainan, Taiwan, Republic of China

Abstract. Recently, researchers consider an approach called the Server Aided Secret Computation (SASC) protocol by using a powerful untrusted auxiliary device to help a smart card for computing a secret function efficiently. However, the computation of their protocol possesses some redundancy. In this paper, we give a new concept called the Secure Addition Sequence and develop an efficient algorithm to construct the Secure Addition Sequence. Based upon the concept of Secure Addition Sequence, performance of the SASC protocol can be enhanced.

1 Introduction

Smart cards under current technology cannot support high speed complicated computations required for most of the public key cryptographic schemes, e.g., the RSA cryptosystem [1] and the key distributions [2] which require modular exponentiation as their fundamental arithmetic operations. Recently, Matsumoto et al. consider the approach by using a powerful untrusted auxiliary device (called the server) to help a smart card (called the client) to perform a secret computation efficiently [3,4]. In this model, the smart card can perform its complicated computations efficiently with the aid of a powerful host computer through an insecure public channel while revealing no secret to the host computer or the eavesdroppers in the public channel. Matsumoto et al. and some other researchers [4,5,6,7] have proposed several protocols to solve the problem. Such protocols are referred to as the *server-aided secret computation* (SASC) protocols. Matsumoto et al. have proposed a solution of the problem which is suitable for the RSA cryptosystem [4]. Kawamura et al. [8] have shown that a client, with a 10^5 times more powerful server's aid, can compute an RSA signature 50 times faster than the case without a server if the communication cost can be ignored. Recently, Pfizmann and Waidner [14] gave passive attacks to reduce the search space and an effective active attack to break the SASC protocol of [4]. Also, Shimbo and Kawamura had proposed an effective attack in [15]. However, a two-round SASC protocol which is free from all the described passive and active attacks with some precautions has been developed by the authors recently [16].

As described in the later section, the computation of the previous protocols possess some redundancy. In order to overcome this disadvantage, we propose a new terminology called the *Secure Addition Sequence* in this paper which is the extended concept of addition sequence with security characteristics. We also develop an efficient algorithm to construct the *secure addition sequence*. Based upon the definition of **secure addition sequence**, the performance of previous SASC protocols for modular exponentiation computations in cryptographic systems, e.g., the RSA decryption and signature, can be improved while the security requirement still holds. Theoretically, security consideration of the *secure addition sequence* and the SASC protocol can be independently solved.

2 The Secure Addition Sequence and the Constructing Algorithm

2.1 Problem description:

In the SASC protocols [4,16], a client has a pair of public keys (e, n) and a secret key d, where n is the product of two secret strong primes p and q. The client wishes to compute the signature $Y \equiv X^d \pmod{n}$ of a message X with the aid of a server under the constraint that it may possibly reveal the client's secret d to the server with the probability lessthan or equal to the security level ϵ.

The computations of the SASC protocols in [4,16] possess some redundancy in their fundamental scheme because all the evaluations of $X^{d_i} \pmod{n}$ are operated independently one after another. Apparently, there are large amount of same operations done repeatedly. This will make the SASC protocols inefficient. To make the computational load of SASC protocols more reasonable for practical applications, addition sequence is applied into the protocol (for the basic idea and more detailed description see reference [13]). Based on the concept of addition sequence the redundancy problem of the previous protocols can be overcome. The first protocol in [13] recommends that the server could construct an addition sequence for D to speed up the modular exponentiations. This protocol will improve the server's efficiency while the performance of the client remains the same. The second protocol in [13] suggested the client to construct an addition sequence, with secrets hidden into it, to speed up the computations for both the client and the server simultaneously.

2.2 The definition of secure addition sequence:

In the protocol-2 proposed by Laih et al. [13], addition sequence is introduced to save both the computations of the client and the server simultaneously while another problem is induced. How can we prevent the secrets from being extracted. This cannot always be the case. The secret numbers cannot always be protected with some required security requirement when they are embedded in a public addition sequence using some kinds of improper addition sequence generation algorithm. To overcome this possible weakness, we introduce a new concept extended from the basic definition of addition sequence and call it the *secure addition sequence*. Here we first give the fundamental definition of **secure addition sequence (SAS)** then a secure and efficient **SAS** generation algorithm is developed.

[Definition-1]

Given an increasing sequence $D = \{d_1, d_2, \cdots, d_t\}$ and one of its corresponding addition sequence$A = \{a_1, a_2, \cdots, a_r\}$ such that

$$a_1 = 1, a_r \geq d_t,$$
$$a_i = a_j + a_k \;\; for \;\; i > j \geq k,$$
$$D \subseteq A.$$

If the probability of finding all elements $d_i, i = 1, 2, ...t$, from A, even the generating algorithm of A is known, is less than or equal to ϵ (> 0) then we say that the addition sequence A is a *secure addition sequence* with security ϵ (abbreviated as SAS-ϵ)for D.

The security level ϵ in the above SAS-ϵ can be any probability required depending on the applications. While it is obvious that the probability ϵ cannot less than the probability of guessing the secret numbers by brute force exhaustive searching method.

[Definition-2]

Given an increasing sequence $D = \{d_1, d_2, \cdots, d_t\}$ and the required security level ϵ . For one of the corresponding addition sequence $A = \{a_1, a_2, \cdots, a_r\}$, if the probability of finding all elements $d_i, i = 1, 2, \cdots, t$, from A is equal to $1/C\binom{r}{t}$ which is less than or equal to ϵ then we say that the addition sequence A is a **perfect secure addition sequence** (abbreviated as PSAS) for D with security level ϵ .

[Lemma-1]

Perfect secure addition sequence always exists for any general sequence D.

Proof:

For any sequence $D = \{d_1, d_2, \cdots, d_t\}$, a PSAS $A = \{1, 2, 3, 4, \cdots, a_r\}$ always exists such that $a_r \geq d_t$ and $1/C\binom{r}{t} \leq \epsilon$. (Q.E.D.)

In the following example, for some special sequences D, there exist corresponding special types of *perfect secure addition sequences.*

[Example-1]

Given $D_1 = \{8, 34\}, D_2 = \{5, 8\}$, and $D_3 = \{16, 64\}$ the corresponding *perfect secure addition sequences* for each of the sequences$D_i, i = 1, 2, 3$, are listed as follows:

$A_1 = \{1, 2, 3, 5, 8, 13, 21, 34, 55, 89\}$ which is a part of the Fibonacci sequence.

$A_2 = \{1, 2, 3, 4, 5, 6, 7, 8, 9, 10\}$.

$A_3 = \{1, 2, 4, 8, 16, 32, 64, 128, 256, 512\}$ which is the series of power of 2. And they all possess the probability ϵ to be $1/C\binom{10}{2}$.

The existence of PSAS cannot be directly applied to the SASC protocols. Since the length of the constructed sequence may be too long to save both the computations of the client and the server simultaneously. In the following, we will give the definition of optimal SAS-ϵ which is needed in the SASC protocol applications.

[Definition-3]

Given an increasing sequence D and a security level ϵ, if one of its corresponding PSAS has the shortest length, then this sequence will be called the **optimal secure addition sequence** (abbreviated as OSAS) for D with the security level ϵ. That is

$$len(A) \leq \{len(A_i)| \text{ all } A_i \in \text{PSAS of } D \text{ with security level } \epsilon\},$$

where *len(A)* denotes the length of sequence A.

A problem induced is that whether the *optimal secure addition sequence* can be easily constructed. To answer this question, we give the following conjecture.

[Conjecture-1]

Given an arbitrary increasing sequence D and the security level ϵ then finding its optimal secure addition sequence may be an **NP-complete** or even harder problem.

The reason for the above conjecture is that finding the shortest addition sequence has been proven to be an NP-complete problem [11] and finding the OSAS takes even more constraints than the previous one. Then, for the OSAS problem of general multiple value sequence D, heuristic algorithm is required to obtain the suboptimal solutions in reasonable time.

[Theorem-1]

Any two or more addition sequences can be combined using the operations of merging, sorting, and deletion into a single addition sequence.

Proof:

Without loss of generality, in this proof we just take the two-sequence combination case for illustration. The extension from two-sequence case to the multi-sequence case is straightforward.

Given any two addition sequences, $A_1 = \{a_{11}, a_{12}, \cdots, a_{1r}\}$ and $A_2 = \{a_{21}, a_{22}, \cdots, a_{2s}\}$. According to the definition of addition sequence, the following equations hold

$$a_{1a} = a_{1b} + a_{1c} \quad for \quad a > b \geq c,$$
$$a_{2d} = a_{2e} + a_{2f} \quad for \quad d > e \geq f.$$

Then we can construct another increasing sequence C' with its elements as $A_1 \bigcup A_2$. After this combination operation, the elements of C' are sorted then the same elements are deleted. The elements of the resulting sequence C may satisfy the following equation

$$c_x = c_y + c_z \quad for \quad x > y \geq z,$$

and therefore C is also an addition sequence. (Q.E.D.)

2.3 The generation algorithm of secure addition sequence with security level ϵ :

From the discussions and properties of SAS-ϵ described in the previous paragraphs, a criteria can be observed. In orderto embed secret numbers into the associated SAS-ϵ , the SAS-ϵ must constitute a long range of duration such that all the elements within this range are distributed homogeneously. The definition of homogeneous distribution is given below.

[Definition-4]

Given an increasing sequence A, we consider a range of numbers in A, i.e., $\{a_x, a_{x+1}, \cdots,$
$a_{y-1}, a_y\}$. The numbers in this range are called **homogeneously distributed** if one of the following conditions holds:

Type-I: $a_i = a_{i-j} + a_{i-k}$ where $1 \leq j, k \leq \delta, \delta$ is a small fixed integer.

Type-II: $a_i = a_{i-j} + a_k$ where j and k are two fixed integers. That is, no matter where you are in the homogeneous range (of course except the least two extreme numbers) the relationship between the number at any position and its former numbers in the range are all the same.

For example, the sequences A_1 and A_3 in the example-1 are the type-I homogeneously distributed sequences while the sequence A_2 is a type-II homogeneously distributed sequence.

Basically, the homogeneous distribution can be divided into the *globally homogeneous distribution* and *locally homogeneous distribution.* In *globally homogeneous distribution* , we mean that the entire range of numbers satisfy the homogeneous property. On the other hand, in *locally homogeneous distribution* , only a fraction of numbers in the sequence A satisfy the homogeneous property. In the following paragraph, two important things about *homogeneous distribution* will be given.

[Theorem-2]

Given an addition sequence A embedding the sequence D of t secret numbers, if the sequence A is *globally homogeneous* distributed then the sequence A is a *perfect secure addition sequence* for D.

Proof:

From the definition of globally homogeneous distribution, every number in the sequence A have the same relationship with the former numbers in the sequence. Then the probabilities for every number to be a secret are all the same even though the generating algorithm is known. Therefore, the sequence A is a PSAS by definition. (Q.E.D.)

The homogeneous relationship can be categorized as the *deterministic type* and the *probabilistic type* . Three examples of the *deterministic type* are given in example-1 while the *probabilistic type* will be demonstrated in the algorithm proposed later. It has been shown that PSAS always exists for a secure sequence D, while in practical applications we prefer the optimal one. It has also been pointed out that finding the OSAS may be an NP-complete problem then we require a heuristic algorithm for practical applications. Previously, we have proved that the *globally homogeneous distributed sequence* containing t secret numbers is a PSAS. In this section, we propose a **probabilistic** algorithm to construct the ***secure addition sequence*** for the secret sequence D containing only one secret number. Based upon this homogeneous SAS generation algorithm, a simple algorithm for constructing the SAS-ϵ of any sequence D with multiple secret numbers can be developed by the ***divide-and-conquer*** like methodology. By probabilistic algorithm, it means that given the same secret number as input data the algorithm may obtain different output at different time.

[Lemma-2]

Given a randomly selected even number E, the probability for $E = O \times 2^k$, where O being an odd number, is equal to $(1/2)^k$.

[Algorithm]:

Given a secret d, this algorithm constructs an addition sequence embedding the secret number d and with the largest element nearest to but less than the number Max and with sequence length Len. Then the obtained addition sequence is called the SAS-ϵ of d where $\epsilon = (1/Len)$.

Initial $A \equiv \phi; Len := 0;$

If (d is even) then $\{DE := d;\ DO := d+1;\}$ else $\{DE := d-1;\ DO := d;\}$

Upward $d_e := DE; d_o := DO;$

While ($d_e < Max$ and $d_o < Max$) do

{ $A := A \bigcup \{d_e, d_o\}; Len := Len + 2;$

Choose d_e or d_o with even probability to generate larger numbers:

If (the choice is d_e) then $\{d_e := 2 \times d_e; d_o := d_e + 1;\}$

else $\{d_e := 2 \times d_o; d_o := d_e + 1;\}$

}

Downward $d_e := DE; d_o := DO;$

While ($d_e > 2$) do

{ $d_e := d_e/2;$

If (d_e is even) then $\{d_o := d_e + 1;\}$ else $\{d_o := d_e; d_e := d_o - 1;\}$

$A := A \bigcup \{d_e, d_o\}; Len := Len + 2;$

}

$A := A \bigcup \{1\}; Len := Len + 1;$

Based upon lemma-2, the above algorithm guarantees to generate an SAS-ϵ for d where ϵ =(1/*Len*).

[Example-2]

Given d=20 and Max=180, one of the SAS-ϵ can be obtained as follows.
$1 \longrightarrow (2,3) \longrightarrow (4,5) \longrightarrow (10,11) \longrightarrow (20,21) \longrightarrow (42,43) \longrightarrow (86,87) \longrightarrow (172,173)$.
The security level ϵ is equal to 1/15.

From the definition of PSAS, it is obvious that the addition sequence for single element sequence D generated by the above algorithm is a PSAS. The following theorem gives the security analysis of the proposed single secret SAS-ϵ generation algorithm.

[Theorem-3]

Given a secure addition sequence generated from the single secret SAS-ϵ generation algorithm with length *Len*, the probability for each element to be the secret number is equal to 1/*Len*.

Proof:

From lemma-2 and theorem-2 it is obvious that all the numbers in the generated addition sequence satisfy the globally homogeneous distribution property. The probabilities for all the numbers to be the secret are all the same. Then the probability for each element to be the secret number is equal to 1/*Len* where *Len* is the length of the generated addition sequence. (Q.E.D)

A critical problem in the proposed algorithm is that why don't we always fix the number d_e or d_o for doubling operation or follow some fixed order to choose d_e or d_o for doubling operation in the upward phase. The reason is obvious, this strategy will leave a weak boundary of upward and downward from the secret number d. And the secret number d may be somewhere near the boundary.

Based on the algorithm described above, we now list the construction algorithm to obtain the SAS-ϵ for the multi-secret sequence $D = \{d_1, d_2, \cdots, d_t\}$ as following:

(Step-1) For each d_i in D, use the algorithm to construct one of its associated SAS-ϵ denoted as A_i with security level ϵ_i where $1/\epsilon_i$ being the length of the addition sequence A_i .

(Step-2) From theorem-1, a combined SAS-ϵ can be obtained from the sequences A_i, i=1,2,...,t, such that

$$\prod_{i=1}^{t} \epsilon_i \leq \epsilon.$$

2.4 Security analysis:

To extract the secret numbers D from the SAS-ϵ , giving the attacker the best benefit, we assume that the attacker can first divide the t sequences $A_i(i = 1, 2, \cdots, t)$ away. Then the attacker applies some best skills to extract the required secret numbers d_i individually. According to theorem-3, all the individual sequences are ***globally homogeneous sequences*** and also PSAS, it may not exist any method to extract the secret with computational complexity less than $(\epsilon_i)^{-1}$. s Under this fact, the computational complexity of extracting all the secret numbers D from the combined ***secure addition sequences*** can no longer less than ϵ^{-1} and the SAS-ϵ generation algorithm is secure for the given security level ϵ .

Here we give an efficient representation method to encode the ***secure addition sequence*** described above. And it is shown to be very useful for the SASC protocol proposed in the next section. In the ***secure addition sequence***, we can use a single bit to represent the selection of either d_e or d_o for doubling from the fixed pair of numbers (2, 3) because all the ***secure addition sequences*** generated from the proposed algorithm always contains the numbers $\{1, 2, 3\}$. If the upward pair of numbers are generated from d_e of the current pair,(d_e, d_o), it is encoded as a binary bit "0" otherwise it is encode as a binary "1". This representation method may be very time and space efficient for data transmission during the SASC protocols.

[Example-3]

Using the encoding method, the addition sequences $A_1 = \{1, 2, 3, 4, 5, 10, 11, 22, 23, 44, 45, \}$ and $A_2 = \{1, 2, 3, 6, 7, 14, 15, 28, 29, 58, 59\}$ can be encoded as: $A_1 = (0, 1, 1, 0)_2$ ***and*** $A_2 = (1, 1, 0, 1)_2$ sectionThe SASC Protocol Based on the Secure Addition Sequence and Its Performance Analysis subsectionThe SASC protocol based on the concept of Secure Addition Sequence: Because the single-round SASC protocol has been shown to posses some weakness, in the following discussion we consider only the two-round SASC protocol proposed in [16]. Using the ***secure addition sequence*** for the RSA-S1 type two-round SASC protocol, in each round the client first divides a secret number d_i into an integer vector $D_i = (d_{i1}, d_{i2}, \cdots, d_{it})$ such that $d_i \equiv \sum_{j=1}^{t} d_{ij} \pmod{\phi(n)}$. Then the client applies the ***secure addition sequence*** generation algorithm to construct a ***combined secure addition sequence*** SAS i for the integer vector D_i. subsubsectionTwo round SASC protocol based on the Secure Addition Sequence

Round-1

Step 1-1 The client sends X, n and $SAS_1 = \{a_{11}, a_{12}, \cdots, a_{1r}\}$ to the server where

$$r = 2 \times |n| - 1.$$

Step 1-2 The server computes the integer vector $Z_1 = (z_{11}, z_{12}, \cdots, z_{1r})$ such that

$$z_{1i} \equiv X^{a_{1i}} \pmod n, \text{ for all } i,$$

and sends Z_1 back to the client.

Step 1-3 The client computes X^{d_1} from Z_1 as follows

$$X^{d_1} \equiv \prod_{for\ a_{1i}=d_{1j}} z_{1i} \pmod n$$

Round-2

Step 2-1 The client sends ($X^{d1} \times R \pmod n$) and $SAS_2 = \{a_{21}, a_{22}, \cdots, a_{2r}\}$ to the server where R is a random number.

Step 2-2 The server computes the integer vector $Z_2 = (z_{21}, z_{22}, \cdots, z_{2r})$ such that

$$z_{2i} \equiv (X^{d1} \times R)^{a_{2i}} \pmod n, \quad for \ \ all \ \ i,$$

and sends Z_2 back to the client.

Step 2-3 The client computes X^d from Z_2 as follows

$$X^d \equiv [\prod_{for\ a_{2i}=d_{2j}} z_{2i}] \times (R^{d_2})^{-1} \pmod n$$

In the above protocol, all the operations of exponent partition and the evalution of $(R^{d_2})^{-1}$ can be performed off-line and it is a reasonable assumption for practical applications. subsectionPerformance analysis and comparisons: In this section, we focus our attention on the performance analysis of the application of *secure addition sequence* to the RSA-S1 type *two-round* SASC *protocol* which is free from the passive and active attacks in [14]. In the following analysis, the exponent d and the modulus n are both 512-bit integers. Under this consideration, we choose the parameter *Max* of the algorithm to be 2^{512}. It is obvious that ϵ_i is equal to 1/1023 because the sequence length of A_i will always be 1023. The performance analysis and security consideration are given in Fig-1 and Fig-2. In Fig-1, conventional implementation of the SASC protocol is considered under the assumption of L=26 and the relationship of security level (in $\log_{10}$ scale) versus parameter M is given. In Fig-2, the relationship of security level (in $\log_{10}$ scale) versus parameter T (where $T = 2 \times t$) of the *secure addition sequence* implementation is sketched. The security level of conventional implementation is approximated by C $\binom{M/2}{L/2}^2$ and the one of the *secure addition sequence* implementation is $(1023)^T$. If the security level is assumed to be 10^{30} then T=10 (for secure addition sequence method) and M=180 (for conventional method) satisfy the requirement. For the *secure addition sequence* implementation, exclusive other operations in the two-round SASC protocol, this takes the client 8 multiplications and the server about $1022 \times 10 = 10220$ multiplications which is about 13.3 exponentiations. For conventional implementation it takes the client 24 multiplications and the server about 180 exponentiations. For the *secure addition sequence* implementation, the communication cost can be reduced using the efficient encoding method proposed previously. sectionConclusion In this paper, a new concept called the secure addition sequence and its construction algorithm is proposed. The new concept has very interesting cryptographic properties. It is shown that the proposed construction algorithm is both simple and efficient. Based upon the concept of secure addition sequence, the SASC protocol can be designed to be very regular and efficient for both parties of the protocol, i.e., the client and the server. Most importantly, the proposed secure addition sequence concept can be used in any SASC protocol, developed or under developing, and any types of cryptosystem using modular exponentiation as their fundamental operations.

Besides the above summary, we want to point out an open problem to conclude this paper. To our best knowledge, it is still an open problem of whether we can construct a near optimal perfect secure addition sequence with general secret sequence D containing multiple secret numbers. If this problem can be solved then we can construct the required secure addition sequence without the aid of divide-and-conquer methodology. Under this condition, the length of the secure addition sequence may be reduced tremendously.

References

1. R.L. Rivest, A. Shamir, and L. Adleman, "A method for obtaining digital signatures and public-key cryptosystem," commun. ACM, Vol. 21, pp.120-126, Feb. 1978.
2. Diffie, W., and Hellman, M.E., "New directions in cryptography," IEEE Trans. on Inform. Theory, vol.IT-22, pp.644-654, 1976.
3. T.Matsumoto and H. Imai, "How to use servers without releasing Privacy-Making IC cards more powerful," IEICE Technical Report, Rep. ISEC88-33.
4. T.Matsumoto, K. Kato, and H. Imai, "Speeding up secret computations with insecure auxiliary devices," Proc. of CRYPTO'88, pp.497-506, 1988.
5. S. Kawamura and A. Shimbo, "Computation methods for RSA with the aid of powerful terminals," 1989 Sym.on Cryptography and Inf. Security, Gotemba, Japan (Feb. 2-4 1989).

6. S. Kawamura and A. Shimbo, "A method for computing an RSA signature with the aid of an auxiliary termimal," 1989 IEICE Autumn Natl. Conv. Rec. A-105.
7. J.J. Quisquater and M. De Soete, "Speeding up smart card RSA computations with insecure coprocessors," Proc. SMART CARD 2000. Amsterdam (Oct. 1989).
8. S. Kawamura and A. Shimbo, "Performance analysis of Server-Aided Secret Computation protocols for the RSA cryptosystem," The Trans. of the IEICE, vol. E73, No. 7, pp. 1073-1080, Jul. 1990.
9. D.E. Knuth, The art of computer programming, Vol. II: Seminumerical algorithms. Reading, Addison Wesley, 1969.
10. J. Bos, M. Coster, "Addition Chain Heuristics," Proceedings CRYPTO'89, Springer-Verlag Lecture Notes in Computer Science, pp.400-407.
11. P. Downey and B. Leony and R. Sethi, "Computing sequences with addition chains," Siam Journal Comput. 3 (1981) pp.638-696.
12. Andrew Yao, "On the evaluation of powers," Siam. J. Comput. 5, (1976).
13. C.S. Laih, S.M. Yen and L. Harn, "Two Efficient Server-Aided Secret Computation Protocols Based on the Addition Sequence," Proc. of the ASIACRYPT'91 and to be appeared in The Lecture Notes in Computer Science by Springer-Verlag.
14. B. Pfitzmann and M. Waidner, "Attacks on Protocols for Server-Aided RSA Computation," Proc. of the EUROCRYPT'92.
15. A. Shimbo and S. Kawamura, "Factorization Attack on Certain Server-Aided Computation Protocols for the RSA Secret Transformation," Electronics Letters, Vol. 26, No. 17, pp. 1387-1388, 1990.
16. T. Matsumoto, H. Imai, C. S. Laih, and S. M. Yen, "On Verifiable Implicit Asking Protocals for RSA Computation, "Proc. of the AUSCRYPT'92.

Fig. 1: Performance of the Conventional Method for the SASC for Parameter L=26

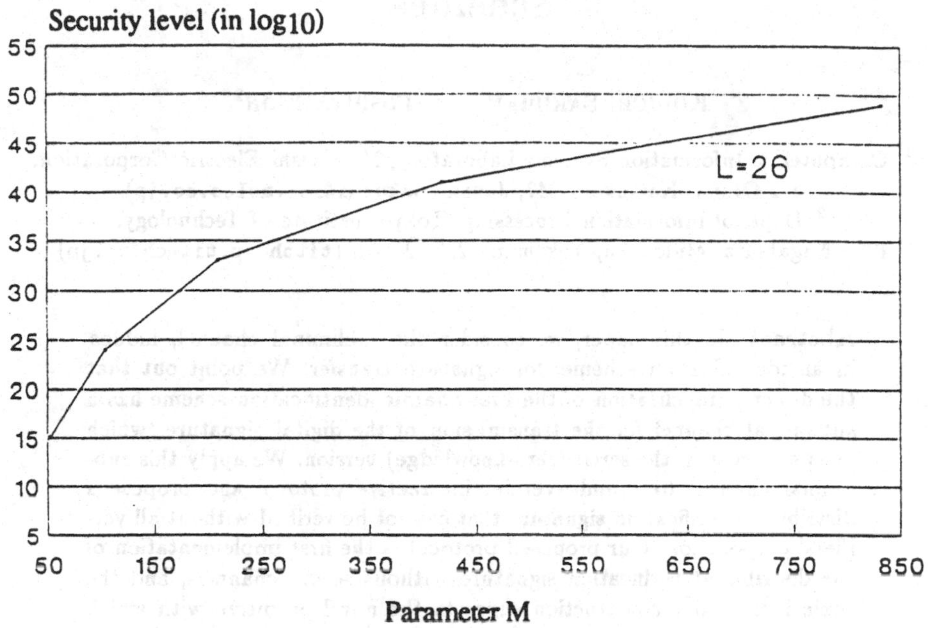

Fig. 2: Performance of the Secure Addition Sequence Method for the SASC

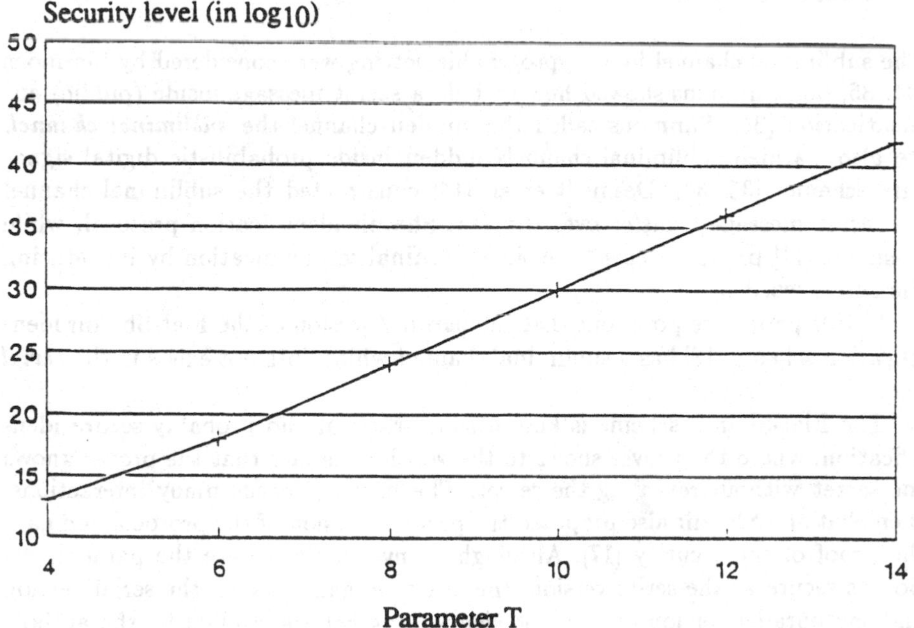

Subliminal Channels for Signature Transfer and Their Application to Signature Distribution Schemes

KOUICHI SAKURAI[1] TOSHIYA ITOH[2]

[1] Computer & Information Systems Laboratory, Mitsubishi Electric Corporation, 5-1-1 Ofuna, Kamakura 247, Japan (sakurai@isl.melco.co.jp).
[2] Dept. of Information Processing, Tokyo Institute of Technology, 4259 Nagatsuta, Midori-ku, Yokohama 227, Japan (titoh@ip.titech.ac.jp).

Abstract. In this paper, we consider the subliminal channel, hidden in an identification scheme, for signature transfer. We point out that the direct parallelization of the Fiat-Shamir identification scheme has a subliminal channel for the transmission of the digital signature, which does not exist in the serial (zero-knowledge) version. We apply this subliminal channel to a multi-verifier interactive protocol and propose a distributed verification signature that cannot be verified without all verifiers' corporation. Our proposed protocol is the first implementation of the distributed verification signature without secure channels, and the basic idea of our construction suggests the novel primitive with which a signature transfer secure against adversary can be constructed using only one-way function (without trapdoor).

1 Introduction

The subliminal channel in a cryptographic setting were considered by Simmons [34, 35, 36]. Simmons showed how to hide a secret message inside (public) authentication [34]. Simmons called this hidden channel the *subliminal channel*. He also clarified subliminal channels hidden inside probabilistic digital signature schemes [35, 36]. Desmedt et al. [14] constructed the subliminal channel for secret message transfer over the Fiat-Shamir identification protocol, while Desmedt [11] proposed how to avoid subliminal communication by introducing the active warden.

In this paper, we point out that the *parallel* version of the Fiat-Shamir identification scheme [18] has a subliminal channel which does not exists in the *serial* one.

The Fiat-Shamir scheme is known as a practical and probably secure identification, where the prover shows to the verifier the fact that the prover knows the secret without revealing the secret. The protocol needs many interactions, then Fiat and Shamir also proposed the parallel version of the protocol and gave the proof of the security [17]. Although many people believe the parallel version as secure as the serial version, the practical gap between the serial version and the parallel version of zero-knowledge protocol are studied by the authors

[33]. However, the previous paper [33] did not mention the gap as the subliminal channel hidden in the parallel version of the Fiat-Shamir's scheme. This subliminal channel is used not for secret message transfer but secret *signature* transfer. Namely, in the parallel version of the Fiat-Shamir scheme., the prover's signature (or prover's authorized object) is transmitted to the verifier over an public channel without being detected by any (polynomial-time bounded) third adversary. Note that our subliminal channel needs no common secret sharing in advance unlike the previous one.

The basic idea of our subliminal channel suggests the novel primitive with which the secure signature transfer is constructed by only one-way function unlike the fact that the the construction of the secure secret transfer is impossible without public-key primitive (or without common key sharing) [22]. In this paper, we suggest that the signature distribution in which the third party cannot keep signatures is possible under the sole assumption of the existence of one-way function. This is a new cryptographic primitive. Our construction is practical than theoretical because the formal characterization of a part of security of our scheme still remains as an open problem. However, authors believe that our approach will effect the theoretical result with probable security just like as Lamport's frontier work on the construction of the digital signature from one-way function [15] induces the theoretical result that the (secure) digital signature is a most fundamental primitive as one way functions [24, 30].

Occasionally, the subliminal channel is considered as an illegal object. However, we positively apply the subliminal channel to the model of single signer and multiple verifiers to get a signature distribution scheme.

A signature with a distributed verification is introduced by De Soete, Quisquater, and Vedder [16]. A $(s,t)-$distributed verification scheme satisfies (1) any t of the verifiers can execute the verification, and (2) the verification cannot be done by any $t-1$ or fewer of the s classes. They constructed a distributed verification scheme using a threshold scheme based on some finite geometric structure. However their implementation heavily depends on some physical assumptions (secure channel, trusted host, etc.), and is not so practical because, as Desmedt [13] pointed out, the available message space in their scheme is small. A practical distributed verification signature is implemented as a part of Pedersen's undeniable signature with distributed provers [29]. Pedersen's scheme is based on the combination of a (selective) convertible undeniable signature [2] and Shamir's secret sharing scheme [31]. Then his implementation needs some secure channel in order to distribute the signer's signature to each agent.

A remaining problem is if the distribution of signatures without secure channel nor secret sharing scheme is possible. Our proposed scheme gives an positive answer to the problem. Our construction is based on the subliminal channel in the Guillou-Quisquater's message authentication [21] with multiple verifiers. The model of the multiple-verifiers (zero-knowledge) interactive protocols was introduced by Burmester and Desmedt [3]. Burmester and Desmedt pointed out the multi-verifier zero-knowledge protocol is efficient than the single-verifier in some broadcast setting. Our proposed scheme is another useful application of multi-verifier interactive protocols.

Our signature distribution scheme satisfies the following conditions.

Validity: *Only* the signer can prove the validity of signer's messages to any verifiers by using signer's public key.

Private distributed recordability: Only the verifiers can keep a peace of signature.

Distributed right of publishing: Without cooperating all verifiers, the signature cannot be published.

2 A new subliminal channel

2.1 The Previous works on the subliminal channel

The subliminal channel in a cryptographic setting is considered by Simmons [34, 35, 36]. Simmons' original subliminal channel is a secret message transfer without secret channel, and Simmons [35, 36] constructs the subliminal channel based on probabilistic digital signature schemes. On the other hand, Desmedt [14] pointed out that the Fiat-Shamir identification scheme has a subliminal channel.

2.2 A New Subliminal Channel without Secret Sharing

All known subliminal channels need a common secret between two parties in advance. The two parties communicate secretly using the shared secret over public channel (e.g. probabilistic digital signatures [35, 36], or the Fiat-Shamir identification scheme [14]). Our question is that: **Does there exist subliminal channel without any shared secret ?** We gives an affirmative answer to the question.

Our basic observation is that the *parallel* version of the Fiat-Shamir's scheme has a subliminal channel which does not exists in the *serial* one, and this subliminal channel is useful for the signature transfer without being detected by any (polynomial-time bounded) third adversary.

The subliminal channel in the Fiat-Shamir identification is pointed out by Desmedt et al. [14]. However, we should note that our subliminal channel exists only in the parallel version.

Note that our subliminal channel needs no common secret sharing in advance unlike the previous ones.

The Fiat-Shamir scheme Recall the Fiat-Shamir identification scheme [18] is provably secure if factoring is difficult.

Fiat-Shamir identification scheme

1. PREPROCESSING STAGE BETWEEN THE TRUSTED CENTER AND EACH USER
   The unique trusted center's secret key in the system is (p, q), and the public key is N, where p, q are distinct large primes, $N = p \times q$. The center generates user A's secret key s_A, where $1/s_A = \sqrt{I_A} \pmod{N}$. I_A is the identity of user A and is published to other users.

2. IDENTIFICATION STAGE BETWEEN USER A AND USER B
   Repeat step (a) to (d) t times.
   (a) The user A picks $r \in_R Z_N^*$, and sends $x \equiv r^2 \pmod N$ to a user B.
   (b) The user B generates $e \in_R \{0,1\}$, and sends e to the user A.
   (c) The user A sends $y \equiv s_A^e r \pmod N$ to the user B.
   (d) The user B checks that $x \equiv y^2 I_A^e \pmod N$. If the check is not valid, the user B quits the procedure.

   The user B accepts A's proof of identity only if all t round checks are successful.

Remark. In the parallel version of the protocol above, A sends B all the x_i ($i = 1,\ldots,t$) simultaneously, then B sends A all the e_i ($i = 1,\ldots,t$), and finally A sends all the y_i ($i = 1,\ldots,t$) to B.

How to secretly transfer authorized objects Consider the following scenario. The prover want to secretly send an authorized objects (e.g. signature) to the verifier over the Fiat-Shamir identification scheme.

Although Desmedt already pointed out the scheme has a subliminal channel [14], we proposed completely different method which is based on the parallel version.

In the parallel version of the Fiat-Shamir scheme, consider the verifier who acts as follows.

> After receiving prover's first message $x_1,\ldots,x_t$, the verifier randomly selects R_V and sends back $e_i (i = 1,\ldots,t)$ which is computed as $(e_1,\ldots,e_t) = h(R_V, x_1,\ldots,x_t)$ by a one-way hash function h. After receiving the prover's answer $y_i (i = 1,\ldots,t)$ for $e_i (i = 1,\ldots,t)$, the verifier gets $H = (x_1,\ldots,x_t, h, R_V, e_1,\ldots,e_t, y_1,\ldots,y_t)$ as the history of the communication with prover A.

Recall the digital signature proposed by Fiat-Shamir [18].

Fiat-Shamir digital signature scheme

1. PREPROCESSING STAGE BETWEEN THE TRUSTED CENTER AND EACH USER
   Same as the preprocessing stage in the Fiat-Shamir identification scheme.
2. TO SIGN A MESSAGE M:
   The user A picks $r_i \in_R Z_N^*$ ($i = 1,\ldots,t$), and calculates $x_i \equiv r_i^2 \pmod N$ ($i = 1,\ldots,t$), $f(M, x_1,\ldots,x_t)$ and sets its first t bits to e_i ($i = 1,\ldots,t$). Furthermore, the user A computes $y_i \equiv s^{e_i} r_i \pmod N$ ($i = 1,\ldots,t$) and sends M, e_i, y_i ($i = 1,\ldots,t$) to the user B.
3. TO VERIFY A'S SIGNATURE ON M :
   The user B calculates $z_i = y_i^2 I_A^{e_i} \pmod N$ ($i = 1,\ldots,t$), $f(M, z_1,\ldots,z_t)$, and checks that its first t bits are equal to e_i ($i = 1,\ldots,t$). If the checks are valid, the user B recognizes that M is A's valid message.

We may regard the history H as A's digital signature on the message R_V. The secrecy of the verifier's random coin R_V implies the following.

Claim: Any (polynomial time bounded) third party cannot detect if the verifier get the history H even if the one-way hash function h is published.

It should be noted that the subliminal channel exists only in the parallel version but not in the serial version.

The community shall convince the subliminal channel hidden in the parallel version of the Fiat-Shamir scheme.

Remark. Desmedt [11] already mentioned that the verifier's random challenge $(e_1, \ldots, e_t)$ is useful for hiding a secret message M: the verifier computes $(e_1, \ldots, e_t) = E_K(M)$, where E is a conventional encryption cipher (e.g. DES) and K is secret key which is shared between the verifier and the prover in advance. We should note that the authors' method above is different from the Desmedt's one: in the authors' proposed subliminal channel, the signature is sent from the prover to the verifier.

General three move Protocols Several variants of the Fiat-Shamir identification and signature scheme are proposed [21, 6]. The technique in our subliminal channel is useful for any three move identification scheme which is the direct parallelization of the zero-knowledge protocol.

Claim: *Suppose the protocol is the iteration of a basic three move protocol (e.g. [4]) and is proven to be zero-knowledge. Then, the direct parallelization has a subliminal channel for secret signature (or authorized object) transfer.*

In the subliminal channel above, the message on the signature is selected by the verifier[3] , and the prover can not send the verifier the signature on the message which the prover selects. We overcome this problem by using the public-key message authentication [21, 10, 27] based on the Fiat-Shamir identification scheme. (We apply this idea to the distributed signature scheme in section 3.)

How to avoid the subliminal channel One protection to the abuse of the subliminal channel is that: *we use the serial version of the Fiat-Shamir scheme, which is proven to be zero-knowledge, instead of the parallel version.*.

The other protection is as follows. (This technique is already proposed by the authors [33] as a countermeasure of the abuse of the divertibility [26] of the parallel Fiat-Shamir scheme.)

[3] Okamoto and Ohta [26] apply this property to blind signature in a sophisticated manner.

After receiving the prover's first message $(x_1, \ldots, x_t)$, the verifier selects a random message R_V, computes $h(R_V, x_1, \ldots, x_t)$ and sets its first t bits to $e_1, \ldots, e_t$. Then the verifier sends the random message R_V to the prover instead of sending $e_1, \ldots, e_t$. The prover sends back the verifier $y_i \equiv u_i(y_i)^{e_i} (i = 1, \ldots, t)$, where $(e_1, \ldots, e_t) = h(R_V, x_1, \ldots, x_t)$ as the ordinary parallel Fiat-Shamir scheme. The verifier accepts the prover only if the checks $(e_1, \ldots, e_t) = h(R_V, x_1, \ldots, x_t)$, and $x_i \equiv y_i^2 (I_A)^{e_i} (i = 1, \ldots, t)$ are passed.

Remark. The modified protocol above still has a subliminal channel: the verifier can hide the secret message M inside the $R_V = E_K(M)$ as we remark in 2.2. However, this subliminal channel exists not only in the parallel version but also in the serial version. It should be noted that our proposed protection above conceals the subliminal channel exists only in the parallel version.

2.3 Yet another cryptographic primitive

The basic idea of the subliminal channel mentioned in the previous subsection suggests the novel primitive with which the secure signature transfer is constructed by only one-way function [32] unlike the fact that the the construction of the secure secret transfer is impossible without public-key primitive (or without common key sharing) [22].

A current research topic in cryptography is to provide constructions of fundamental primitives under assumptions that are as general (or weak) as possible. One of the most general (and popular) cryptographic primitives is the one-way function with no trapdoor. Another primitive is secure message transmissions. However, the secure message transmission seems to be indispensable to public-key primitive with trapdoor primitive (or common secret key sharing). Impagliazzo and Rudich [22] observed that public-key primitive (common secret key sharing) is a stronger assumption than the existence of one-way function.

In the case of the transfer of the signatures, the condition to be considered is mild. Namely, it is enough for the signer and the verifier to avoid being recorded the signatures by the third party even if the signature is convinced.

Okamoto and Ohta [27] proposed non-transitive signature based on the Fiat-Shamir scheme with a modification on the prover's randomness. The same idea is already inspired by Desmedt [11]. Desmedt's construction of one-time public-key message authentication is based on the zero-knowledge protocol with a tool of GMR[20]'s signature scheme. Their implementation depends on the the structure of underlying problems. However, Okamoto and Ohta also constructs non-transitive signature based on the zero-knowledge protocol and bit-commitment scheme. The assumption in the Okamoto-Ohta's general construction is only the existence of the one-way function.

We consider the subliminal channel in the 3 move protocol based on the (possibly) general assumption. Our technique, the modification of verifier's randomness based on the prover's first message in the (3 move) parallelization of zero-knowledge protocol, is useful for Desmedt's scheme and the general construction by Okamoto and Ohta.

This implies a remarkable fact that signatures can be transferred securely using only one-way function (without public key), because the secret message transfer is believed to be indispensable to public key primitive with trapdoor primitive (or common secret key sharing) [22].

A remaining problem is the security of the scheme. The security against the cheating verifier in the direct parallelization of zero-knowledge protocol, which is a three move protocol, is still unclear. Indeed Feige, Fiat, and Shamir have proved the security for their scheme (based on Rabin scheme). However their proof heavily depends upon the underlying computational problem (factorization). However, if the verifier is honest, the prover can send his signature to the verifier without any third party's recording.

Authors believe that our idea will be the first attempt to give the effect on the theoretical result with probable security just like as Lamport's frontier work on the construction of the digital signature from the one-way function [23] induces the fact that the digital signature is the most fundamental primitive as one-way functions [24, 30].

3 Signature distribution scheme

The subliminal channel is regarded as a negative object in cryptographic setting if the community does not identifies the existence of the channel and if and the cheating entity abuses the channel. In this section, however, we discuss a positive application of our subliminal channel. We consider the Fiat-Shamir identification scheme with multiple verifiers and apply the subliminal channel to the signature distribution scheme.

3.1 Previous works on multi-party signatures

The signature scheme with the multiple signers has been investigated by some researchers [25, 28, 13], and some simultaneous multiplesignature schemes are proposed [5, 21]. These schemes consist of multiple signers and a single verifier: the signature can be verified by only one verifier with each signer's public key.

The distributed verification signature, which is generated by a single signer, and verified by multiple verifier, is firstly proposed by De Soete, Quisquater, and Veder [16]. They constructed a distributed verification scheme using a threshold scheme based on some finite geometric structure. However their implementation heavily depends on some physical assumptions (secure channel, trusted host, etc.), and is not so practical because, as Desmedt [13] points out, the available message space in their scheme is small.

A practical distributed verification signature is implemented as a part of Pedersen's undeniable signature with distributed provers [29]. Pedersen's scheme is based on the combination of a (selective) convertible undeniable signature [2] and Shamir's secret sharing scheme [31]. Then his implementation needs some secure channel in order to distribute the signer's signature to each agent. A remaining problem is if the distribution of signatures without secure channel nor secret sharing scheme is possible. The author gives an positive answer to the problem.

3.2 The model of the multiple verifier interactive protocol

The model of multiple-verifiers (zero-knowledge) interactive protocols was introduced by Burmester and Desmedt [3] from the point of view of the efficiency of a broadcasting setting.

We give a definition of our model of the multiple verifier interactive protocol which is simpler than the Burmester and Desmedt[3]'s one.

Definition 1. A Turing machine which is equipped with a read-only input tape, a work tape, a random tape, a history tape, i read-only communication tapes and j write-only communication tapes, is called an *(i, j)-interactive Turing machine*(ITM).

Definition 2. A multi-verifier interactive protocol $(P, V_1, \ldots, V_v)$ consists of v ITM's $V_1, \ldots, V_v$, and one (v, v)-ITM denoted by P, which satisfies the following conditions:

1. P and $V_1, \ldots, V_v$ share a common communication tape which is write-only for P and read-only for all the V_i. When P writes a string on this tape, we say that P broadcasts that string.
2. P and $V_1, \ldots, V_v$ share a common communication tape which is read-only for P and write-only for all the V_i.

Remark. The model above has the assumption that there exists a *broadcasts channel* where every verifier can check that any other verifier receives the same information from the prover.

3.3 Definition of Signature Distribution Scheme

In digital signatures by Diffie and Helmann [15], the validity of a signature can be tested with signer's public key. The original notion of digital signature does not take account of how the signature is distributed, because the digital signature is transitive and everyone can check the validity of a signature with signer's public key. With the development of interactive protocols [19, 1], the notion of digital signature is extended. Chaum's undeniable signature [9, 8] cannot be verified without signer's corporation. In the undeniable signature, the signature is verified in the interactive protocol between the signer and the verifier.

We consider the signature generated in the interactive protocol between single signer and multiple verifiers.

Definition 3. The interactively generated signature $IGS(P, V_1, , \ldots, V_v)$ has the following components:

1. A security parameter k, determining the security, running time, and the length of messages.

2. A message space, which we will assume to be strings in $\{0,1\}^{poly(k)}$.
3. A polynomial S_B called the signature bound. The value $S_B(k)$ represents a bound on the number of messages that can be signed when the security parameter is k.
4. A probabilistic polynomial-time key generation algorithm KG, which on input 1^k, outputs a signer P's public key PK_P, and a matching signer P's secret key SK_P.
5. A *multiple verifier interactive* signing algorithm ISP, which given a message m and a matching pair of keys $< PK_P, SK_P >$, outputs a distributed signature $S_1, \ldots, S_v$ of m with respect to PK_P, where $S_i = IPS(P(PK_P, SK_P,), V_i(PK_P))[m]$.
6. A polynomial-time verification algorithm VER, which given $S_1, \ldots, S_v$, m, PK_P, tests whether or not $S, \ldots, S_v$ is a valid signature of m with respect to PK_P.

Definition 4. We call an interactively generated signature $IGS(P, V_1, \ldots, V_v)$ distributed signature scheme if it satisfies the following conditions:

Validity: *Only* the signer can prove the validity of signer's messages to any verifiers by using signer's public key.

Private distributed recordability: Only the verifiers can keep a peace of signature.

Distributed right of publishing: Without cooperating all verifiers, the signature cannot be published.

3.4 Proposed scheme

The basic idea of our construction of a signature distribution scheme is to divide the verifier's private coin in the parallel version of the Fiat-Shamir scheme. The following protocol is based on the public-key message authentication in [21].

Proposed signature distribution scheme with k verifiers based on the Guillou-Quisquater's message authentication scheme

1. PREPROCESSING STAGE
In the system, the center's secret key is p, q (primes), the center's public key is $n = pq$ and $L = O(|n|)$. The center generates the secret key s of the Guillou-Quisquater scheme for a signer P, where $1/s = (I_P)^{1/L} \pmod n$, and I_P is the identity of the signer P. Furthermore, two one-way hash functions h and g are published to each user.
2. DISTRIBUTION STAGE
   (a) The signer P broadcasts his identity I_P and a message M_P to all verifiers $V_i (i = 1, \ldots, k)$.
   (b) The signer P picks $r \in_R Z_n^*$, and computes $x \equiv r^L \pmod N$ and $u = g(M_P, x)$. The prover broadcasts x and u to all verifiers $V_i (i = 1, \ldots, k)$.
   (c) Each verifier V_i chooses R_{V_i} and sends back e_{V_i} which are computed as $e_{V_i} = h(R_{V_i}, x) \in Z_L$.

(d) The signer P broadcasts (e, y) such that $e = e_{V_1} \times \cdots \times e_{V_k} \pmod{L}$ and $y \equiv s^e r \pmod{n}$ to all verifiers $V_i (i = 1, \ldots, k)$.

(e) Each verifier V_i checks that $x \equiv y^L (I_P)^e \pmod{n}$, and $u = g(M_P, x)$. If all checks are valid, the verifiers $V_i (i = 1, \ldots, k)$ recognize that M_P is P's valid message.

3. PUBLISHING STAGE
When the verifiers publish the signature, each verifier V_i publishes his private coin R_{V_i} with the public view $(I_P, M_P, u, R_{V_i}, e_{V_i}, y)$.

4. VERIFICATION STAGE
Once all of verifier view are published, anyone can check the validity and the origin of message M_P by calculating $e = e_{V_1} \times \cdots \times e_{V_k} \pmod{L}$, $x \equiv y^L (I_P)^e \pmod{n}$ and checking that $u = g(M_P, x)$ and $e_{V_i} = h(R_{V_i}, x)$.

Remark. In the protocol above, the condition that $L = O(|n|)$ is important. If we adapt the serial version with $L = O(1)$, then the verifier can easily forge the signature because the protocol is zero-knowledge.

3.5 Properties of the proposed scheme

We consider properties of the proposed scheme, especially for the security. (We will describe more detail in the final version of this paper.) Our proposed protocols satisfies the following conditions on interactive protocols with multiple-verifiers [3].

Completeness: When all the verifiers are honest, the honest prover will answer correctly every verifier with overwhelming probability.

Soundness: The probability that a dishonest prover will answer correctly to a challenge of an honest verifier is negligible even when all the other verifiers cheat.

Attack by the dishonest verifier An important assumption of our model on the signature distribution is that all verifiers are honest. However, there is a possibility that an "active" dishonest verifier would deny the corporation on the verification stage if the message is unfavorable to him. (Consider the last dying statement on the succession to property.) Although a solution that overcomes such a dishonest verifier's cheating may be to construct some threshold, it is not a perfect solution. It remains as an open problem.

Security against passive attacker Our proposed signature distribution scheme is secure against the (polynomial bounded) third adversary if all verifiers are honest, because the secrecy of each verifier's random coin R_{V_i}. Then we consider passive attack by the verifiers who act honestly to get pieces of the prover's signature. We can show that the difficulty of forging a (total) signature from any $k-1$ or fewer verifiers' pieces of the signature is as hard as the difficulty of forging a signature in single verifier signature transfer. Note that the security of a signature in single verifier signature transfers is as same as the original Fiat-Shamir digital signature. Thus the security against passive (verifier's) attack is reducible to the security of the original Fiat-Shamir digital signature scheme. (We will give more precise statement and formal proof in the final version.)

4 Concluding remarks

In this paper we have pointed out a new subliminal channel hidden in the parallel version of the Fiat-Shamir identification scheme (or general three move protocols), which does not exists in the serial (zero-knowledge) version.

We have also considered a positive aspect on the subliminal channel. We have constructed a signature distribution scheme based on the public-key message authentication scheme with multiple verifiers. Desmedt investigated several subliminal channels in cryptographic protocol and proposed how to avoid the abuse [12]. We conclude with the following topics.

Find other subliminal channels hidden in the cryptographic objects, and

1. Consider the countermeasure to avoid the abuse.
2. Positively apply the subliminal channels to the cryptographic systems.

Acknowledgments

The authors wish to thank Dr. Tatsuaki Okamoto for the discussion with him on the cryptographic primitive of secure message transfer.

References

1. Brassard, G., Chaum, D., and Crépeau, C., "Minimum Disclosure Proof of Knowledge," JCSS Vol.37, pp.156-189 (1989).
2. Boyor, J., Chaum, D., Damgård, I., and Pedersen, T., "Convertible Undeniable Signature," Proc. of Crypto'90.
3. Burmester, M. and Desmedt, Y., "Broadcast Interactive Proofs," Proc. of Eurocrypt'91.
4. Burmester, M. and Desmedt, Yvo., Piper, F., and Walker, M., "A general zero-knowledge scheme," Proc. of Eurocrypt'89.
5. Brickell, E.F., Lee, P.J., and Yacobi, Y., "Secure Audio Teleconference," Proc. of Crypto'87.
6. Brickell, E. F. and McCurley, K.S "An Interactive Identification Scheme Based on Discrete Logarithms and Factoring," *Journal of Cryptology*, Vol.5, pp.29-40 (1992).
7. Chaum,D., Damgård, I., and van de Graaf, J. "Multiparty computations ensuring privacy of each party's input and correctness of the result," Proc. of Crypto'87.
8. Chaum, D., "Zero-knowledge undeniable signatures," Proc. of Eurocrypt'90.
9. Chaum, D. and van Antwepen, H., "Undeniable signature," Proc. of Crypto'89.

10. Desmedt, Y., "Major security problems with the "unforgeable" (Feige-)Fiat-Shamir proofs of identity and how to overcome them," *Proc. of Securicom'88*.
11. Desmedt, Y.: "Subliminal-free Authentication and Signature," Proc. of Eurocrypt'88.
12. Desmedt, Y.: "Abuse in cryptography and how to fight them," Proc. of Crypto'88.
13. Desmedt, Y. and Frankel, Y.: "Shared generation of authenticators and signatures," Proc. of Crypto'91.
14. Desmedt, Y., Goutier, C. and Bengio,S.: "Special Uses and abuses of the Fiat-Shamir Passport Protocol," Proc. of Crypto'87.
15. Diffie, W., and Helmann, M. "New Directions in Cryptology", IEEE Trans. on Info. Technology, vol. IT-22, 6(1976) pp.644-654 (1976).
16. De Soete,M., Quisquater,J., and Vedder, K., "A signature with shared verification scheme," Proc. of Crypto'89.
17. Feige, U., Fiat, A., and Shamir, A., "Zero-Knowledge Proofs of Identity," *Journal of Cryptology*, Vol.1, pp.179-194 (1988).
18. Fiat, A. and Shamir, A., "How to Prove Yourself," Proc. of Crypto'86.
19. Goldwasser, S., Micali, S., and Rackoff, C., "The Knowledge Complexity of Interactive Proof Systems," *SIAM Journal on Computing*, Vol.18, No.1, pp.186-208 (February 1989).
20. Goldwasser,M., Micali, S., and Rivest, R., "A digital signature scheme secure against adaptive chosen-message attacks," Proc. of FOCS'84. *IEEE Annual Symposium on Foundations of Computer Science*, pp.441-448 (October 1984).
21. Guillou,L.C. and Quisquater,J.J. "A "Paradoxical" Identity-Based Signature Scheme Resulting from Zero-Knowledge ", Proc. of Crypto'88.
22. Impagliazzo, R. and Rudich, S., "Limits on the Provable Consequences of One-way Permutations ," Proc. of STOC'89. *ACM Annual Symposium on Theory on Computing*, pp.44-61 (May 1989).
23. Lamport, L., "Constructing digital signatures from one-way functions," *SRI intl. CSL-98*, pp.33-43 (Oct. 1979).
24. Naor, M. and Yung, M. "Universal One-Way Hash Functions and their Cryptographic Applications," Proc. of STOC'89. *ACM Annual Symposium on Theory on Computing*, pp.33-43 (May 1989).
25. Okamoto, T., "A digital Multisignature Scheme Using Bijective Public-Key Cryptosystems," *ACM Trans. on Comp. Systems,* Vol.6, No.8, pp.432-441 (1988).
26. Okamoto, T. and Ohta, K., "Divertible Zero-Knowledge Interactive Proofs and Commutative Random Self-Reducibility," Proc. of Eurocrypt'89.
27. Okamoto, T. and Ohta, K., "How to utilize the randomness of Zero-Knowledge Proofs," Proc. of Crypto'90.
28. Ohta, K. and Okamoto, T., "A Digital Multisignature Scheme Based on the Fiat-Shamir Scheme," Abstracts of Asiacrypt'91 (1991).
29. Pedersen, P.T., "Distributed Provers with Applications to Undeniable Signatures," Proc. of Eurocrypt'91.
30. Rompel, J., "One-way functions are necessary and sufficient for secure signature," Proc. of STOC'90.
31. Shamir, A., "How to share a secret," *CACM*, 22, pp.612-613 (1979).
32. Sakurai, I., and Itoh,T., "Privately recordable signatures and signature sharing scheme," *Proc. of 1992 SCIS*, **6C**, (Japan, 1992).
33. Sakurai, I., and Itoh,T., "On the discrepancy between Serial and Parallel of Zero-Knowledge Protocols," Abstracts of Crypto'92.

34. Simmons, G. J., "The Prisoner's Problem and the Subliminal Channel," Proc. of Crypto'83.
35. Simmons, G. J., "The Subliminal Channel and Digital Signature," Proc. of Eurocrypt'84.
36. Simmons, G.J., "The Secure subliminal Channel (?)," Proc. of Crypto'85.

A Practical Secret Voting Scheme for Large Scale Elections

Atsushi Fujioka Tatsuaki Okamoto Kazuo Ohta

NTT Laboratories
Nippon Telegraph and Telephone Corporation
1-2356 Take, Yokosuka-shi, Kanagawa-ken, 238-03 Japan

Abstract. This paper proposes a practical secret voting scheme for large scale elections. The participants of the scheme are voters, an administrator, and a counter. The scheme ensures the privacy of the voters even if both the administrator and the counter conspire, and realizes voting fairness, i.e., no one can know even intermediate result of the voting. Furthermore fraud by either the voter or the administrator is prohibited.

1 Introduction

Secret voting schemes have been proposed by many researchers from both the theoretical and practical points of view.

In the scheme using the multi-party protocol [GMW87, BGW88, CCD88], all procedures are managed by just the voters, however, it takes many communication acts to prove that all acts were performed correctly. Therefore, this approach is interesting theoretically, but is impractical. A practical secret voting scheme requires additional participants, e.g., a trusted center or an administrator.

There are two types of this approach: one is the voter sends the ballot in an encrypted form and the other is voter sends the ballot through an anonymous communication channel.

The first type have been proposed by Benaloh (Cohen) *et al.* and Iversen [CF85, BY86, Iv91], and their schemes utilize the higher degree residue encryption technique. The scheme in [CF85] needs distributed centers to protect voter privacy, however, the voter must prove that the distributed ballot is valid, so all voting must be done at the same time. Iversen [Iv91] evades this problem by using the technique proposed to realize electronic money [CFN90]. However, the essential drawback of this approach is that if all centers conspire, the privacy of voters is violated. Moreover, these schemes are less practical for large scale elections, since it takes a lot of communication and computation overhead when the number of voters is large.

As a scheme of the second type, Chaum proposed a voting scheme that used an anonymous communication channel, and it provides unconditionally security against tracing the voting [Ch88b]. Independently Ohta proposed a practical secret voting scheme using only one administrator in similar manner [Oh88]. These schemes are more suitable for large scale elections, since the communication and computation overheads are reasonable even if the number of voters is large.

However, both schemes have the same drawbacks: the problems of fairness and privacy. As the fairness problem, these schemes don't ensure voting fairness, i.e., the center knows the intermediate result of opening the ballot, so he can affect the voting by leaking the result. As the privacy problem, voter's privacy is violated, when a voter notices that his voting was not counted correctly, and claims it by showing his voting.

Recently, Asano *et al.* proposed a scheme which overcomes the fairness problem [AMI91]. This scheme, unfortunately, is not secure against disruption by the administrator. Subsequently, Sako proposed a scheme to solve the privacy problem [Sa92]. This scheme, however, doesn't overcome the fairness problem and voting is limited to only yes/no.

Therefore, no practical and secure secret voting scheme has been proposed which is suitable for large scale elections (or based on the second type) and solves the privacy and fairness problems at the same time.

This paper solves these problems: we propose practical and secure secret voting scheme which is suitable for large scale elections and solves the privacy and fairness problems at the same time. That is, our scheme ensures the privacy of the voters even if both the administrator and the counter conspire, and voting fairness, i.e., no one can know even intermediate result of the voting. Furthermore fraud by either the voter or the administrator is prohibited.

Section 2 defines the security needed by a practical secret voting scheme. In **Section 3**, a practical secret voting scheme is proposed and we prove that the proposed scheme is secure, and we conclude this paper in **Section 4**.

2 Security of Secret Voting Scheme

In this paper, we discuss the security of the secret voting scheme using the following definition.

Definition 1. We say that the secret voting scheme is *secure* if we have the following:

- *Completeness* ... All valid votes are counted correctly.
- *Soundness* ...... The dishonest voter cannot disrupt the voting.
- *Privacy* .... All votes must be secret.
- *Unreusability* .... No voter can vote twice.
- *Eligibility* .... No one who isn't allowed to vote can vote.
- *Fairness* .... Nothing must affect the voting.
- *Verifiability* ..... No one can falsify the result of the voting.

3 Proposed Voting Scheme

3.1 Model of Proposed Scheme

Our model consists of voters, an administrator, and a counter (the counter can be replaced with a public board), and the voters and the counter communicate through an anonymous communication channel [Ch81, Pf84, Ch88a]. The scheme requires the bit-commitment scheme [Na90], the ordinary digital signature scheme [DH76], and the blind signature scheme [Ch85].

Every voter has his own ordinary digital signature scheme, and the administrator has blind signature scheme. The counter only creates a list of ballots, and publishes it.

3.2 Notations

In this paper, we use the following notations.

V_i: Voter i
A: Administrator
C: Counter
$\xi(v,k)$: Bit-commitment scheme for message v using key k
$\sigma_i(m)$: Voter V_i's signature scheme
$\sigma_A(m)$: Administrator's signature scheme
$\chi_A(m,r)$: Blinding technique for message m and random number r
$\delta_A(s,r)$: Retrieving technique of blind signature
ID_i: Voter V_i's identification
v_i: Vote of voter V_i

3.3 Structure of Proposed Scheme

In this subsection, we propose a practical secret voting scheme based on the model described in **Subsection 3.1**. The proposed scheme is secure in the sense of **Definition 1**.

First we outline the proposed scheme. The scheme consists of the following stages executed by the voters, the administrator, and the counter.

PREPARATION: Voter fills in a ballot, makes the message using the blind signature technique to get the administrator's signature, and sends it to the administrator.
ADMINISTRATION: Administrator signs the message in which the voter's ballot is hidden, and returns the signature to the voter.
VOTING: The voter gets the ballot signed by administrator, and sends it to the counter anonymously.
COLLECTING: Counter publishes a list of the received ballots.
OPENING: The voter opens his vote by sending his encryption key anonymously.
COUNTING: Counter counts the voting and announces the result.

Roughly speaking, the voter prepares a ballot, get a administration, and vote anonymously. Administrator gives a administration to an eligible voter, and counter only collects the ballots and published a list.

On the communication between the voter and the administrator, voter gets qualification, and after it, voter acts anonymously. The blinding signature technique provides the separation between the identification and anonymous communication.

Here we explain the scheme in detail.

- PREPARATION
  - Voter V_i selects vote v_i and completes the ballot $x_i = \xi(v_i, k_i)$ using a key k_i randomly chosen.
  - V_i computes the message e_i using blinding technique $e_i = \chi(x_i, r_i)$.
  - V_i signs $s_i = \sigma_i(e_i)$ to e_i and sends $\langle ID_i, e_i, s_i \rangle$ to administrator.
- ADMINISTRATION
  - Administrator A checks that the voter V_i has the right to vote. If V_i doesn't have the right, A rejects the administration.
  - A checks that V_i has not already applied for a signature. If V_i has already applied, A rejects the administration.
  - A checks the signature s_i of message e_i. If they are valid, then A signs $d_i = \sigma_A(e_i)$ to e_i and sends d_i as A's certificate to V_i.
  - At the end of the ADMINISTRATION stage, A announces the number of voters who were given the administrator's signature, and publishes a list that contains $\langle ID_i, e_i, s_i \rangle$.
- VOTING
  - Voter V_i retrieves the desired signature y_i of the ballot x_i by $y_i = \delta(d_i, r_i)$.
  - V_i checks that y_i is the administrator's signature of x_i. If the check fails, V_i claims it by showing that $\langle x_i, y_i \rangle$ is invalid.
  - V_i sends $\langle x_i, y_i \rangle$ to the counter through the anonymous communication channel.
- COLLECTING
  - Counter C checks the signature y_i of the ballot x_i using the administrator's verification key. If the check succeeds, C enters $\langle l, x_i, y_i \rangle$ onto a list with number l.
  - After all voters vote, C publishes the list. (The list can be accessed by all voters.)

Table 1. List of the ballots (in the COLLECTING stage)

Entry	Ballot & additional information
1	x_j, y_j
⋮	⋮
l	x_i, y_i
⋮	⋮

- **OPENING**
  - Voter Vi checks that the number of ballots on list is equal to the number of voters. If the check fails, voters claim this by opening r_i used in encryption.
  - V_i checks that his ballot is listed on the list. If his vote is not listed, then V_i claims this by opening $\langle x_i, y_i \rangle$, the valid ballot and its signature.
  - V_i sends key k_i with number l, i.e., $\langle l, k_i \rangle$ to C through an anonymous communication channel.
- **COUNTING**
  - Counter C opens the commitment of the ballot x_i, retrieves the vote v_i (or C adds k_i and v_i to the list), and checks that v_i is valid voting.
  - C counts the voting and announces the voting results.

Table 2. List of the ballots (in the COUNTING stage)

Entry	Ballot & additional information
1	x_j, y_j, k_j, v_j
⋮	⋮
l	x_i, y_i, k_i, v_i
⋮	⋮

3.4 Security

Theorem 1 (Completeness). *If every participant (the voters, the administrator, and the counter) is honest, the result of the voting is trustable.*

Sketch of Proof. It is clear. □

Theorem 2 (Soundness). *Even if a voter intends to disrupt the election, there is no way to do it.*

Sketch of Proof. The only way to disrupt the elections is for the voter to keep sending invalid ballots, however, this can be detected in the COUNTING stage. Furthermore, the votes are bound by using the bit-commitment scheme, so the voter cannot change his mind. □

Remark. It is possible to assume that the voter sends an illegal key which cannot open the vote. In this situation, there is no way to distinguish between a dishonest voter or a dishonest counter. To prevent this, the voter should send his key to several independent parties, e.g., the candidates of the election, who are assumed not to collaborate.

Theorem 3 (Privacy). *Even if the administrator and the counter conspire, they cannot detect the relation between vote v_i and voter V_i.*

Sketch of Proof. The relation between the voter's identity ID_i and the ballot x_i is hidden by the blind signature scheme. The ballot x_i and the key k_i are sent through the anonymous communication channel. So no one can trace the communication and violet the privacy of the voters. It is unconditionally secure against tracing the voting.

In addition, when the voter claim the disruption by the administrator or counter, he need not release his vote v_i.

In VOTING stage, when the administrator sends the invalid signature, the voter only show the pair $\langle x_i, y_i \rangle$ to claim the cheating.

In OPENING stage, when the counter doesn't list the voter's ballot, the voter only show the pair $\langle x_i, y_i \rangle$ to claim the disruption.

So he can claim the disruption with keeping his vote v_i secret. This ensures the voter's privacy. □

Theorem 4 (Unreusability). *Assume that no voter can break the blind signature scheme. Then, the voter cannot reuse the right to vote.*

Sketch of Proof. To vote twice, voter must have two valid pairs of the ballot and the signature. He can get one signature by right procedure, however, he has to create another pair by himself. This means that he can break the blind signature scheme, and it contradicts the assumption. □

Theorem 5 (Eligibility). *Assume that no one can break the ordinary digital signature scheme. Then, the dishonest person cannot vote.*

Sketch of Proof. In the opposite direction, assume the dishonest person can vote. The administrator checks the list of voters who have the right to vote. So the dishonest person must create a valid pair of the ballot and the signature by himself. This contradicts that no one can break the ordinary digital signature scheme. □

Theorem 6 (Fairness). *The counting of ballots doesn't affect the voting.*

Sketch of Proof. The COUNTING stage is done after the VOTING stage, and the votes are hidden by using the bit-commitment scheme. So it is impossible that the counting of ballots affects the voting. □

Theorem 7 (Verifiability). *Assume that there is no voter who abstains from the voting and no one can forge the ordinary digital signature scheme. Then, even if the administrator and the counter conspire, they can not change the result of the voting.*

Sketch of Proof. It is clear that only disruption is for the counter not to list a valid ballot to the list. However, this disruption can be easily proved by showing the valid pair of the ballot and the signature by the valid voter. So we only consider the disruption by the administrator in the following.

If there is no voter who abstains from the voting (i.e., he sends a ballot even if he abstains), there is no way for the administrator to dummy vote. So only valid voters can vote, and the result is trustable.

If the list overflows, every voter claims that he is an eligible voter and he was given a valid signature by the administrator. To claim it, the voter open the number r_i which he used in the blinding technique, and requires the administrator to show the voter's signature. By opening r_i, the message e_i is fixed, so the signature which the administrator must show is determined. If the administrator is honest, he can show the signatures for all requests. However, when he dummy voted, there remains the ballots which were not shown the signature because he cannot forge the digital signature scheme. The fraud is detected here. □

Remark. The following procedure is followed after the ADMINISTRATION stage if fraud by the administrator or counter is proved by one or more voters. First, the usual voting protocol is followed and disputed votes are omitted. If the number of omitted votes changes the result, the voting process is invalidated and is restarted. If the number of omitted votes fails to change the result, the voting process is accepted. In any case, the fraudulent party should be appropriately punished.

4 Conclusion

This paper proposed a practical secret voting scheme for large scale elections. The scheme ensures the privacy of the voters and prevents any disruption by voters or the administrator. Furthermore, voting fairness is ensured.

Acknowledgement

The authors would like to thank Choonsik Park for pointing out the illegal key problem. We would also like to thank Kazue Sako, Tsutomu Matsumoto, and Tomoyuki Asano for their valuable comments about the dishonesty of the administrator and the counter.

References

[AMI91] T. Asano, T. Matsumoto, and H. Imai, "A Study on Some Schemes for Fair Electronic Secret Voting" (in Japanese), The Proceedings of the 1991 Symposium on Cryptography and Information Security, SCIS91-12A (Feb., 1991).

[BGW88] M. Ben-Or, S. Goldwasser, and A. Wigderson, "Completeness Theorems for Non-Cryptographic Fault-Tolerant Distributed Computation", Proceedings of the 20th Annual ACM Symposium on Theory of Computing, pp.1–10 (May, 1988).

[BY86] J. Benaloh and M. Yung, "Distributing the Power of a Government to Enhance the Privacy of Votes", Proceedings of the 5th ACM Symposium on Principles of Distributed Computing, pp.52–62 (Aug., 1986).

[Ch81] D. L. Chaum, "Untraceable Electronic Mail, Return Addresses, and Digital Pseudonyms", *Communications of the ACM*, Vol.24, No.2, pp.84–88 (Feb., 1981).

[Ch85] D. Chaum, "Security without Identification: Transaction Systems to Make Big Brother Obsolete", *Communications of the ACM*, Vol.28, No.10, pp.1030–1044 (Oct., 1985).

[Ch88a] D. Chaum, "The Dining Cryptographers Problem: Unconditional Sender and Recipient Untraceability", *Journal of Cryptology*, Vol.1, No.1, pp.65–75 (1988).

[Ch88b] D. Chaum, "Elections with Unconditionally-Secret Ballots and Disruption Equivalent to Breaking RSA", in *Advances in Cryptology — EUROCRYPT '88*, Lecture Notes in Computer Science 330, Springer-Verlag, Berlin, pp.177–182 (1988).

[CCD88] D. Chaum, C. Crépeau, and I. Damgård, "Multiparty Unconditionally Secure Protocols", Proceedings of the 20th Annual ACM Symposium on Theory of Computing, pp.11–19 (May, 1988).

[CF85] J. Cohen and M. Fisher, "A Robust and Verifiable Cryptographically Secure Election Scheme", 26th Annual Symposium on Foundations of Computer Science, IEEE, pp.372–382 (Oct., 1985).

[CFN90] D. Chaum, A. Fiat, and M. Naor, "Untraceable Electronic Cash", in *Advances in Cryptology — CRYPTO '88*, Lecture Notes in Computer Science 403, Springer-Verlag, Berlin, pp.319–327 (1990).

[DH76] W. Diffie and M. E. Hellman, "New Directions in Cryptography", *IEEE Transactions on Information Theory*, Vol.IT-22, No.6, pp.644–654 (Nov., 1976).

[GMW87] O. Goldreich, S. Micali, and A. Wigderson, "How to Play Any Mental Game or a Completeness Theorem for Protocols with Honest Majority", Proceedings of the 19th Annual ACM Symposium on Theory of Computing, pp.218–229 (May, 1987).

[Iv91] K. R. Iversen, "A Cryptographic Scheme for Computerized General Elections", in *Advances in Cryptology — CRYPTO '91*, Lecture Notes in Computer Science 576, Springer-Verlag, Berlin, pp.405–419 (1992).

[Na90] M. Naor, "Bit Commitment Using Pseudo-Randomness", in *Advances in Cryptology — CRYPTO '89*, Lecture Notes in Computer Science 435, Springer-Verlag, Berlin, pp.128–136 (1990).

[Pf84] A. Pfitzmann, "A Switched/Broadcast ISDN to Decrease User Observability", 1984 International Zurich Seminar on Digital Communications, IEEE, pp.183–190 (Mar., 1984).

[Oh88] K. Ohta, "An Electrical Voting Scheme using a Single Administrator" (in Japanese), 1988 Spring National Convention Record, IEICE, A-294 (Mar., 1988).

[Sa92] K. Sako, "Electronic Voting System with Objection to the Center" (in Japanese), The Proceedings of the 1992 Symposium on Cryptography and Information Security, SCIS92-13C (Apr., 1992).

Privacy for Multi-Party Protocols

Takashi Satoh, Kaoru Kurosawa and Shigeo Tsujii

Department of Electrical and Electronic Engineering,
Faculty of Engineering, Tokyo Institute of Technology
2-12-1 O-okayama, Meguro-ku, Tokyo, 152 Japan

Email: {tsato,kkurosaw}@ss.titech.ac.jp

Abstract. We initiate information theoretic investigation of the amount of leaked information in multi-party protocols. It is shown that mutual information is a satisfactory measure of the amount of the leaked information. For more than two parties, we apply multi-terminal information theory and give a formulation of the leaked information.

1 Introduction

A multi-party protocol is a set of n interactive probabilistic Turing machines, $\{P_i\}$. Party P_i has a private input x_i. Multi-party protocols require two conditions: correctness and privacy. Correctness is that each party P_i can compute the value of $f(x_1, x_2, \cdots, x_n)$ correctly. Privacy is that dishonest parties cannot compute any information on the private input of any other party.

Many researchers studied multi-party protocols in various models. Our model is as follows: (i) There are no cryptographic assumptions. (ii) There exists a secret channel between any two parties. (iii) Each party is honest or semi-honest. (iv) Semi-honest parties can be infinitely powerful. The meanings of an honest party and a semi-honest party are as follows. An honest party follows the protocol faithfully, and a faulty party behaves arbitrarily. A semi-honest party follows the protocol, but try to compute something more than the value of f.

We say that a protocol is t-private if any t parties coalition cannot compute the local input of any other party. Ben-Or, Goldwasser and Wigderson [2] showed the following interesting result. (i) For any function f and for any $t < n/2$, there exists a t-private protocol. (ii) There is a function f for which there are no $n/2$-private protocols. For two parties ($n = 2$), there exists a not privately computable function as a result. Then, what functions are privately computable ? Chor and Kushilevitz [3, 6] gave a characterization of private computable functions for two parties. Their results suggest that "most" functions are not privately computable. Therefore what we can do is to minimize the amount of leaked information in the execution of a protocol. Then, the next question is how to formulate the amount of the leaked information. Bar-Yehuda, Chor and Kushilevitz [1] gave a formulation by using an artificial function $g(x_1, x_2)$ (We adopt g instead of h, which they used). However, they implicitly assumed that the private inputs of parties are probabilistically distributed independently and uniformly.

In this paper, we investigate the amount of leaked information in the execution of multi-party protocols from an information theoretic point of view. It is shown that mutual information is a satisfactory measure of the amount of leaked information. Our formulation assumes nothing on the probability distribution of the private input of each party. We also show a relationship between our definitions and the definition of [1]. For more than two parties, we apply multi-terminal information theory and give a formulation of the leaked information.

2 Preliminaries

2.1 Notation

At the end of a protocol, each party can compute the value of $f(x, y, \cdots)$ from his inputs and the ***communicated string*** c, that is all messages passed in the protocol. We assume that each party is honest or semi-honest.

- x denotes a private input to party P_x. It is distributed over a finite set X.
- y denotes a private input to party P_y. It is distributed over a finite set Y.
- r_x denotes a random input to party P_x. It is distributed over a finite set R_x.
- r_y denotes a random input to party P_y. It is distributed over a finite set R_y.
- $f(x, y)$ is distributed over a finite set F.
- $g(x, y)$ is distributed over a finite set G.
- c is a communicated string (all messages sent by party P_x and party P_y in the protocol execution). It is distributed over a finite set C.

2.2 Related Work

Bar-Yehuda, Chor and Kushilevitz [1] gave a formulation of the leaked information for two party protocols by using an artificial function g. For any y, consider x_i and x_j such that $f(x_i, y) = f(x_j, y)$. Let c_i and c_j be the communicated string for x_i and x_j, respectively. If $c_i = c_j$, then it is clear that party P_y cannot know whether party P_x has x_i or x_j. Based on this observation, they formulated the amount of the leaked information as follows.

Definition 1. A protocol which computes a function f is weakly g-private for party P_x if for every two inputs (x, y) and (x', y) satisfying $f(x, y) = f(x', y)$ and $g(x, y) = g(x', y)$, and for every communicated string c,

$$\Pr(c|x, y) = \Pr(c|x', y) ,$$

where the probability is taken over the random inputs of both parties.

Weak g-privacy with respect to party P_y is defined symmetrically.

Definition 2. A protocol which computes a function f is strongly g-private for party P_x if for every two inputs (x, y) and (x', y) satisfying $f(x, y) = f(x', y)$

and $g(x,y) = g(x',y)$, for any random input r_y of party P_y, and for every communicated string c,

$$\Pr(c|x,y,r_y) = \Pr(c|x',y,r_y) \ ,$$

where the probability is taken over the random input r_x of party P_x.

Definition 3. A protocol is weakly g-private (strongly g-private) if it is weakly g-private (strongly g-private) for both parties.

Definition 4. A function f is weakly g-private (strongly g-private) if there exists a weakly g-private (strongly g-private) protocol which computes f.

Definition 5. A function f is weakly private (strongly private) if it is weakly g-private (strongly g-private) with respect to the constant function $g(x,y) \equiv 0$.

They showed that the weak and the strong g-privacy are equivalent requirements. The amount of leaked information required to compute f is defined as the minimum of $\log_2 |range(g)|$, where the minimum is taken over all functions g such that f is weakly g-private.

However, they implicitly assumed that the private inputs of parties are probabilistically distributed independently and uniformly. Their formulation cannot treat any other probability distribution. For example, if x_1 and x_2 are correlated, party P_2 can know some information on x_1 before a run of the protocol. Then, how much information is leaked by executing the protocol ? Their formulation cannot answer this question.

3 Proposed Formulation of Leaked Information (Two Parties Model)

For convenience, we denote the random variable distributed a finite set X by the same capital X. $H(X)$,$H(X|Y)$,$I(X;Y)$,$I(X;Y|Z)$ denote the standard information quantities as defined in Gallager [5].

3.1 Proposed Formulation

Party P_y knows at least his private input y, the random input r_y and the value v of a function f. We define p_i and q_i as follows:

$$p_i = \Pr(X = x_i|Y = y, R_y = r_y, F = v, C = c) \ ,$$
$$q_i = \Pr(X = x_i|Y = y, R_y = r_y, F = v) \ .$$

It is clear that if the communicated string c leaks no information on x, then $p_i = q_i$ for all i. Thus, we obtain the following formulation.

Definition 6. A function f is *type-1 privately computable* for party P_x iff $H(X|Y, R_y, F) = H(X|Y, R_y, F, C)$, that is $I(X;C|Y, R_y, F) = 0$.

From this definition, we see that it is natural to define the leaked information as $I(X;C|Y,R_y,F)$. This formulation can be applied to any probability distributions of x and y.

Definition 7. The amount of leaked information from party P_x to party P_y is defined as $I(X;C|Y,R_y,F) = H(X|Y,R_y,F) - H(X|Y,R_y,F,C)$.

Notice that this quantity equals 0 if f is type–1 private for party P_x. Leaked information with respect to party P_y is defined similarly.

3.2 Relations among Perfect Privacy

From $H(X|Y,R_y,F) = H(X,Y,R_y,F,C)$, however, we cannot conclude that $p_i = q_i$. In this sense, our formulation of the perfect privacy is somewhat weaker. To obtain a stronger definition, we adopt Kullback-Leibler's information.

Definition 8. A function f is *type–2 privately computable* for party P_x iff for $\forall y, \forall r_y, \forall v, \forall c$,

$$KL[\boldsymbol{p}||\boldsymbol{q}] = \sum_i p_i \log \frac{p_i}{q_i} = 0 ,$$

where $\boldsymbol{p}$ and $\boldsymbol{q}$ denotes the vector of $\{p_i\}$ and $\{q_i\}$, respectively. i.e. $\boldsymbol{p} = (p_1, p_2, \cdots)$ and $\boldsymbol{q} = (q_1, q_2, \cdots)$.

$KL[\boldsymbol{p}||\boldsymbol{q}]$ is known as *Kullback–Leibler's information*. Note that $KL[\boldsymbol{p}||\boldsymbol{q}]$ =0 iff $\boldsymbol{p} = \boldsymbol{q}$.

Theorem 9. *Definition 8 (type–2 private) is equivalent to Definition 5.*

Proof. From Definition 5, $\Pr(c|x,y,r_y) = \Pr(c|x',y,r_y)$ for every two inputs (x,y) and (x',y) satisfying $f(x,y) = f(x',y)$. Therefore $\Pr(c|x,y,r_y,v) = \Pr(c|y,r_y,v)$ for every x, y, r_y and v satisfying $f(x,y) = v$. Thus,

$$\begin{aligned} p_i &= \Pr(x_i|y,r_y,v,c) \\ &= \frac{\Pr(c,x_i|y,r_y,v)}{\Pr(c|y,r_y,v)} \\ &= \frac{\Pr(c|x_i,y,r_y,v)}{\Pr(c|y,r_y,v)} \cdot \Pr(x_i|y,r_y,v) \\ &= q_i \ . \end{aligned}$$

We get $KL[\boldsymbol{p}||\boldsymbol{q}] = 0$. Vice versa. □

We define type–3 perfect privacy as non-averaged version of type–1 perfect privacy.

Definition 10. A function f is *type–3 privately computable* for party P_x iff $\forall y, \forall r_y, \forall v, \forall c$,

$$H(X|Y=y,R_y=r_y,F=v) = H(X|Y=y,R_y=r_y,F=v,C=c) \ .$$

Type–i privacy for party P_y is defined similarly for $i = 1, 2, 3$.

Theorem 11. *If x is uniformly distributed, Definition 10 (type–3 private) is equivalent to Definition 5 and Definition 8 (type–2 private).*

Proof. We show that Definition 10 is equivalent to Definition 8 if x is uniformly distributed. If f is type–2 privately computable, then p=q. Thus, $H(X|Y = y, R_y = r_y, F = v) = H(X|Y = y, R_y = r_y, F = v, C = c)$. This means that f is type–3 private.

Next, we assume that f is type–3 private. It is easy to see that x is also uniformly distributed for $\forall y, \forall r_y, \forall v, \forall c$. Fix y,r_y,v and c arbitrarily. It is clear that $0 \leq \Pr(x_i|y, r_y, v) \leq \Pr(x_i|y, r_y, v, c)$. Without loss of generality, we can set that

$$\Pr(x_i|y, r_y, v) = \begin{cases} 1/n & (1 \leq i \leq n) \quad \text{for some } n \\ 0 & \text{otherwise} \end{cases},$$

$$\Pr(x_i|y, r_y, v, c) = \begin{cases} 1/m & 1 \leq i \leq m (\leq n) \\ 0 & \text{otherwise} \end{cases}.$$

Because f is type–3 private, $\log_2 n = \log_2 m$, and then $n = m$. Thus, $\Pr(x_i|y, r_y, v) = \Pr(x_i|y, r_y, v, c)$ for all i. Hence, we obtain p=q and $KL[p||q] = 0$. □

Theorem 12. *If f is type–3 privately computable then it is type–1 privately computable.*

Proof. Clear from the definition of conditional entropy. □

3.3 Comparison on Leaked Information

Next, we make a comparison between the amount of leaked information by our definition and that by Bar-Yehuda et al.'s definition.

Lemma 13. *For random variables U,V,W, if $H(U|V) = H(U|V, W)$ then $I(U; W) = H(U) - H(U|W) \leq H(V)$.*

Proof. The following identity always holds:

$$H(V) = H(V|U, W) + I(U; V|W) + I(V; W|U) + I(U; V; W) .$$

Noticing that entropy and conditional mutual information are always non-negative, we have $I(U; V; W) \leq H(V)$. Thus, from $I(U; W|V) = H(U|V) - H(U|V, W) = 0$,

$$\begin{aligned} I(U; V; W) &= I(U; W; V) = I(U; W) - I(U; W|V) \\ &= I(U; W) = H(U) - H(U|W) \leq H(V) . \end{aligned}$$

□

Theorem 14. *If a function f is weak g-privately (strong g-privately) computable for party P_x, then $I(X;C|Y,R_y,F) \leq H(G) \leq \log_2 |g(x,y)|$.*

Notice that $I(X;C|Y,R_y,F)$ and $\log_2 |g(x,y)|$ are the amount of leaked information by our definition and [1]'s definition, respectively.

Proof. If f is g-privately computable for party P_x, then we obtain $H(X|Y,R_y,F,G) = H(X|Y,R_y,F,G,C)$ by modifying our type-1 definition. From Lemma 13, $I(X;C|Y,R_y,F) \leq H(G)$ (substituting $U = (X|Y,R_y,F)$, $V = G$ and $W = C$). Clearly, $H(G) \leq \log_2 |g(x,y)|$. □

3.4 Deterministic Case v.s. Probabilistic Case

Theorem 15. *If x and y are independent and the protocol is deterministic, the following statements are equivalent.*

1. *f is type-1 privately computable for both parties.*
2. *$I(F;C) = H(F) = H(C)$.*

Proof. Since a communicated string c is determined by their private inputs x and y, $H(C|X,Y) = 0$. Both parties can compute the value of f from their private inputs and the communicated string. Therefore $H(F|X,C) = 0$ and $H(F|Y,C) = 0$. From $H(C) = H(C|X,Y,F) + I(C;F) + I(C;X,Y|F)$, we get $H(C) = I(C;F)$ because $H(C|X,Y) = H(F|X,C) = H(F|Y,C) = 0$. From the definition of a privately computable function, $I(X;C|Y,F) = 0$ and $I(Y;C|X,F) = 0$.

If X and Y are independent, the following equations hold:

$$H(F|C) = I(X;F|C) + I(Y;F|C) + H(F|X,Y,C) \ ,$$
$$H(F|X,C) = I(F;Y|C) + H(F|X,Y,C) \ ,$$
$$H(F|Y,C) = I(F;X|C) + H(F|X,Y,C) \ .$$

From $H(F|X,C) = 0$ and $H(F|Y,C) = 0$, $I(F;X|C) \leq H(F|Y,C) = 0$ and $I(F;Y|C) \leq H(F|X,C) = 0$. Thus, we obtain $H(F|C) = 0$, that is $H(F) = I(F;C)$. □

Corollary 16. *Assuming that a communicated string corresponds to a value of a function f, $I(F;C) = H(F) = H(C)$ states that the distribution of f and c is identical.*

Proof. Let P_{f_i} be the probability that the value of f equals to i. Let $P_{f_{i,j}}$ be the probability that the value of f equals to i with a communicated string c_j. Obviously, $P_{f_i} = \sum_j P_{f_{i,j}}$. Then, $-P_{f_i} \log P_{f_{i,j}} = -(\sum_j P_{f_{i,j}}) \log \sum_j (P_{f_{i,j}}) \leq -\sum_j P_{f_{i,j}} \log P_{f_{i,j}}$ with equality if and only if $P_{f_i} = P_{f_{i,j}}$ for a single j. From $H(F) = H(C)$, we obtain that a communicated string corresponds one-to-one to a value of f. □

Theorem 17. *If x, y, r_x, r_y are independent and the protocol is probabilistic, the following statements are equivalent.*

1. *f is type-1 privately computable for both parties.*
2. *$I(F;C) = I(F;C|X,Y)$.*

Proof. By using a similar arguments to Theorem 15, we have this theorem. (details omitted). □

4 More than Two Parties Case

In this section we consider the privacy in four party protocols. The amount of leaked information in more than two party protocols has not been formulated so far. We give a formulation of the leaked information by extending our formulation for two party protocols and applying multi-terminal information theory.

Recall the definition of the amount of leaked information for two party protocols. It is defined as $I(X;C|Y,R_y,F) = H(X|Y,R_y,F) - H(X|Y,R_y,F,C)$. $H(X|Y,R_y,F)$ and $H(X|Y,R_y,F,C)$ represents the amount of information necessary for party P_y to compute x before the execution of the protocol and after the execution of the protocol, respectively.

Suppose that P_w,P_x,P_y and P_z compute $f(w,x,y,z)$. P_w and P_x are semi-honest and they collaborate to get some information on the private inputs of P_y and P_z. We define the *view* of P_w and P_z as whatever they can see in the run of the protocol. We consider the following problem: After a run of the protocol, P_w and P_z knows the value of $f(w,x,y,z)$ and the view. Suppose that P_y sends m_y and P_z sends m_z after the protocol execution so that P_w and P_x can compute y and z. Let $R_y = |m_y|/|y|$ and $R_z = |m_z|/|z|$.

Theorem 18 (Cover's theorem). [4]
M jointly ergodic countable-alphabet stochastic processes can be sent separately at rates $R_1, R_2, \ldots, R_M$ to a common receiver with arbitrarily small probability of error, if and only if

$$\sum_{i \in S} R_i \geq H(X_i, i \in S | X_i, i \in \overline{S})$$

for all subsets $S \subseteq \{1, 2, \cdots, M\}$.

Then, by Cover's theorem, there exists a coding scheme so that P_w and P_x can compute y and z if and only if R_y and R_z satisfy the conditions of the following theorem.

Theorem 19. *After the execution of the protocol, there exists a coding scheme so that P_w and P_x can compute y and z with arbitrary small probability of error, if and only if*

$$R_y \geq H(Y|Z,W,X,View,F)\ ,$$
$$R_z \geq H(Z|Y,W,X,View,F)\ ,$$
$$R_y + R_z \geq H(Y,Z|W,X,View,F)\ ,$$

where View denotes the random variable of the view.

Definition 20. Given R_y and R_z, we define L_y and L_z as following. We define the leaked information as this L_y and this L_z.

$$\begin{aligned} R_y + L_y &= H(Y|Z,W,X,F) \ , \\ R_z + L_z &= H(Z|Y,W,X,F) \ , \\ R_y + R_z + L_y + L_z &= H(Y,Z|W,X,F) \ . \end{aligned}$$

Remark. The first equation shows that the leaked information on y plus the message to be sent after the protocol on y equals the conditional entropy of y. The condition is what dishonest parties know after a run of the protocol. The other two equations have similar meanings. We think that this is a natural formulation of the leaked information.

From this definition and the previous theorem, we obtain the following corollary.

Corollary 21.

$$\begin{aligned} L_y &\leq I(Y;View|Z,W,X,F) \ , \\ L_z &\leq I(Z;View|Y,W,X,F) \ , \\ L_y + L_z &\leq I(Y,Z;View|W,X,F) \ . \end{aligned}$$

5 Conclusion

In this paper we have shown a new formulation of the leaked information from an information theoretic point of view. We have shown a relationship between our definitions and the definition of [1]. For more than two party protocols, we have applied multi-terminal information theory and have given a formulation of the leaked information by extending our formulation for two party protocols. It is a further work to design the protocol which retains privacy as much as possible.

References

1. Bar-Yehuda, R., Chor, B., Kushilevitz, E.: Privacy, additional information, and communication. Proceedings 5 Annual Structure in Complexity Theory Conference (1990) 55–65
2. Ben-or, M., Goldwasser, S., Wigderson, A.: Completeness theorems for non-cryptographic fault-tolerant distributed computation. Proc. 20th ACM Annual Symposium on Theory of Computing (1988) 1–10
3. Chor, B., Kushilevitz, E.: A zero-one law for boolean privacy. Proc. 21st ACM Annual Symposium on Theory of Computing (1989) 62–72
4. Cover, T. M.: A proof of the data compression theorem of Slepian and Wolf for ergodic sources. IEEE Trans. Inform. Theory IT–21 (1975) 226–228

5. Gallager, R. G.: Information Theory and Reliable Communication. John Wiley & Sons, New York (1968).
6. Kushilevitz, E.: Privacy and communication complexity. SIAM Journal on Discrete Mathematics **5** (1992) 273–284 (preliminary version in Proc. 30th FOCS (1989) 416–421)

Session 7

PROTOCOLS II

Chair: Ed Dawson
(Queensland University of Technology, Australia)

NEW PROTOCOLS FOR ELECTRONIC MONEY

Jean Claude PAILLES
Service d'Etudes communes des Postes et Télécommunications (SEPT)
42 Rue des Coutures, 14040 CAEN, FRANCE

ABSTRACT
This paper proposes 2 new protocols for electronic cash system, which give in a simple way, three main features :

- non reusability of the coins
- anonymity of the payments
- divisibility of the coins, in order to be able to use a coin C in its totality, or subdividing it into several pieces.

This simplicity seems to be an advantage over other previous work on this issue. In this paper, the way to address divisibility, and to use a variant of Schnorr scheme is original (§3.2,3.3,3.4)
The last section gives information on practical aspects, like amount of computation, of storage , and of transmission.

1 - INTRODUCTION :Prepayment systems

A model of prepayment system is shown on fig. 1. Only 3 types of actors are required : the bank, the user, the shop. We consider in this paper the approach of prepayment based on "coins which are numbers" which has been first described by Chaum [CHAU 83]. We do not consider the approach of prepayment based on "purse cards" which are in fact a counter inside an object (a Smart Card) providing physical security, and protocols for secured increase or decrease operations between 2 cards exchanging money. This approach does not present any difficulty (except for the problem of physical security) and is in the scope of European Standardization Work [CEN].

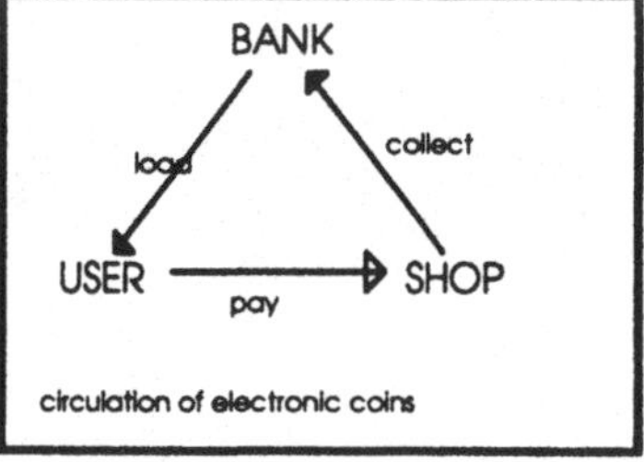

circulation of electronic coins

The approach "coins which are numbers" leads to the following principles.
a/ * An electronic coin is data of the form C = [x, S(x)] where S is the bank signature function.
b/ * Anonymity
This is an important feature, which now is offered by the traditionnals coin and bank note systems. So it is important that, when the user buys a coin from the bank (loading transaction), there are no means for the bank to link the identity of the user

and the bits of informations of the coin ; if not, it would be possible to trace the coin and its uses.
Fortunately, blind signature schemes exist, using the multiplicative properties of RSA [RSA] :to obtain the signature of x, the user asks the bank for the signature of P(z) x where P is the public verification function of the bank, and z a random number choosen by the user. The signature given by the bank is then S(P(z).x) = z S(x), and the user computes easily S(x).

c/ * During the payment protocol, the user must not reveal C = (x, S(x)) to the shop, in order to prevent some missuse of the coin by the shopkeeper himself. So, fake money could be created. Therefore, the user must prove that he has a coin C without giving all the informations of C.
Another problem appears with the duplication of C by the user himself who owns the coin . Nothing is easier than the duplication of data in a computer ! Of course, if the device which contains the electronic money is a physically secured one, like a Smart Card, there is no problem. Without this assumption, and with an off line system, the solution is to use a scheme to prove the knowledge of C which is "one time", ie which can be used only one time, unless giving to the bank the possibility of revealing for example the identity of the user who has duplicated his coin.
The alternative of the on line system, where for each transaction, the shop has to ask to the bank "have you already seen this coin C ", is completely ineffective in terms of cost and time.
d/ * Several schemes for "one time" proof of knowledge or signature exist : [CHFN 88, OKOH 90]. In all cases, data of the coin C have to be structured in such a way that these kinds of schemes can work. For example, the identity of the user must be present in these data. But this is contradictory with.b) and the blind signature methodology. Fortunately, there is a solution based on the "cut and choose" method [RA 77] : so it is possible to conciliate blinding for signature by the bank and control by the bank of some sensitive information in the data which are signed. This method is used in the protocols presented below. The principle of this method is simple : the user sends to the bank several candidates, blinded, for signature. The bank chooses those it wants to control. Then the user gives all the information to the bank to remove the blinding and control the candidates. If controls are successful, bank signs the others (blinded) candidates. Then, if the user wants to fraud, and if K candidates are proposed to the Bank which controls K/2 of these candidates, the user has to build K/2 candidates with fraudulent information, unless he gives to the bank a means to reveal his own identity; so, he has to hope that the bank chooses exactly the others K/2 candidates which will be controled successfully by the bank. For K = 10, the probability of such an event is $1/C_5^{10}=1/252$.
e/ * If a coin C can be used as a whole, or as smaller pieces, untill the total value of the pieces used equals the value of C, then we have divisibility of the coin. This feature is interesting on a practical point of view if the amount of information and the protocol overhead for a dividible coin are reasonable compared with the solution of having several smaller undivisible coins.

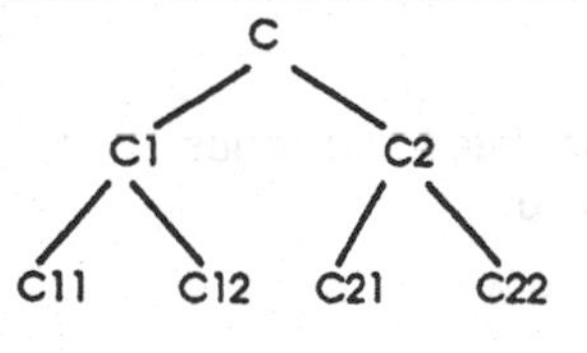

Tree organisation of a coin

As in [OKOH 91], the principle taken in this paper is to divide the coin in 2 parts, which are themselves divided recursively. The problem of non reusability is slightly changed by such an organisation. Taking the example shown in the figure, reuse a coin may mean to use 2 times C_1, or use one time C_1 and C_{12} for example, since C_{12} is a part of C_1.

2 - PREVIOUS WORK

Chaum, in [Chau 83] gave the bases of on line untraceable Electronic Cash Systems. [CHFN 88] described protocols for Electronic Cash working off line . Unreusability is possible, thanks to the use of collision free hash functions. [OKOH 90], for the purpose of unreusability, proposed the use of the GQ scheme [GQ 88]. [BH 90] followed the construction of [CHFN 88] and improved the mechanisms of "hot listing". [OKOH 91] introduced the concept of divisibility with a binary tree, and based the unreusability and divisibility on a square-rooting technique.

3. DETAILED PROTOCOLS

3.1: SCHEMES FOR ONE TIME PROOF OF KNOWLEDGE

a/ scheme based on GQ [GQ 88]

Let **U** and **B** be a prover and a verifier; let n be a composite number (whose factorization is secret), and e a prime, $e < n$. In order to prove that **U** knows two eth-roots of I and J, $u = \sqrt[e]{I}, v = \sqrt[e]{J}$, the following protocol is used, where computations are made modulo n:

U	**B**
sends I and J	
	chooses c, random, $0 < c < e$ sends c
computes and sends $y=v.u^c$	
	verifies that $y^e=J.I^c$

This protocol, used one time, doesn't give any information on v nor u. Nor if it is used several time, with v changing randomly.

If it is executed twice with the same u and v, but c and c' different, then it is easy to compute u: if $y=vu^c$ and $y'=vu^{c'}$, then $y/y'=u^{c-c'}$. But c-c' and e being relatively primes, then one can find r and s such that $(c-c')s+er=1$, and $v=(y/y')^s I^r$. The probability of having c and c' different is 1-1/e-1.

b/ Scheme based on Schnorr scheme [SCH89].

Let n be a prime, and e, a generator of the multiplicative group of Z/nZ.
U wants to prove that he knows u and v, the logarithms of I and J: $I=e^u$, $J=e^v$.

U	V
sends I and J	
	sends c random, $0< c < n-1$
computes and sends $y=v+cu \mod n-1$	
	verifies that $J.I^c=e^y \mod n$

The one-timeness of this scheme is obvious, since having $y=v+cu$, and $y'=v+c'u$, c and c', it is easy to obtain u and v by solving this linear system.

3.2: PRINCIPLES OF THE NEW PROTOCOLS FOR ELECTRONIC MONEY

In this paragraph, we use the first scheme (3.1,a) . Computations are done modulo n, where n is a composite number whose factorization is only known by the user of identity idU.

a/ with the scheme of § 3.1a/ , v will contain information related to the coin C, and u the identity of the user idU. This scheme being "one time", v must change with the piece of the coin which is used: so $v, v_1, v_2, v_{11}, \ldots, v_l \ldots$ are defined for each node of the coin tree. Let l and m be two positions in this tree (for example l=121)

b/ The use of the same piece of the coin twice, or of a part of a piece already used must allow to find the relevant u, and then idU. Then, the v_l must be structured in such a way that $v_{l'}/v_l$ must be computable using v_l^e if and only if l' is in the subtree of l.

For this purpose, 2 functions f and g from Z/nZ into Z/nZ are used, and the sons of v_l are computed by $v_{l1}=v_l f(v_l^e)$, for the left one, and $v_{l2}=v_l g(v_l^e)$ for the right one. v_{lm} may be written as $q.v_l$, where q is computable only with v_l^e.

Indeed, if $v_{lm}=q(v_l^e)\, v_l$, then $v_{lm1}=v_{lm} f(v_{lm}^e)=q(v_l^e) v_l f((q(v_l^e))^e v_l^e)$, which can be written $v_l.q'(v_l^e)$, where $q'(v_l^e)$ is computable only with v_l^e ; for v_{lm2}, the same applies. Then, this statement can be proven by induction.

Then, for a payment with v_l and v_{lm} , the bank, with v_l^e , may compute v_{lm}/v_l .. Then, using the method of §3.1a, it can compute u easily, and obtain idU.

Requirements for functions f and g are simply that f/g has a good statistical repartition (for example, not a constant), in order that v_{l1}/v_{l2} is not computable without the knowlege of v_l^e.

c/ In order for the bank to verify the authenticity of the v_l used for a transaction, and to prevent the need of one signature per v_l , a vector T is defined by : $T=(j(v^e),...,j(v_l^e),........)$. T has the same number of components than the number of nodes in the coin tree; j is a hash function, with result on a few bytes, in order to limit the size of memory required. Twill be part of the data describing the coin C, which will be signed by the bank.

d/ Finally, coin data are:

u, containing idU, concatenated with some random value;

n, choosen by **U**, who keeps secret its factorization in 2 primes ($\phi(n)$ must be prime with e);

T , defined previously

All this information is defined for each candidate sent to the bank (cf §1.1.d) in the cut and choose procedure which allows blinding and signature of well formed messages to be conciliated; this signature has the form $SigC=S(\prod_{i\in g} h(n_i,u_i^e,T_i))$, where h is a hashing function relevant to RSA function, and g the set of candidates choosen by the bank.

3.3: DETAILS

Here is considered the protocol between the user **U**, (identity idU), and the bank **B**. **B** has a couple of RSA functions S (signature), and P corresponding to a moduli n_B; **U**, who is a customer of **B**, has his own RSA functions Su, Pu, and knows P and n_B; The bank knows also idU et Pu. The means by which the secured knowledge of the public key of the corresponding entity is possible, is out of the scope of this paper. X509 may be used.

Bank and users know also the functions:

h hashing, transforming a message of any length into a data field relevant for signature by S or Su.

f and g for the computations mentioned § 3.2.b

j mentioned § 3.2.c

e, prime

a/ loading a coin C

User U / **Bank B**

User U: idU,Su,Pu,P,n_B

Bank B: S,P , n_B , Pu

For i=1 à K, do:

chooses randoms: v_i , b_i , a_i
chooses a composite number n_i
computes $v_{i,1}, v_{i,2}, v_{i,11}, \ldots, v_{i,l}, \ldots$
with v_i, (rules of § 3.2.b);
$u_i=(idU,b_i)$

computes :
$T_i=(j(v_i^e),j(v_{i,1}^e),\ldots j(v_{i,l}^e),\ldots.)$ (calculations mod n_i)
$Z_i=P(a_i)h(n_i,u_i^e,T_i)$ mod n_B

sends
idU
$Z_1,Z_2,\ldots,Z_k$
$s=Su(h(Z_1,Z_2,\ldots,Z_k))$

For each i $\in$g, U et B do :

User U: *sends a_i, b_i, n_i, v_i*

Bank B:
computes $v_{i\,1}, v_{i,2}, v_{i,11}, \ldots, v_{i,l}, \ldots$
with v_i, (rules of § 3.2.b)
computes :
$T_i=(j(v_i^e),j(v_{i,1}^e),\ldots j(v_{i,l}^e),\ldots.)$
verifies that Z_i =
$P(a_i)h(n_i,u_i^e,T_i)$mod n_B

Bank B:

verifies that $Pu(s)=h(Z_1,Z_2,\ldots Z_k)$

If all the verifications are positive:

debit of the account of U

computes $y = S(\prod_{i \notin g} Z_i \bmod n_B)$

sends y

User U:

computes $y / \prod_{i \notin g} a_i \bmod n_B$ = sigC

U stores the data of the Coin C:
sigC;
for each i$\notin$g; n_i,b_i,v_i

Notes:
1. The index of g has not to be stored: with some reordering, we now consider index between 1 and K/2.
2. The signature with Su of $Z_1,..,Z_K$ allows the bank to authenticate the user ; masquerade is then impossible. No one but the user knowing the a_i, for $i \notin g$, can use y; then y has not to be protected when sent from the bank to the user.

b) payment

The shop terminal $\underline{T}$ knows P and n_B. Z is a working variable which is used for the verification of SigC, and set to 1.

User $\underline{U}$	**Shop $\underline{T}$**
chooses among the not yet used coins or coin-pieces the relevant one for the current transaction. Let C be the choosen coin, and l the relevant position in the tree, and (n_i,v_i,b_i,T_i) i=1...K/2 the data of C, as computed during the loading transaction.	
For i =1..K/2 $\underline{U}$ and $\underline{T}$ do:	
computes T_i from v_i, and $v_{i,l}^e \bmod n_i = t$, *and* $(idU \| b_i)^e \bmod n_i = s$ *sends* t, s, n_i, T_i	
	verifies that $j(t)$ *is the l-th component of* T_i *computes* $Z := Z \cdot h(n_i, s, T_i) \bmod n_B$ *chooses random* c_i $0 < c_i < e$ *sends* c_i
computes and sends $y_i = v_{i,l} u_i^{c_i} \bmod n_i$	
	verifies that $y_i^e = t.s^{c_i} \bmod n_i$
----------------------------	
sends SigC	
	verifies that P(SigC) =Z If yes, the payment is completed
keeps in memory that part l of coin C has been used.	stores for i=1.. K/2 $(u_i^e, v_{i,l}^e, n_i, T_i, y_i, c_i)$

c/ Collect:

Batch of data stored by the shop-terminal is sent to the Bank. The security of this file transfer is out of the scope of this paper. But the Bank has to perform the following tasks, specific to this protocol.

3.4 USE OF SCHNORR SCHEME:

Adaptations are straightforward.

-e is here a generator of the multiplicative group of Zn.

-As previously, u contains idU, and v depends on the piece of the coin which is used.

The v_l are defined in the following way:

- f and g are two functions of Zn into Zn-1.

- v, corresponding to the root of the coin tree, is choosen randomly, 0<=v<n-1, and the various nodes in the tree are computed by $v_{l1}=v_l f(e^{v_l})$ mod n-1; and $v_{l2}=v_l\, g(e^{v_l})$ mod n-1;

With this choice, if l' is a descendant of l, then $v_{l'}/v_l$ may be computed with e^{v_l} .

Indeed, if $z=yf(e^y)$ mod n-1, and $y=xf(e^x)$ mod n-1, then z/x is computable with e^x

- T becomes $(j(e^v),...., j(e^{v_l}),.....)$

Other adaptations are obvious.

3.5 PRACTICAL ASPECTS, AND CONCLUSION.

An example of practical choices for the various parameters of the protocols described in 3.3, based on GQ scheme, is given here. The consequences in security level, amount of storage, amount of data to be exchanged, and computation to be done, are described below. The values chosen are:

- 512 bits for the different moduli;

- 10 candidates for the "cut and choose " methodology. This number can be considered as sufficient, since in the loading protocol, the user is identified by the bank, and any attempt of the user to fraud is detected with a probability of 251/252.

- parameter e of GQ (cf 3.1.A) : if chosen too low, the probability to not detect a coin used twice or more becomes not neglectable; but if too large, the number of modular multiplications becomes too high; so we take e a prime between 2^7 and 2^8. With 5 repetitions for the 5 chosen candidates in the cut and choose metod,, the security level is then 2^{40} .

- Coin tree has 8 levels; it then contains 255 nodes, and a coin is divisible up to 1/128 of the coin value;

- function j hashs 64 bytes onto 8 bytes; Then the vector T (cf § 3.2.c) is 2k-bytes long; this is the reason why it is not kept in memory, but recomputed when required. Then the size of a coin in memory doesn't depend on divisibility level.

- idU is 32 bytes long

<u>Storage requirements:</u>

A coin , like defined in §3.3.a, needs 864 bytes of memory .

<u>Computation requirements:</u>

We consider here that the computation time required for functions f,g,h,j is low compared with the time spent in modular exponantiations.

We only consider the payment transaction, which is certainly the most critical in time. The case where the amount to be paid doesn't fit with the divisibility, and then requires several coins is not adressed here. Refering to §3.3.b, and since the parameter of exponantiations e is on 8 bits, we obtain:

- computation of T:	255x8x1.5	3000
- computation of t and s		24
- computation of y 13x8		104

Verification of payment transaction authenticity:

The bank, in order to counter frauds like a shopkeeper sending fake payment transactions, must perform the same control as in §b.

Tests of non reusability

Bank may sort the collected transactions by same n_i, T_i, i=1..K/2

When several transactions with same n_i, T_i are found, the test is simply:

- the same I (position in the coin tree) must not appear twice.
- for each I used, it must not appear an I' being a descendant of I.

In case of reuse, with the mechanisms described in §32b, the bank obtains easily idU, the identity of the fraudulent user.

d/ Black lists

Having found a fraudulent user idU, the problem is to prevent him from doing payment transactions in the shops. But in the payment protocol of §32b, idU evidently does not appear, since anonimity is the main goal of these protocols. Then, the following modifications are to be made, in order to keep anonymity.

Loading:

U adds data d_i et r_i to each of the k candidates.

r_i is choosen randomly by U; d_i is computed using a one-way function O, by d_i=O(idU,r_i). r_i and d_i are part of the computation of Z_i.

d_i and r_i are not given to the bank, except for $i \in g$; r_i and d_i, $i \notin g$, take part to the computation of SigC, since being in Z_i;

d_i and r_i, $i \notin g$, being unknown by the bank, they can't be used to link each payment transaction with a user.

Payment:

The black list of the shop terminal contains all the idU of the fraudulent users. After verification of SigC, the authenticity of the d_i and r_i, i=1..K/2, sent by the user, may be verified.

The test which has to be done to see if the current user is in the black-list is then:

```
For each idU in the blacklist,
    for i =1..k/2
                test if di=O(idU,ri)
                if yes, refuse the payment.;
```

Collect:

d_i and r_i must be transmitted with the other data of the payment transactions, in order to allow the bank to verify the authenticity of the collected transactions .

Note: It is assumed in this paragraph that the Bank will not perform an exhaustive search of IdU on each collected transaction, unless anonymity would disappear.

For 5 candidates, the total is then about 15000 modular-multiplications, or equivalent to 20 full 512 bits exponentiations.

Data exchanges

transfer of T,t,s,n and y is 2.2 k bytes; with 5 candidates, this gives about 11 k bytes.

Use of Schnorr scheme: if T is recomputed at each transaction, the computation load becomes too high (about 10^6); if T is stored the number of computations is very low, but 16 kbytes per coin are needed for its storage.

CONCLUSIONS

The above results are not too far from the possibilities of existing or announced chips for smart-cards, for RSA computations (0.8 s at a clock of 3.57 MHZ, but the clock can be accelerated). However, in this paper, have been considered only the payment transaction and the cards. With this kind of protocols, the load transaction is certainly more time consuming, but has not the same time constraints. Also, the cost of the central bank system would be very high, since it must perform a huge amount of controls and computations.

Another aspect is that an implementation of such a protocol doesn't imply necessarily the use of smart-cards, as described above, at least for paying. In the previously described protocols, the possibility to create fake money is not related to any physical security, and then the storage and computations for the payment transaction could be done mainly by a portable device (PC, or pocket calculator). Then, in this case, the storage and computation requirements become much more easy to reach. The smart card could be used only as a key to access to the portable device, and would play an important role mainly in the loading procedure.

ACKNOWLEDGEMENTS:

The author would like to thank Marc Girault and Jacques Traoere, who are ingeneers at the SEPT, for their helpful comments.

ANNEX: Transferability.

Here are given indications on how to modify the above protocols (in the case of GQ scheme) in order to obtain transferability.

Transferability is the ability for one user to pay an other user with one or several coins, or parts of coins. Then, the model of circulation of electronic money of §1 is changed accordingly. It is clear that transferability has to be limited to a few numbers of payments between users, in order to keep for the Bank the possibility of controling reuse of coins, which need the Bank to collect all the transactions , and to perform the tests as defined in §33c.

To afford transferability, we take the following simplifications:

-n (moduli) is the same for all the users and all the coins; n is a composite number which factorization is secret, including for the Bank, unless anonimity disappears; e is a prime, which size depends on the security level required.

-u=(IdU,random) is unique for each User. During the payment transactions, only u^e is visible, and then anonimity is preseved (but in a weaker sense than previously).

-Each user has a licencence $L=S_{bank}[h(u^e$, others informations...)]. It is obtained by a suscribtion transaction, that the user has to do before to load, to pay, to reload, etc.... Anonimity of the user is preserved if the bank is not able to make any link between IdU, and u^e nor L. This can be the case if a blinding combined with a cut and choose methodolgy is used , like in §3.3.a.

-Signature of a coin by the Bank becomes S=Sbank[$\prod$ h(T)] , where T has the form defined in §3.3.a; $\prod$ is the product of K/2 candidates, and this signature is obtained in the same way than in 3.3.a. Pbank is the public verification function.

Now we consider payment transactions with a coin which data are for K/2 candidates: v_l for the value attached to node l of the coin tree, T, and S. Different users U,U',U",..., $U^{(k)}$ having licences L,L',L",...,$L^{(k)}$ are considerd .

1/ U pays U' with part l of a coin

The protocol is basically the same than in 3.3.b, with the adaptations implied by the modifications of T and L. After the transaction U' obtains from U:

L,u^e
for each K/2 candidates;
T
v_l^e, v_lu^c — where c is the challenge chosen by U'
S

U' has to perform the following verifications:

verif. of L and u^e with P_{bank}
verif. of v_l^e, with T
verif of v_lu^c with v_l^e and u^{ec}
verif. of S with P_{bank}

U' keeps in his memory history (all the data obtained by U' from U) of this transaction.

2/ U' pays U" with part m of a coin (node m descendant of node l)

U' computes $v'_m = v_lu^c(v_m/v_l)$ where v_m/v_l can be computed from v_l^e using formulas of §3.2. Then U" obtains from U':

L,u^e,L',u'^e
for each K/2 candidates
T,c
$v'_m{}^e$, $v'_mu'^d$ — where d is the challenge chosen by U"
S

U" performs same controls as in 1/; more precisely, U" verifies $v'_m{}^e$ by computing $v_m{}^e = v'_m{}^e/u^{ec}$, and checking with T.
U" keeps in his memory history of this transaction.

3/ U" pays Shop with part q of a coin (node q descendant of node m)

U" computes $v''_q = v'_mu'^d(v_q/vm)$ where v_q/v_m can be computed from $v_m{}^e$ using formulas of §3.2.

Then U" obtains from U':

L,u^e,L',u'^e, L",u''^e
for each K/2 candidates
T,c,d
$v''_q{}^e$,$v''_qu''^f$ — where f is the challenge chosen by the Shop
S

Shop performs same kind of controls as in 2/; $v''_q{}^e$ is verified by computing $v_q{}^e=v''_q{}^e/u^{ec}u'^{ed}$ .Shop keeps in its memory history of this transaction. These data are collected by the bank during the collect transaction.
note: this process can be generalized on any number of transfers of the same coin.

This protocol offers the same functionnalities that protocol described in §3.
Anonimity is preserved, except in the case of collusion between bank and users: eg, user U, if he gives v_lu^c to the Bank, could allow Bank to find idU' in transaction 2.
Divisibility is unchanged, since based on the fact that $v_l^{(k)}/v_m^{(k)}$ is computable with $v_l^{(k)e}$ if node m is a descendant of node l.

Reusability is slightly changed, since different parts of the same coin can be transfered to different users.
For example, Bank will receive after the collect 2 expressions of the form

(1) $v_l u^a u'^b u''^c$

(2) $v_m u^a u'^d u'''^f u''''^g$

(for each of K/2 candidates) if user U' has used twice the same part l of a coin (or a part of it); one part has circulated from U' to U" and then to a Shop; the other part has circulated from U' to U''' to U'''' and then to a Shop. As vl/vm is known, then by dividing (1) by (2) , Bank obtain K/2 expressions of the form $u'^p u''^q u'''^r u''''^t$ with different p,q,r,t, for each candidate. Then it is possible to compute by successive elimination u'^z , and then, u' and IdU' (like in 3.1.a).

Note: a same coin can't be used by the same user, to pay and to be payed. For example, in steps 2 and 3, U" must be different from U, since U could discover IdU', and then, inpersonate U'. Then, to cover this case, in this protocol, the payer must control that the payee is not in the history of the transfered coin: U' controls that u''^e different from u^e.

REFERENCES

Chau 83 : David Chaum : Blind Signatures for untraceable payments ; Crypto 82, Plenum Press, New York 1983, 199-203.

Chau 90 : David Chaum : Showing credentials without identification : Transferring signatures between unconditionally unlinkable pseudonyme ; Auscrypt 90, LNCS 453, Springer-Verlag, Berlin 1990, 246-264.

CEN: "Intersector electronic purse" : standardization work of the european organization CEN/TC224/WG10; to be published in 93.

CHFN 88 : David Chaum, Amos Flat, Moni Naor Untraceable Electronic Cash ; Crypto 88, LNCS 403, Springer-Verlag, Berlin 1990, 319-327.

RA 77 : Rabin M.O. Digitalized Signature. "Fundations of Secure Computation" Academic Press, NY, 1978.

GQ 88 : LC. Guillou and JJ. Quisquatter "A practical zero knowledge protocole fitted to security microprocessor minimizing both transmission and memory".
The Proc of Eurocrypt 88, PP 123-1

BH 90 : Barry Hayes : Anonymous one time Signatures flexible untraceable Electronic Cash ; the Proc of Auscrypt 90.

OKOH 91 : Tatsuati Okamoto, Kazuo Ohta : Universal Electronic Cash, Proc of Crypto 91.

RSA 78 : Ronald Rives, Adi Shamir, Leonard Adleman. A method for obtaining digital signature and public key cryptosystems, com. of the ACM 120-126, Feb 78.

SCH89 : C.P. Schnorr, Efficient identification and signatures for smart cards, Advances in Cryptology, Proceedings of CRYPTO 89, pp235-251.

Modelling and Analyzing Cryptographic Protocols Using Petri Nets

Benjamin B. Nieh† and Stafford E. Tavares
Department of Electrical Engineering
Queen's University
Kingston, Ontario, Canada K7L 3N6

Abstract — In this paper, we present a Petri net based methodology for the formal modelling and analysis of cryptographic protocols. We set up modelling rules that represent the protocols in terms of Petri nets. The modelling produces formal descriptions for the protocols with good visibility and layered abstraction. In particular, the descriptions clearly visualize the causal relations and constraints among the data flows in the protocols. An intruder model is introduced to formulate intruder attacks and to generate test cases against the cryptographic protocols. A procedure that exhaustively generates the test cases and searches for states that violate specified security criteria, is also proposed. We demonstrate the value of this methodology by applying it to a number of published protocols. In this way, we are able to reveal security flaws of these protocols. This methodology is applicable to both public-key and private-key based cryptographic protocols.

1. Introduction

When using cryptographic techniques to achieve network security, one must consider both the cryptoalgorithms and the cryptographic protocols that utilize the algorithms. An improperly designed protocol may provide potential loopholes for an intruder to compromise the system security no matter how strong the cryptoalgorithm deployed is. Therefore, the security of these applications depends not only on the strength of the cryptoalgorithms but also on the correctness and security of the protocols. Since informal descriptions are prone to subtle security errors, formal methods, including formal modelling and analysis, provide a crucial means to ensure the correctness and security of cryptographic protocol designs [1][2].

Progress has been made by various researchers to model and analyze cryptographic protocols, using state machines [2][3], term-rewriting [1][4], rule based expert systems [5], and modal logic [6][7]. However, these methodologies have been developed either to fit specific classes of protocols, or to study security issues with no explicit assumptions about intruder attacks. As a graphical and mathematical formalism, Petri nets [8][9][10] have been widely applied to the modeling and analysis of conventional communication protocols [11][12]. But there is no published work on using Petri nets to study cryptographic protocols.

The methodology presented in this paper is to use Petri nets to model and analyze cryptographic protocols. We build a bridge between data security and Petri net formalism through modelling to formally describe cryptographic protocols in terms of Petri nets. Formulated in terms of Petri nets, security problems can be investigated by using analytical tools that are available in this formalism. Illustrated by its application to a number of

† Presently with Department of Electrical and Computer Engineering, University of Waterloo, Waterloo, Ontario, Canada N2L 3G1.

protocols, including both public-key and private-key protocols, this methodology shows its promise by providing unambiguous descriptions and by discovering security flaws. Since no restrictions are placed on the types of cryptoalgorithms and security goals employed, this methodology is applicable to a wide spectrum of protocols.

2. Petri Nets

The modelling and analytical tool employed in our approach is Petri nets. Specifically, we use ordinary Petri nets [9], and Coloured Petri nets [10]. In this section, we provide a brief introduction to these two types of Petri nets.

An ordinary Petri net is a particular type of directed graph with two kinds of nodes: places and transitions, connected by directed arcs. In graphical representation, places are drawn as circles, and transitions as rectangles. An integer number of markers, called tokens, which are drawn as black dots, are allowed to occupy the places. The distribution of tokens over the net is called a marking which represents the state of the net. A transition is enabled and can fire when each of its input places contains at least one token. When a transition fires, it removes one token from each of its input places and adds one token to each of its output places. A place with no input (output) is called a source (drain).

From a condition-event point of view, a place represents a condition, a transition represents a set of actions, and the movement of tokens though the net represents the execution of actions — an event taking place after being enabled by conditions. The presence of a token in a place is interpreted as holding the truth of the condition associated with the place. Tokens that appear in the same place are viewed as *identical* to each other. Figure 1 shows an ordinary Petri net and its change of marking (state) resulting from a transition firing. It also illustrates the encryption of plaintext P with key K using the encryption algorithm E. Very often, ordinary Petri nets are simply referred as "Petri nets".

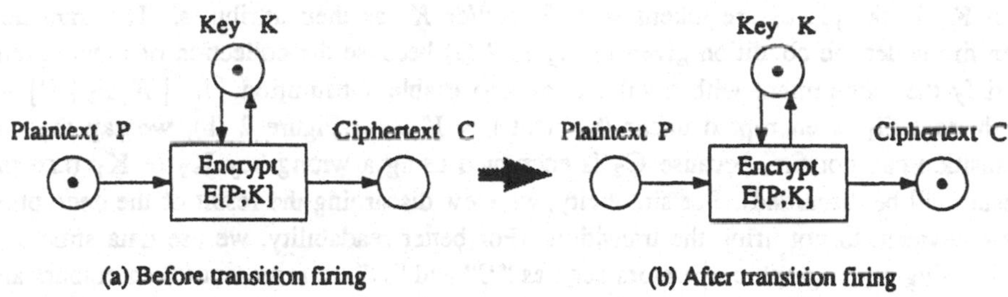

Figure 1. Transition Firing in a Petri Net

As an extension of (ordinary) Petri nets, Coloured Petri nets have been developed to produce more compact and manageable descriptions than those produced in terms of ordinary Petri nets [10]. By means of arc expressions with *variables*, a Coloured Petri net can merge separate sets of Petri net places, transitions and arcs that are of certain common characteristics to a simple structure with greatly reduced redundancy. In a Coloured Petri net, a place can host tokens with *different* meanings. Each type of token is distinguished from others by the information it carries — a *colour* defined by a set of *attributes*. These attributes specify the contents of the token and the operations that

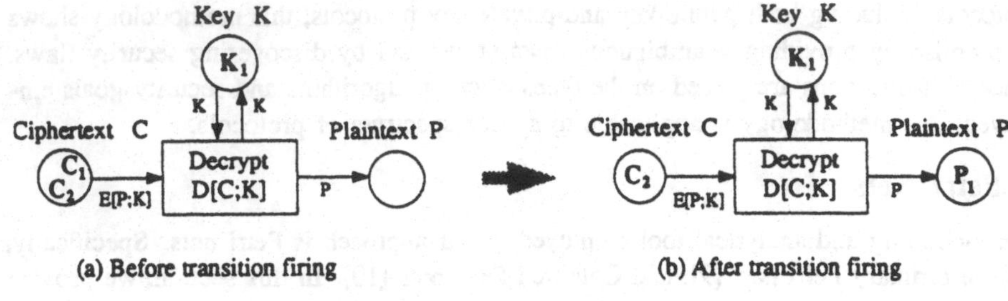

Note: (i) D[C;K]: Decryption of ciphertext C under key K.
(ii) $C_i = E[P_1; K_i], i=1,2$ and $D[C_1; K_1] = P_1$.

Figure 2. Transition Firing in a Coloured Petri Net

have been performed on them. These attributes can be inspected and modified when a transition fires. The condition for attributes of the tokens in an input place to enable a transition, is defined by a *colour function* attached to the arc between the place and the transition. The *variables* in this function correspond to the attributes of the token(s) allowed to pass through the arc. The collection of input colour functions for a transition define the relation among the attributes of input tokens that must be satisfied in order to enable the transition. The output colour functions define the way to assign attributes to output tokens according to the attributes of input tokens. A transition will not fire unless the attributes of the input tokens constitute a valid substitution of the variables in the colour functions inscribing the input arcs.

Figure 2 shows a Coloured Petri net and its change of marking (state) that results from a transition firing. It also illustrates the decryption of ciphertext C with key K using the decryption algorithm D. In this net, P, K and $E[P; K]$ inscribing the arcs are colour functions with variables P and K, while $C_1 = E[P_1; K_1]$, $C_2 = E[P_1; K_2]$, P_1 and K_1 in the places are tokens with P_1 and/or K_1 as their attributes. The transition can fire under the condition given in Figure 2 (a) because the collection of input tokens satisfy the requirement with a valid attribute-variable substitution $\{K_1 \mid K, P_1 \mid P\}$ — ciphertext C_1 is encrypted under the right key K_1. In Figure 2 (b), we say that the transition can not fire, because C_2 is encrypted using a wrong key $K_2 \neq K_1$ (and the result will be discarded). For simplicity, we view discarding the result of the decryption as equivalent to not firing the transition. (For better readability, we use data structures containing cryptographic operators such as "E" and "D", to represent token colours and colour functions. These operators are neither attributes nor variables.)

Petri nets have a strong mathematical basis for analyzing general net properties, such as deadlock and livelock freedom, boundedness and proper termination, and for analyzing specific properties of individual nets. For more details about mathematical definitions, properties, analysis and applications of Petri nets, we refer the reader to [9] and [10].

3. Example Protocols

We will use the Burns-Mitchell private-key resource sharing protocol [13] and the Meyer-Matyas public-key distribution protocol [14, pp. 578–583] as the main examples to explain our methodology. In this section, we briefly introduce these two protocols. More details about the two protocols can be found in the quoted references.

3.1 The Burns-Mitchell Protocol

This protocol is designed for sharing the resources provided by suppliers, through a trusted broker who arbitrates the resources with users who have resource needs.

As shown in Figure 3, we select an arbitrary pair of supplier *S* (with id *S*) and user *U* (with id *U*), and broker *B* for the discussion. The private key shared between supplier and broker, and that between user and broker are denoted by K_S, and K_U, respectively. We denote by SK the session key; $E[I;K]$ the encryption of the data I under key K; and $MAC[I;K]$ the *M*essage *A*uthentication *C*ode (MAC) computed on I using K. The use of commas in message notations denotes concatenation of data, while "$\longrightarrow$" represents the direction in which a message is transferred. Notations in *italic* are used to denote both entity id's appearing in data structures, and entity names, which can be differentiated from each other through their contexts. Let N be the serial number and V_o the value (amount of resource) offered by S; V_r be the value required by U (assuming $V_r \leq V_o$); and T the task to be finished by S who in turn returns the result denoted by R. Informally, the protocol can be described as follows (see Figure 3):

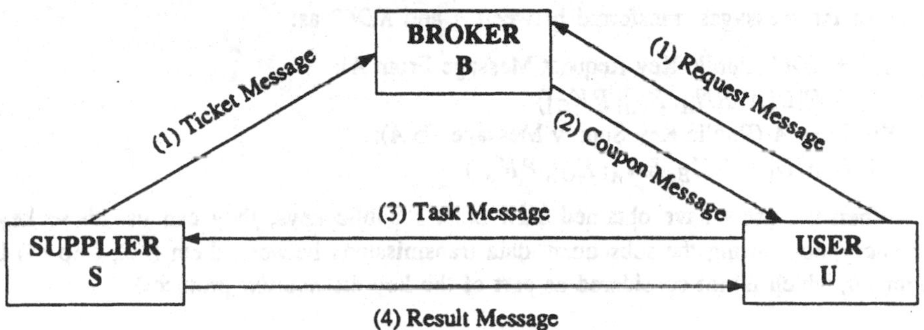

Figure 3. Message Transmissions in the Burns-Mitchell Protocol

(1) $S \longrightarrow B$ (Ticket Message): $(S, N, V_o, MAC[S, N, V_o; K_S])$
$U \longrightarrow B$ (Request Message): $(U, V_r, MAC[U, V_r; K_U])$
(2) $B \longrightarrow U$ (Coupon Message): $(S, N, V_o, U, MAC[S, N, V_o, U, SK; K_S],$
$MAC[S, N, V_o, U, SK; K_U], E[SK; K_U], E[SK; K_S])$
(3) $U \longrightarrow S$ (Task Message): $(S, N, V_o, U, MAC[S, N, V_o, U, SK; K_S],$
$E[SK; K_S], T, MAC[T; SK])$
(4) $S \longrightarrow U$ (Result Message): $(R, S, N, MAC[R, S, N; SK])$.

3.2 The Meyer-Matyas Protocol

This protocol is designed to distribute authenticated public keys to users who request other users' public keys from a key distribution centre (KDC). In this protocol, users and the KDC are initially assumed to each have a pair of secret and public keys. A secret key is known only to each individual owner. Only the KDC's public key is initially known to the public.

Again we choose the *KDC*, and an arbitrary pair of users *A* and *B* who want to get each other's public key through *KDC* (see Figure 4). In addition to the notations we have been using, we denote by PK_A, PK_B and PK_C the public keys of *A*, *B* and *KDC*,

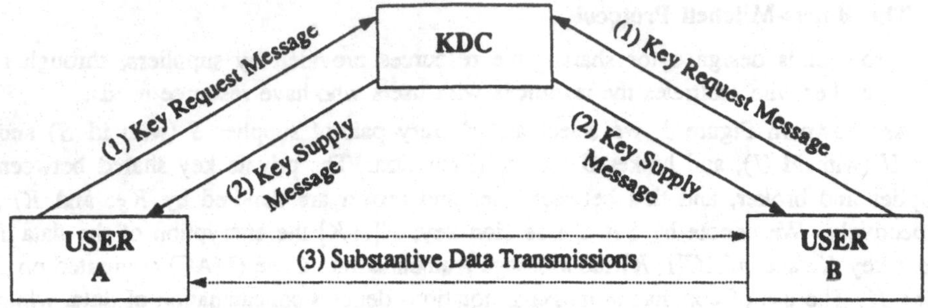

Note: In the message names, "Key" stands for public key.

Figure 4. Message Transmissions in the Meyer-Matyas Protocol

respectively; and RN_A a random number generated by A. K_A and K_C are the secret keys of A and KDC. $D[I; K]$ is used to denote the decryption of I under K.

Because this protocol is symmetric for all the users, we only need to specify the connection between one user and the KDC. Then the protocol can be described (in terms of the messages transferred between A and KDC) as:

(1) $A \longrightarrow KDC$ (Public Key Request Message From A):
$(A, B, E[D[A, RN_A; K_A]; PK_C])$

(2) $KDC \longrightarrow A$ (Public Key Supply Message To A):
$(A, B, E[D[A, PK_B, RN_A; K_C]; PK_A])$

When the users have obtained each other's public keys, they can use these keys for encryption during the subsequent data transmissions between them (i.e., step (3) in Figure 4, which is not considered as part of the key distribution protocol).

4. Modelling of Protocols

While informal protocol descriptions in natural language are readily accessible to human readers, they have two disadvantages [15]: (i) they often contain ambiguities and are difficult to check for their correctness; and (ii) they can not be processed by automatic tools in conformance testing processes. Formal protocol modelling is the process to represent a protocol system in terms of a formalism. The result of the modelling is a set of formal descriptions that provide an unambiguous reference for understanding the protocol, and a basis for automated formal analysis.

Two types of entities are identified in our modelling. They are protocol entities, and an intruder[1] who can eavesdrop and interfere with the communication between the protocol entities. In Figure 5, we present a general model of a protocol system and an intruder. We complete the modelling in two steps. First, we model the protocol without

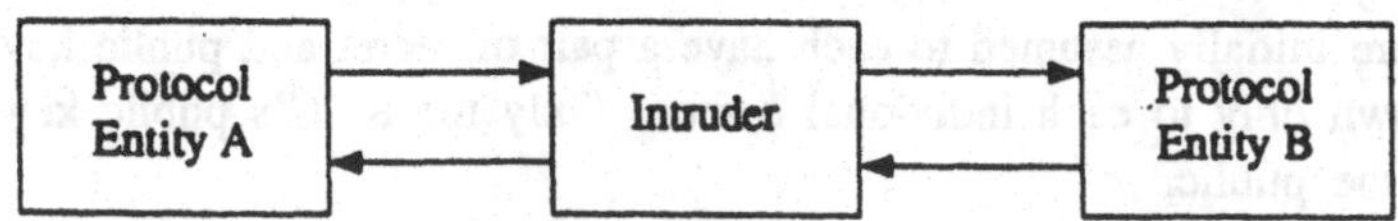

Figure 5. Protocol Entities with an Intruder — A General Model

[1] Individual intruders can have the maximum power by joining their efforts as a single one.

intruder present, to provide easy access to the modelling procedure. Second we model the protocol with the intruder present, to obtain insights into the design semantics, which are of fundamental importance for our investigation of cryptographic protocols.

4.1 Protocol Modelling without Intruder

In this part of protocol modelling, we assume that all functions performed by protocol entities are error free, and messages are exchanged between protocol entities over lossless, error-free channels. In addition, we assume each protocol entity is in a desired initial state at the beginning of each protocol run.

To model a protocol in terms of a formalism, we need to establish a relationship between protocol components and the elements in the formalism. A cryptographic protocol is a system with defined inputs and outputs, which obeys a set of well-defined rules to use cryptographic transformations to achieve security protection of information. Governed by these rules, each protocol entity has the capability of transmitting and receiving messages, and performing a set of operations on the data items in the messages, such as generation, retrieval, encryption and decryption of data. Thus, all the actions in a cryptographic protocol system can be abstracted as the ones that follow the same paradigm: a process which performs operations on a set of input data and produces a set of output data by modifying and changing the input data. Hence four types of basic protocol components can be identified. They are processes, data input/output media (such as buffers, memories and channels), data items, and data flow relationships. The relationship between these components precisely conforms the semantics of Petri nets introduced in §2. Therefore, we can establish the mapping between the basic protocol components and the elements of (ordinary) Petri nets, as shown in Table 1.

Table 1. Mapping between Protocol Components and Petri Net Elements

Protocol Component	**Petri Net Element**
Process	Transition
Data input/output medium	Place
Data item	Token
Data flow relationship	Directed arc

Based on this relation, we can represent the components of each protocol in terms of Petri net elements and then represent the complete protocols in terms of Petri nets. We model the protocols with Petri nets at two structurally equivalent (guaranteed by structural transformation rules of Petri nets [16]) levels of abstraction, namely, the *conceptual level* and the *functional level*.

In [17] we proposed a systematic modelling framework to identify objects at these two levels. This framework defines a number of modelling rules to develop a unique representation for each protocol at these levels from its informal description. Following this framework, we can develop the descriptions for the two protocols we have introduced. In Figures 6 and 7, we give the descriptions of the Burns-Mitchell protocol at both conceptual and functional levels with markings indicating the initial states. And we present in Figures 8 and 9 the descriptions of the Meyer-Matyas protocol at the same

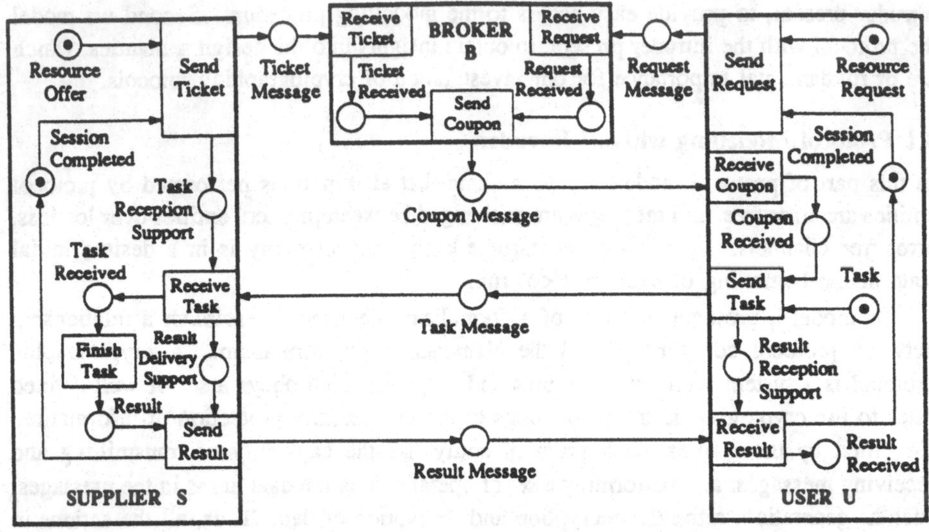

Figure 6. Conceptual Level Description of the Burns-Mitchell Protocol (Intruder not Present)

levels. The descriptions of the Meyer-Matyas protocol include only one user (User A) and the KDC because of the symmetric nature of this protocol for all users. (Note that in these descriptions, a single notation is used to denote both a data medium/place and the data or information/token it holds; and each bidirectional arc is equivalent to a pair of directed arcs in opposite directions.)

The conceptual level description provides a concept about major processes and information flows in a protocol. At this level, "inherent data" which are constantly

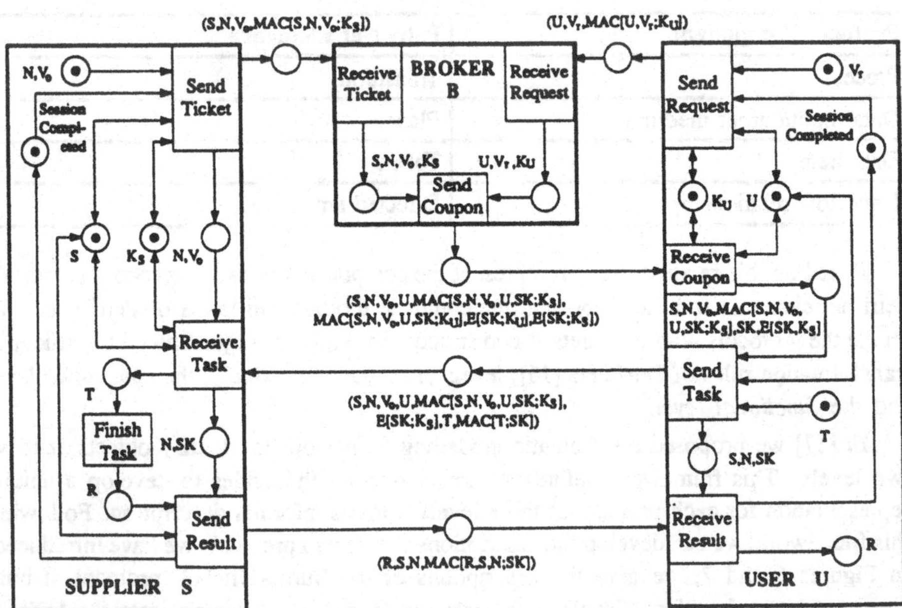

Figure 7. Functional Level Description of the Burns-Mitchell Protocol (Intruder not Present)

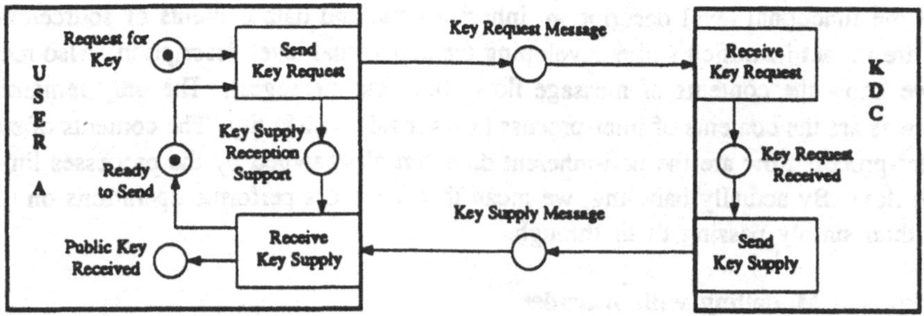

Figure 8. Conceptual Level Description of the Meyer-Matyas Protocol (Intruder not Present)

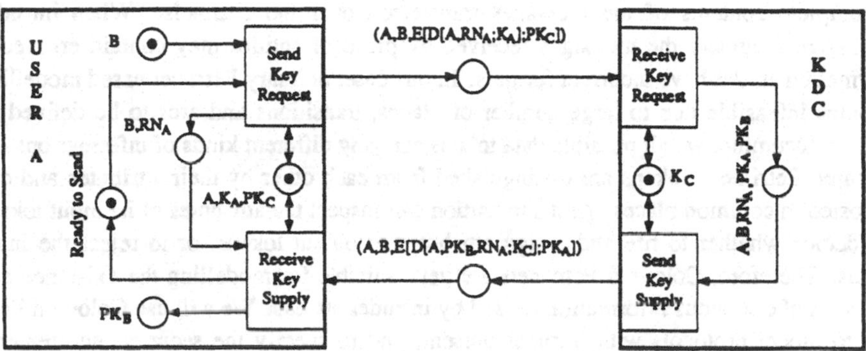

Figure 9. Functional Level Description of the Meyer-Matyas Protocol (Intruder not Present)

available for a protocol entity, such as the entity's id and private key, are considered inherently attached to the processes of the entity and therefore invisible.

The functional level description has the same processes identified at the conceptual level. It visualizes the inherent data (not seen in the conceptual level description) and the data contents/data structures of information flows. In this way, this description specifies the functions performed on the data items by the processes.

The above two levels of descriptions clearly visualize the information/data flows between and inside protocol entities, and the causal relationships among these flows. The detailed steps and complete set of modelling rules to develop these descriptions can be found in [17]. In the following, we discuss only the guideline to develop these descriptions that can be easily justified by referring to Figures 6–9.

In the conceptual level description, for each message transferred, we define two boundary processes, one at the entity sending the message and the other at the entity receiving the message. The input/output interfaces (including the start and termination points of a protocol run), where the protocol exchanges information with the "*outside world*" (i.e., the client of the protocol), are modelled as information sources/drains. An internal process (e.g., "Finish Task" in Figure 6) is defined within an entity when a process taking place in the "outside world" needs to be treated as a process in the protocol to ease the protocol analysis [17]. There is an information flow from one process to a subsequent process, if any non-inherent information is handled in both processes and not in any process taking place between them.

In the functional level description, inherent data, and data contents of source/drain flows are in fact identified while developing the conceptual level description. Also recall that we know the contents of message flows (i.e., the messages). The only remaining unknowns are the contents of inter-process flows inside each entity. The contents of such an inter-process flow are the non-inherent data *actually* handled by the processes linked by the flow. By actually handling, we mean that a process performs operations on data other than simply passing them through.

4.2 Protocol Modelling with Intruder

A cryptographic protocol is designed to combat attacks from an intruder, who has the control of the communication medium and tries to insert, modify or delete partial or complete contents of the messages transferred over the channels. When intruder attacks are mounted, the messages received by protocol entities may contain erroneous information and/or have incorrect formats. In this case, ordinary Petri net based modelling becomes infeasible due to large number of places, transitions and arcs to be defined in order to accommodate all possible data tokens carrying different kinds of informations. In Coloured Petri nets, tokens are distinguished from each other by their attributes and can be hosted in common places. And a transition can inspect the attributes of its input tokens and decide whether to fire and assign attributes to output tokens, or to reject the input tokens. Therefore, Coloured Petri nets are very suitable for modelling the existence and detection of erroneous information caused by intruder attacks. We will use Coloured Petri nets to model protocols with intruder present, and to specify the security requirements enforced by processes protocol entities.

With the details about the intruder attacks temporarily unspecified, we consider the simple fact that the messages coming from the channels may not have the correct contents and formats. Thus we propose a general model for protocol entities and an intruder that is represented by a modified Petri net[2] [18] as shown in Figure 10. Assume there are n message transfers in a protocol run. Since each message from a protocol entity will be "unconditionally accepted" (for manipulation) by the intruder, we use a single colour function F_i to inscribe the place holding the ith message output $(i = 1, 2, \cdots, n)$ from a protocol entity and its surrounding arcs. We use two different colour functions F_j' *and* F_j'' to inscribe the two arcs associated with the place holding the jth output message $(j = 1, 2, \cdots, n)$ from the intruder (see Figure 10). This is because intruder attacks may produce erroneous format and contents of a message (as represented by F_j'), and a protocol entity has its own requirements on the format and attributes of each message coming from the channel (as represented by F_j'').

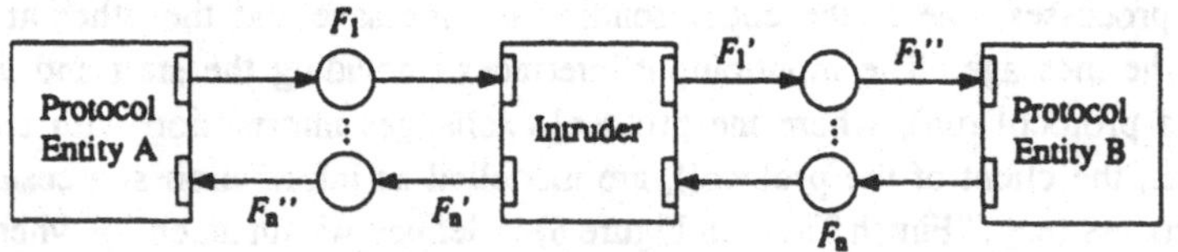

Figure 10. Protocol Entities with an Intruder — A General Petri Net Model

[2] A modified Petri net transition may contain boundary transitions whose detailed interconnections are masked and need to be specified only when lower levels of abstraction are of concern.

We complete the modelling by preserving the objects identified in ordinary Petri net modelling at the *functional level*, assigning attributes to data tokens, and attaching colour functions to arcs. In specific, the attributes are defined according to elementary data items, such as keys, user id's, random numbers, pieces of plaintext information, etc. When defining colour functions, one variable appearing in different colour functions attached to the arcs of a transition requires that the tokens allowed by those arcs must have the same attribute to constitute a valid substitution of this variable. Otherwise, different variables must be used. When defining a colour function to specify the requirement that a key is retrieved according to an entity's id, this key (e.g., K_A) must be viewed as a function of the id (e.g., A). (This is reflected through the consistency between the subscript of a key notation and the id.) For simplicity, a single notation is used to denote the input/output colour functions around a place inside a protocol entity and to denote colour of the token (if any) initially marking the place. Black dots are used to represent tokens that carry determined information and are initially held in places inside protocol entities. ***We use notations identical to the ones in Petri net protocol descriptions to denote only the data items that are originally generated by protocol entities.*** When a process is executed, actual values are assigned to the attributes of output data tokens according to the values of attributes attached to input data tokens. Again we emphasize that cryptographic operators such as "E", "D" and "MAC" are not counted as attributes or variables.

As a result of the modelling, Figure 11 shows the Coloured Petri net description for the user entity in the Burns-Mitchell protocol with intruder present. In this description, the colour functions surrounding each process precisely define the restraints among the attributes of the data inputs and outputs. For example, the appearance of variable K_U in the colour functions associated with the process "Receive Coupon" specifies one of the conditions for the coupon message to be accepted: the private key used to compute MAC and to encrypt the session key for the user in the message must be identical to

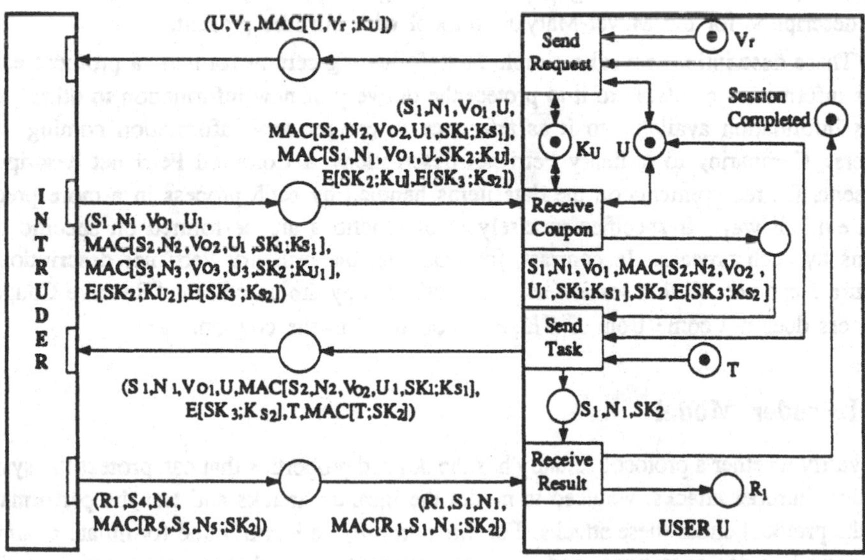

Figure 11. Description of User U in the Burns-Mitchell Protocol with Intruder Present

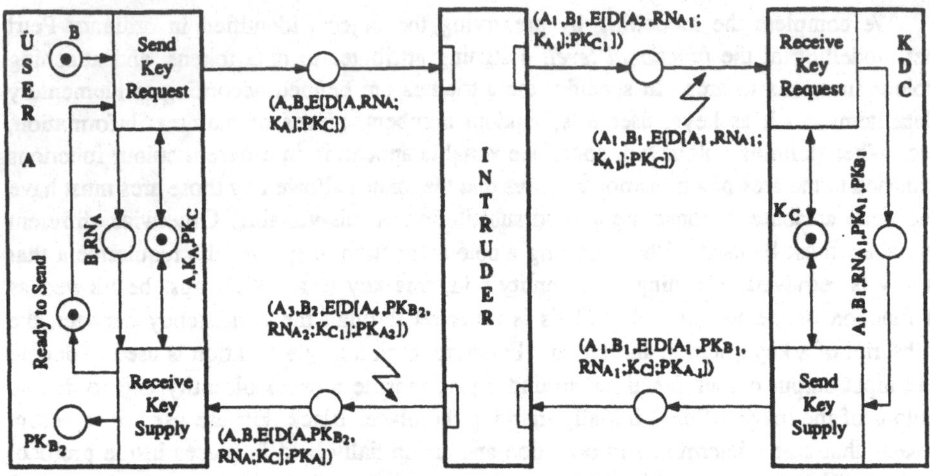

Figure 12. Description of the Meyer-Matyas Protocol with Intruder Present

the private key owned by the user. The colour functions surrounding this process also indicate that the process has no specific requirements on the contents of the MAC and encrypted session key that are to be forwarded to the supplier.

The only difference between modelling of private-key protocols and that of public-key protocols is the treatment of keys. Since a pair of public key PK and secret key K are complementary to each other, we look at them as functions of each other though in fact PK is a one-way function of K. This can be expressed as $PK = [K]^C$ and $K = [PK]^C$, where $[k]^C$ stands for the complementary value of k. This observation is based on the fact that the plaintext encrypted using PK (K) can only be recovered by decrypting the ciphertext using K (PK). In Figure 12, we present the Coloured Petri net description for the Meyer-Matyas protocol with intruder present.

These descriptions reveal a simple fact: following certain formats, a protocol entity uses information available to it to protect the delivery of new information to others; and uses information available to it as reference to extract new information coming from others. Comparing to ordinary Petri net descriptions, a Coloured Petri net description presents the requirements on the data items handled by each process in a more precise and explicit way. It specifies precisely what functions are performed on specific data items by each process. In contrast, for example, the ordinary Petri net description in Figure 7 can not tell that session key SK endorsed by the user in the "Receive Coupon" process does not come from $E[SK; K_S]$ contained in the coupon.

5. Intruder Model

To verify whether a protocol actually has the desired properties that can protect the system against intruder attacks, we need to model the intruder attacks and test the performance of the protocol under these attacks. For this purpose, we identify and formulate a number of basic intruder operations, and develop an intruder model that uses these operations to produce all possible sequences of intruder attacks. The generation of these intruder attacks are governed by a number of defined rules.

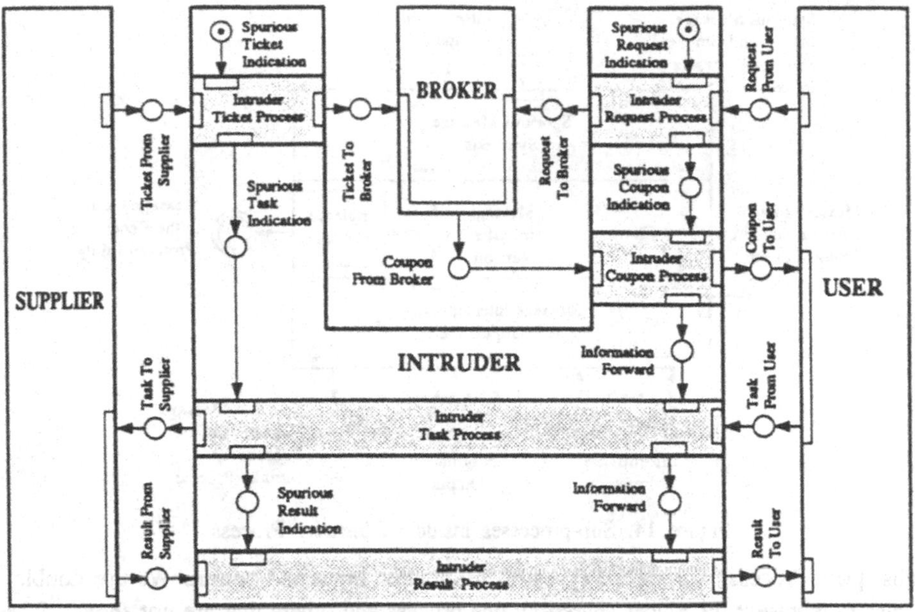

Figure 13. Intruder Model for the Burns-Mitchell Protocol

We assume that the intruder has a full knowledge of the protocol, a record of previous transactions between the protocol entities, and the facility that can perform any operation defined in the protocol. Further, the intruder can be a legitimate user in the system. In addition, the intruder is assumed to have old session keys obtained through some means, e.g., theft or cryptanalysis. However, we put a restriction on the intruder behaviour: it performs operations following the format of each message. We do not consider the cases that the intruder alters the order of message contents, because it can be easily detected.

The proposed intruder model is composed of a number of intruder processes that each corresponds to a message transferred over the channel. Figure 13 illustrates a simplified intruder model for the Burns-Mitchell protocol described at the conceptual level with a modified Petri net.

In each intruder process, the intruder can choose to intercept a message from the channel, delete it, or return a manipulated version of the message back to the channel; or to synthesize a spurious message and inject it into the channel. Thus we define two sub-processes inside each intruder process, namely, "Message Interception & Manipulation" and "Spurious Message Synthesis". To perform operations in an intruder process, the intruder receives information inputs from intruder processes handling preceding message transfers, and provides new information extracted from the intercepted message to its subsequent processes.

Figure 14 presents the structure inside a general intruder process composed of these two sub-processes and the connections between this process and the channel and other intruder processes. An additional "Release Message" operation is defined to send the message of the intruder version back to the channel. Note that "Spurious Message Indication from the Preceding Process" and "Spurious Message Indication to the

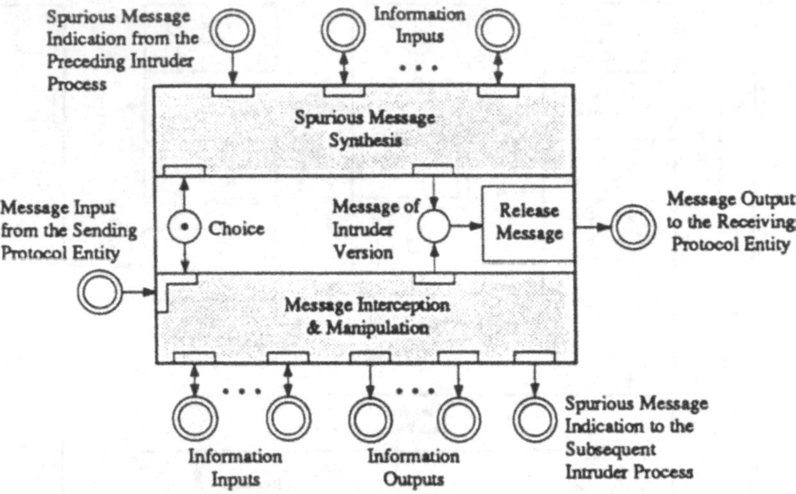

Figure 14. Sub-processes inside an Intruder Process

Subsequent Process" are special types of information input and output. We use double-circles to represent the places shared by one process and others that are not shown. The place "Choice" is only marked at the beginning of each protocol run, so that the intruder can not employ two sub-processes at same time and produce more than one version of the message in a process. Bidirectional arcs are used to indicate that the information inputs to this process may be used by other intruder processes.

Before proceeding further, we define that an *individual item* is a visible field that are not dividable without performing cryptographic transformation. For instance, message $(A, B, E[D[A, RN_A; K_A]; PK_C])$ in the Meyer-Matyas protocol contains three individual items, A, B and $E[D[A, RN_A; K_A]; PK_C]$.

The information inputs to an intruder process may include: (1) a correct version and a wrong version of every piece of predictable plaintext information in the message; (2) a wrong version of every piece of unpredictable plaintext information in the message, such as a random number; (3) compromised secret information in the message in a previous session belonging to the association, such as an old session key; (4) the messages and their individual fields obtained in previous sessions that belong to the association under the attack and an association other than the one under the attack; (5) a wrong key for every key used in the message; and (6) information outputs from preceding processes. Since the intruder can be a legitimate user in the system, it is reasonable to assume that the wrong version of a data item uniquely associated with the identity of the entity under attack (e.g., a user id, a public key, etc.) corresponds to the true identity of the intruder.

The information outputs from an intruder process provide new information that is not available to the intruder until the process is engaged. They may include (1) unpredictable plaintext information contained in the intercepted messages; and (2) intercepted secret information generated by protocol entities in the protocol run.

At a lower layer, we define the modules inside each sub-process. Figure 15 shows the internal configuration of the "Message Interception & Manipulation" sub-process, assuming that the intruder process handles a message composed of m individual items.

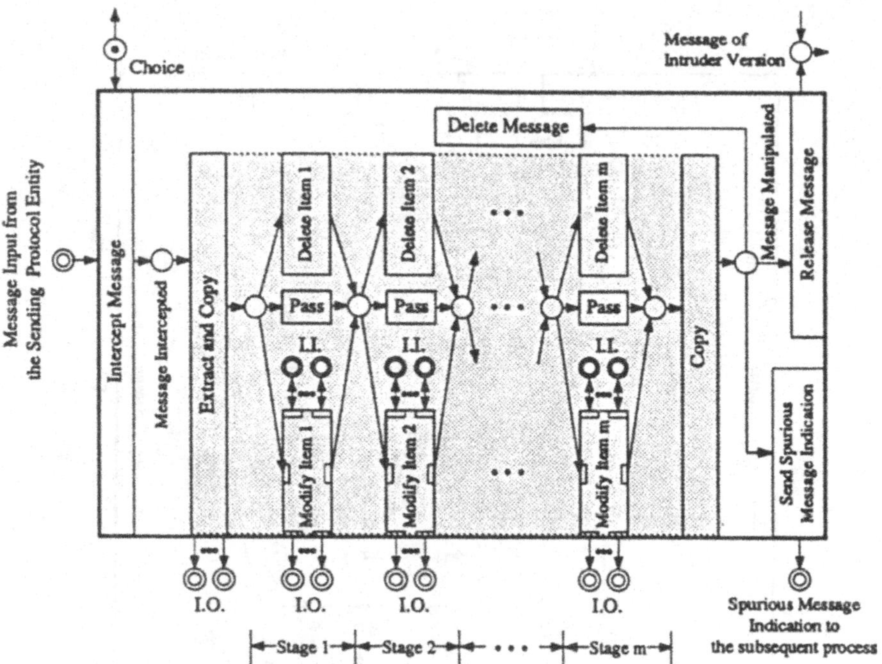

Legend: I.I.: Information inputs.
I.O.: Information outputs.
Stage i : Manipulation stage for Item i, where i=1, ... , m.

Figure 15. Internal Configuration of Message Interception & Manipulation Sub-process

In this sub-process, individual items are first extracted as information outputs, and then go through *m* manipulation stages that each corresponds to an individual item. Each stage provides three exclusive choices to handle the message: "Pass" the item without changing anything, "Delete" the item corresponding to that stage, or "Modify" the item with available information inputs and extract possible new information as information outputs. The stages are concatenated in such a way that every stage is placed behind the ones from which the stage receives possible information inputs. Then a choice is made to send the manipulated message back to the channel, to delete it, or to simply trigger a spurious message indication to the next intruder process (note that the latter two choices are provide after some new information has been extracted at the manipulation stages).

Figure 16 shows the internal configuration of a "Modify Item" module (for Item r), which can be determined according to the type of a given item. We identify three basic item types. They are: Type I, a simple plaintext item; Type II, a ciphertext item; and Type III, a cryptographic check-sum such as the MAC discussed in §3.

As shown in Figure 16, the item to be modified is first separated from the others. No matter what type the item belongs to, the intruder can always make a choice to substitute the item with any of the versions that are maybe available. If the item is of Type III, the intruder can also choose to remove the original item, synthesize the data and compute the check-sum with any of the keys that are maybe available. If the item belongs to Type II, the intruder can also choose to either decrypt the item using any of the keys that are maybe available, if it succeeds, manipulate the plaintext contents, and then

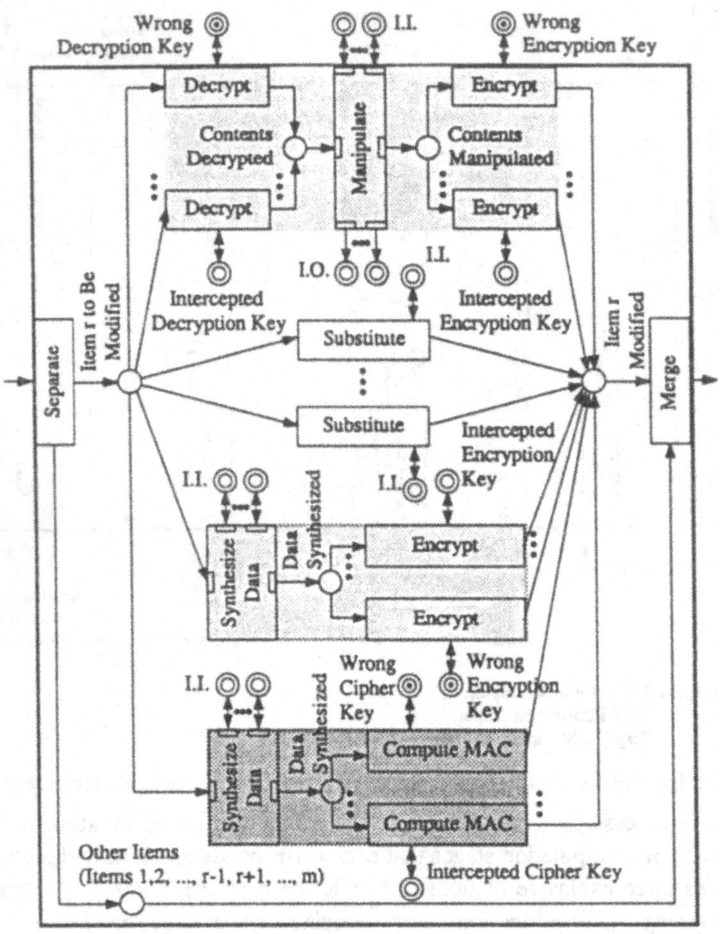

Figure 16. Internal Configuration of a Modify Item Module

encrypt it using any of the keys; or to synthesize the plaintext portion of the item and then encrypt it. Finally the modified item is merged with the other items of the message to complete the operation of the "Modify Item" module. (Note that the en/decryption keys are special type of information inputs. In addition, the contents in some defined information in/outputs may never be available.)

The internal configuration of the "Manipulate" sub-module in Figure 16 is identical to the collection of modules covered by the shaded area in Figure 15. If we eliminate the "Decrypt" and "Manipulate" sub-modules in Figure 16, replace each "Substitute" with a "Paste" operation that produces an item with any of the versions that are available, and change the "Separate" and "Merge" operations to simple "Copy" operations, we can obtain the internal configuration of the "Synthesize Data" sub-module in Figure 16.

The "Spurious Message Synthesis" sub-process (in Figure 14) is primarily composed of a number of concatenated modules that are similar to the "Synthesize Data" sub-module. Each of these modules synthesizes one individual item in the message, and appends the item to the ones that are synthesized by its preceding modules.

Depending on the complexity of a message format, the modules and sub-modules may be further decomposed to the modules and sub-modules we have defined. For an individual item of a complex form, the module partitioning can be repeated as many times as the number of layers of en/decryption employed in this item. The lowest layer of module partitioning should contain only simple operations such as "Encrypt", "Decrypt", "Substitute", "Paste", "Pass", "Delete", etc.

It should be emphasized that all the components in an intruder model are modelled by Coloured Petri net elements. Intruder operations will change the attributes corresponding to the data items on which the operations are performed. In an intruder process, the intruder always "expects" to perform operations with certain information attained through manipulations in preceding processes (such as to use an intercepted key to decrypt a ciphertext). But the intruder can succeed only if such information can actually be obtained. For instance, the intruder will get nothing if it tries to decrypt a ciphertext item (token with certain attributes) using a wrong key (token with wrong attribute).

Using the above modular approach, we can develop the complete set of intruder processes according to the specific message formats given in a protocol. By coupling the processes together through the information inputs/outputs, we can construct the complete intruder model for the protocol. This model explicitly formulates and specifies the behaviour of an intruder. The detailed intruder models for the two example protocols have been developed in [17], and will not be presented here due to space limitation.

6. Security Property Analysis of Protocols

The purpose of protocol modelling is not only to provide a formal representation of protocol systems, but also to facilitate formal analysis of these systems. The Coloured Petri net-based protocol descriptions provide a formal basis on which we can investigate security properties of these protocols, and search for possible security flaws.

6.1 Security Property Analysis

In general, secrecy and authenticity are two basic security goals in designing cryptographic protocols. They are also the security properties expected in most protocol designs. To decide whether a protocol possesses its desired security properties, we need to set up criteria that characterize these properties. The security criteria for a protocol are defined as the requirements that must be satisfied in order to achieve the security goals. The satisfaction of these properties can be judged by verifying whether the system's performance meets these criteria. Our analysis of security properties of cryptographic protocols is based on an exhaustive penetration test. This test utilizes the intruder model, and searches for the states of the global system that violate certain specified security criteria.

We explain the analysis procedure as follows. First, we couple the Coloured Petri net description of the protocol and the associated intruder model to construct the global system. Specific security criteria which guarantees the protection of protocol entities against the modelled threats are defined and then translated to requirements on Coloured

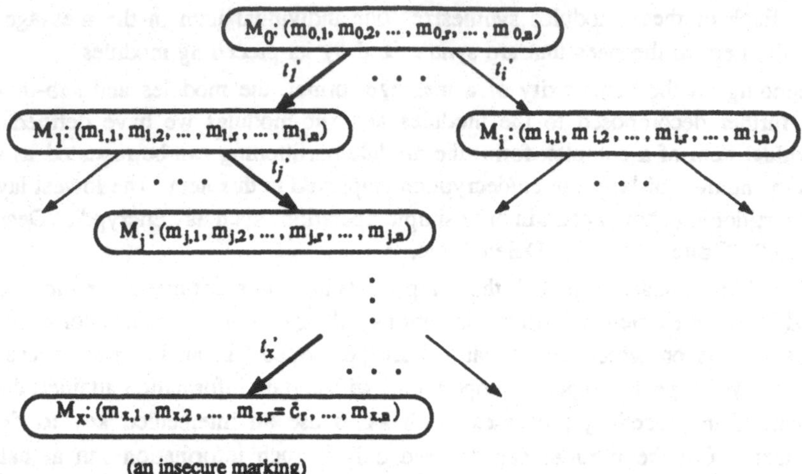

Legend: M_j : The jth global marking.
$m_{j,s}$: The marking in place s, also the sth element of M_j, with $1 \leq s \leq n$, where n is the total number of places in a protocol description and its intruder model.

Note: (1) We assume that $\bar{c}_r$ is an insecure marking in place r, which violates a security criterion, with $1 \leq r \leq n$.
(2) Attribute-variable substitution attached to each transition firing (denoted by t_k') is not shown.

Figure 17. Security Flaw Search in a Coloured Petri Net Reachability Tree

Petri net markings (states) of the global system. We use the intruder model to generate all possible sequences of attacks, and construct a state reachability tree [9][10] by listing all possible markings (states) reached through transition firings in the global system. Each intruder attack sequence is a sequence of specific choices made among the operations defined in the intruder processes. We then search the reachability tree for (insecure) markings that violate the security criteria. When a violation of the criteria is discovered, the protocol is asserted as flawed. As shown in Figure 17, we can find the path of penetration used by the intruder that leads to the insecure marking, by tracing the marking sequence (indicated by bold arrows) in the reachability tree.

6.2 Analysis of the Burns-Mitchell Protocol

In the Burns-Mitchell protocol, the ultimate goals are to have a genuine task being accepted by the supplier and to have a genuine result being accepted by the user. Both the task and the result are provided with verified authenticity. We can then define the security criteria for this protocol as:

Criterion 1: The task endorsed by the supplier must be the one that was issued by the user in the current protocol run; and

Criterion 2: The result endorsed by the user must be the one that was produced by the supplier based on the task issued by the user in the current protocol run.

These criteria can be easily translated to requirements on Coloured Petri net markings of the global system. For example, Criterion 2 can be translated as: " the colour of the result data token inside the user entity must never be any colour other than *R*."

Using the proposed exhaustive search strategy, we have found a security flaw in the Burns-Mitchell protocol, which violates the second criterion [17]. The flaw occurs when the intruder plays back an old coupon message with a compromised session key upon intercepting the user's request. This session key will be accepted by the user and therefore the intruder can send a false result (or even simply the result message in the previous session where the old coupon belongs) back to the user without being detected. Such weakness is caused by the user's lack of capability of detecting, at the "Receive Coupon" process, whether the message is a current one or is a replay.

This security flaw was discovered and first published by Meadows in [19], and was also independently identified by the authors in [20].

6.3 Analysis of the Meyer-Matyas Protocol

For this protocol, the ultimate goal is the correct public key being received as requested by each user. Thus the security criterion can be defined as follows:

Criterion: The public key endorsed by a user must be the genuine one belonging to the user whose public key is requested by this user.

In terms of requirement on Coloured Petri net markings, the criterion can be specified as: "the colour of the public key endorsed by user *A* must never be any colour other thanPK_B."

Our test result reveals a security flaw in this protocol design, which is demonstrated by a branch of the reachability tree shown in Figure 18. The processes and data medium engaged in this successful penetration is shown in Figure 19. In each of the global markings in Figure 18, we only show the markings of the marked places that are denoted as collections of "token / place" pairs (with the numbering of places shown in Figure 19) of protocol entities and the channels. The attribute-variable substitution engaged in a transition firing is denoted as a set of "attribute | variable" pairs (see Figure 18). In this penetration, a wrong public key $(P\bar{K}_B)$ is endorsed by *A* as *B*'s public key.

This intruder penetration can be explained as follows. Upon intercepting a "Key Request Message" from *A*, the intruder substitutes *B*'s id in the message with a wrong id $\bar{B}$, and sends the modified message to *KDC*. *KDC* disassembles this message, and endorses the plaintext part including *A*'s id and $\bar{B}$, since the ciphertext part of this message is encrypted using the right keys and contains the correct id *A*. $P\bar{K}_B$ (= $PK_{\bar{B}}$), the public key for the user "$\bar{B}$", is retrieved by *KDC* and then included in the "Key Supply Message" for *A*. The intruder intercepts this message, changes the wrong id $\bar{B}$ in the plaintext back to *B*'s id, and sends the modified message to *A*. This erroneous key supply message can pass through the detection of *A*. As a result, $P\bar{K}_B$ will be endorsed by *A* as *B*'s public key. Assuming that the intruder is a user with id $\bar{B}$, it can then impersonate *B* to communicate with *A* using the wrong public key $P\bar{K}_B$. The cause of this security flaw is the lack of cryptographic protection for the id of the user whose public key is requested.

We have also applied the above methodology to model and analyze the well-known Needham-Schroeder protocol [21]. We have identified the same security flaw in the Needham-Schroeder protocol originally discovered in [22]. The flaw is caused by its vulnerability to playback attacks. More details about the analysis of this and the above two example protocols can be found in [17].

The protocols analyzed are designed to achieve two common security goals — secrecy and authenticity. However, thanks to the modelling and analytical power of Coloured

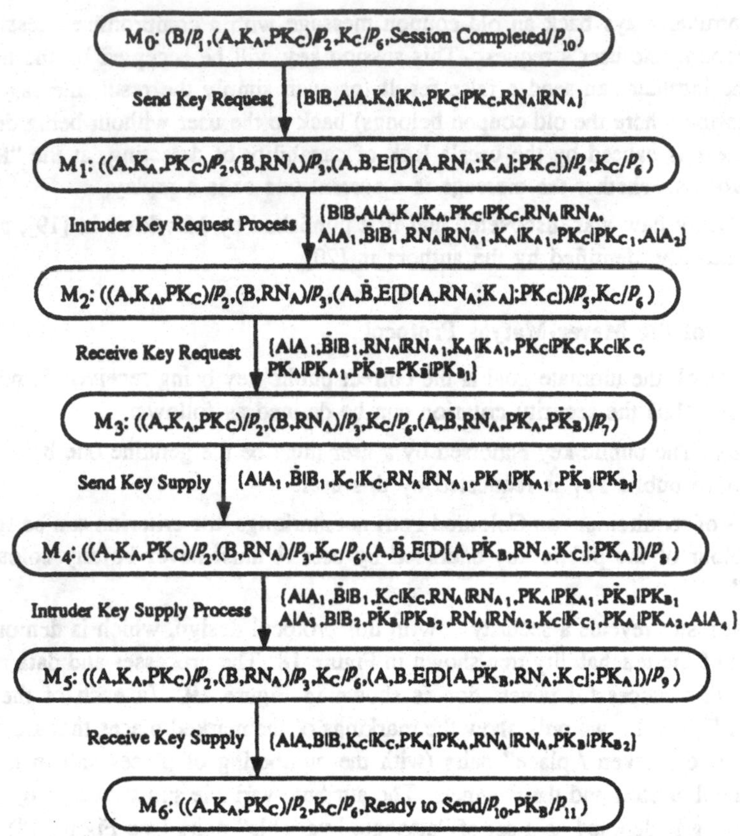

Figure 18. A Branch Leading to an Insecure Marking in the Meyer-Matyas Protocol

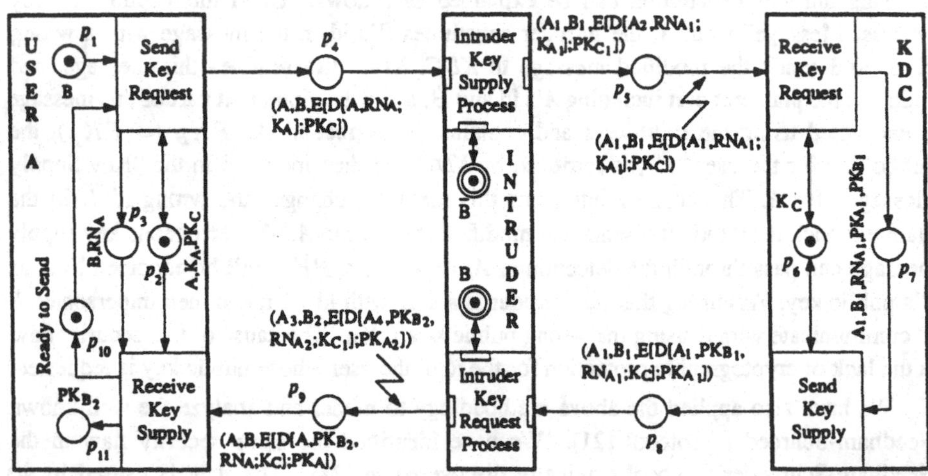

Figure 19. Processes and Inputs/outputs Related to the Reachability Tree Branch in Figure 18

Petri nets, we can also analyze protocols with other security goals. This is because many other security criteria, such as the one for a protocol that can detect denial of service proposed in [23], can also be specified as requirements on the markings in their global Coloured Petri net descriptions [17].

A flawed protocol can be strengthened by fixing the discovered flaws and performing the analysis proposed here iteratively until no new flaws are discovered. The final strengthened version of the protocol will be secure against the threats specified in developing the intruder model.

7. Conclusions

This research has applied Petri net formalism to the modelling and analysis of cryptographic protocols. The methodology we have developed is capable of providing precise descriptions for cryptographic protocols, analyzing their security properties, and detecting possible security flaws. Although informal methods may reveal obvious flaws, a careful formal approach is required to eliminate the more subtle flaws.

We have developed a framework of cryptographic protocol modelling using Coloured Petri nets, explicitly including an intruder. Using this framework, we are able to develop precise protocol descriptions. These descriptions explicitly specify the operations performed by protocol entities, and the relations and constraints among the data flows. Based on a set of typical threats, we have proposed a Coloured Petri net based intruder model that can be constructed according to each specific protocol. We have also proposed a security property analysis procedure that utilizes the intruder model and the protocol description to perform exhaustive search for insecure states in the global system. Following this analysis procedure, we are able to determine whether a protocol is secure against the specified threats, and to trace the paths of intruder penetration.

Using this methodology, we have investigated three published protocols, including both public-key and private-key protocols, and have identified their security flaws. Regardless of the cryptoalgorithm(s) deployed, a protocol can be analyzed as long as its security goals can be specified as requirements on the markings in its global Coloured Petri net description. Therefore, this methodology is applicable to a wide range of cryptographic protocols.

We suggest that Petri nets, a formalism with strong modelling power and theoretical foundation, is a powerful tool for the study of security systems. It is hoped that the work reported here will interest researchers in this new application of Petri net theory — using Petri nets to study cryptographic protocols.

Acknowledgment

The authors would like to thank Cynthia Morton in our department for her assistance in editing the manuscript of this paper.

References

[1] D. Dolev and A. C. Yao, "On the security of public key protocols," *IEEE Trans. Infor. Theory*, vol. IT-29, no. 2, pp. 198–208, March 1983.

[2] R. A. Kemmerer, "Analyzing encryption protocols using formal verification techniques," *IEEE J. Selected Areas in Commun.*, vol. 7, pp. 448–457, May, 1989.

[3] J. K. Millen, S. C. Clark and S. B. Freedman, "The interrogator: Protocol security analysis," *IEEE Trans. Software Eng.*, vol. SE-13, no. 2, pp. 274–288, 1987.

[4] C. Meadows, "Using narrowing in the analysis of key management protocols," in *Proc. IEEE Symp. Security and Privacy, Oakland, California*, pp. 138–147, 1989.

[5] D. Longley and S. Rigby, "An automatic search for security flaws in key management schemes," *Computers & Security*, vol. 11, no. 1, pp. 75–89, 1992.

[6] M. Burrows, M. Abadi and R. Needham, "A logic of authentication," *ACM Trans. Computer Systems*, vol. 8, no. 1, pp. 18–36, 1990.

[7] P. V. Rangan, "An axiomatic theory of trust in secure communication protocols," *Computers & Security*, vol. 11, no. 2, pp. 163–172, 1992.

[8] C. A. Petri, *Kommunikation mit Automaten*. PhD thesis, Bonn: Institut für Instrumentelle Mathematik, Schriften des IIM Nr. 3, 1962.

[9] T. Murata, "Petri nets: Properties, analysis and applications," *Proc. IEEE*, vol. 77, no. 4, pp. 541–580, April 1989.

[10] K. Jensen, "Coloured Petri nets: A high level language for system design and analysis," in *Advances in Petri Nets 1990*, vol. LNCS–483, pp. 342–416, Springer-Verlag, 1991.

[11] T. Suzuki, S. M. Shatz and T. Murata, "A protocol modeling and verification approach based on a specification language and Petri nets," *IEEE Trans. Software Eng.*, vol. 16, no. 5, pp. 523–536, May, 1990.

[12] M. Diaz, "Modeling and analysis of communication and cooperation protocols using Petri net based models," *Computer Networks*, vol. 6, no. 6, pp. 419–441, 1982.

[13] J. Burns and C. J. Mitchell, "A security scheme for resource sharing over a network," *Computers & Security*, vol. 9, no. 1, pp. 67–75, 1990.

[14] C. H. Meyer and S. M. Matyas, *Cryptography: A New Dimension in Computer Data Security*. New York: John Wiley & Sons, 1982.

[15] G. V. Bochmann, "Protocol specification for OSI," *Computer Networks and ISDN Systems*, vol. 18, pp. 167–184, 1989/90.

[16] G. Berthelot, "Checking properties of nets using transformations," in *Advances in Petri Nets 1985*, vol. LNCS–222, pp. 19–40, Springer-Verlag, 1985.

[17] B. B. Nieh, "Modelling and analysis of cryptographic protocols using Petri nets," Master's thesis, Electrical Engineering, Queen's University, Kingston, Canada, 1992.

[18] N. Behki and S. E. Tavares, "An integrated approach to protocol design," in *Proc. IEEE Pacific Rim Conf. on Commun., Computers and Signal Processing*, pp. 244–248, June 2–4 1989.

[19] C. Meadows, "A system for the specification and analysis of key management protocols," in *Proc. IEEE Symp. Research in Security and Privacy, Oakland, California*, pp. 182–195, May, 1991.

[20] B. B. Nieh and S. E. Tavares, "Modelling and analysis of secure communication protocols using Petri net formalism," in *Proc. Canadian Conference on Electrical and Computer Engineering, Quebec, P.Q., Canada*, pp. 37.3.1–37.3.4, Sept. 1991.

[21] R. M. Needham and M. D. Schroeder, "Using encryption for authentication in large networks of computers," *Commun. ACM*, vol. 21, no. 12, pp. 993–999, 1978.

[22] D. E. Denning and G. M. Sacco, "Timestamps in key distribution protocols," *Commun. ACM*, vol. 24, no. 8, pp. 533–536, 1982.

[23] V. L. Voydock and S. T. Kent, "Security mechanisms in a transport layer protocol," *Computers & Security*, no. 5, pp. 325–341, 1985.

On Verifiable Implicit Asking Protocols For RSA Computation

Tsutomu Matsumoto[1]
Hideki Imai[2]
Chi-Sung Laih[3]
Sung-Ming Yen[3]

[1] Division of Electrical and Computer Engineering
YOKOHAMA NATIONAL UNIVERSITY
156 Tokiwadai, Hodogaya, Yokohama, 240 Japan

[2] Institute of Industrial Science
UNIVERSITY OF TOKYO
7-22-1 Roppongi, Minato-ku, Tokyo, 106 Japan

[3] Department of Electrical Engineering
NATIONAL CHENG KUNG UNIVERSITY
Tainan, Taiwan, 70101 Republic of China

Abstract. The verifiable implicit asking is to speed up a certain feasible computation (e.g., $y = x^d \bmod n$) based on a secret (d) stored in a relatively powerless device (called Client) with the help of powerful device(s) (called Server(s)) in such a way that Client can check the behavior of Server(s) and that the leakage of the secret to Server(s) should be suppressed as much as possible. Possible attacks to obtain Client's secret are classified into passive and active attacks. Passive attacks can be completely nullified by dividing the target computation into two parts so that one depends on d but the other does not and then by asking Server to do only the latter part. However since such a method brings relatively low speed-up performance, we discuss a method to obtain verifiable implicit asking protocols highly secure against passive attacks by modifying some base protocols which are fast enough but not completely free from passive attacks since sending to Server some information not independent from d.

1 Introduction

As an approach to enhance the utilization of portable devices such as smart cards or electronic pocket-books, a course of study has been actively conducted under the subject called the server-aided secret computation, the verifiable implicit asking, or *iraikeisan* initiated by Matsumoto et al. [1][2]. The aim of the verifiable implicit asking is to speed up a certain (polynomial-time) computation based on a secret stored in a relatively powerless device (called *Client*) with the assistance of powerful auxiliary device(s) (called *Server*(s)) in such a way that Client can check the behavior of Server(s) and that the leakage of the secret to Server(s) should be suppressed as much as possible.

The secret transformation of the RSA cryptosystem, e.g., the RSA signature generation, is to get y from x, d, n such as

$$y = x^d \bmod n$$

where n is the product of different primes p and q of 256 bits or more, and d is an integer relatively prime to $\lambda(n) = \mathrm{lcm}(p-1, q-1)$, and x and y are non-negative integers less than n. Since this transformation is quite important in practical application of smart cards, many verifiable implicit asking (or, server-aided secret computation) protocols for it, namely the protocols to utilize Server to correctly evaluate y without releasing secret d, have been extensively devised and analyzed by Matsumoto-Kato-Imai[1][2], Matsumoto-Imai[3][12], Kawamura-Shimbo [4][6][7][9][14], Quisquater-De Soete [5], Laih-Yen-Harn [8], Laih-Yen [13][15], Pfitzmann-Waidner [10], Anderson [11]...

Typical speed-up achieved by such a protocol with an ordinary smart card client and an available RSA-hardware server is roughly 10 to 100 times faster execution compared to one solely done by an ordinary smart card. Recently there have been developed smart cards with special circuits which can perform a single RSA computation with 512 bit modulus in around one second. Note that this fact never loses the practical importance of the verifiable implicit asking protocols for RSA computation, because the protocols can speed up even such novel smart cards and both items can be harmonized, particularly in the case when multiple evaluation of RSA transformation is required such as possible retail banking applications.

This paper studies how to resist a type of possible attacks on such protocols. Let Client and Server be the entities who conduct the protocols. The attacks by Server to obtain Client's secret can be classified into the following two categories:

Passive Attack: While honestly conducting the protocol the attacker (Server) makes some computation based on the messages obtained through the normal execution.

Active Attack: The attacker (Server) deviates from the protocol and may send false messages to Client to extract some information about the secret by observing the replies for them in a single round or in multiple rounds.

Previous studies have clarified the followings:

As for active attacks the situation changes whether the final result (y) is given to Server or not. In the scenario where Server cannot see the final result, it is known [3][4][5][6][7] that if the target protocol has appropriate verification function then the irregular action by Server can be ineffective. Thus the current interest of researchers is how to achieve effective verification with lower computational and/ or communication costs. In the other scenario, the interest focuses on how to efficiently avoid the multi-round active attacks such as one given in [10].

On the other hand it is known that the passive attacks can be completely nullified by dividing the target computation into two parts so that one depends on d but the other part does not and then by asking Server to do only the latter part. However the obtained protocols by such a method have relatively low speed-up performance [5][9][14].

In this paper we concentrate on the last problem. That is, we discuss a method to obtain verifiable implicit asking protocols highly secure against passive attacks by modifying some base protocols which are fast enough but not completely free from passive attacks since sending to Server some information not independent from d.

2 Basic Protocol and Passive Attack

2.1 Protocol RSA-S1

As a simplest example of the protocols which cannot completely avoid passive attacks but have high speed-up performance, we describe here protocol RSA-S1 proposed by Matsumoto-Kato-Imai[2]. Let Weight(A) denote the Hamming weight of a vector A.

Let M and L be positive integers.

0. Client randomly selects integer vectors $D=[d_1,\ldots,d_M]$ and $F=[f_1,\ldots,f_M]$ such that
$$d \equiv d_1 f_1 + \cdots + d_M f_M \pmod{\lambda(n)}$$
and $1 \leq d_i < n$ and F is binary and Weight(F) $\leq L$.

1. Client sends n, D, and x to Server.
2. Server computes $Z = [z_1,\ldots,z_M]$, $z_i = x^{d_i} \bmod n$, and sends Z back to Client.
3. Client computes y as
$$y = (\prod_{i=1}^{M} {z_i}^{f_i}) \bmod n.$$

In RSA-S1 the maximum on-line computational complexity of Client is $L-1$ times [mod n-multiplications] and that of Server is M times [mod n-powerings (exponentiations)]. The communication complexity between Client and Server is $2(M+1)$ [blocks of log n-bit data]. For simplicity, we use the following notation to indicate the above characteristics:

$$\begin{cases} \mathrm{COMP_{Client}(RSA-S1)} = L-1, \\ \mathrm{COMP_{Server}(RSA-S1)} = M, \\ \mathrm{COMMUN(RSA-S1)} = 2(M+1). \end{cases}$$

Besides $\mathrm{COMP_{Client}(RSA-S1)}$ Client should spend $\mathrm{COMP^{offline}_{Client}(RSA-S1)}$, the computational complexity for step 0, but this computation can be done prior to getting input x.

Note that while this protocol has no verification function for the behavior of Server it is easy to add such function by techniques in [3]. However, to focus on the resistance against passive attacks, we treat RSA-S1 as it is. Also note that RSA-S1 can be generalized [2] by modifying the condition on F so that the computational complexity of obtaining $y = (\prod_{i=1}^{M} z_i{}^{f_i}) \bmod n$ given F and Z and n is upperbounded by some constant. Such a generalized protocol is called a non-binary version of RSA-S1.

2.2 Passive Attacks on RSA-S1

Simple Exhaustive Search: It was pointed out in [2] that d can be determined by an exhaustive search of every binary vector F of Hamming weight less than or equal to L. The number of such vectors is

$$K_{S1}(M,L) = \sum_{i=1}^{L} \binom{M}{i},$$

which represents the complexity of this passive attack. To check if a candidate vector is correct or not the attacker should use a pair $(x, x^d \bmod n)$ for the unknown exponent d. Such a pair can be available if the attacker knows the public exponent $e = d^{-1} \bmod \lambda(n)$ corresponding to d, or if he can observe the execution of the protocol to have the final result y corresponding to x.

For non-binary versions of RSA-S1 similar simple exhaustive searches can be applied.

Bilateral Exhaustive Search: Recently Pfitzmann-Waidner[10] found a more effective passive attack, which we call here the bilateral exhaustive search.

1. For a message x with known $y = x^d \bmod n$, and for every $i = 1, \ldots, M$, compute $z_i = x^{d_i} \bmod n$.
2. For every binary vector F with Weight(F) $\leq$ $\lceil L/2 \rceil$ compute $y_F = (\prod_{i=1}^{M} z_i{}^{f_i}) \bmod n$ and $y{*}_F = (y \cdot y_F{}^{-1}) \bmod n$.
3. Search the domains $\{(F, y_F)\}$ and $\{(F, y{*}_F)\}$ for a pair (F_1, F_2) of vectors such that $y_{F_1} = y{*}_{F_2}$ and that for every coordinate F_1 and F_2 have no common 1. For such a matched pair, $F = F_1 + F_2$ is a candidate of the secret vector F.
4. If there is only one such F found in step 3 then it is the desired one. If plural F's are found, for these try steps 2 and 3 on another x until the number of candidates becomes one.

Both the number of y_F's and that of $y{*}_F$'s are equal to

$$N_{S1}(M,L) = \sum_{i=0}^{\lceil L/2 \rceil} \binom{M}{i},$$

and the computational complexity of sorting and searching required in the above procedure is order of $N_{S1}(M,L)\log N_{S1}(M,L)$, each of which is slightly greater than $\sqrt{K_{S1}(M,L)}$ and is much smaller than $K_{S1}(M,L)$.

For non-binary versions of protocol RSA-S1 similar bilateral exhaustive searches can be applied [10].

3 Countermeasure

We demonstrate a key idea to enhance the resistance of certain protocols against passive attacks.

3.1 Protocol RSA-S1M

The following protocol (RSA-S1M) has been proposed by Matsumoto-Imai [12] to gain resistance against the bilateral exhaustive search. Essentially the same protocol has been independently reported by Laih-Yen [13].

Let R and S be positive integers.

0. Client randomly selects integer vectors $D = [d_1, \ldots, d_R]$, $G = [g_1, \ldots, g_R]$, and $H = [h_1, \ldots, h_R]$ such that

$$d \equiv g \cdot h \pmod{\lambda(n)}$$

$$g \equiv d_1 g_1 + \cdots + d_R g_R \pmod{\lambda(n)}$$

$$h \equiv d_1 h_1 + \cdots + d_R h_R \pmod{\lambda(n)}$$

where $1 \leq d_i < n$ and both G and H are binary and Weight(G) $\leq S$ and Weight(H) $\leq S$. Client randomly prepares a pair $(t, t^{-h} \bmod n)$ of integers such that $0 < t < n$.

1. Client sends n, D, and x to Server.
2. Server computes $Z = [z_1, \ldots, z_R]$, $z_i = x^{d_i} \bmod n$, and sends Z back to Client.
3. Client computes $x^g \bmod n$ as

$$x^g \bmod n = \left(\prod_{i=1}^{R} z_i^{g_i}\right) \bmod n$$

and makes $u = (x^g \cdot t) \bmod n$ and sends it to Server.
4. Server computes $V = [v_1, \ldots, v_R]$, $v_i = u^{d_i} \bmod n$, and sends V back to Client.

5. Client computes $u^h \bmod n$ as

$$u^h \bmod n = (\prod_{i=1}^{R} v_i^{h_i}) \bmod n$$

and obtains $y = (u^h \cdot t^{-h}) \bmod n$.

This protocol is in the form of two-phase serial execution of RSA-S1. Thus, the maximum on-line computational complexities and the communication complexity of RSA-S1M are evaluated as follows. We use the similar notation as for RSA-S1.

$$\begin{cases} \mathrm{COMP_{Client}(RSA-S1M)} = 2S, \\ \mathrm{COMP_{Server}(RSA-S1M)} = 2R, \\ \mathrm{COMMUN(RSA-S1M)} = 3(R+1). \end{cases}$$

By $\mathrm{COMP^{offline}_{Client}(RSA-S1M)}$ we denote the off-line computational complexity required for step 0. The term 'off-line' means that D, G, H, and $(t, t^{-h} \bmod n)$ can be precomputed before getting the input message x. Note that the pair $(t, t^{-h} \bmod n)$ can be generated by using a pair $(u, u^h \bmod n)$ obtained in earlier execution of steps 3, 4, and 5.

What dominates the real execution time of a verifiable implicit asking protocol (say, P) is the maximum on-line computational complexity of Client $\mathrm{COMP_{Client}(P)}$ and the communication complexity between Client and Server $\mathrm{COMMUN(P)}$. Thus, we adjust parameters so that $\mathrm{COMP_{Client}(RSA-S1M)}$ and $\mathrm{COMMUN(RSA-S1M)}$ are approximately equal to $\mathrm{COMP_{Client}(RSA-S1)}$ and $\mathrm{COMMUN(RSA-S1)}$, respectively.

If parameters R and S are set as

$$R = \lfloor (2M-1)/3 \rfloor, \quad S = \lfloor (L-1)/2 \rfloor \qquad (\#)$$

then

$$0 \le \mathrm{COMP_{Client}(RSA-S1)} - \mathrm{COMP_{Client}(RSA-S1M)} \le 1,$$
$$0 \le \mathrm{COMMUN(RSA-S1)} - \mathrm{COMMUN(RSA-S1M)} \le 1,$$

while

$$\mathrm{COMP_{Server}(RSA-S1M)} = 2\lfloor (2M-1)/3 \rfloor \le \tfrac{4}{3}\mathrm{COMP_{Server}(RSA-S1)},$$
$$\mathrm{COMP^{offline}_{Client}(RSA-S1M)} = O(\mathrm{COMP^{offline}_{Client}(RSA-S1)}).$$

3.2 Resistance of RSA-S1M

We evaluate the complexity of passive attacks on RSA-S1M.

Simple Exhaustive Search: Given a pair $(x, x^d \bmod n)$, the secret exponent d can be determined by an exhaustive search of every pair of binary vectors G and H of Hamming weight less than or equal to S. The number of such pairs is

$$K_{\mathrm{S1M}}(R,S) = (\sum_{i=1}^{S} \binom{R}{i})^2,$$

which represents the complexity of this passive attack. It is easy to see that the actual computational complexity of the simple exhaustive search on RSA-S1M is upper bounded by

$$\frac{R(R+3)}{2} \text{ [mod}n - \text{exponentiations]}$$
$$+\sum_{\alpha=1}^{S}\sum_{\beta=1}^{S}\binom{R}{\alpha}\binom{R}{\beta}(\alpha\beta-1) \text{ [mod}n - \text{multiplications]}.$$

Bilateral Exhaustive Search: Consider the applicability of the passive attack by Pfitzmann-Waidner[10] on RSA-S1M. To apply the bilateral exhaustive search the attacker should have a pair $(x, x^d \bmod n)$ for the unknown exponent d. In the case of RSA-S1, if the attacker knows the public exponent $e = d^{-1} \bmod \lambda(n)$ corresponding to d then he can readily generate such a pair, and if he can observe the execution of the protocol to have the final result y he can also get such a pair.

Trying to apply the bilateral exhaustive search on steps 1, 2, 3 or on steps 3, 4, 5 of RSA-S1M to get an unknown exponent g or h will fail since the output value of each part is masked by t or $t^{-h} \bmod n$. The way to apply this attack to the full protocol is to get an R^2-dimensional binary vector of Hamming weight less than or equal to S^2 observing that

$$d \equiv \sum_{i=1}^{R}\sum_{j=1}^{R}(d_i d_j)(g_i h_j) \pmod{\lambda(n)}.$$

The complexity measure of this application of the bilateral exhaustive search,

$$N_{\mathrm{S1M}}(R,S) = \sum_{i=0}^{\lceil S^2/2 \rceil}\binom{R^2}{i},$$

becomes greater than $K_{\mathrm{S1M}}(R,S)$ on most of reasonable values of the parameters, which means that the bilateral exhaustive search is not effective for RSA-S1M.

Therefore the best known passive attack on RSA-S1M is the simple exhaustive search. When we select parameters set as (#) the computational complexity of the search is estimated as $K_{\mathrm{S1M}}(\lfloor(2M-1)/3\rfloor, \lfloor(L-1)/2\rfloor)$, which is greater than $K_{\mathrm{S1}}(M,L)$ on most of reasonable values of the parameters. Thus it can be recognized that protocol RSA-S1M acquires much greater resistance against passive attacks than RSA-S1 by introducing 33% increase of the maximum on-line computational complexity required for Server and some increase of the off-line computational complexity required for Client.

Besides passive attacks, there have been proposed active attacks on RSA-S1. The single-round active attack by Anderson [11] can be prevented [10]. This attack can be modified for RSA-S1M. However, Yen-Laih [15] shows that RSA-S1M is highly resistant against the modified attack.

Multiple-round active attacks will be the last threat to verifiable implicit asking. Protocol RSA-S1M also has certain resistance against the multiple-round active attacks although precise security evaluation is not finished.

4 Generalization

4.1 Increase of Interaction

A possible generalization of RSA-S1M is to introduce more interaction. However within the same computational and communication resources this seems not to increase the resistance drastically.

4.2 Non-binary RSA-S1M

It is easy to have non-binary versions of RSA-S1M when we exclude 'binary' from the condition to G and H. The obtained protocols have the same advantage as RSA-S1M on the resistance against the corresponding attacks.

4.3 Protocol RSA-S2

As the second example which cannot completely avoid passive attacks but has high speed-up performance, we describe here RSA-S2, also proposed by Matsumoto-Kato-Imai[2]. The following is a non-binary version of RSA-S2 adopting the parameter restriction suggested by Shimbo-Kawamura [7].

0. Client randomly selects integer vectors $D = [d_1, \ldots, d_M]$, $F = [f_1, \ldots, f_M]$ and $G = [g_1, \ldots, g_M]$ such that

$$\begin{cases} d \equiv d_1 f_1 + \cdots + d_M f_M \pmod{p-1}, \\ d \equiv d_1 g_1 + \cdots + d_M g_M \pmod{q-1}, \end{cases}$$

and $1 \leq d_i < n$ and $\quad f_i \equiv g_i \pmod 2$.

1. Client sends n, D, and x to Server.
2. Server computes $Z = [z_1, \ldots, z_M]$, $z_i = x^{d_i} \bmod n$, and sends Z back to Client.
3. Client computes y as

$$< y >_p = (\prod_{i=1}^{M} z_i^{f_i}) \bmod p, \quad < y >_q = (\prod_{i=1}^{M} z_i^{g_i}) \bmod q,$$

$$y = (< y >_p w_p + < y >_q w_q) \bmod n$$

where $w_p = q(q^{-1} \bmod p)$ and $w_q = p(p^{-1} \bmod q)$ are precomputed values.

4.4 Protocol RSA-S2M

The following protocol (RSA-S2M) can be derived from RSA-S2 using the same technique as RSA-S1M was derived.

0. Client randomly selects integer vectors $D = [d_1, \ldots, d_R]$, $F = [f_1, \ldots, f_R]$, $G = [g_1, \ldots, g_R]$, $H = [h_1, \ldots, h_R]$, and $K = [k_1, \ldots, k_R]$ such that

$$d \equiv a \cdot b \pmod{\lambda(n)},$$

$$\begin{cases} a \equiv d_1 f_1 + \cdots + d_R f_R \pmod{p-1}, \\ a \equiv d_1 g_1 + \cdots + d_R g_R \pmod{q-1}, \end{cases}$$

$$\begin{cases} b \equiv d_1 h_1 + \cdots + d_R h_R \pmod{p-1}, \\ b \equiv d_1 k_1 + \cdots + d_R k_R \pmod{q-1}, \end{cases}$$

and $1 \leq d_i < n$, $f_i \equiv g_i \pmod 2$, and $h_i \equiv k_i \pmod 2$.

Client prepares a pair $(t, t^{-h} \bmod n)$ of integers such that $0 < t < n$.

1. Client sends n, D, and x to Server.
2. Server computes $Z = [z_1, \ldots, z_R]$, $z_i = x^{d_i} \bmod n$, and sends Z back to Client.
3. Client computes $x^a \bmod n$ as

$$< x^a >_p = (\prod_{i=1}^{R} z_i^{f_i}) \bmod p, \quad < x^a >_q = (\prod_{i=1}^{R} z_i^{g_i}) \bmod q,$$

$$x^a \bmod n = (< x^a >_p w_p + < x^a >_q w_q) \bmod n$$

and makes $u = (x^a \cdot t) \bmod n$ and sends it to Server.
4. Server computes $V = [v_1, \ldots, v_R]$, $v_i = u^{d_i} \bmod n$, and sends V back to Client.
5. Client computes $u^b \bmod n$ as

$$< u^b >_p = (\prod_{i=1}^{R} v_i^{h_i}) \bmod p, \quad < u^b >_q = (\prod_{i=1}^{R} v_i^{k_i}) \bmod q,$$

$$u^b \bmod n = (< u^b >_p w_p + < u^b >_q w_q) \bmod n,$$

where $w_p = q(q^{-1} \bmod p)$ and $w_q = p(p^{-1} \bmod q)$ are precomputed values. Then Client obtains $y = (u^b \cdot t^{-b}) \bmod n$.

4.5 Resistance of RSA-S2M

Let C be a positive integer such that the number of mod p or mod q multiplications in steps 3 and 5 of RSA-S2M is limited by C. We derive the complexities of exhaustive searches of every candidate of (F, G, H, K) constrained by R and C, which provide the evaluation of the resistance of RSA-S2M. Assume that the bit length of f_i, g_i, h_i, and k_i is k.

Simple Exhaustive Search: The exponent d can be determined by an exhaustive search of every candidate of (F, G, H, K) constrained by R and C. Let $\mathbf{V}$ denote the set of all such candidates:

$$\mathbf{V} = \{(F, G, H, K) | \sum_{i=1}^{R} \chi(f_i) + \sum_{i=1}^{R} \chi(g_i) + \sum_{i=1}^{R} \chi(h_i) + \sum_{i=1}^{R} \chi(k_i) \leq C,$$

$$0 \leq f_i, g_i, h_i, k_i \leq 2^k - 1,\ f_i \equiv g_i \pmod 2,\ h_i \equiv k_i \pmod 2\},$$

where $\chi(x)$ is the number of modular multiplications required to compute the x-th power for a given number. Using a powering algorithm realized by successive square and multiply operations, which is referred to as the *fast exponentiation algorithm*, we have

$$\chi(x) = k + w - 2,$$

where k and w are the bit length and the Hamming weight of the binary representation of x, respectively. We then have

$$\sum_{i=1}^{R} \mathrm{Weight}(f_i) + \sum_{i=1}^{R} \mathrm{Weight}(g_i) + \sum_{i=1}^{R} \mathrm{Weight}(h_i) + \sum_{i=1}^{R} \mathrm{Weight}(k_i) \leq C - 4R(k-2).$$

The least significant bit (LSB) of f_i is equal to that of g_i since $f_i \equiv g_i \pmod 2$. Similarly, the LSB of h_i is equal to that of k_i. Therefore, the number of candidates $K_{\mathrm{S2M}}(R, C) = \#\mathbf{V}$ is

$$K_{\mathrm{S2M}}(R, C) = \sum_{i=1}^{C-4R(k-2)} \sum_{j=0}^{\lfloor i/2 \rfloor} \binom{4R(k-1)}{i-2j} \binom{2R}{j}.$$

Bilateral Exhaustive Search: Let $N_{\mathrm{S2M}}(R, C)$ be the number of candidates of (F, G, H, K) in the bilateral exhaustive search, suggested in [10], for RSA-S2. Trying to apply this attack to get an unknown exponent a or b will fail. The way to apply this attack to RSA-S2M is to get $2R^2(2k-1)$-dimensional vectors of Hamming weight less than or equal to $(C - 4R(k-2))^2$ observing that

$$d \equiv \sum_{i=1}^{R} \sum_{j=1}^{R} (d_i d_j)(f_i h_j) \pmod{p-1},$$

$$d \equiv \sum_{i=1}^{R} \sum_{j=1}^{R} (d_i d_j)(g_i k_j) \pmod{q-1},$$

$$f_i \equiv g_i \pmod 2,$$

$$h_i \equiv k_i \pmod 2.$$

Thus we have

$$N_{\mathrm{S2M}}(R,C)=\sum_{i=1}^{\lceil (C-4R(k-2))^2/2\rceil}\ \sum_{j=0}^{\lfloor i/2\rfloor}\binom{2R^2(2k-2)}{i-2j}\binom{R^2}{j},$$

which is much greater than $K_{\mathrm{S2M}}(R,C)$ on most of reasonable values of the parameters. When $k=4$, we have Table 1 showing $N_{\mathrm{S2M}}(R,C) \gg K_{\mathrm{S2M}}(R,C)$.

Table 1. Examples of $K_{\mathrm{S2M}}(R,C)$ and $N_{\mathrm{S2M}}(R,C)$.

k	R	C	$K_{\mathrm{S2M}}(R,C)$	$N_{\mathrm{S2M}}(R,C)$
4	7	100	8.14×10^{27}	5.70×10^{191}
4	8	100	2.35×10^{28}	2.86×10^{250}
4	8	103	2.16×10^{29}	2.86×10^{250}
4	8	108	4.85×10^{30}	2.86×10^{250}
4	8	110	1.39×10^{31}	2.86×10^{250}

In this section, the complexities of the passive attacks for RSA-S2M have been evaluated. As the result, it has been found that the bilateral exhaustive search is not effective for RSA-S2M.

5 Conclusion

We have demonstrated how to enhance verifiable implicit asking protocols or server-aided secret computation protocols with respect to the resistance against passive attacks. The essential technique can be applied to other implicit asking protocols. Next research problems are to extend the technique in the most general setting and to apply its generalized version into previously known protocols for obtaining much secure (against both passive and active attacks) and practical verifiable implicit asking protocols.

Acknowledgment

The first and second authors, T. Matsumoto and H. Imai, have been partially supported by Ministry of Education, Science, and Culture under Grant-in-Aid for Scientific Researches No. 04402033. Also they thank Yuich Saitoh for discussing the analysis in section 4.5.

References

[1] Tsutomu Matsumoto, Koki Kato, Hideki Imai, "Smart cards can compute secret heavy functions with powerful terminals," Proc. of the 10th Symposium on Information Theory and Its Applications, pp.17-22, (1987-11).

[2] Tsutomu Matsumoto, Koki Kato, Hideki Imai, "Speeding up secret computation with insecure auxiliary devices," Advances in Cryptology — CRYPTO'88, Santa Barbara (1988-08), Lecture Notes in Computer Science No.403, pp.497-506, Springer-Verlag, (1990).

[3] Tsutomu Matsumoto, Hideki Imai, "How to ask and verify oracles for speeding up secret computations (Part 2)," IEICE Technical Report, IT89-24, (1989-07).

[4] Shin-ichi Kawamura, Atsushi Shimbo, "A note on checking the faithfulness of the server in client-server systems (II)," IEICE Technical Report, ISEC89-17, (1989-09).

[5] Jean-Jacques Quisquater, Marijk De Soete, "Speeding up smart card RSA computation with insecure coprocessors," Smart Card 2000, Amsterdam (1989-10).

[6] Shin-ichi Kawamura, Atsushi Shimbo, "Performance analysis of server-aided secret computation protocols," Transactions of IEICE, Vol.E73, No.7, pp.1073-1080, (1990-17).

[7] Atsushi Shimbo, Shin-ichi Kawamura, "Factorisation attack on certain server-aided secret computation protocols for the RSA secret transformation," IEE Electronics Letters, Vol.26, No.17, pp.1387-1388, (1990-08).

[8] Chi-Sung Laih, Sung-Ming Yen, Lein Harn, "Two efficient server-aided secret computation protocols based on addition chain sequence," ASIACRYPT'91 — Abstracts, pp.270-274, (1991-11).

[9] Shin-ichi Kawamura, Atsushi Shimbo, "A server-aided secret computation based on the addition chain," Proc. 1992 Symp. on Cryptography and Information Security, SCIS92-12A, (1992-04).

[10] Birgit Pfitzmann, Michael Waidner, "Attacks on protocols for server-aided RSA computation," to appear in Advances in Cryptology — EUROCRYPT'92, Balatonfüred (1992-05), Lecture Notes in Computer Science, Springer-Verlag.

[11] R. J. Anderson, "Attack on server assisted authentication protocols," IEE Electronics Letters, Vol.28, No.15, p.1473, (1992-07).

[12] Tsutomu Matsumoto, Hideki Imai, "On verifiable implicit asking — or server-aided secret computation," IEICE Technical Report, ISEC92-15, (1992-08).

[13] Chi-Sung Laih, Sung-Ming Yen, "Two phase server-aided secret computation for RSA," Technical Report, E.E. Dept. of National Cheng Kung University, (1992).

[14] Atsushi Shimbo, Shin-ichi Kawamura, "Efficient server-aided RSA computation protocols," Proc. of the 15th Symposium on Information Theory and Its Applications, pp.269-272, (1992-09).

[15] Sung-Ming Yen, Chi-Sung Laih, "More about the active attack on the server-aided secret computation protocol," submitted to IEE Electronics Letters, (1992).

Modified Maurer-Yacobi's scheme and its applications*

Chae Hoon Lim and Pil Joong Lee

Department of Electrical Engineering, Pohang Institute of Science and Technology (POSTECH), Pohang, 790-784, KOREA

Abstract. In Eurocrypt'91, Maurer and Yacobi developed a method for building a trapdoor into the one-way function of exponentiation modulo a composite number which enables an identity-based non-interactive key distribution system. In this paper, we provide some improvements of their scheme and then present a modified trapdoor one-way function by combining Maurer-Yacobi's scheme and RSA scheme. We demonstrate that a lot of applications can be constructed based on this modified scheme which are impossible in the original scheme. As examples, we present several protocols based on it, such as identifications, key distributions and signature schemes. We have implemented the Pohlig-Hellman and Pollard's ρ-methods for computing discrete logarithms modulo a composite number, which shows that average running time for computing logarithms is too large to be realizable in practice. Therefore, considering current algorithms and technology, we maintain that it is more efficient and practical to take a certificate-based scheme on which all protocols presented in this paper can be based as well.

1 Introduction

In Eurocrypt'91, Maurer and Yacobi [MY] proposed an identity-based non-interactive key distribution system based on a novel method for building a trapdoor into the modular exponentiation one-way function which allows a trusted authority to compute a discrete logarithm modulo a composite number m while this is completely infeasible for an adversary not knowing the factorization of m. The public key of a user is defined to be his identity number derived from publicly known identification information such as name, address, physical description, etc., and the corresponding secret key is computed by a trusted authority as the discrete logarithm of his identity number modulo a composite number m. Since not every number possesses a discrete logarithm modulo a composite number, it is necessary to derive an identity number from publicly known identification information such that its discrete logarithm modulo a composite number is guaranteed to exist. They presented two such methods, the squaring method and the Jacobi symbol method.

* This work was supported in part by the Ministry of Science and Technology (MOST) of the Korea.

In this paper, we point out that the squaring method for deriving an identity number has a critical defect that all or part of trapdoor information (i.e., the factorization of m) is easily disclosed in most cases. We present one possible way to circumvent the defect when it is for use in non-interactive key distribution systems. It is also shown that the Jacobi symbol method can be adapted to the case where the modulus is composed of more than two prime factors at the expense of a few bits interactive communication. Furthermore we present a modified trapdoor one-way function by combining the Maurer-Yacobi's trapdoor one-way function with the RSA scheme [RSA]. It is shown that many interesting applications can be constructed based on this modified scheme, such as mutual identification scheme with key distribution, conference key distttribution scheme, directed signatures and undeniable signatures.

It is main drawback of the Maurer-Yacobi's scheme that a trusted authority needs great computing power to compute secret keys of subscribers. Implementation of the Pohlig-Hellman and Pollard's ρ-methods shows that a trusted authority should have several giga flops computing power for computing logarithms in a few minutes on the average in case where the largest prime factors of p-1 is around 15 decimal digits. Therefore we think that a certificate-based scheme is a better choice in practice and we show that all the applications of the modified scheme presented in this paper can be based on very flexible certificate-based schemes as well.

Here we remark that we have been informed after our independent work that the weakness of the squaring method and the countermeasure in non-interactive key distribution systems have been reported by Murakami and Kasahara in Japan in 1990 and by Y.Yacobi in the rump session of Eurocrypt'92, respectively. However, these results are briefly described in this paper for the convenience of subsequent explanation of other related results.

This paper is organized as follows. Chapter 2 is devoted to brief description of Maurer-Yacobi's scheme and its weakness and countermeasures including some improvement. The modified scheme and several protocols based on it are given in chapter 3. We discuss the implementation result of Pohlig-Hellman and Pollard's ρ-methods in chapter 4 and finally conclude in chapter 5.

2 Analysis and improvement of Maurer-Yacobi's scheme

2.1 Description of Maurer-Yacobi's scheme

Maurer and Yacobi's scheme is based on the following two facts. First, by choosing the prime factors of m appropriately such that discrete logarithms modulo each prime factor can feasibly be computed while factoring their product m is still completely infeasible, a trusted authority can compute the discrete logarithm of a user's identity number modulo a composite number m whereas this is completely infeasible for an adversary not knowing the factorization of m. This idea comes from the discrepancy of the running time between factoring algorithms and algorithms for computing discrete logarithms. Second, though there exists no element in the multiplicative group Z_m^* that generates the entire group, an identity number can be derived in a publicly known way from publicly known identification information such that it is guaranteed to have a discrete logarithm modulo a composite number m. The following two paragraphs briefly describe the systen setup and the methods for guaranteeing the existence of logarithms modulo a composite, respectively.

For the system setup, a trusted authority can construct a composite modulus m in the following two ways. The first way is to choose the prime factors of m as 3 to 4 primes of between 55 and 70 decimal digits where for the prime factors p_i the numbers $(p_i - 1)/2$ are odd and pairwise relatively prime, which is based on the discrepancy between the running time of the best known factoring algorithm for finding factors of moderate size, the elliptic curve method [L], and the fastest known algorithm for computing discrete logarithms in GF(p) due to Coppersmith-Odlyzko-Schroepell [COS]. An alternative approach is to choose m as the product of two primes p_1 and p_2 of about 100 decimal digits where $(p_1 - 1)/2$ and $(p_2 - 1)/2$ are odd and relatively prime and both contain several prime factors of 13 to 15 decimal digits but no larger prime factor, which is based on the factoring algorithm of Pollard's p-1 [P1] and the discrete logarithm algorithm of Pohlig-Hellman [PH], both of which are most efficient in case that all prime factors of p-1 have only moderate size. Of course these figures can vary according to the progress in computer technology and number theoretic algorithms.

Two different methods were presented in [MY] for solving the problem that not every number is guaranteed to have a discrete logarithm modulo a composite number m. Throughout this paper, we assume that D_i, a descriptor of a user i, denotes the number describing a user i uniquely by his name, address, physical description, etc., with publicly known format and that ID_i, an identity number of a user i, denotes the number derived from D_i so that it is guaranteed to have a discrete logarithm modulo a composite number m. The first solution presented by Maurer and Yacobi is based on the fact that the squaring of any element in Z_m^* has a discrete logarithm modulo m. Therefore, by defining an identity number of user i as $ID_i \equiv D_i^2 \bmod m$, we can be assured that the discrete logarithm of the number ID_i modulo m does exist since discrete logarithms modulo m exist if and only if logarithms modulo each prime factors of m have the same parity (See the original paper [MY] for more details). The other solution, which is restricted to the latter case of constructing a composite modulus m, is to define a user i's identity number ID_i as the smallest integer x greater than or equal to the number D_i such that the Jacobi symbol (x/m) is equal to 1. It can be shown that $(x/m) = 1$ if and only if x has a discrete logarithm modulo m in case of $m = p_1 p_2$. Since the Jacobi symbol can be easily computed without knowing the factorization of m, a user i can obtain his counterpart user j's identity number ID_j from his descriptor D_j.

Once the system is set up, each user i wishing to join the system visits the trusted authority, presents his identity information together with an appropriate proof of identity and receives the secret key S_i corresponding to his identity number ID_i such that $ID_i \equiv g^{S_i} \bmod m$, which can be computed in either way described in the second paragraph. Now any two users i and j in the system can compute the common key between them by

$$K_{ij} \equiv ID_j^{S_i} \equiv g^{S_i S_j} \equiv ID_i^{S_j} \equiv K_{ji} \bmod m.$$

2.2 Failure of the squaring method

In the squaring method, each user i in the system will be given his secret key S_i satisfying $g^{S_i} \equiv D_i^2 \equiv ID_i \bmod m$ and this S_i is always even due to the squaring. This means that he can create the congruence of the form $x^2 \equiv y^2 \bmod m$ commonly used in many factoring algorithms, i.e., $(g^{S_i/2})^2 \equiv D_i^2 \bmod m$. Moreover if the secret key S_i happens to be the discrete logarithm of D_i^2 modulo m for such a D_i that the D_i itself does not have a discrete logarithm modulo m, then this fact implies that the numbers x and y are not trivial solutions of the congruence $x^2 \equiv y^2 \bmod m$, i.e., $x \neq \pm y$. Therefore any user who has such an identity number will be able to obtain all or part of prime factors of m by computing the greatest common divisor of the number $g^{S_i/2} - D_i \bmod m$ and the modulus m. Unfortunately the probability that a randomly selected number does not have a discrete logarithm modulo m is overwhelming with average probability $1 - 2/2^r$ where r is the number of prime factors of m and thus the trapdoor information will be released in most cases. In summary, though for the number D_i not having a discrete logarithm modulo m its square D_i^2 is guaranteed to have a discrete logarithm modulo m, paradoxically the resulting logarithm (the secret key of user i) makes it possible to compute all or part of prime factors of m since it is always even. This is demonstrated by the following small-size example.

Example :

Let's take $m = p_1 p_2 p_3 = 227.347.467 = 36785123$, where $p_1 - 1 = 2.113$, $p_2 - 1 = 2.173$, and $p_3 - 1 = 2.233$. Then $\lambda(m)$ is given by $2.113.173.233 = 9109834$. The base element g is simply chosen as 2, the smallest element in Z_m^* that is primitive in every GF(p_i) for $i = 1, 2, 3$. Now the descriptor of a user A in the system is assumed to be D_A = LCH = 4C434816 (ASCII value in hexadecimal) = 4213496. Then the trusted authority will compute a user A's secret key S_A by first computing x_i for $i = 1, 2, 3$ such that $D_A \equiv g^{x_i} \bmod p_i$ and then solving the system of congruences $S_A \equiv 2x_i \bmod p_i - 1$ for $i = 1, 2, 3$ by the Chinese remainder technique.

$$D_A \equiv 4213496 \equiv 149 \equiv 2^{x_1} \bmod 227 ---> x_1 \equiv 221 \bmod 226,$$

$$D_A \equiv 4213496 \equiv 222 \equiv 2^{x_2} \bmod 347 ---> x_2 \equiv 312 \bmod 346,$$

$$D_A \equiv 4213496 \equiv 222 \equiv 2^{x_3} \bmod 467 ---> x_3 \equiv 104 \bmod 466.$$

Note that the numbers x_i for $i = 1, 2, 3$ are neither all even nor all odd and thus D_A itself does not have a discrete logarithm modulo m. The secret key S_A, the discrete logarithm of $D_A^2 \equiv ID_A$ modulo m, is given by $S_A \equiv 2.221.173.233.60 + 2.312.113.233.21 + 2.104.113.173.81 \equiv 3393154 \bmod 9109834$. Now that $D_A^2 (= 4213496^2) \equiv g^{S_A} (= 2^{1696577.2} \equiv 6805836^2) \bmod 36785123$, a user A receiving the number S_A will be able to compute part of factors of m by $GCD(6805836 - 4213496, 36785123) = GCD(2592340, 36785123) = 227$.

2.3 Countermeasures against the failure and some improvements

If we are only interested in generating a common secret key between two users, then there exists a simple way to counter the abovementioned attack. A trusted authority chooses a random secret number S and computes each user i's secret key S_i as $S_i \equiv S.\ X_i \bmod \lambda(m)$ where X_i is the discrete logarithm of $ID_i \equiv D_i^2$ modulo m as computed in the preceding section. Now no user can create the congruence of the form $x^2 \equiv y^2 \bmod m$ as before, since he knows nothing about the trusted authority's secret S, but he can still share a common key with any other user j as before :

$$\begin{aligned} K_{ij} &\equiv ID_j^{S_i} \equiv (D_j^2)^{S_i} \equiv g^{X_j S_i} \equiv g^{X_j X_i S} \\ &\equiv g^{X_i S_j} \equiv (D_i^2)^{S_j} \equiv ID_i^{S_j} \equiv K_{ji} \bmod m \end{aligned}$$

The main drawback of this countermeasure is that though it allows two users to share a common secret key, there exists no explicit relationship between a user's secret key and his identity number such that $ID_i \equiv D_i^2 \equiv g^{S_i} \bmod m$, which can be used for other purposes.

Jacobi symbol method when $m = p_1 p_2$:

In the case where m is the product of only two prime factors, i.e., $m = p_1 p_2$ with $GCD(p_1 - 1, p_2 - 1) = 2$, we can safely use the Jacobi symbol method proposed by Maurer and Yacobi, since the Jacobi symbol $(x/m) = (x/p_1).(x/p_2)$ is equal to 1 if and only if x has a discrete logarithm modulo m. This is followed by the fact that $(x/p_1) = (x/p_2) = 1$ if and only if the discrete logarithms of x modulo p_i for $i = 1, 2$ are all even and that $(x/p_1) = (x/p_2) = -1$ if and only if the discrete logarithms of x modulo p_i for $i = 1, 2$ are all odd.

Maurer and Yacobi defined a user i's identity number ID_i as the smallest integer x such that $(x/m) = 1$ for $x \geq D_i$. However, if a trusted authority chooses the prime factors p_1 and p_2 such that $p_1 \equiv \pm 1 \bmod 8$ and $p_2 \equiv \pm 3 \bmod 8$ so that the Jacobi symbol $(2/m)$ is equal to -1, then only a single computation of the Jacobi symbol (D_i/m) is enough in order to determine a user i's identity number, since if $(D_i/m) = -1$, then $(2D_i/m)$ will be sure to be equal to 1. Therefore a user i's identity number ID_i will be definitely either D_i or $2D_i$ according to the value of the Jacobi symbol (D_i/m). Now it is clear that the Jacobi symbol approach becomes more attractive than the squaring method, since it preserves the relationship between a user's secret key and his identity number with just a single computation of Jacobi symbol. However the Jacobi symbol method is restricted to the case where the modulus is a product of two prime factors. We will show in the following that this method can be also adapted to the case of the modulus with more than two prime factors if we introduce an interactive communication of a few bits via more general media such as telephones.

Jacobi symbol method when $m = p_1p_2p_3$:

We assume that the modulus m of about 200 decimal digits is composed of three prime factors p_1, p_2 and p_3 of around 65 digits where $(p_i - 1)/2 = p_i'$ for i = 1, 2, 3 are all distinct primes. Besides publishing the modulus m and the base element g as before, the trusted authority computes and publishes the set of small positive integers $G_i, i = 0, 1, \ldots, 7$ such that the set of values of the Legendre symbol $\{(G_i/p_1), (G_i/p_2), (G_i/p_3)\}$ for i = 0, 1, ...,7 covers all eight possible combinations from {1, 1, 1} to {-1, -1, -1}. Here we can take $G_0 = 1$. Then we can define the identity number of user i as $ID_i = D_iG_k$ where the multiplier $G_k, k \in [0, 7]$, is computed by the authority so that the set of values of the Legendre symbols $\{(ID_i/p_1), (ID_i/p_2), (ID_i/p_3)\}$ is equal to {1, 1, 1}. Of course, for the uniqueness of user's identity number it would be better that the bit-length of D_i be restricted to be less than that of the modulus by the maximum bit-length among the G_i, i = 0, 1, ..., 7. Here note that the 3-bit index k of the multiplier G_k of user i plays a role of group identifier which identifies the group that user i's identity number belongs to. Now the authority can compute the discrete logarithm of ID_i modulo m as before. Note that the 3-bit group identifier of user i should be available to all the users wishing to communicate with him in order to determine his identity number. This introduction of interactiveness seems inevitable in case where the prime factors of the modulus are more than two. However, since the 3-bit index can be communicated via more common media such as telephones and we may assume that most users will keep or access to the telephone number of the counterpart they want to communicate with, we think that this method practically does not lose the non-interactiveness of the identity-based scheme.

3 Modified scheme and its applications

In this chapter we present a modified trapdoor one-way function which is obtained by combining the Maurer and Yacobi's scheme with the RSA scheme. We demonstrate through several examples that this simple modification allows very useful applications.

3.1 System setup

As before, the trusted authority generates two prime numbers p_1 and p_2 of about 100 decimal digits such that both $p_1 - 1$ and $p_2 - 1$ contain several prime factors of 13 - 15 decimal digits but no larger one and that $(p_1 - 1)/2$ and $(p_2 - 1)/2$ are relatively prime. As pointed out earlier, the prime factors p_1 and p_2 can be chosen to satisfy $p_1 \equiv \pm 1 \bmod 8$ and $p_2 \equiv \pm 3 \bmod 8$ so that the Jacobi symbol

$(2/m)$ is equal to -1. The authority computes $m = p_1 p_2$, $\lambda(m) = (p_1-1)(p_2-1)/2$ and the base element g which is primitive in both $GF(p_1)$ and $GF(p_2)$. Next it computes the RSA keys e and d where e is a t-bit odd integer relatively prime to $\lambda(m)$ and d satisfies $e.d \equiv 1 \bmod \lambda(m)$. The security parameter t can be flexibly determined according to the applications, lying between 20 - 70 bits. Now the authority publishes m, g and e but it keeps secret the prime factors p_1 and p_2 and d. An one-way hash function h is also made public which produces positive integers smaller than e for arbitrary input bit strings.

For each verified user i, the authority computes his secret key S_i corresponding to his identity number ID_i as follows, where the identity number ID_i is determined from his descriptor D_i using the Jacobi symbol method as explained before. It first computes the number X_i satisfying $ID_i \equiv g^{X_i} \bmod m$ using the prime factorization of m and the Pohlig-Hellman algorithm, and then computes his secret key S_i as $S_i \equiv d.X_i \bmod \lambda(m)$. Now the user i receiving his secret key S_i from the authority can verify its validity by checking that the following equation holds :

$$ID_i \equiv g^{eS_i} \bmod m$$

For convenience, we define G and G_i as $G \equiv g^e \bmod m$ and $G_i \equiv g^{S_i} \bmod m$, respectively. Then the equation $ID_i \equiv G_i^e \equiv G^{S_i} \bmod m$ holds. Moreover we note that user i may use as his secret information the number G_i as well as the number S_i, which is very useful in many applications as explained in the following sections. Since G has the same order as the base element g and thus there can be no security problem on the equation $ID_i \equiv G^{S_i} \bmod m$, it suffices to consider the security of the modified scheme on the equation $ID_i \equiv G_i^e \bmod m$.

Security of the modified scheme :

Up to present, solving the equation $ID_i \equiv G_i^e \bmod m$ for the secret number G_i remains intractable for an appropriately chosen parameters. Since the modulus m is chosen to be infeasible to factor, most specialized attacks applicable to the RSA scheme are ineffective in the above scheme. However proper precautions should be taken to prevent the iterative exponentiation attack in which an attacker attempts to find the secret G_i without factoring m by successively powering the public exponent e [SN] [R].

In order that such an attack could be made futile , the authority should carefully select the public exponent e. Let $p_i - 1 = q_{i0} \cdot q_{i1} \cdots q_{ir}$, $i = 1, 2$ with the primes q_{ij}, $j = 1, 2, \ldots, r$ and $q_{i0} = 2$. Then $\lambda(m)$ is given by $q_{10} \cdot q_{11} \cdots q_{1r} \cdot q_{21} \cdots q_{2r}$. If the secret S_i of user i is relatively prime to $\lambda(m)$, the order of G_i modulo m will be the same as that of the base element g. Now suppose that the greatest common divisor of S_i and $\lambda(m)$ is not equal to 1. Then the order of G_i modulo m will be given by $\lambda(m)' = \lambda(m)/GCD(S_i, \lambda(m))$ and we require the order of e modulo all possible values of $\lambda(m)'$ to be large enough to defeat the iterative attack. For this, the authority had better choose the public exponent e as the t-bit odd integer which is primitive in $GF(q_{ij})$ for all q_{ij} greater than or equal to certain limit, say 10 decimal digits. Then the number e

will have large enough order modulo $\lambda(m)'$ to defeat the iterative attack, since it is unlikely in practice that $GCD(S_i, \lambda(m))$ consists of several large prime factors of $\lambda(m)$. Finally we remark that the above modification can be applicable equally well to the case where the modulus is composed of more than two prime factors as is suggested in the section 2.3, if 3 to 4 bit group identifiers of users in the system can be available in an efficient way.

3.2 Identification, signature and key distribution

The basic idea for identification and signature protocols based on the modified scheme is that each user i can also utilize the number $G_i \equiv g^{S_i} \bmod m$ as his secret key in the GQ-like scheme [GQ1][GQ2][OT][OO1]. Then the identification and signature protocols can be constructed in the the same way as in the GQ scheme [GQ1][GQ2]. In the followings we explain the usefulness of the modified trapdoor one-way function through three applications : the combined scheme of signature and encryption, the mutual identification protocol with key distribution and the conference key distribution scheme.

Signature with encryption :

Suppose that user i wants to send an encrypted signed message M to user j. Then he chooses a random number ρ over Z_m, computes $R \equiv g^{\rho} \bmod m$ and generates the signature $\{X, Y\}$ as follows :

$$X \equiv R^e \bmod m, D = h(X, M), Y \equiv R.\ G_i^D \bmod m$$

Next he computes the session key K for the conventional cryptosystem to be used and then produces the ciphertext C for the message M using the session key K and the encryption algorithm E_K as follows :

$$K \equiv ID_j^{\rho} \equiv g^{e\rho S_j} \bmod m, C = E_K(M)$$

Now he sends the set of integers $\{X, Y, C\}$ to user j.

Upon receiving them, user j computes the common secret key K and recovers the message M by the decryption algorithm D_K:

$$K \equiv X^{S_j} \equiv g^{e\rho S_j} \bmod m, M = D_K(C)$$

Then he computes $\underline{D} = h(X, M)$ and checks if the equation $Y^e.\ ID_i^{-\underline{D}} \equiv X \bmod m$ holds. We note that the above combined scheme provides an efficient encryption capability incorporated into the GQ signature scheme with neither interaction nor access to the public files.

The security of the signature part is identical to that of the GQ scheme, since the number $R \equiv g^{\rho} \bmod m$ computed with a randomly selected number ρ over Z_m is indistinguishable from the number randomly generated over Z_m. The common key in the encryption part is directly related to the signature of a specific message via a random number ρ, and thus we claim that the entire scheme remains as secure as the original GQ scheme.

3-move mutual identification with key distribution :

We now present a 3-move mutual identification scheme with key distribution, which is obtained by combining the GQ identification scheme and the Diffie-Hellman scheme [DH]. We assume that user i initiates the protocol.

i) User i selects a random number ρ_i over Z_m, computes $R_i \equiv g^{\rho_i} \bmod m$ and $X_i \equiv R_i^e \bmod m$ and sends the number X_i to user j.
ii) Using the number X_i received from user i in step i), user j computes $K_{ij} \equiv X_i^{S_j} \bmod m$. Then he chooses a random number R_j over Z_m, computes $h(K_{ij}, R_j) = D_i$ and sends the numbers D_i and R_j to user i.
iii) User i computes $\underline{K}_{ij} \equiv ID_j^{\rho_i} \bmod m$ and $h(\underline{K}_{ij}, R_j) = \underline{D}_i$ and checks if the computed number $\underline{D}_i$ is equal to the number D_i received from user j in step ii). If they are equal, he is assured that the connection is correct. As a response, he computes $Y_i \equiv R_i . G_i^{D_i} \bmod m$ and sends it to user j. Now he computes the common key to be used between them in a subsequent communication as $K \equiv K_{ij}^{D_i} \bmod m$. If the verification fails, he stops the protocol.
iv) User j computes $Y_i^e . ID_i^{-D_i} \equiv \underline{X}_i \bmod m$ and compares it with the number X_i received from user i in step i). If they are equal, he computes the common key with user i as user i does in step iii). Otherwise he stops the protocol.

In this protocol, the GQ identification scheme is applied to user i when he is acting as a prover and the Diffie-Hellman scheme is applied to user j when he is acting as a prover. It is of great importance to take proper precautions for preventing the middleperson attack or the divertibility which is known as a major problem in most schemes based on zero-knowledge technique [DGB][OO2][BBDGQ]. In this point of view it is essential in the above protocol that user j's random challenge in step ii) should be dependent not only on the number he chooses randomly by himself but also on the number user i transmitted in step i). The number D_i computed by $D_i = h(K_{ij}, R_j)$ plays both the role of user j's challenge to user i which is due to the random number R_j generated by user j and publicly transmitted and the role of user j's response to user i's challenge in step i) which is due to the secret number K_{ij} computed only by user i and user j. We believe that this effectively defeats any on-line real-time attack.

All transmissions that outsiders can observe in the above protocol are random numbers and so no useful information about the secrets is revealed. The mutual identification protocol remains secure as long as both solving the Diffie-Hellman problem and breaking the GQ identification scheme remains intractable. The session key is computed using the common secret number K_{ij} and it depends both on the random number ρ_i generated by user i and on the random number R_j generated by user j.

Conference key distribution :

Based on the modified trapdoor one-way function, we can construct a conference key distribution system for generating a common secret conference key for two or more users. Here we assume a star-connected network so that messages are transmitted from one user to the others. The basic idea for the proposed scheme is to extract a secret key from the key-embedding random number authenticated by the GQ signature of the initiator. We assume that n users are involved and that user 1 initiates the protocol.

- (user 1) User 1 prepares the following data D_i, Y_i, Z_i, T_i and sends them to the conference attendants user i $(2 \leq i \leq n)$: User 1 generates n random numbers $\gamma, \rho_i (2 \leq i \leq n)$, computes $X_i \equiv R_i^e \bmod m$ where $R_i \equiv g^{\rho_i} \bmod m$, $Z_i \equiv g^{\gamma}.ID_i^{\rho_i} \equiv g^{\gamma + e\rho_i S_i} \bmod m$. He also generates the signature on the message Z_i and T_i by computing $D_i = h(X_i, Z_i, T_i)$ where T_i is a timestamp and $Y_i \equiv R_i.\, G_1^{D_i} \equiv R_i.g^{S_1 D_i} \bmod m$. Now the conference key to be used among n users will be $K \equiv g^{\gamma} \bmod m$.
- (user i) $(2 \leq i \leq n)$ Upon receiving D_i, Y_i, Z_i and T_i, user i checks the timeliness of a timestamp T_i, computes $Y_i^e.ID_1^{-D_i} \equiv \underline{X}_i \bmod m$ and $\underline{D}_i = h(\underline{X}_i, Z_i, T_i)$ and compares it with the number D_i received from user 1. If $\underline{D_i}$ is equal to D_i, he can verify that user 1 sent D_i, Y_i, Z_i and T_i. After successful verification, he computes $K \equiv X_i^{-S_i}.\, Z_i \equiv g^{\gamma} \bmod m$ as a conference key.

In the above scheme an initiator (user 1) utilizes GQ signature scheme to authenticate the message Z_i into which a conference key is embedded by ElGamal cryptosystem [E]. The timestamp protects against a replay attack. The validity of signature implies that the number Z_i is actually transmitted by the initiator user 1. Finally only the designated receiver user i can extract the correct conference key K from Z_i. A conference key K may be embedded directly into Z_i by computing $Z_i \equiv K.ID_i^{\rho_i} \bmod m$. We remark that similar approach for constructing conference key distribution schemes can be applied to the case where we use the Shnorr's signature scheme [S] or its variants and the Diffie-Hellman scheme. In this case a conference key $K \equiv g^{\gamma} \bmod p$ can be embedded into Z_i by $Z_i \equiv P_i^{\gamma} \bmod p$, where P_i is a public key of user i.

3.3 Directed signature and undeniable signature

In this section, we consider a signature scheme with the following properties, assuming user i sends a message M to user j :

i) Only user j can directly verify the signature on user i's message M while the others know nothing on the origin and validity of the message M without the help of user i and user j.
ii) User j can prove, if necessary, to any third party that the signature is a valid signature on the message M issued to him by user i and conversely user i can also prove, if necessary, that he issued the signature on the message M to user j.

This signature is different from an ordinary signature in its directedness to a specific receiver and in its verifiability to a third party (transitivity) only by the issuer or the designated receiver and thus we call it as a verifiable directed signature scheme. Due to these properties, this signature is suitable for the applications in which a signed message is personally or commercially sensitive to and/or somewhat obligatory on the designated receiver, for example as in a bill of tax, a bill of health, a writ of summons, etc. The second property enables a dispute resolution in case that the signer tries to deny his signature or the receiver tries to deny the directedness of the signature. Compared with undeniable signatures introduced by D.Chaum which is a signature that can be verified only with the help of the signer [CA] [C1], both signatures have the advantage that unlike ordinary signatures they cannot be transferred without the help of the signer or the designated receiver, but an undeniable signature scheme is intended for the applications in which a signed message is sensitive to the signer's privacy or interests as in a signature on the software releases.

Verifiable directed signature :

Now we present an example of signature construction based on the modified trap-door one-way function. Suppose that user i wants to send a signature of message M to user j so that only user j can directly verify it. Then he proceeds as follows. First, he chooses a random number ρ over Z_m and computes $R \equiv g^{\rho} \bmod m$ which plays a role of a random number in the GQ scheme. The signature on a message M that he generates consists of $\{X, Y\}$ computed as follows :

$$X \equiv ID_j^{\rho} \equiv g^{e\rho S_j} \bmod m, D = h(X, M), Y \equiv R.\ G_i^D \bmod m$$

Now he sends the set of integers $\{X, Y, M\}$ to user j. If necessary for later dispute resolution, he stores securely the secret random number ρ related to that signature. More efficiently, he can generate these random numbers using some pseudo-random function f with a secret seed K by $\rho = f(K, ID_j, M)$. Then he will be able to recover the random number related to the specific signature, when needed, using the secret number K.

Upon receiving them, user j can verify the validity of the signature on user i's message M as follows. First he computes $\underline{D} = h(X, M)$ and $V \equiv Y^e.ID_i^{-\underline{D}} \bmod m$ and checks if the equation $X \equiv V^{S_j} \bmod m$ holds. It is clear that the verification, the reconstruction of X from the signature, can be done only by the designated receiver user j.

In times of trouble or if necessary, user j should be able to prove that the set of integers $\{X, Y\}$ is a valid signature on the message M which is issued to him by user i. For this he can prove that he knows the secret number S_j such that $X \equiv V^{S_j} \bmod m$ and $ID_j \equiv G^{S_j} \bmod m$ where $V \equiv Y^e.\ ID_i^{-D} \bmod m$ which can be computed by anyone from the signature. This can be achieved through the confirmation protocol in [C1] [BCDP]. Similarly user i can prove that he issued that signature to user j by showing that he knows the secret ranfiom number ρ such that $X \equiv ID_j^{\rho} \bmod m$ and $V \equiv G^{\rho} \bmod m$.

The weaknesses of Chaum's confirmation protocol is pointed out in [DY] [C2] where it is shown that without isolation of the verifier such as physical isolation or temporal restriction on the verifier's responses, the abuses of the protocol, the multi-verifier attack, is always possible assuming the majority of the verifiers are honest. Under some restricted scenario satisfying non-detectivity and the dishonest majority of the verifiers, the solution to break the divertibility is proposed in [OOF] by using a secure bit commitment function against divertibility. But under similar assumption the multi-verifier attack is easy to prevent by simply breaking the homomorphic property of the challenge generating function through one-way hash function commonly used in many protocols presented in this paper (for example, the challenge value can be generated by $X \equiv m^{a+t} g^b \bmod p$ with random numbers a and b and the hashing result $t = h(a, b)$ in step 1 of Chaum's protocol in [C1]). Of course these methods are not effective when the majority of the collaborating verifiers are honest, but we think that as for directed signatures the prevention of divertibility is sufficient in most cases since the protocol is used by the signer or the receiver for the purpose of dispute resolutions or proofs of the validity and directedness of the signature.

Selectively convertible undeniable signature :

Finally we show that a selectively convertible undeniable signature can be easily constructed based on the modified trap-door one-way function. Convertible undeniable signatures are undeniable signatures with the added property that the signer can convert all the undeniable signatures into ordinary signatures by revealing a part of his secret information, and selectively convertible undeniable signatures are convertible undeniable signatures in which a specific signature can be selectively converted into ordinary signatures without affecting other undeniable signatures [BCDP].

In this scheme we assume that the signer possesses another secret key K for conventional cryptosystems which will be used as a secret key for the MAC (Message Authentication Code)-like algorithm f which is assumed to be publicly available. The signature generation procedures are as follows. First the signer computes $R \equiv g^{\rho} \bmod m$ with a random number ρ over Z_m and then $Z \equiv R^e \bmod m$. He next generates a secret number γ using the public algorithm f and his secret key K by $\gamma = f(K, Z, M)$ and then computes the followings :

$$X \equiv g^{e\gamma} \bmod m, D = h(X, M), Y \equiv R^D \cdot G_i^{\gamma} \bmod m$$

Now the signature on the message M is given as $\{Z, X, Y\}$.

The triple $\{Z, X, Y\}$ is a legal signature on the message M if and only if $V \equiv ID_i^{\gamma} \equiv X^{S_i} \bmod m$, where $V \equiv Y^e \cdot Z^{-D} \bmod m$. Here observe that a single signature can be converted into an ordinary signature by releasing the secret number γ specific to the signature. Moreover if the secret key K itself is released, all the undeniable signatures will be converted into ordinary signatures, since everyone can compute the number γ from the signature and the released K. As is clear in construction of Y, the release of γ and/or K does not affect the security of the signer's secret key. The verification (confirmation or disavowal) can be done via those protocols in [C1] [BCDP].

4 Implementation of Pohlig-Hellman and Pollard's ρ-methods

We have implemented the method for computing logarithms modulo $m = p_1p_2$ which is a combination of Pohlig-Hellman algorithm and Pollard's ρ-method [P2]. We have chosen prime factors p_i of 100 decimal digits such that $p_i - 1$ have 5-6 prime factors of 10-15 decimal digits as follows :

$$p_1 - 1 = 2.163.87083.17313839.705313823.35491269587.$$
$$343765241003.2965638692243.5694860095979.$$
$$45583609221803.473542954399787,$$
$$p_2 - 1 = 2.179.204599.1168859.19800239.67241159.$$
$$115897319.68770525127.236514898199.2821358425703.$$
$$20826680869859.546499152259883$$

All prime factors q_{ij} of $p_i - 1$ except for the first two small factors are of the form $q_{ij} - 1 = 2q'_{ij}$, where q'_{ij} is also prime. The base element g is chosen as $g = 5$, which is primitive in both $GF(p_1)$ and $GF(p_2)$. The public exponent e for the modified scheme can be computed so that it is primitive in all $GF(q_{ij})$ for the q_{ij}'s greater than or equal to 10 decimal digits. For example, we have computed 65-bit size odd integer e as $e = 18446744073709622793$ which has a Hamming weight 7.

We used the Brent's cycle-finding algorithm [B] for computing the index modulo each prime factor q_{ij} of $p_i - 1$. In the Brent's algorithm, one iteration consists of one multiplication modulo p_i, one addition modulo q_{ij} and one comparison. At the expense of storage, we used much precomputation for modular multiplications when constant is multiplied and thus about 285 iterations per second could be performed in SUN SPARC station 1+. For the largest prime factor q of $p_i - 1$, the Brent's algorithm requires on the average $2\sqrt{q}$ iterations in order to find the cycle on which the whole running time mainly depends. We have run the program several times in 5 workstations independently. The running times for computing logarithms modulo p_i varied from about 30 to 150 hours in CPU time.

Here we present an instance of computation. The identity information and its decimal equivalent are as follows :

ID = Chae Hoon, Lim, 630311-1925515, Dept. Elect. Eng.,
POSTECH, Pohang, Korea.
= 1628159180138416771842818454554241515077144141 2616486
9197606408240222770738330293006683818608139580849 2937
5337284683844075187505680587922404505212651915923 6790
68310942867758 (173 digits).

Parameters g, p_1, p_2 and $m = p_1 p_2$ are the same as given above. Since $(ID/m) = 1$, the ID itself has a discrete logarithm modulo m, which can be obtained by first computing x_i, $i = 1, 2$, such that $y \equiv y_i \equiv g^{x_i} \bmod p_i$ and then solving the congruences $x \equiv x_i \bmod p_i$.The followings are the results:

$$x_1 = 1319710947706104836178493618152145378567958010555572 1\\ 2007471950183683944854509404162919336118 2514973$$

(100 digits : 28 hours 55 minutes),

$$x_2 = 2772588890972151494269646886785692846146822771063857 5\\ 916483788782210073820478201733392691573 26370841$$

(100 digits : 87 hours 8 minutes),

$$x = 4954462606290161666747909510131614042994072237853200 5\\ 8569195515977196853486255508858258147278714251338367 4\\ 1174251649895004053754418291090178349753808686928067 9\\ 2454911161251274109090619626205642588911 \text{ (199 digits).}$$

This running time shows that the scheme with the parameters given above seems not suitable for practical use even with giga flops computer. Of course, since the Pollard's p-1 method, the most efficient factoring algorithm at present for this kind of modulus, requires the computational amount proportional to the largest prime factor of $p_i - 1$ and thus increasing the computational effort of the trusted authority by a factor k forces an adversary to increase his computational effort by a factor k^2, the number and size of prime factors of $p_i - 1$ may be more sophisticatedly chosen so that the scheme becomes more practical while keeping the security level of some predetermined limit.

On the other hand, if we are willing to give up the advantage of non-interactiveness of the identity-based scheme, much more efficient certificate-based scheme can be obtained as follows. First the trusted authority prepares RSA key parameters, makes public (m, e, g) and keeps secret d. Each user i wishing to join the system visits the trusted authority and presents his identity information. After verifying user i's physical identity, the authority prepares user i's ID with public format, computes $ID_i^d \bmod m$ and gives it to user i securely. On receiving $ID_i^d \bmod m$, user i chooses his secret key S_i by himself and computes $C_i \equiv g^{S_i} . ID_i^{-d} \bmod m$. Then the equation $P_i \equiv g^{eS_i} \equiv C_i^e . ID_i \bmod m$ holds and this P_i can be used as his public key whose certificate is C_i. Note that each user i in the system can change his public key without the help of the authority by first choosing another secret key and then computing its certificate using the secret number $ID_i^d \bmod m$ received from the authority at the subscription time. Now it is enough to exchange and verify a pair (P_i, C_i) at the start of protocols for the schemes described in the preceding chapter to be the equivalent certificate-based schemes. Of course since we are going to use the number $g^{S_i} \bmod m$ as another secret information, it should be kept secret as well. This certificate-based scheme is much more efficient and practical than the identity-based scheme since only moderate computing power is enough for the authority and a secret key can be selected by each user.

5 Conclusion

We have shown that the squaring method for building a trapdoor exponentiation one-way function proposed by Maurer and Yacobi is not secure and provided some countermeasures to overcome the defect. With a slight penalty of 3 to 4 bit interactive communication via more general media such as telephones, the Jacobi symbol method was shown to be applicable to the case where the modulus has more than two prime factors. We also presented a modified trapdoor one-way function in which the RSA scheme is incorporated into the Maurer-Yacobi's scheme. Since in this modified scheme we can make full use of the benefits of both the RSA scheme and the Maurer-Yacobi's scheme, much more applications can be provided such as identifications, key distributions and signatures. We presented several such applications including mutual identification with key distribution, conference key distribution, directed signatures and undeniable signatures.

For the Maurer-Yacobi's scheme to be practical, the authority should have great computing power. Implementation of Pohlig-Hellman and Pollard's ρ-methods shows that it takes about 30 to 150 CPU hours in SUN SPARC station 1+ to compute logarithms mod p when p is around 100 decimal digit prime and $p-1$ has several prime factors of 13-15 decimal digits. From this experiment, we conclude that with the parameter setting suggested by Maurer and Yacobi the authority can hardly do its role properly even with giga flops computer. It is shown that a certificate-based scheme is much more efficient and flexible, which can be used instead of the identity-based scheme in any protocols presented in this paper.

References

[B] R.Brent, "An improved Monte Carlo factoring algorithm," *BIT*, 20, 1980, pp.176-184.

[BBDGQ] S.Bengio, G.Brassard, T.G.Desmedt, C.Goutier and J.J.Quisquater, "Secure implementation of identification systems," *J. Cryptology*, 4, 3, 1991, pp.175-183.

[BCDP] J.Boyar, D.Chaum, I.Damgard and T.Pedersen, "Convertible undeniable signatures," *Advances in Cryptology - Crypto'90*, Lecture Notes in Computer Science (LNCS), Vol.537, Springer-Verlag, 1991.

[C1] D.Chaum, "Zero-knowledge undeniable signatures," *Advances in Cryptology - Eurocrypt'90*, LNCS, Vol.473, Springer-Verlag, 1991, pp.458-464.

[C2] ——, "Some weaknesses of 'Weaknesses of undeniable signatures'," *Advances in Cryptology - Eurocrypt'91*, LNCS, Vol.547, 1991, pp.554-556.

[CA] D.Chaum and H.Antwerpen, "Undeniable signatures," *Advances in Cryptology - Crypto'89*, LNCS, Vol.435, Springer-Verlag, 1990, pp.212-216.

[COS] D.Coppersmith, A.M.Odlyzko and R.Schroeppel, "Discrete logarithms in GF(p)," *Algorithmica*, Vol.1, 1986, pp.1-15.

[DGB] Y.Desmedt, C.Goutier and S.Bengio, "Special uses and abuses of the Fiat-Shamir passport protocol," *Advances in Cryptology - Crypto'87*, LNCS, Vol.293, Springer-Verlag, 1988, pp.21-39.

[DH] W.Diffie and M.E.Hellman, "New directions in cryptography," *IEEE Trans. Inform. Theory*, IT-22, 6, 1976, pp.644-654.

[DY] Y.Desmedt and M.Yung, "Weaknesses of undeniable signature schemes," *Advances in Cryptology - Eurocrypt'91*, LNCS, Vol.547, 1991, pp.205-220.

[GQ1] L.S.Guillou and J.J.Quisquater, "A practical zero-knowledge protocol fitted to security microprocessor minimizing both transmission and memory," *Advances in Cryptology - Eurocrypt'88*, LNCS, Vol.330, Springer-Verlag, 1988, pp.123-128.

[GQ2] —— , "A paradoxical identity-based signature scheme resulting from zero-knowledge," *Advances in Cryptology - Crypto'88*, LNCS, Vol.403, Springer-Verlag, 1989, pp.216-231.

[E] T.ElGamal, "A public key cryptosystem and a signature scheme based on discrete logarithm," *IEEE Trans. Inform. Theory*, IT-31, 1985, pp.469-472.

[L] H.W.Lenstra, "Factoring integers with elliptic curves," *Ann. Math.*, Vol.126, 1987, pp.649-673.

[MY] U.M.Maurer and Y.Yacobi, "Non-interactive public key cryptography," *Advances in Cryptology - Eurocrypt'91*, LNCS, Vol.547, 1991, pp.498-507.

[OO1] T.Okamoto and K.Otha, "How to utilize the randomness of the zero-knowledge proofs," *Advances in Cryptology - Crypto'90*, LNCS, Vol.537, Springer-Verlag, 1991.

[OO2] —— , "Divertible zero-knowledge interactive proofs and commutative random self-reducibility," *Advances in Cryptology - Eurocrypt'89*, LNCS, Vol.434, Springer-Verlag, 1990, pp.134-149.

[OOF] K.Ohta, T.Okamoto and A.Fujioka, "Secure bit commitment function against divertibility," *Proc. Eurocrypt'92*.

[OT] E.Okamoto and K.Tanaka, "Identity-based information security management system for personal computer networks," *IEEE JSAC*, Vol.7, No.2, 1989, pp.290-294.

[P1] J.M.Pollard, "Theorems on factorization and primality testing," *Proc. Cambridge Philos. Soc.*, Vol.76, 1974, pp.521-528.

[P2] —— , "Monte Carlo methods for index computation (mod p)," *Math. Comp.*, 32, 1978, pp.918-924.

[PH] S.C.Pohlig and M.E.Hellman, "An improved algorithm for computing logarithms over GF(p) and its cryptographic significance," *IEEE Trans. Inform. Theory*, Vol.IT-24, 1978, pp.106-110.

[R] R.L.Rivest, "Remarks on a proposed cryptanalytic attack on the M.I.T. public key cryptosystem," *Cryptologia*, Vol.2, No.1, 1978, pp.62-65.

[RSA] R.L.Rivest, A.Shamir and L.Adleman, "A method of obtaining digital signatures and public key cryptosystem," *Comm. ACM*, 21, 2, 1978, pp.120-126.

[S] C.P.Schnorr, "Efficient identification and signatures for smart cards," *Advances in Cryptology - Crypto'89*, LNCS, Vol.435, Springer-Verlag, 1990, pp.239-252.

[SN] G.J.Simmons and M.J.Norris, "Preliminary comments on the M.I.T. public key cryptosystem," *Cryptologia*, Vol.1, No.4, 1977, pp.406-414.

Session 8

SEQUENCES

Chair: Rudi Lidl
(University of Tasmania, Australia)

The Vulnerability of Geometric Sequences Based on Fields of Odd Characteristic
(Extended Abstract)

Andrew Klapper*

University of Manitoba
Department of Computer Science
Winnipeg, Manitoba
Canada R3T 2N2
e-mail: klapper@cs.umanitoba.ca

Abstract. A new method of cryptologic attack on binary sequences is given, using their linear complexities relative to odd prime numbers. We show that, relative to a particular prime number p, the linear complexity of a binary geometric sequences is low. It is also shown that the prime p, can be determined with high probability by a randomized algorithm if a number of bits much smaller than the linear complexity is known. This determination is made by exploiting the imbalance in the number of zeros and ones in the sequences in question, and uses a new statistical measure, the partial imbalance.

1 Introduction

In several applications in modern communication systems, periodic binary sequences are used that must be difficult for an adversary to determine when a short subsequence is known, and must be easy to generate given a secret key. In particular, sequences must have large linear complexities, ensuring resistence to attack by the Berlekamp-Massey algorithm [7]. This is true in stream cipher systems, in which the binary sequence is used as a pseudo-one-time-pad [9], and in secure spread spectrum systems, in which the sequence is used to spread a signal over a large range of frequencies [8]. Much effort has gone into finding ways of modifying linear feedback shift registers, typically by adding some nonlinearity, so that the sequences they generate have large linear complexities. One way to do this is to apply a nonlinear feedforward function to an m-sequence over a finite field, giving rise to a *geometric sequence.*

In this paper we describe a cryptologic attack on geometric sequences based on m-sequences over fields $GF(q)$ with q odd. Previously, because geometric sequences are binary, they have been thought of as sequences over $GF(2)$ and

* Project sponsored by the National Science and Engineering Research Council under grant number OGP0121648, and by the National Security Agency under Grant Number MDA904-91-H-0012. The United States Government is authorized to reproduce and distribute reprints notwithstanding any copyright notation hereon.

only their $GF(2)$-linear complexity has been considered. This linear complexity has been shown by Chan and Games to be high when q is odd [3], and thus geometric sequences have been thought to be relatively secure. Here we take the novel approach of considering binary sequences as sequences over $GF(q)$, and showing that their $GF(q)$-linear complexity is considerably lower. We show that if $q = p^e$, p prime, and the geometric sequence $\mathbf{S}$ is based on an m-sequence over $GF(q)$ of period $q^n - 1$, then the $GF(q)$-linear complexity of $\mathbf{S}$ is at most

$$\binom{n+p-1}{p-1}^e.$$

To make geometric sequences easy to generate, n is often chosen to be close to q. For example, with $n = q = 27$, we can produce geometric sequences with period about $1.4\text{x}2^{128}$, but their $GF(q)$-linear complexity is at most 2^{26}. Thus high $GF(2)$-linear complexity is inadequate for security, even when one is only concerned with attacks using the Berlekamp-Massey algorithm.

In order to exploit the low linear complexity, we must first determine q. This is done by a probabilistic algorithm based on a lack of balance in these sequences (when q is odd). We compute the imbalance of a known small subsequence, and use this statistic to determine q. By exhibiting a bound on the variance of such a subsequence (with the length of the subsequence fixed and the start position varying), we show that with high probability q can be determined from this partial imbalance. In general the number of bits required to make this probability overwhelming is far smaller than the $GF(q)$-linear complexity. This algorithm is similar to one described in an earlier paper by Klapper and Goresky based on the partial period autocorrelation [5], but requires far fewer bits for success.

2 Definitions

If $\mathbf{S} = (\mathbf{S}_i)$ is a sequence over a field F, then the *linear complexity* of $\mathbf{S}$ over F, denoted by $\lambda_F(\mathbf{S})$, is the smallest n such that $\mathbf{S}$ satisfies a linear recurrence .

$$\forall k \geq 1 : \sum_{i=0}^{n} c_i \mathbf{S}_{k+i} = 0 \tag{1}$$

with coefficients c_i in F, not all zero (n is the length of the linear recurrence). Any sequence over F that satisfies such a recurrence is called a *linear recurrent sequence*. It is well known [7] that $2\lambda_F(\mathbf{S})$ consecutive elements of $\mathbf{S}$ suffice to determine the smallest linear recurrence satisfied by $\mathbf{S}$ (or, equivalently, to synthesize a linear feedback shift register which generates $\mathbf{S}$). Moreover, the algorithm for this determination is highly efficient. Thus for purposes of security, sequences must be found which have large linear complexity.

Consider a binary sequence $\mathbf{S}$. While it is natural to think of $\mathbf{S}$ as a sequence of elements of $GF(2)$ and concern ourselves only with the linear complexity over $GF(2)$, we can in fact think of $\mathbf{S}$ as a sequence over any prime field $GF(p)$. We denote by $\lambda_p(\mathbf{S})$ the linear complexity of $\mathbf{S}$ considered as a sequence of elements of $GF(p)$. It is our purpose to show that, despite the fact that $\lambda_2(\mathbf{S})$ is large, $\lambda_p(\mathbf{S})$ may be small for some efficiently computable p.

The sequences we consider are known as *geometric sequences* and have been investigated by several authors with respect to their linear complexity and correlation properties [2, 3, 4]. Chan and Games showed that for q odd, geometric sequences can have high linear complexities (over $GF(2)$) [3]. For this reason, these sequences are candidates for applications that require easily generated sequences that resist chosen plaintext attacks.

Let p be a prime integer, $q = p^e$ be a fixed power of p, and $GF(q)$ denote the Galois field with q elements. For any $n \geq 1$, we denote the *trace function* from $GF(q^n)$ to $GF(q)$ by $Tr_q^{q^n}(x) = \sum_{i=0}^{n-1} x^{q^i}$. Let α be a primitive element of $GF(q^n)$. The infinite periodic sequence $\mathbf{U}$ whose ith term is $\mathbf{U}_i = Tr_q^{q^n}(\alpha^i)$ is known as an *m-sequence* over $GF(q)$ of span n [6]. The sequence whose ith term is $Tr_q^{q^n}(A\alpha^i)$, $A \in GF(q^n)$, is a cyclic shift of the sequence $\mathbf{U}$, and we do not consider it to be a distinct sequence here. Every m-sequence of span n can be generated by a linear feedback shift register of length n and has period $q^n - 1$, the maximum possible period for a sequence generated by a linear feedback shift register of length n over $GF(q)$. Every such maximal period sequence is (a shift of) an m-sequence [6, pp. 394-410].

Definition 1. Let n be a positive integer, α be a primitive element of $GF(q^n)$, and $g : GF(q) \rightarrow GF(2)$. The binary sequence $\mathbf{S}$ whose ith term is $\mathbf{S}_i = g(Tr_q^{q^n}(\alpha^i))$ is a geometric sequence based on feedforward function g.

If p is odd, then g cannot be defined algebraically. We can, however, think of g as a function into $\{0, 1\} \subseteq GF(r)$ for any prime r. If $r = p$, then $g(x)$ can be expressed as a polynomial in x with coefficients in $GF(q)$. $\mathbf{S}$ is easy to generate if the feedforward function g is easy to compute. This is always the case, for example, if q is small. A geometric sequence is a $(q^n - 1)$-periodic binary sequence. Geometric sequences with q odd have been used in applications where easily generated sequences with large linear complexities are needed, due to the following theorem of Chan and Games [3], in which we let $\nu = (q^n - 1)/(q - 1)$.

Theorem 2 Chan and Games. *Let* $\mathbf{S}$ *be a geometric sequence satisfying* $\mathbf{S}_i = g(Tr_q^{q^n}(\alpha^i))$ *and suppose* $g(0) = 0$. *Let* $\mathbf{T}$ *be the sequence* $\mathbf{T}_i = g(\alpha^{i\nu})$ *(note that* α^ν *is a primitive element of* $GF(q)$*). Then* $\lambda_2(\mathbf{S}) = \nu\lambda_2(\mathbf{T})$.

Thus, if g maximizes $\lambda_2(\mathbf{T})$ at $q - 1$, then $\lambda_2(\mathbf{S})$ will be $q^n - 1$, i.e., maximal. This theorem generalizes to the linear complexity λ_r over an arbitrary prime number $r \neq p$.

Theorem 3. *Let* $\mathbf{S}$ *be a geometric sequence,* $\mathbf{S}_i = g(Tr_q^{q^n}(\alpha^i))$ *and suppose* $g(0) = 0$. *Let* $\mathbf{T}$ *be the sequence* $\mathbf{T}_i = g(\alpha^{i\nu})$. *Let* r *be a prime number,* $r \neq p$. *Then* $\lambda_r(\mathbf{S}) = \nu\lambda_r(\mathbf{T})$.

In particular, as long as $g(x) \neq 0$ for some $x \in GF(q)$, then $\lambda_r(\mathbf{S}) \geq (q^n - 1)/(q - 1)$ for all primes $r \neq p$. Thus the only prime over which $\mathbf{S}$ can have small linear complexity is p, and we show in the next section that this is indeed the case.

3 Bounding λ_p for Geometric Sequences

We show that if **S** a geometric sequence based on an m-sequence over a finite field of characteristic p, then $\lambda_p(\mathbf{S})$ is far from maximal. Our results follow from the work of Zierler and Mills on algebraic combinations of periodic sequences over fields [10], and generalize Brynielsson's theorem on the linear complexity of a geometric sequence based on even q [2].

If F is a field, we consider the variable t as acting on the F-algebra of infinite sequences over F (with componentwise addition and multiplication) as the shift operator: $t(\mathbf{S}_0, \mathbf{S}_1, \mathbf{S}_2, \cdots) = (\mathbf{S}_1, \mathbf{S}_2, \mathbf{S}_3, \cdots)$. This extends to a linear action of the set of polynomials over F on the vector space of infinite sequences over F. The condition in equation (1) is equivalent to $f(t)\mathbf{S} = 0$, where $f(t) = \sum c_i t^i$. If $f_{\mathbf{S}}(t)$ is the nonzero polynomial of smallest degree such that $f_{\mathbf{S}}(t)\mathbf{S} = 0$, and $g(t)$ is another polynomial such that $g(t)\mathbf{S} = 0$, then $f_{\mathbf{S}}(t)$ divides $g(t)$. The degree of $f_{\mathbf{S}}$ is $\lambda_F(\mathbf{S})$.

The following proposition is a consequence of Zierler and Mills' results relating the algebra of infinite sequences over F to the algebra of polynomials which annhilate them. If f and g are two polynomials over F, we denote by $f \vee g$ the monic poynomial whose roots (in an algebraic closure of F) are the distinct elements of the form $\gamma\delta$ where γ is a root of f and γ is a root of g ($f \vee g$ is necessarily defined over F).

Proposition 4.

1. *If* **S** *and* **T** *are linearly recurrent sequences over F, then $f_{\mathbf{S}+\mathbf{T}}$ divides the least common multiple of $f_{\mathbf{S}}$ and $f_{\mathbf{T}}$, and $\lambda_F(\mathbf{S}+\mathbf{T}) \leq \lambda_F(\mathbf{S}) + \lambda_F(\mathbf{T})$. If, moreover, $f_{\mathbf{S}}$ and $f_{\mathbf{T}}$ have no roots in common, then $f_{\mathbf{S}+\mathbf{T}} = f_{\mathbf{S}} f_{\mathbf{T}}$ and $\lambda_F(\mathbf{S}+\mathbf{T}) = \lambda_F(\mathbf{S}) + \lambda_F(\mathbf{T})$.*
2. *If* **S** *and* **T** *are linearly recurrent sequences over F, then $f_{\mathbf{ST}}$ divides $f_{\mathbf{S}} \vee f_{\mathbf{T}}$ and $\lambda_F(\mathbf{ST}) \leq \lambda_F(\mathbf{S})\lambda_F(\mathbf{T}) =$ the number of distinct root products $\gamma\delta$, γ a root of $f_{\mathbf{S}}$, δ a root of $f_{\mathbf{T}}$. If, moreover, all the root products from $f_{\mathbf{S}}$ and $f_{\mathbf{T}}$ are distinct, then $f_{\mathbf{ST}} = f_{\mathbf{S}} \vee f_{\mathbf{T}}$ and $\lambda_F(\mathbf{ST}) = \lambda_F(\mathbf{S})\lambda_F(\mathbf{T})$.*

We will also need

Lemma 5. *If $f_{\mathbf{S}}(t) = \prod_{j=1}^{n}(t - \gamma_j)$ with distinct $\gamma_j \neq 0$, then* **S** *can be written uniquely as $\mathbf{S}_i = \sum_{j=1}^{n} c_j \gamma_j^i$. (The γ_j may lie in an extension field of F. The c_j will lie in $F[\gamma_1, \cdots, \gamma_n]$.)*

We have an m-sequence **U** of period $q^n - 1$ over $GF(q)$, hence of linear complexity n over $GF(q)$. To this sequence we apply a feedforward function $g : GF(q) \to GF(q)$. We eventually specialize to the case where the image of g is in $\{0, 1\}$. We can express g as a polynomial

$$g(x) = \sum_{i=0}^{p^e - 1} a_i x^i$$

with $a_i \in GF(q)$ (recall $q = p^e$). The resulting sequence **S** can be treated as a sum of constant multiples of products of the original sequence **U** with itself. We will use the above results to express $\lambda_q(\mathbf{S})$ in terms of p, e, n, and $\{a_i\}$. Note that $f_{\mathbf{U}}$ is the minimal polynomial of α over $GF(q)$, and α is a primitive element of $GF(q^n)$. Hence $f_{\mathbf{U}}$ has distinct roots $\alpha, \alpha^q, \cdots, \alpha^{q^{n-1}}$. Moreover,

$$\mathbf{S}_i = Tr_q^{q^n}(\alpha^i) = \sum_{j=0}^{n-1} \alpha^{q^j i}$$

is the representation of **S** as in Lemma 5.

If $0 \le k < q$, then we can express k in base p, $k = \sum_{i=0}^{e-1} b_i p^i$ where $0 \le b_i < p$. Thus

$$x^k = \prod_{i=0}^{e-1} (x^{p^i})^{b_i}.$$

We derive $\lambda_q(\mathbf{S})$ and the roots of $f_{\mathbf{S}}$ in stages.

Theorem 6.

1. *Suppose $g(x) = x^{p^i}$. Then $\lambda_q(\mathbf{S}) = n$ and $f_{\mathbf{S}}$ has distinct roots*

$$\alpha^{p^i}, \alpha^{p^i q}, \cdots, \alpha^{p^i q^{n-1}}.$$

2. *Suppose $g(x) = x^{bp^i}$, where $1 \le b < p$. Then*

$$\lambda_q(\mathbf{S}) = \binom{n+b-1}{b}$$

and $f_{\mathbf{S}}$ has distinct roots

$$\{\alpha^{p^i m} : \ where\ m = \sum_{j=0}^{n-1} r_j q^j, \sum_{j=0}^{n-1} r_j = b, \ and\ \forall j : r_j \ge 0\}.$$

3. *Suppose $g(x) = x^k$, $k = \sum_{i=0}^{e-1} p^i b_i$, where $0 \le b_i < p$. Then*

$$\lambda_q(\mathbf{S}) = \prod_{i=0}^{e-1} \binom{n+b_i-1}{b_i}$$

and $f_{\mathbf{S}}$ has distinct roots

$$\{\alpha^{\sum_{i=0}^{e-1} p^i \sum_{j=0}^{n-1} r_{i,j} q^j} : \sum r_{i,j} = b_i \wedge \forall j : r_{i,j} \ge 0\}.$$

4. *Suppose $g(x) = \sum_{k=0}^{p^e-1} a_k x^k$. For each k, let $k = \sum_{i=0}^{e-1} b_{k,i} p^i$ where $0 \le b_{k,i} < p$. Then*

$$\lambda_q(\mathbf{S}) = \sum_{a_k \neq 0} \prod_{i=0}^{e-1} \binom{n+b_{k,i}-1}{b_{k,i}}.$$

Proof. (Sketch) In the first part the roots of f_S are the p^ith powers of the roots of f_U. Each remaining part is proved by applying Proposition 4 to the preceeding part. Properties of p-ary expansions of integers are used to show the root products remain distinct.

If S is a sequence over a field F, and F is a subfield of E, then the linear complexity of S over F equals its linear complexity over E. Thus we have:

Corollary 7. *If, for all $x \in GF(q)$, $g(x) \in GF(p)$, then*

$$\lambda_p(\mathbf{S}) = \sum_{a_k \neq 0} \prod_{i=0}^{e-1} \binom{n + b_{k,i} - 1}{b_{k,i}}.$$

The maximum possible linear complexity occurs when the a_k are all nonzero. This gives the following upper bound.

Corollary 8. *The linear complexity of* S *(over $GF(q)$ or $GF(p)$) is at most*

$$\binom{n+p-1}{p-1}^e.$$

For a given period $p^r - 1$, we can maximize $\lambda_p(\mathbf{S})$ by taking $e = r$ and $n = 1$, but S will not be efficiently computable. A more practical approach is to balance the work involved in computing g and the m-sequence by choosing n approximately equal to $q = p^e$. We can generate a sequence of period $q^q - 1$ with a relatively easily defined feedforward function on q elements and a q stage linear feedback shift register over $GF(q)$. We must take care, however, that $g(x)$ cannot be written as $g(x) = h(Tr^{p^e}_{p^f}(x))$ for some f dividing e and $h : GF(p^e) \rightarrow \{0, 1\}$. With relatively small q we can obtain sequences of enormous period that are easily generated. The largest linear complexity we can achieve with such a sequence is

$$\lambda_p(\mathbf{S}) = \binom{p^e+p-1}{p-1}^e.$$

This expression is largest if e is small. Thus choosing q to be a larger prime rather than a power of a small prime gives a larger linear complexity. If $p = 3$ and $q = 27$, we have a sequence of period about 1.4x2^{128}, while $\lambda_p(\mathbf{S}) < 2^{26}$. If $p = q = 17$, the period is only 1.4x2^{69}, but $\lambda_p(\mathbf{S})$ can be made greater than 2^{30}. Even for the latter sequence, if one bit is generated per microsecond, then the sequence can be cracked in approximately one hour.

4 Finding p

The fact that $\lambda_p(\mathbf{S})$ is low is of no use to a cryptanalyst unless p is known. In this section we describe an algorithm that determines p with high probability when S is optimally balanced. It was recently shown that, for a geometric sequence S based on an m-sequence over $GF(q)$, q can be determined with high probability if at least $2q^8$ bits of $\mathbf{S}$ are known [5]. This attack is based on the calculation of a partial period autocorrelation function. More available bits give a higher probability of success. We show that the same information can be obtained with far fewer bits of the sequence available – approximately q^4 bits – using a simpler statistical measure, the partial imbalance.

An attack on a system using geometric sequences proceeds in two stages. First, use the partial imbalance to determine q (with high probability). Knowing q tells us p, and this allows us to use the Berlekamp-Massey algorithm to synthesize a linear feedback shift register over $GF(p)$ that generates $\mathbf{S}$. The number of bits needed for this last stage to work is determined by the linear complexity calculation in Section 3. Our basic tool is the following.

Definition 9.

1. The imbalance of a binary sequence $\mathbf{S}$ of period N is: $\mathcal{I}_{\mathbf{S}} = \sum_{i=1}^{N}(-1)^{\mathbf{S}_i} = |\{i : \mathbf{S}_i = 0\}| - |\{i : \mathbf{S}_i = 1\}| = 2|\{i : \mathbf{S}_i = 0\}| - N$.
2. The imbalance of a function $g : GF(q) \to GF(2)$ is: $\mathcal{I}_g = \sum_{x \in GF(q)} G(x) = |\{x : g(x) = 0\}| - |\{x : g(x) = 1\}| = 2|\{x : g(x) = 0\}| - q$.
3. The partial imbalance of a sequence is defined by limiting the range of values in the sum defining the imbalance to a fixed window, parametrized by the start position k and length D of the window:

$$\mathcal{I}_{\mathbf{S}}(k, D) = \sum_{i=k}^{D+k-1} (-1)^{\mathbf{S}_i}.$$

Proposition 10. *The imbalance of a geometric sequence* $\mathbf{S}$ *of period* $q^n - 1$ *with feedforward function* $g : GF(q) \to GF(2)$ *is* $\mathcal{I}_{\mathbf{S}} = q^{n-1}\mathcal{I}_g - G(0)$.

If q is even, then $\mathcal{I}_{\mathbf{S}}$ can be ± 1, and we can learn nothing about $\mathbf{S}$ by computing its imbalance. If q is odd, then the smallest imbalance of g that we can achieve is ± 1 so $\mathcal{I}_{\mathbf{S}} = \pm q^{n-1} \pm 1$. In this case we can determine q^{n-1} if we can compute $\mathcal{I}_{\mathbf{S}}$. This is hopeless – we would need to know the entire sequence $\mathbf{S}$ to compute $\mathcal{I}_{\mathbf{S}}$. The most we can hope to compute is a partial imbalance.

We show that, if q is odd and $\mathbf{S}$ is as balanced as possible, the expected imbalance (keeping D fixed and letting k vary) is approximately D/q, and that the variance is small. Thus Chebyshev's inequality [1] implies that the partial imbalance is close to its expectation with high probability. One way to view our results is that we have introduced a new statistical test that a sequence must satisfy in order to be secure – the variance of the partial imbalance must be high for small subsequences if the imbalance is high.

We first show that the expected partial imbalance of any sequence can be determined from its full imbalance. We denote expectations by $E[-]$. Expectations are taken for fixed window size D, assuming a uniform distribution on all start positions k.

Theorem 11. *If* **S** *is a binary sequences with period* N*, then* $E[\mathcal{I}_{\mathbf{S}}(k,D)] = D\mathcal{I}_{\mathbf{S}}/N$.

Corollary 12. *Suppose* **S** *is a geometric sequence based on an m-sequence of span n with elements in* $GF(q)$*, q odd, with feedforward function* $g : GF(q) \to GF(2)$*. Assume that* **S** *is as balanced as possible, i.e.,* $|\mathcal{I}_g| = 1$*. Then the expected partial imbalance of* **S** *for odd q is* $E[\mathcal{I}_{\mathbf{S}}(k,D)] = D(\pm q^{n-1} \pm 1)/(q^n - 1)$.

If q is odd, then the expected partial imbalance is approximately D/q, and we can hope to learn q from it. If the partial imbalances (for varying k) lie close to the expected partial imbalance, then we can compute a partial imbalance (with q and k unknown), determine for which q the partial imbalance is closest to $D(\pm q^{n-1} \pm 1)/(q^n - 1)$, and conclude that q was used to generate the sequence. That is, we use the following algorithm:

Algorithm for Finding q**:**

1. **Establish** disjoint intervals $U_3, U_5, \cdots$ in the positive real line (one for each power of each odd integer, to avoiding having to recognize prime powers).
2. **Input** a subsequence $\mathbf{S}_k, \mathbf{S}_{k+1}, \cdots, \mathbf{S}_{k+D-1}$ of **S** (Determined, say, by a known plaintext attack on a stream cipher system. The index k is unknown).
3. **Compute** $x = \sum_{i=k}^{k+D-1}(-1)^{\mathbf{S}_i}$ $(= \mathcal{I}_{\mathbf{S}}(k,D))$.
4. **Output** q such that $|x| \in U_q$.

Is this algorithm likely to succeed? Only if we can choose disjoint intervals U_q so that $\mathcal{I}_{\mathbf{S}}(k,D)$ is in U_q with high probability whenever q was used as the parameter for generating **S**. If we can show that the partial imbalances do not deviate too much from their expectations, then it will be possible to choose the intervals U_q. Chebyshev's inequality bounds the deviation of a random variable from its expectation in terms of its variance.

Proposition 13 Chebyshev's Inequality [1]. *If* X *is a random variable with expectation* $E[X]$ *and variance* $V(X)$*, then for any* $\epsilon > 0$*,* $Prob\{|X - E[X]| > \epsilon\} < V(X)/\epsilon^2$.

As it turns out, the algorithm needs at least q^4 bits for success, more than are available if $n < 4$, so we assume $n \geq 4$. If q is an odd prime power, then for $n \geq 4$

$$\frac{D(q^3-1)}{q^4-1} \leq |E[\mathcal{I}_{\mathbf{S}}(k,D)]| \leq \frac{D(q^3+1)}{q^4-1}.$$

U_q should contain all these points. For each odd q, we pick $\epsilon_q > 0$, such that

$$\frac{D(q^3+1)}{q^4-1} + \epsilon_q = \frac{D((q-2)^3-1)}{(q-2)^4-1} - \epsilon_{q-2}.$$

Let $U_q = [D(q^3-1)/(q^4-1) - \epsilon_q, D(q^3+1)/(q^4-1) + \epsilon_q]$. If we can choose the ϵ_q so that $V(\mathcal{I}_{\mathbf{S}}(k,D))/\epsilon_q^2$ is small whenever **S** is a geometric sequence of period $q^n - 1$ based on a $g : GF(q) \to GF(2)$, and $n \geq 4$, then Chebyshev's inequality can be applied to show that the algorithm is successful with high probability. We next show that this is the case if enough bits are available, i.e., if D is large enough. Moreover, ϵ_q and the bound on $V(\mathcal{I}_{\mathbf{S}}(k,D))$ are proportional to D, so the more bits that are available, the higher the probability of success.

Theorem 14. *If $D \le \nu = (q^n-1)/(q-1)$, then the variance of the partial imbalance of a geometric sequence with window D is bounded above by D.*

Proof. The variance of a random variable X is $E[(X-E[X])^2] = E[X^2]-E[X]^2$, so we must determine $E[\mathcal{I}_{\mathbf{S}}(k,D)^2]$. We can reduce this to the determination of the cardinalities of certain sets, as stated in the following proposition. If $s \in GF(q)$, and $A \in GF(q^n)$, then we denote by H_A^s the set $\{x : Tr_q^{q^n}(Ax) = s\}$.

Proposition 15. *If $\mathbf{S}$ is a geometric sequence, then*

$$E[\mathcal{I}_{\mathbf{S}}(k,D)^2] = \frac{1}{q^n-1}\sum_{i,j=0}^{D-1}\Big(\sum_{s,t\in GF(q)} N_{i,j}(s,t)(-1)^{g(s)}(-1)^{g(t)} - 1\Big)$$

where $N_{i,j}(s,t) = |H^s_{\alpha^i} \cap H^t_{\alpha^j}| \in \{q^{n-1}, q^{n-2}, 0\}$.

Proposition 16. *For any $0 \le i,j \le q^n-2$, and $s,t \in GF(q)$,*

1. *If $\alpha^{i-j} \notin GF(q)$, then $N_{i,j}(s,t) = q^{n-1}$.*
2. *If $\alpha^{i-j} \in GF(q)$ (i.e, there is an integer m such that $i-j = m(q^n-1)/(q-1)$) then $N_{i,j}(s,t) = q^{n-2}$ if $\alpha^i t = \alpha^j s$, and $N_{i,j}(s,t) = 0$ otherwise.*

Proof. (Sketch) We consider $GF(q^n)$ to be affine n-space over $GF(q)$, and H_A^s to be a hyperplane in $GF(q^n)$. The proposition follows from an analysis of the conditions under which two hyperplanes are parallel.

If the window size D is less than ν, then $N_{i,j}(s,t) = q^{n-2}$ if $i \ne j$; $N_{i,i}(s,t) = 0$ if $s \ne t$; and $N_{i,i}(s,s) = q^{n-1}$. Thus

$$\begin{aligned}
E[\mathcal{I}_{\mathbf{S}}(k,D)^2] &= \frac{1}{q^n-1}\sum_{i,j=0}^{D-1}\Big(\sum_{s,t\in GF(q)} N_{i,j}(s,t)(-1)^{g(s)}(-1)^{g(t)} - 1\Big)\\
&= \frac{1}{q^n-1}\Big(\sum_{0\le i\ne j<D}\Big(\sum_{s,t\in GF(q)} q^{n-2}(-1)^{g(s)}(-1)^{g(t)} - 1\Big)\\
&\quad + \sum_{i=0}^{D-1}\Big(\sum_{s\in GF(q)} q^{n-1} - 1\Big)\Big)\\
&= \frac{1}{q^n-1}\Big(\sum_{0\le i\ne j<D}(q^{n-2}I_g^2 - 1) + \sum_{i=0}^{D-1}(q^n-1)\Big)\\
&= \frac{(D^2-D)(q^{n-2}I_g^2-1)}{q^n-1} + D.
\end{aligned}$$

It follows that $V(\mathcal{I}_{\mathbf{S}}(k,D)) = E[\mathcal{I}_{\mathbf{S}}(k,D)^2] - E[\mathcal{I}_{\mathbf{S}}(k,D)]^2 < D$, proving Theorem 14.

The last step is to choose the intervals U_q so that the bound on the variance, combined with Chebyshev's inequality, imply that the algorithm for finding q is successful with high probability. We assume that g is as balanced as possible (that is, $I(g) = \pm 1$). We find a positive number ϵ_q for each odd q.

Proposition 17. *Let* $\epsilon_q = (q^2 - 3)/(q^4 - 1)$ *and*

$$U_q = [\frac{q^3 - 1}{q^4 - 1} - \epsilon_q, \frac{q^3 + 1}{q^4 - 1} + \epsilon_q] = [\frac{q^3 - q^2 + 2}{q^4 - 1}, \frac{q^3 + q^2 - 2}{q^4 - 1}].$$

Then $\{U_q\}$ *are pairwise disjoint intervals. If* $\mathbf{S}$ *is based on an m-sequence over* $GF(q)$ *of span* ≥ 4, *then the interval of radius* ϵ_q *centered at* $E[I_{\mathbf{S}}(k, D)]$ *is contained in* U_q.

We can combine this with Chebyshev's inequality and our results on the variance of the partial imbalance to obtain the following theorem.

Theorem 18. *Let* $n \geq 4$, *and let* $\mathbf{S}$ *be an optimally balanced geometric sequence based on an m-sequence of span* n *over* $GF(q)$. *Then the algorithm for determining* q *using partial imbalances with a window* $D \leq \nu = (q^n - 1)/(q - 1)$ *will succeed with probability at least*

$$1 - \frac{(q^4 - 1)^2}{D(q^2 - 3)^2}.$$

Proof. The probability that the algorithm is successful is the probability that $I_{\mathbf{S}}(k, D)$ is in U_q. We have

$$\begin{aligned} Prob_k\{I_{\mathbf{S}}(k, D) \in U_q\} &\geq Prob_k\{|I_{\mathbf{S}}(k, D) - E[I_{\mathbf{S}}(k, D)]| < \epsilon_q\} \\ &\geq 1 - V_k(I_{\mathbf{S}}(k, D))/\epsilon_q^2 \text{ (by Chebyshev's inequality)} \\ &\geq 1 - D/\epsilon_q^2 \text{ (by Theorem 14)} \\ &\geq 1 - (q^4 - 1)^2/D(q^2 - 3)^2 \text{ (by the definition of } \epsilon_q). \end{aligned}$$

If $n = 5$ and $q \leq 5$, or if $n = 4$ this probability will be negative for all D, so Chebyshev's inequality does not tell us whether the attack has a positive probability of determining q. However, we have

Corollary 19. *If* $n = 5$ *and* $q \geq 7$, *or* $n \geq 6$, *then using a window of size*

$$D > \frac{(q^4 - 1)^2}{(q^2 - 3)^2}$$

gives a positive probability of successfully determining q.

For example, if $q = 27$, then 535,841 bits suffice to determine q with positive probability. Of course we want to determine q with high probability.

Corollary 20. *The probability that the algorithm is successful is greater than* δ *if we use a window size of*

$$D > \frac{(q^4-1)^2}{(1-\delta)(q^2-3)^2}.$$

This is possible if n *is large enough that*

$$\frac{q^n-1}{q-1} > \frac{(q^4-1)^2}{(1-\delta)(q^2-3)^2}.$$

For example, if $q = 27$ and $n \geq 6$, then we can determine q with probability at least 1/2 if 1,071,682 bits are known, a relatively small number.

5 Conclusions

Perhaps the most important aspect of this paper is the consideration, for a binary sequence, of linear complexity relative to an odd prime number. We have demonstrated that this linear complexity can be far smaller than the period of the sequence, even when the usual linear complexity is quite large. This can be exploited in a cryptologic attack. The belief that high linear complexity gives a degree of security is falacious. At the very least, the linear complexity must be high relative to all small primes.

We have shown that geometric sequences based on m-sequences over a finite field $GF(q)$ of odd characteristic p can be cracked if enough bits are known. By finding upper bounds on the linear complexity *relative to p* we show that these sequences are vulnerable to a Berlekamp-Massey type attack. The number of bits required depends on the parameters of the sequence (not simply the period). If geometric sequences of this type continue to be used, this dependence and considerations of efficiency should influence the choice of parameters. It seems that it is best to choose p fairly large and generate a sequence of period $p^p - 1$. This gives an easily generated sequence with linear complexity relative to p as large as possible for easily generated geometric sequences with approximately this period.

We have also shown that if p is not known, then it can be discovered with high probability if enough bits are known. The algorithm for determining p exploits the lack of balance in geometric sequences and uses a new statistical measure, the partial imbalance. In general far fewer bits are required to determine p than are required for the Berlekamp-Massey attack. For example, if we use a geometric sequence $\mathbf{S}$ based on an m-sequence of span 17 over $GF(17)$, so $n = p = q = 17$, then $\lambda_{17}(\mathbf{S})$ is approximately 1.1x2^{30}. The period of the sequence is approximately 1.4x2^{69}. Thus with 2.2x2^{30} bits available we can determine p with probability at least $1 - 2^{-14}$, and then determine a linear feedback shift register over $GF(17)$ that outputs $\mathbf{S}$. The drawback is that if $\lambda_{17}(\mathbf{S})$ is close to 2^{30}, then this feedback register will have span close to 2^{30} and will generate the sequence much more slowly than the original device using comparable hardware.

It is an interesting question whether the information we have acquired can be used to synthesize a faster device for generating the sequence.

The moral of this paper is that it is dangerous to rely on linear complexity as a measure of cryptographic security. There are many other statistical tests a sequence must pass – in this paper we have shown that the linear complexity relative to other primes must be high and the variance of the partial imbalance must be high if the imbalance is large.

6 Acknowledgements

The author would like to thank Mark Goresky for many helpful discussions, and Judy Goldsmith for valuable comments on the manuscript.

References

1. Bauer, H.: Probability Theory and Elements of Measure Theory, Holt, Rinehart and Winston, New York, 1972
2. Brynielsson, L.: On the Linear Complexity of Combined Shift Registers. Proceedings of Eurocrypt 1984 (1984) 156-160
3. Chan, A. and Games, R.: On the linear span of binary sequences from finite geometries, q odd. Advances in Cryptology: Proceedings of Crypto 1986, Springer-Verlag (1987) 405-417
4. Klapper, A., Chan, A. H., and Goresky, M.: Cross-Correlations of Linearly and Quadratically Related Geometric Sequences and GMW Sequences. Discrete Applied Mathematics (to appear)
5. Klapper, A. and Goresky, M.: Revealing information with partial period autocorrelations. Proceedings of Asiacrypt '91, Fujyoshida, Japan (1991)
6. Lidl, R. and Niederreiter, H.: Finite Fields, Encyclopedia of Mathematics vol. 20, Cambridge University Press, Cambridge, 1983
7. Massey, J.L.: Shift register sequences and BCH decoding. IEEE Trans. Info. Thy. **IT-15** (1969), 122-127
8. Simon, M., Omura, J., Scholtz, R., and Levitt, B.: Spread-Spectrum Communications, Vol. 1, Computer Science Press, 1985.
9. Welsh, D.: Codes and Cryptography, Clarendon Press, Oxford, 1988
10. Zierler, N. and Mills, W.: Products of linearly recurring sequences. Journal of Algebra **27** (1973) 147-157

A Fast Cryptographic Checksum Algorithm Based on Stream Ciphers

Xuejia Lai
Signal and Information Processing Laboratory
Swiss Federal Institute of Technology Zürich

Rainer A. Rueppel
R^3 Security Engineering
Bahnhofstr. 242, 8623 Wetzikon, Switzerland

Jack Woollven
Neuschwändi 2, 8496 Steg, Switzerland

Abstract. A design principle for the computation of a cryptographic checksum is proposed. Unlike most of the existing message authentication algorithms, the proposed scheme is based on stream cipher techniques and is non-iterative. In this scheme, a key stream sequence is used to demultiplex the message into two subsequences, which are then fed into two accumulating feedback shift registers to produce the checksum (also called message authentication code). The scheme is suitable for high-speed implementation and possesses valuable properties such as "perfect hashing", "perfect MAC" and complete key diffusion.

1 Introduction

A **cryptographic checksum algorithm** A_K is a hash function, parameterized by a secret key. It maps variable-length input strings (the messages) into fixed-length output strings (the authenticators). Before transmission of a message, its authenticator is computed using the secret key. Then both the message and the authenticator are sent to the intended recipient which then verifies the integrity of the message by again computing the authenticator using his local secret key and comparing it with the authenticator received. To offer cryptographic protection it must be computationally infeasible, given any number of messages and corresponding authenticators, to determine the value $A_K(x)$ for any new x. Cryptographic checksum algorithms are used to produce a fixed-length key-dependent external redundancy to any data whose integrity is to be protected. No additional protection of the checksum is required, since its computation requires knowledge of a secret key. Cryptographic checksum algorithms are also referred to as **keyed hash functions** to distinguish them from one-way hash functions which are publicly known (i.e. do not make use of a secret key). Integrity protection by means of a cryptographic checksum algorithm has been standardized (e.g., see American National Standards Institute (ANSI) X9.9 [1], ISO 8731 [8] and ISO 9797 [9]). In this context the cryptographic checksum is called a **message authentication code (MAC)**.

Most of the published keyed and non-keyed hash functions are iterative and belong to two main categories [13]. The first category, apparently motivated by the available secret-key block ciphers, consists of schemes based on block ciphers (see, e.g., [4, 9, 11, 13, 18]). The second category, apparently motivated by the available public-key cryptosystems [6, 14], consists of schemes based on modular arithmetic (see, e.g., [4, 10, 13]).

In this paper, we propose a design principle for the computation of a cryptographic checksum. This proposal differs from existing schemes in the following aspects: it is based on stream cipher techniques (see, e.g., [15, 16]) and it is non-iterative. A related scheme has been presented at Crypto'85 [5]. In our scheme, a running key sequence generated from the secret key is used to demultiplex a message sequence into two subsequences, which then are fed into two accumulating feedback shift registers to produce the checksum (MAC). We show that under certain conditions the proposed scheme achieves the following security properties:

- Perfect hashing: exactly the same number of messages hash to any value of the authenticator (Theorem 3).
- Perfect MAC property: the authenticator is statistically independent of the message for a "one-time" key (Theorem 6).
- Complete key diffusion: each bit of the running key sequence influences every bit of the authenticator in a highly nonlinear way (Lemma 8).
- In a known plaintext attack, determining the secret key for the proposed scheme is at least as hard as determining the secret key in a conventional stream cipher.

In general, the use of stream-cipher techniques leads to efficient implementations. The cryptographic checksum algorithm is described in Section 2. Section 3 gives an introduction to message authentication. The basic properties of the proposed scheme are discussed in Section 4. The security of the scheme against various attacks is considered in Section 5.

2 The Cryptographic Checksum Algorithm

The model of the cryptographic checksum algorithm is shown in Fig.1.

Let $x^n = x_0, x_1, \ldots, x_{n-1}$ be the message whose cryptographic checksum (MAC) we wish to compute. Let k denote the key which is used as the seed of a keystream generator (KSG) [15, 16]. The resulting keystream $z^n = z_0, z_1, \ldots, z_{n-1}$ is used to demultiplex the message into two subsequences $u_1^{n_1}$ and $u_2^{n_2}$ which are fed into two accumulating feedback shift registers FSR_1 and FSR_2. If $z_i = 1$ then x_i is routed to the upper FSR_1, otherwise x_i is routed to the lower FSR_2. The two FSRs have lengths m_1 and m_2, feedback functions f_1 and f_2, and initial states $y_0^{(1)}$ and $y_0^{(2)}$, respectively. They implement the recursions

$$y_i^{(1)} = u_i^{(1)} \oplus f_1(y_{i-1}^{(1)}, \ldots, y_{i-m_1}^{(1)}), \quad 0 \le i \le n_1 - 1; \tag{1}$$

$$y_i^{(2)} = u_i^{(2)} \oplus f_2(y_{i-1}^{(2)}, \ldots, y_{i-m_2}^{(2)}), \quad 0 \le i \le n_2 - 1. \tag{2}$$

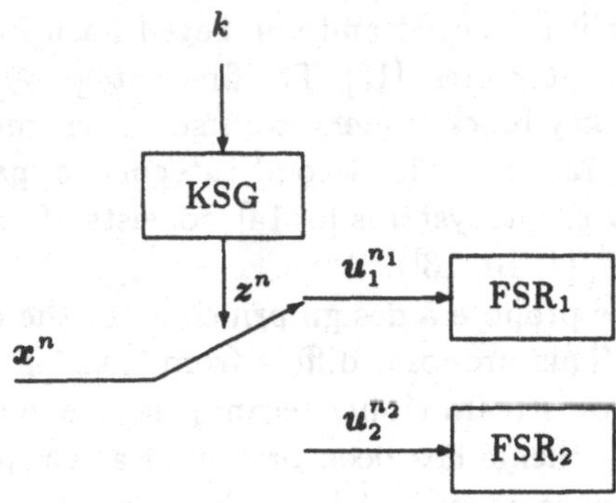

Fig. 1. Computation of the cryptographic checksum using a keystream generator and two feedback shift registers

The final states of the FSRs,

$$w_1^{m_1} = (y_{n_1-1}^{(1)}, \ldots, y_{n_1-m_1}^{(1)}) \text{ and } w_2^{m_2} = (y_{n_2-1}^{(2)}, \ldots, y_{n_2-m_2}^{(2)})$$

serve as the cryptographic checksum (MAC).

The initial states of the two FSRs are used as part of the secret key and we require that the keystream generator KSG be used securely as in a conventional stream cipher system, i.e., no keystream z^n is to be used more than once.

3 Message Authentication

In this section a short description is given of what a message authentication scheme is, how it is used, and what the relevant attacks are (see also [2]).

The purpose of message authentication is to *detect the presence of an active opponent.* A message authentication scheme consists of a key space $\mathcal{K}$, a message space $\mathcal{M}$ and an authenticator (tag) space $\mathcal{T}$ such that for each $k \in \mathcal{K}$ there is an authenticator function $A_k : \mathcal{M} \to \mathcal{T}$. [Note that the $\mathcal{M}$ and $\mathcal{T}$ are independent of k for the reason that the choices of the message space and authenticator space should not release information about the key.] Given any key k, it must be easy to obtain an efficient algorithm for computing A_k.

Use of the system

1. Two parties A and B agree on a secret key k.
2. When party A wants to provide authentication for a message $x \in \mathcal{M}$ then A computes the message authenticator $w = A_k(x)$ and sends it along with the message x.
3. To verify the authenticity of the received message x' B computes $A_k(x')$ and compares it with the w' received.

Attacks against the system
It is assumed that sender and receiver trust each other and want to protect against outsider attacks. The relevant attacks are:

1. Known message attack: the opponent has seen $(x_1, w_1),(x_2, w_2), \ldots,(x_i, w_i)$ such that $w_j = A_k(x_j)$ for $1 \leq j \leq i$ for an unknown key k. He succeeds if he can
   (a) find the secret key k or, lacking this ability,
   (b) can determine $w_{i+1} = A_k(x_{i+1})$ for a message of his choice x_{i+1} or, lacking this ability,
   (c) can determine $w_{i+1} = A_k(x_{i+1})$ for any message x_{i+1} distinct from x_1, ..., x_i or, lacking this ability,
   (d) can figure out a pair $(x_{i+1}. w_{i+1})$ for a new x_{i+1} such that w_{i+1} has non-negligible probability of being $A_k(x_{i+1})$.
2. Chosen message attack: the opponent is allowed to choose $x_1, \ldots, x_i$ and is given the corresponding authenticators $w_j = A_k(x_j)$ for $1 \leq j \leq i$. Perform any of the above attacks.

4 Properties of the proposed scheme

Definition. Let the binary FSR of length m have feedback function $f : \mathbb{Z}_2^m \rightarrow \mathbb{Z}_2$ and initial state $y_0 = (y_{-1}, \ldots, y_{-m})$. Consider its use as a scrambler as shown in Fig.2, satisfying the recursion

$$y_i = u_i \oplus f(y_{i-1}, \ldots, y_{1-m}) \quad 0 \leq i \leq n-1 \tag{3}$$

where $u_0, u_1, \ldots, u_{n-1}$ denotes the data sequence to be scrambled. Let

$$F_{y_0} : \mathbb{Z}_2^n \rightarrow \mathbb{Z}_2^m$$

be the hashing function which maps the data sequence u^n into the fixed-length digest value $w^m = (y_{n-1}, \ldots, y_{n-m})$, the final state of the FSR.

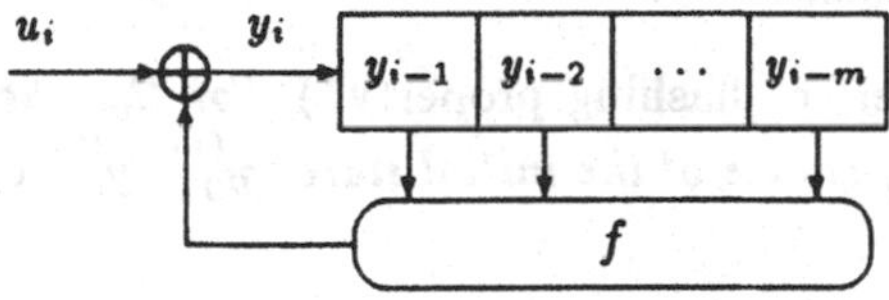

Fig. 2. The scrambler obtained from a FSR.

Lemma 1. *For each choice of the initial state y_0 and $n \geq m$, every digest value w^m in $\mathbb{Z}_2^m$ has exactly 2^{n-m} preimages, that is,*

$$|\{u^n : F_{y_0}(u^n) = w^m\}| = 2^{n-m}.$$

In other words, this scrambler produces a "perfect hashing". Note that we did not make any assumption about the feedback function.

Proof: Obviously, for every data sequence u^n there is a scrambled sequence $y^n = y_0, \dots, y_{n-1}$. Now suppose two different data sequences u^n and u'^n yield the same scrambled sequence y^n. Let i be the first position where u^n differs from u'^n, that is, $u^i = u'^i, u_i \neq u'_i$. Hence y^i is equal for u^n and u'^n, and

$$\begin{aligned} y_i &= u_i \oplus f(y_{i-1}, \dots, y_{i-m}) \\ &= u'_i \oplus f(y_{i-1}, \dots, y_{i-m}) \end{aligned}$$

implies that $u_i = u'_i$, which is a contradiction. Therefore, for any initial state y_0, there is a one-to-one relation between data sequence u^n and scrambled sequence y^n. Thus, 2^n distinct sequences u^n yield 2^n distinct sequences y^n, among them there are exactly 2^{n-m} (for $n \geq m$) sequences y^n with the last m bits $y_{n-1}, \dots, y_{n-m}$ being the given digest w^m. It then follows that every digest w^m has 2^{n-m} preimage sequences u^n. □

Lemma 2. *Given the digest w^m in $\mathbb{Z}_2^m$ and initial state y_0 in $\mathbb{Z}_2^m$, it is easy to find a sequence u^n such that*

$$F_{y_0}(u^n) = w^m.$$

In other words, the hashing function obtained from such a scrambler is not "one-way".

Proof: Pick an arbitrary $n-m$ bit sequence $y_0, \dots, y_{n-m-1}$. Append the digest w^m to obtain an n-bit sequence y^n. Initialize the descrambler with y_0 and "descramble" y^n by computing

$$u_i = y_i \oplus f(y_{i-1}, \dots, y_{i-m}) \quad 0 \leq i \leq n-1.$$

The resulting sequence u^n, when fed into the scrambler, produces the given digest value w^m. Moreover, for any given initial state y_0 and digest w^m, from the 2^{n-m} choices of the sequence $y_0, \dots, y_{n-m-1}$ one can obtain all 2^{n-m} possible u^n that yield the same w^m. □

Theorem 3. ("Perfect hashing property") *For the checksum algorithm shown in Fig.1, for every choice of the initial states $y_0^{(1)}$, $y_0^{(2)}$ of the two FSRs of length m_1, m_2,*

$$\#\{x^n;\quad x^n \text{ produces authenticator } w^m\} = \frac{2^n}{2^m} \text{ for all } w^m$$

if $n_1 =$ the Hamming weight of $z^n \geq m_1$ and $n - n_1 \geq m_2$.

Proof: With the given keystream z^n with Hamming weight $W_H(z^n) = n_1$, we know that n_1 digits are routed to the upper register, and $n_2 = n - n_1$ digits are routed to the lower register. For a given keystream, it is easy to multiplex the two data sequences to produce a message x^n. The Theorem then follows by invoking the above two Lemmas. □

Lemma 4. *For the hashing function $F_{y_0}(\cdot)$ with feedback function f, the initial state y_0 is uniquely determined by the data sequence u^n and the digest w^m $(n > m)$ if and only if the state transition diagram of the FSR shown in Fig.3 contains no transients (or roots) with length more than m, that is, if and only if there is no choice of $(y_0, ... y_{m-1})$ and y_0, y'_0, such that*

$$\begin{aligned} f(y_{-1}, y_{-2}, ..., y_{-m}) &\neq f(y'_{-1}, y'_{-2}, ..., y'_{-m}) \\ f(y_0, y_{-1}, ..., y_{-m+1}) &= f(\overline{y_0}, y'_{-1}, ..., y'_{-m+1}) \\ f(y_1, y_0, y_{-1}, ..., y_{-m+2}) &= f(y_1, \overline{y_0}, y'_{-1}, ..., y'_{-m+2}) \\ &\vdots \\ f(y_{m-2}, ..., y_0, y_{-1}) &= f(y_{m-2}, ..., \overline{y_0}, y'_{-1}) \\ f(y_{m-1}, ..., y_1, y_0) &= f(y_{m-1}, ..., y_1, \overline{y_0}). \end{aligned} \tag{4}$$

Note that the first inequality implies $y_0 \neq y'_0$.

Proof: If there exist such y_0, y'_0 and $y_0, ..., y_{m-1}$ satisfying (4), then the data sequence u^m, obtained from

$$u_i = y_i \oplus f(y_{i-1}, ..., y_{i-m}) \quad 0 \le i \le m-1,$$

will yield the same sequence $y_1, ..., y_{m-1}$ for both initial states y_0 and y'_0. Any sequence u^n, $(n > m)$ with the first m bits as above will yields the same digest w^m from these two different initial states. Conversely, if there are two distinct initial states y_0 and y'_0 and u^n $(n > m)$ such that the resulting digests are equal, then for the two sequences y^n and y'^n, there exists an n_0, $0 \le n_0 < n-m$, such that $y_{n_0} \neq y'_{n_0}$ but $y_j = y'_j$, for $j = n_0+1, ..., n_0+m, ..., n$. It follows from (3) that

$$\begin{aligned} f(y_{n_0-1}, y_{n_0-2}, ..., y_{n_0-m}) &\neq f(y'_{n_0-1}, y'_{n_0-2}, ..., y'_{n_0-m}) \\ f(y_{n_0}, y_{n_0-1}, ..., y_{n_0-m+1}) &= f(y'_{n_0}, y'_{n_0-1}, ..., y'_{n_0-m+1}) \\ f(y_{n_0+1}, y_{n_0}, ..., y_{n_0-m+2}) &= f(y_{n_0+1}, y'_{n_0}, ..., y'_{n_0-m+2}) \\ &\vdots \\ f(y_{n_0+m-1}, ..., y_{n_0+1}, y_{n_0}) &= f(y_{n_0+m-1}, ..., y_{n_0+1}, y'_{n_0}), \end{aligned} \tag{5}$$

that is, we have found a solution for (4). □

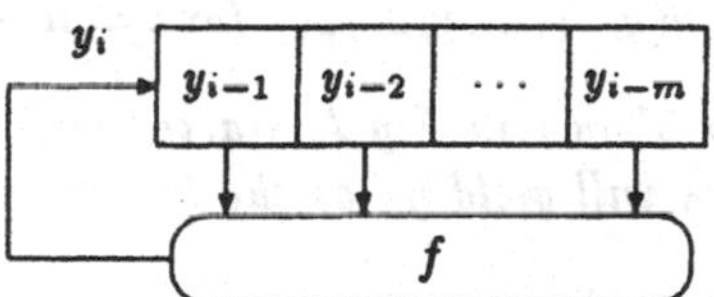

Fig. 3. Feedback shift register with feedback function f.

The condition of Lemma 4 is equivalent to the condition that, for any y_0, a periodic data sequence u^n always yields a periodic scrambled sequence y^n (see [12]). We will call a scrambler satisfying the conditions of Lemma 4 a **p-scrambler**.

Lemma 5. *For the checksum algorithm shown in Fig.1, for any given* $u_1^{n_1}$ *and* $u_2^{n_2}$, *if the initial state* $(y_0^{(1)}, y_0^{(2)})$ *is chosen with uniform probability from* $\mathbb{Z}_2^{m_1+m_2}$ *and if the two FSRs are p-scramblers, then the checksum* $(w_1^{m_1}, w_2^{m_2})$ *takes on values in* $\mathbb{Z}_2^{m_1+m_2}$ *with uniform probability.*

Proof: It follows from Lemma 4 that for any choice of $u_1^{n_1}$ and $u_2^{n_2}$ the two p-scramblers determine a bijective mapping between $(y_0^{(1)}, y_0^{(2)})$ and $(w_1^{m_1}, w_2^{m_2})$. □

We will call a checksum algorithm *"perfect MAC"* if the authenticator is statistically independent of the message. [This is similar to the "perfect secrecy" property for a cipher system, which is defined by Shannon in [17] as that the ciphertext is statistically independent of the plaintext.] Note that for a given z^n, there is a one-to-one correspondence between message x^n and $(u_1^{n_1}, u_2^{n_2})$, we obtain the following result from the above Lemma.

Theorem 6. ("Perfect MAC" property) *If the initial states of the two FSRs are chosen independently and uniformly at random and the two FSRs are p-scramblers, then the authenticator is statistically independent of the message .*

Note that if the initial state is kept secret, then even knowledge of the keystream z^n to be used for the computation of the authenticator does not provide any clue whatsoever about the resulting value of the authenticator.

A sufficient condition for a scrambler to be p-scrambler is that the state transition diagram of the FSR in Fig.3 consists of only cycles, i.e., all transients have length zero. This is further equivalent to the condition stated in the following lemma.

Lemma 7. *If the feedback function f is* non-singular *[7], i.e., if f has the form*

$$f(x_m, ..., x_2, x_1) = g(x_m, ..., x_2) \oplus x_1$$

where $g : \mathbb{Z}_2^{m-1} \mapsto \mathbb{Z}_2$ *is a binary function with* $m-1$ *variables, then the initial state* y_0 *can be computed from* u^n *and* w^m $(n \geq m)$ *recursively as follows:*

$$y_{i-m} = u_i \oplus y_i \oplus g(y_{i-1}, ..., y_{i-m+1}) \text{ for } i = n-1, n-2, \ldots, 1, 0.$$

That is, the descrambler shown in Fig.4, started with initial state w^m *and fed with sequence* $u_{n-1}, ..., u_0$ *will yield* y_0 *as the last m bits of output.*

The following result implies that each bit of the keystream will influence every bit of the authenticator in a "highly-nonlinear" way.

Lemma 8. ("Complete key diffusion")
Each bit of the sequence $u_1^{n_1}$ *depends on every bit of* z^n.
Let n_1 *be the Hamming weight of the keystream* z^n, *then the dependence of* $u_1^{n_1} = (u_0, u_1, \cdots, u_{n_1-1})$ *on* z^n *can be described in the following way: if* $n_1 \geq 1$, *then*

$$u_0 = x_0 z_0 \oplus x_1 \bar{z}_0 z_1 \oplus x_1 \bar{z}_0 \bar{z}_1 z_2 \oplus \cdots \oplus x_{n-1} \bar{z}_0 \bar{z}_1 \cdots \bar{z}_{n-2} z_{n-1};$$

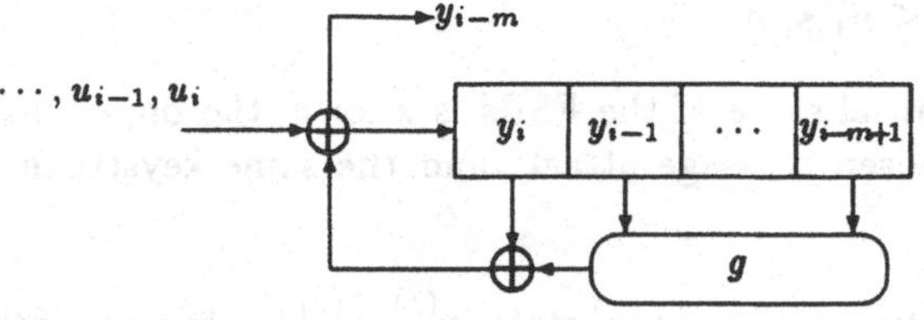

Fig. 4. The descrambler computing y_0 from u^n and w^m.

in general, if $n_1 \geq i+1$ $(1 \leq i \leq n-1)$, *then*

$$u_i = x_i z_0 z_1 \cdots z_i \oplus x_{i+1} s_i(z_0 z_1 \cdots z_i) z_{i+1} \oplus \cdots x_{n-1} s_i(z_0 z_1 \cdots z_{n-2}) z_{n-1},$$

where

$$s_i(z_0 z_1 \cdots z_l) = \begin{cases} 1 & \text{if} \quad W_H(z_0 z_1 \cdots z_l) = i \\ 0 & \text{otherwise.} \end{cases}$$

Proof: Note that the $(l+1)$-th bit of x^n, x_l, becomes the $(i+1)$-th bit of $u_1^{n_1}$, u_i $(i \leq l)$, if and only if the first l bits of keystream z^n contain i ones and $z_l = 1$.

5 Attacks

In this section, we consider some attacks on the proposed scheme. In the following discussion, we assume that the attacker knows the feedback functions f_1 and f_2 of the two FSRs. To simplify the analysis we will often assume that the same keystream sequence is used repeatedly, although the secure use of the system (as well as of any stream cipher) requires that any used portion of the keystream sequence is to be discarded immediately (or, at least prevented from further use).

Suppose the keystream generator is always started in the same position, that is, the keystream sequence z^n is used repeatedly. From a known message x^n and its checksum $(w_1^{m_1}, w_2^{m_2})$, one can find a new valid message and checksum with probability 1/2 by changing the last bit x_{n-1} in x^n and the last bit in either $w_1^{m_1}$ or $w_2^{m_2}$.

Suppose the initial state of the FSRs is known and the opponent is given the possibility of a chosen message attack.

Lemma 9. *Suppose the initial state* $(y_0^{(1)}, y_0^{(2)})$ *is known. Choose* $x^n = o^n$. *Then, from* $(y_0^{(1)}, y_0^{(2)})$ *and* $(w_1^{m_1}, w_2^{m_2})$ *one can determine the Hamming weight of the keystream, i.e., the number of 1's in the keystream* z^n, $n_1 = W_H(z^n)$.

Proof: With $x^n = o^n$ there is nothing to scramble. The Hamming weight of the key, $W_H(z^n)$, simply determines the number of iterations of FSR$_1$. With given initial and final state one can determine the number of steps by searching the state diagram beginning with $y_0^{(1)}$ and testing for occurrence of the state $w_1^{m_1}$

in the range $0 \leq n_1 \leq n$. □

Suppose the initial state of the FSRs is known, the opponent is given the possibility of a chosen message attack, and the same keystream sequence is used repeatedly.

Lemma 10. *Suppose the initial state $(y_0^{(1)}, y_0^{(2)})$ is known. If the same keystream is used repeatedly for different messages, then, in a chosen message attack, one can unravel the key sequence bit by bit.*

Proof: Choose $x^n = o^n$ and compute $n_1 = W_H(z^n)$ from the obtained cryptographic checksum $(w_1^{m_1}, w_2^{m_2})$ by using Lemma 9.
Now choose $x^{n-1} = o^{n-1}$. One of the digests will change (with high probability) since the corresponding FSR did one step less. If it is $w_1^{m_1}$ then $z_{n-1} = 1$, otherwise $z_{n-1} = 0$.
Now choose $x^{n-2} = o^{n-2}$. As compared to the checksum of the message $x^{n-1} = o^{n-1}$, one of the digests will change (with high probability) since the corresponding FSR did one step less. If it is $w_1^{m_1}$ then $z_{n-2} = 1$, otherwise $z_{n-2} = 0$.
Continue this procedure down to the minimum allowable length of a message. From then on, by setting $x_i = 1$ for $i = 0, \ldots, l-1$, one can determine the ith bit of the key sequence. □

Suppose the initial state of the FSRs is known, the opponent is given the possibility of a chosen message attack, but the keystream generator is used securely, that is, different portions of the keystream are used for the computation of subsequent authenticators. In this case, the best attack found so far is to choose $x^n = 10...0$ and then to determine the Hamming weight of z^n and the last bit of z^n from the known checksum and initial state. But provided that a cryptographically secure keystream generator is employed, no advantage can be gained in predicting future keystream bits based on the available information.

Note that determining the secret key in a known plaintext (i.e., the message) attack is at least as hard for the proposed scheme as for a conventional stream cipher. In our scheme the opponent cannot uniquely determine the keystream, while in a conventional stream cipher system knowledge of the plaintext implies knowledge of the keystream.

6 Summary

In this paper a principle for computing cryptographic checksums has been proposed. Unlike most of the existing proposals, this scheme is based on stream cipher techniques and is non-iterative. It allows for very high speed implementations, in particular, since the keystream can be generated offline. The properties and the security of the scheme have been analyzed under various conditions. The scheme has been shown to possess valuable properties such as "perfect hashing", "perfect MAC" and complete key diffusion.

References

1. ANSI X9.9-1986, *Financial Institution Message Authentication* (Wholesale), American Bankers Association, Washington DC, 1986.
2. G. Brassard, *Modern Cryptology*, LNCS 325, Springer-Verlag, 1988.
3. *Data Encryption Standard*, FIPS PUB 46, National Tech. Info. Service, Springfield, VA, 1977.
4. I. B. Damgaard, "A Design Principle for Hash Functions", Advances in Cryptology-CRYPTO'89, LNCS 435, pp. 416-427, Springer-Verlag, 1990.
5. Yvo Desmedt, "Unconditional secure authentication schemes and practical and theoretical consequences", Advances in Cryptology-CRYPTO'85, LNCS 218 pp. 42-55, Springer-Verlag, 1986.
6. T. El Gamal, "A Public Key Cryptosystem and a Signature Scheme based on Discrete Logarithms", IEEE Trans. on Info. Th., Vol. IT-31, pp. 469–472, 1985.
7. S. W. Golomb, *Shift Register Sequences*, Holden-Day, San Francisco, 1967.
8. ISO 8731-1, "Banking - Approved Algorithms for Message Authentication - Part 1 : DEA", International Organization for Standardization, 1987.
9. ISO/IEC 9797, "Data cryptographic techniques - Data integrity mechanism using a cryptographic check function employing a block cipher algorithm", International Organization for Standardization, 1-st. edition, 1989.
10. R. R. Jueneman, "A High-Speed Manipulation Detection Code", Advances in Cryptology-CRYPTO'86, Proceedings, pp. 327–346, Springer-Verlag, 1987.
11. X. Lai, "On the Design and Security of Block Ciphers", ETH series in Information Processing (Ed. J.L.Massey), Vol.1, Hartung-Gorre Verlag, Konstanz, 1992.
12. X. Lai and J. L. Massey, "Some Connections between Scramblers and Invertible Automata", Proceedings of Beijing International Workshop on Information Theory (BIWIT'88), p. DI-5.1, International Academic Publishers, July, 1988.
13. C. J. Mitchell, F. Piper and P. Wild, "Digital Signatures", *Contemporary Cryptology* (Ed. G. Simmons), pp. 325-378, IEEE Press, 1991.
14. R. L. Rivest, A. Shamir and L. Adleman, "A Method for obtaining digital Signatures and Public-Key Cryptosystems", Communications of the ACM 21 (1978), pp. 120–126.
15. R. A. Rueppel, *Analysis and Design of Stream Ciphers*, New York, NY, Springer-Verlag, 1986.
16. R. A. Rueppel, "Stream Ciphers", *Contemporary Cryptology: The Science of Information Integrity* (Ed. G. Simmons), pp. 65-134, IEEE Press, 1991.
17. C. E. Shannon, "Communication Theory of Secrecy Systems", Bell. System Technical Journal, Vol. 28, pp. 656-715, Oct. 1949.
18. R. S. Winternitz, "Producing One-Way Hash Function from DES", Advances in Cryptology-CRYPTO'83, Proceedings, pp. 203-207, Plenum Press, New York, 1984.

An Approach to the Initial State Reconstruction of a Clock-Controlled Shift Register Based on a Novel Distance Measure

Miodrag J. Mihaljevic*

Institute of Applied Mathematics and Electronics, Institute of Mathematics
Academy of Arts and Sciences, 11001 Belgrade, Yugoslavia
Mailing Address: Solina 4, 11040 Belgrade, Yugoslavia

Abstract. The initial state reconstruction problem of a clock-controlled shift register is considered when the characteristic polynomial, a segment of the output sequence and the probability of ones in the clock sequence are known. This problem is more general than the considered one (in [2]), and it is solved using a quite different approach. A novel distance measure for comparison of two different length binary sequences is proposed and its main characteristics relevant for the cryptanalysis is derived. An algorithm for the cryptanalysis based on the proposed distance measure is presented and its main characteristics are pointed out. Expected minimal length of the observed sequence for the unique initial state solution is estimated. Illustrative numerical examples are included.

Key words and phrases: Cryptanalysis, Key-stream generators, Clock-controlled shift register, Correlation attack, Sequence comparison, Distance measures, Algorithms.

1 Introduction

Clock-controlled shift registers have become popular building blocks for the key-stream generators. A review of the clock-controlled shift registers is presented in [1]. The object of this paper is the initial state, X_0, reconstruction of a given clock-controlled linear feedback shift register ($LFSR$) when it is clocked by an unknown sequence. Let $\{x_n\}$ be an output sequence segment from the regularly clocked $LFSR$ assuming that the initial state is X_0, and let $\{y_n\}$ be the clock-controlled $LFSR$ output sequence

$$y_n = x_{f(n)}, \qquad f(n) = n + \sum_{i=1}^{f(n)} s_i, \qquad n = 1, 2, \cdots. \tag{1}$$

where $\{s_n\}$ is the clock-controlled binary sequence such that each one in this sequences produces an extra non-observable $LFSR$ shift. An equivalent model of the clock-controlled $LFSR$ sequences we are considering is displayed in Figure

* This research was supported by the Science Fund, grant #0403, through the Institute of Mathematics, Academy of Arts and Sciences.

1 where $\{s_n\}$ plays the role of the decimation sequence. The decimation box forwards the input bit to the output only when $s_n = 0$. Consequently, $\{y_n\}$ can be seen as a sequences obtained by deleting a number of bits, in arbitrary positions, from $\{x_n\}$. In the statistical model, $\{s_n\}$ is regarded as a realization of the sequence of independent identically distributed (i.i.d.) binary variables $\{S_n\}$ such that $Pr(S_n = 1) = p$ for each n.

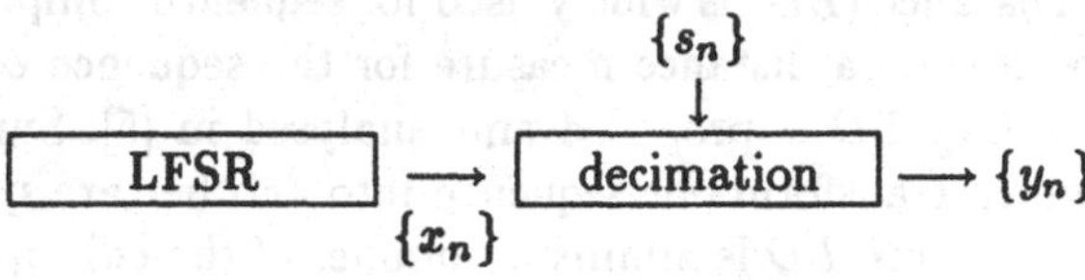

Fig. 1. A model of the clock-controlled shift register sequence.

In this paper, the cryptanalytical problem is to reconstruct the initial state X_0, reconstruction assuming that the $LFSR$ characteristic polynomial, the probability $p \leq 0.5$ of ones in $\{s_n\}$, and the segment $\{y_n\}_{n=1}^{N}$ are known.

A similar problem is considered in [2] but only in the constrained case when successive deletions in the decimation box are not allowed except for an arbitrary number of deletions at the end of $\{x_n\}_{n=1}^{M}$. Here we address a more general problem. The algorithm [2] for the $LFSR$ initial state reconstruction is based on the estimation of an upper bound on the number of different sequences which can be embedded into a given sequence. In this paper a quite different and general approach is employed. Note that the modified procedure [3] could also be employed to solve the problem, but it is impractical here, due to its complexity, because it is related to a more complicated situation where not only deletions but also complementations are involved to produce a transformation of an ordinary $LFSR$ sequence.

In this paper an algorithm for the cryptanalysis is proposed based on a novel sequence comparison approach which is relevant for the comparison of two binary sequences when the shorter sequence is obtained from the longer one by deletion of a number of bits, or both sequences are uncorrelated. This approach ensures analytical determination of the relevant probability distributions which are important for the cryptanalysis, and it enables estimation of the minimal length of the observed sequence $\{y_n\}_{n=1}^{N}$ required for the unique reconstruction of the initial state.

The basis for the problem solution is given in Section 2. A novel sequence comparison approach is proposed, and its relevant characteristics are derived in Section 3. An algorithm, based on the proposed sequence comparison approach, for the clock-controlled shift register initial state reconstruction is presented in Section 4. The novel algorithm characteristics are discussed in Section 5, and the conclusions are given in Section 6.

2 Preliminaries

Suppose we have defined a suitable distance measure d between two binary sequences of different length, which reflects the transformation of the $LFSR$ sequence $\{x_n\}$ to the output sequence $\{y_n\}$. Here, the word *distance* is not restricted only to indicating a function which satisfies the metric conditions. The Levenshtein Distance (LD) is widely used for sequence comparison purposes (see [4], for example). As a distance measure for the sequence comparison and error-correction coding LD is proposed and analysed in [5]. Suppose that the edit operations which transform one sequence into another are symbol deletion, insertion and substitution. LD is minimum number of the edit operations which transform one sequence into another.

Given an observed segment $\{y_n\}_{n=1}^N$, the optimal decision strategy (yielding the minimum probability of decision error) is to decide on the initial state with the maximum posterior probability. However, when the $LFSR$ is clocked irregularly it is not clear how to find an optimum decision rule. Anyway, given an appropriate distance measure, we can define a decision procedure that is close to being optimal.

Let $\{\hat{x}_n\}_{n=1}^M$ be an $LFSR$ sequence corresponding to an initial state $\hat{X}_0$, where $M \geq N$. Let d be the distance between $\{\hat{x}_n\}_{n=1}^M$ and $\{y_n\}_{n=1}^N$. Then two hypothetical cases are possible:

H_0 : the observed sequence $\{y_n\}_{n=1}^N$ is not produced by $\hat{X}_0$;

H_1 : the observed sequence $\{y_n\}_{n=1}^N$ is produced by $\hat{X}_0$.

Consequently, d is a realisation of a random variable D with two possible probability distributions (statistically averaged over the set of all the initial states): $\{Pr(D \mid H_0)\}$ and $\{Pr(D \mid H_1)\}$.

Suppose that they are known. Note that they depend on N, assuming that $M = M(N)$. First determine the threshold t and length N so as to achieve the given probabilities of *the missing event* P_m and the *false alarm* P_f. As in [3] P_m could be chosen close to zero (for example 10^{-3}) and P_f picked very close to zero (for example $P_f \cong 2^{-L}$), so that the expected number of false alarms is very small $\cong 1$). Then, the decision procedure goes through the following steps, for every possible initial state $\hat{X}_0$:

- generate $\{\hat{x}_n\}_{n=1}^M$,
- calculate the distance d between $\{\hat{x}_n\}_{n=1}^M$ and $\{y_n\}_{n=1}^N$,
- using the threshold t accept H_0 or H_1.

The output of the procedure is the set of the most probable candidates for the true initial state.

Accordingly, the distance measure should be defined so that it enables statistical discrimination between the two cases: the first, when $\{\hat{x}_n\}_{n=1}^M$ and $\{y_n\}_{n=1}^N$ are picked at random, uniformly and independently (which is a reasonable model for H_0), and the second, when $\{y_n\}_{n=1}^N$ is obtained from $\{\hat{x}_n\}_{n=1}^M$ according to (1), that is, by the deletion of some bits.

3 A Sequence Comparison Approach

In this section a sequence comparison approach which employs a novel distance measure is proposed. The distance measure definition originates from the *LD* calculation procedure proposed and discussed in [6, 7].

Denote by $LD(\{a_i\}_{i=1}^{M}, \{b_i\}_{i=1}^{N})$ the *LD* between two arbitrary binary sequences $\{a_i\}_{i=1}^{M}$ and $\{b_i\}_{i=1}^{N}$, $M \geq N$.

The novel distance measure d^*, for a given positive integer Δ, is defined by the following.

Definition 1. The distance d^* is given by:

$$d^* = max\{\ell \mid LD(\{a_i\}_{i=1}^{\Delta+\ell}, \{b_j\}_{j=1}^{\ell}) = \Delta\}.$$

Note that d^* can be efficiently calculated using the procedure for the Levenshtein distance calculation proposed in [6]. In the binary case when only deletions and substitutions of the symbols are allowed, the main part of the algorithm [6] is generation of a $(M-N)\times N$-dimensional matrix $u = [u(i,j)]_{i=0}^{M-N}\,_{j=0}^{N}$ such that:

$$\begin{aligned}
u(i,0) &= i & i &= 0,1,\cdots,M-N,\\
u(0,j) &= u(0,j-1) + (a_j \oplus b_j), & j &= 1,2,\cdots,N,\\
u(i,j) &= min\{[u(i-1,j)+1],\\
&\quad [u(i,j-1) + (a_{i+j} \oplus b_j)]\}, & i &= 1,2,\cdots,M-N,\ j = 1,2,\cdots,N.
\end{aligned} \tag{2}$$

It can be directly shown that

$$d^* = \sum_{j=1}^{N} \delta_{u(M-N,0),u(M-N,j)}, \tag{3}$$

where δ denotes Kronecker symbol $\delta_{k,\ell} = \begin{cases} 1 \text{ for } k=\ell \\ 0 \text{ for } k \neq \ell. \end{cases}$

On the other hand d^* could be considered as the realisation of the stochastic integer variables D^*. The following theorems establish the relevant probability distributions of D^*.

Theorem 2. *Suppose that $\{a_i\}_{i=1}^{M}$ and $\{b_i\}_{i=1}^{N}$ are realisations of the sequences of i. i. d. equally probably binary variables, and denotes this case as H_0. Then*

$$Pr(D^* = d \mid H_0) =$$

$$\frac{f(d)}{\left[\sum_{i=0}^{N-1} 2^{-(M-N+i)} \binom{M-N+i}{M-n}\right] + \left[\sum_{i=0}^{M-N} 2^{-(N+i)} \binom{N+i}{i}\right]}, \tag{4}$$

where

$$f(d) = \begin{cases} 2^{-(M-N+d)} \binom{M-N+d}{d}, & d \leq N-1 \\ \sum_{i=0}^{M-N} 2^{-(N-i)} \binom{N+i}{i}, & d = N \end{cases} \tag{5}$$

The proof of this Theorem is given in the appendix. On the other hand it can also be shown that the following theorem holds.

Theorem 3. *Let $\{b_i\}_{i=1}^N$ be produced from $\{a_i\}_{i=1}^M$ by deletion of $M-N$ bits and denote this case as H_1. then*

$$Pr(D^* = d \mid H_1) = \begin{cases} 1 \text{ for } d = N \\ 0 \text{ for } 0 < d \leq N-1. \end{cases} \tag{6}$$

4 Algorithm for the Cryptanalysis

In this section an algorithm for the cryptanalysis based on the defined distance measure d^* is proposed.

Assume that the following decision rule is applied for a shift register initial state reconstruction:

- accept H_1 if $d^* = N$,
- otherwise accept H_0.

Then, note that the missing event probability P_m is given by

$$P_m = Pr(\sum_{i=1}^{M} S_i > M-N) = \sum_{i=M-N+1}^{M} \binom{M}{i} p^i (1-p)^{M-i}. \tag{7}$$

Algorithm

INPUT: $\{y_n\}_{n=1}^N$, p, and the $LFSR$ characteristic polynomial.

Phase I Initialisation:

For given N, P, and adopted P_m using (7), determine minimal M.

Phase II Hypothesis testing:

For every possible initial state repeat the following steps:

Algorithm 1 *9*

Assume a new initial state $\hat{X}_0$ different from those previously considered and generate the corresponding sequence $\{\hat{x}_m\}_{m=1}^M$;
Using the definition calculate the distance measure d^* between $\{\hat{x}_m\}_{m=1}^M$ and $\{y_n\}_{n=1}^N$
If $d^* = N$ preserve the current $\hat{X}_0$ as a possible candidate for the solution.

end

Phase III Final processing:

Using an appropriately adjusted method choose a solution from the candidates selected or decide that the solution has not been found.

OUTPUT: The solution X_0, or the decision that the unique solution has not been found.

5 Discussion

In this section the main characteristics of the proposed algorithm are pointed out, and illustrative numerical examples are presented.

The number of hypotheses which had to be tested in the algorithm Phase II equalled 2^L, where L is the shift register length. On the other hand, according to the definition and [6], for given N, the number of elementary operations required for hypothesis testing is proportional to $M - N$. So, the algorithm employs the total number of elementary operations which is proportional to $2^L(M - N)$.

Let us now consider the expected number $\overline{No_f}$ of false candidates that have to be processed in the Phase III. Obviously,

$$\overline{No_f} = (2^L - 1)P_f \tag{8}$$

where P_f is the false alarm probability related to the decision rule,

$$P_f = Pr(D^* = N \mid H_0) \tag{9}$$

and $Pr(D^* = N \mid H_0)$ is given by (4)–(5).

$\overline{No_f}$ as a function of N for $M = 1.6\,N$, $p = 1/3$, and $L = 30, 50$ is displayed in Table 1.

Table 1. $\overline{No_f}$ as a function of N when d^* is employed for $M = 1.6\,N$, $p = 1/3$, and $L = 30,\ 50$.

N	$L = 30$	$L = 50$
150	5.5×10^4	5.7×10^{10}
200	3867	4.0×10^9
250	279	2.9×10^8
300	20	2.1×10^7
350	1.51	1.5×10^6
400	0.11	1.1×10^5
450	0.01	9016
500	0.00	685
550	0.00	52
600	0.00	4.0

When $\overline{No_f} \ll 1$ we expect to have the unique solution, and also, $\overline{No_f} =$ constant < 1 determines the sequence lengths which lead to a unique solution. Accordingly, we have the critical length of the observed sequence $\{y_n\}_{n=1}^{N}$ for which we expect the unique solution is typically less than $15L$ (assuming a reasonably small P_m) in the general-unconstrained case. The experiments with the algorithm for cryptanalysis confirm this estimation.

6 Conclusions

The initial state reconstruction problem of a clock-controlled shift register is considered when the characteristic polynomial, a segment of the output sequence and the probability of ones in the clock- controlled sequence are know, assuming that each one in this sequence produces a non-observable bit. The problem addressed here is more general than the problem considered in [2] which focused only on the constrained case when successive non-observable bits are not allowed except for an arbitrary number at the end of the sequence. In this paper the problem solution is based on a quite different approach to that previously published. A novel distance measure for the problem of comparison of binary sequences is defined and studied and probability distributions derived for the cryptanalysis. An algorithm for the initial state reconstruction which employs the defined distance measure is proposed. Its main characteristics are pointed out, and illustrative numerical examples are presented. The proposed approach enables prior estimation of the minimal observed sequence length which is required for the unique initial state solution and it is shown that in the general-unconstrained case this length is typically less than 15 × (the shift register length).

7 Appendix

Proof of Theorem 1. Let $\{e_i\}_{i=1}^{M-N+\ell}$ be the edit-sequence which reflects the transformation of a maximal part or whole sequence $\{b_i\}_{i=1}^{N}$ into the corresponding part of sequence $\{a_i\}_{i=1}^{M}$ by $M - N$ insertions such that:

$$\begin{aligned} e_i &= 0 \text{ if } a_i = b_{i - \sum_{j=1}^{i} e_j} \\ e_i &= 1 \text{ otherwise,} \end{aligned}$$

where $0 \leq \ell \leq N$. So ℓ could be considered as a realisation of a stochastic integer variable L.

On the other hand, the *LD*-calculation procedure [6] gives that $\{e_i\}_{i=1}^{M-N+\ell}$ could be considered as a realisation of the stochastic path in the matrix u (see (2)) starting from the upper-left corner, and ending in the point $(M-N, \ell)$ when, in each step, only progressive right or down movements are allowed, assuming that the zeros in $\{e_i\}$ correspond to the right (horizontal) movements and the ones correspond to the down (vertical) movements.

When $\{a_i\}$ and $\{b_i\}$ are realisations of the sequences of i.i.d. binary variables $\{A_i\}$ and $\{B_i\}$ respectively, such that for any i $Pr(A_i = 1) = Pr(B_i = 1) = 0.5$, the LD-calculation procedure [6] yields that $\{e_i\}_{i=1}^{M-N+\ell}$ is a realisation of the sequence of i.i.d. binary variables $\{E_i\}_{i=1}^{M-N+\ell}$ such that $\Pr(E_i = 1) = 0.5$. Let a stochastic integer variable W be defined by $W = \sum_{i=1}^{M-N+\ell} E_i$. Accordingly we have

$$Pr(L = \ell \mid W = w) = Pr(L = \ell, W = w) \mid Pr(W = w), \tag{10}$$

where

$$Pr(L = \ell, W = w) = Pr(W = w \mid L = \ell) \;\; Pr(L = \ell), \tag{11}$$

$$Pr(W = w) = \sum_{L=0}^{N} Pr(L = \ell)Pr(W = w \mid L = \ell), \tag{12}$$

and

$$Pr(W = M - N \mid L = \ell) = 2^{-(M-N+\ell)}\binom{M-N+\ell}{M-N}, \; 0 \leq \ell \leq N-1, \tag{13}$$

$$Pr(W = M - N \mid L = N) \; = \; \sum_{i=0}^{M-N} 2^{(N+i)}\binom{N+i}{i}, \tag{14}$$

and

$$Pr(L = \ell) = \text{constant}, \; 0 \leq \ell \leq N, \tag{15}$$

because when the number of insertions is not limited, all the lengths are equally probable. Finally, (10)–(15) yield the theorem statement. □

References

1. D. Gollman and W.G. Chambers, Clock-controlled shift registers: A review, IEEE Journal on Selected Areas in Communications, Volume SAC-7, May (1989) 525-533
2. M. Zivkovic, An algorithm for the initial state reconstruction of the clock-controlled shift register, IEEE Trans. Information Theory, Volume 37, September (1991) 1488-1490
3. J. Golic and M. Mihaljevic, A generalised correlation attack on a class of stream ciphers based on the Levenshtein distance, Journal of Cryptology, Volume 3 (3), (1991) 201-212
4. D. Sankoff and J.B. Kruskal, Time Warps, String Edits and Macro Molecules: The Theory and Practice of Sequence Comparison. Reading, MA: Addison-Wesley, 1983.
5. A Levenshtein, Binary codes capable of correcting deletions, insertions, and reversals. Sov. Phy. Dokl., Volume 10, (1966) 707-710
6. B.J. Oommen, Recognition of noisy subsequences using constrained edit distance. IEEE Trans Pattern Analysis Mach. Intell., Volume PAMI-9, September (1987) 636-685
7. B.J. Oommen, Correction to recognition of noisy subsequences using constrained edit distance, IEEE Trans, Pattern Analysis Mach. Intell., Volume PAMI-10, November (1988) 983-984, November 1988.

Construction of m-ary de Bruijn Sequences (Extended abstract)

Jun-Hui Yang[1,2] and Zong-Duo Dai[2]

[1] Computing Center, Academia Sinica, 100080, Beijing, China.
[2] State Key Laboratory of Information Security, Academia Sinica, 100039-08, Beijing, China.

Abstract. A new method for constructing large numbers of m-ary de Bruijn sequences is given.

Key words: m-ary de Bruijn sequences, Shift-register sequence.

1 Introduction

An m-ary feed back shift-register sequence is a recursion sequence defined by

$$x_n = f(x_0, x_1, \ldots, x_{n-1}),$$

where $f(x_0, x_1, \ldots, x_{n-1})$ is a function mapping from $\mathcal{A}^n$ to $\mathcal{A}$ and $\mathcal{A}$ is a set of size m. As we know, the maximum period of linear shift-register sequences over $\mathcal{A}$ of degree n is less than or equal to $m^n - 1$. The m-ary shift-register sequences with period m^n are called m-ary de Bruijn sequences, on which n-tuples run over all m^n states of $\mathcal{A}^n$. The de Bruijn sequences of degree n are the maximum length nonlinear shift-register sequences of degree n. With high linear complexities and some special properties, the sequences can be applied in cryptography and other fields. A comprehensive survey about construction of binary de Bruijn sequences can be found in [5]. For $m > 2$ a few methods for constructing the sequences have been proposed [3,7,11,12]. However, these methods are not efficient in practical use. In [1,2,10], the authors suggested some efficient methods for construction of large numbers of binary de Bruijn sequences. In this paper we develop their ideas to deal with the problem of construction of m-ary de Bruijn sequences for the general $m(\geqq 2)$, and a new method for constructing these sequences will be given.

We start with any given nonsingular recursion function $f(x_0, x_1, \cdots, x_{n-1})$ and construct large numbers of modification sets. Each modification set is made of some elements (call states) in $\mathcal{A}^n$. With any given modification set we design some ways to modify the values of $f(x_0, x_1, \ldots, x_{n-1})$ at the states in the modification set to get a wide variety of recursion functions of m-ary de Bruijn sequences. It will be shown that large numbers of m-ary de Bruijn cycles can be constructed from any one of the modification sets, and that different modification sets will provide totally different m-ary de Bruijn cycles.

2 Construction

Let $\mathcal{A} = \{0, 1, \ldots, m-1\}$ be a set of m elements. By a nonsingular function $f(x_0, x_1, \ldots, x_{n-1})$, we mean a mapping from $\mathcal{A}^n$ to $\mathcal{A}$, and satisfying $f(a_0, a_1, \ldots, a_{n-1}) \neq f(a_0', a_1, \ldots, a_{n-1})$, for all $(a_1, \ldots, a_{n-1}) \in \mathcal{A}^{n-1}$ and all distinct elements a_0 and a_0'. A sequence $\{a_i\}_{i=0}^{\infty}$ of elements $a_i \in \mathcal{A}$ and satisfying the following recursion equation

$$a_{n+t} = f(a_t, a_{t+1}, \ldots, a_{t+n-1}), \forall t \geqq 0,$$

is called a recursion sequence of degree n generated by $f(x_0, x_1, \ldots, x_{n-1})$ over $\mathcal{A}$. It is known that any sequence generated by a nonsingular function must be periodic. An m-ary de Bruijn sequence of degree n is just a recursion sequence for degree n with period m^n over $\mathcal{A}$.

For any sequence $\{a_i\}_{i=0}^{\infty}$ generated by a nonsingular recursion function $f(x_0, x_1, \ldots, x_{n-1})$, the n-tuple $\mathbf{s}_t = (a_t, a_{t+1}, \ldots, a_{t+n-1})$ is called the t-th state. Assume that the sequence $\{a_i\}_{i=0}^{\infty}$ is of period T, then its first T consecutive states will make a cycle of length T as $[\mathbf{s}_0 \mathbf{s}_1 \ldots \mathbf{s}_{T-1} \mathbf{s}_0]$, which will be called a cycle of degree n generated by $f(x_0, x_1, \ldots, x_{n-1})$. The cycle $[\mathbf{s}_i \ldots \mathbf{s}_{T-1} \mathbf{s}_0 \ldots \mathbf{s}_{i-1} \mathbf{s}_i]$ will be considered as the same cycle as $[\mathbf{s}_0 \mathbf{s}_1 \ldots \mathbf{s}_{T-1} \mathbf{s}_0]$. The fact that a state $\mathbf{s}$ is in a cycle σ will be written as $\mathbf{s} \in \sigma$.

In order to construct m-ary de Bruijn sequences of degree n, we start with any given nonsingular function $f(x_0, x_1, \ldots, x_{n-1})$, and assume that the function generates N cycles: $\sigma_0, \sigma_1, \ldots, \sigma_{N-1}$. It is clear that any n-tuple $(a_0, a_1, \ldots, a_{n-1}) \in \mathcal{A}^n$ will be in one and only one of these N cycles. If we could join all these cycles together, then we would get an m-ary de Bruijn cycle of length m^n.

2.1 Joining Cycles

Two states $(a_0, a_1, \ldots, a_{n-1})$ and $(b_0, b_1, \ldots, b_{n-1})$ are said to be conjugate to each other if their last $n-1$ components are the same, i.e. $a_i = b_i, 1 \leqq i \leqq n-1$.

Lemma 1. *Let $\sigma_0, \sigma_1, \ldots, \sigma_{N-1}$ be all the cycles generated by $f(x_0, x_1, \ldots, x_{n-1})$, and let the s conjugate states $(a_0^{(i)}, a_1, \ldots, a_{n-1}), 1 \leqq i \leqq s$, be in the s distinct cycles $\sigma_i, 1 \leqq i \leqq s$, respectively, and let*

$$F(\mathbf{x}) = \begin{cases} f(\mathbf{x}) & \textit{if } \mathbf{x} \neq (a_0^{(i)}, a_1, \cdots, a_{n-1}), 1 \leqq ix \leqq s, \\ f(a_0^{(i+1)}, a_1, \cdots, a_{n-1}) & \textit{if } \mathbf{x} = (a_0^{(i)}, a_1, \cdots, a_{n-1}), 1 \leqq i < s, \\ f(a_0^{(1)}, a_1, \cdots, a_{n-1}) & \textit{if } \mathbf{x} = (a_0^{(i)}, a_1, \cdots, a_{n-1}), \end{cases}$$

where $\mathbf{x} = (x_0, x_1, \ldots, x_{n-1})$. Then $F(\mathbf{x})$ will join all the cycles $\sigma_i, 1 \leqq i \leqq s$, together to become a new cycle and will keep all other cycles unchanged.

2.2 Order Function

We choose state $\mathbf{p} = (p_0, \ldots, p_{n-1})$ and assume that $\mathbf{p} \in \sigma_0$. Let S_m be the symmetric group on $\mathcal{A} = \{0, 1, \ldots, m-1\}$, and for any $a \in \mathcal{A}$ let H_a be the subset of S_m consisting of all permutations τ such that $\tau(a) = 0$. Let $\boldsymbol{\tau} = (r_0, \ldots, r_{n-1})$, where $r_i \in H_{p_i}$, then $\boldsymbol{\tau}$ will specify an order on $\mathcal{A}^n$, as well as an order on all these N cycles $\sigma_i, 0 \leqq i < N$. In fact, for any $\mathbf{a} = (a_0, \ldots, a_{n-1}) \in \mathcal{A}^n$, let

$$Ord_\tau(\mathbf{a}) = \sum_{i=0}^{n-1} \tau_i(a_i) m^i,$$

then the function $Ord_\tau(\mathbf{a})$ provides an order on the set $\mathcal{A}^n$. And for any cycle $\sigma = \sigma_i, 0 \leqq i < N$, let

$$Ord_\tau(\sigma) = min\{Ord_\tau(\mathbf{a}) \mid \mathbf{a} \in \sigma\},$$

then the function $Ord_\tau(\sigma)$ provides an order on all the cycles generated by $f(x_0, x_1, \ldots, x_{n-1})$. It is easy to see that $Ord_\tau(\mathbf{a}) \neq Ord_\tau(\mathbf{b})$ if $\mathbf{a} \neq \mathbf{b}$, so for any cycle $\sigma_i, 0 \leqq i < N$, there exists an unique state $\mathbf{a} = (a_0, a_1, \ldots, a_{n-1})$ in σ_i such that $Ord(\mathbf{a}) = Ord(\sigma_i)$. We call $Ord_\tau(\sigma)$ an order function. It is clear that the function $Ord_\tau(\sigma)$ depends not only on $\boldsymbol{\tau}$, and also on the chosen state $\mathbf{p} = (p_0, \ldots, p_{n-1})$. From now on we shall fix both of $\mathbf{p}$ and $\boldsymbol{\tau}$, and simply write $Ord_\tau(\sigma)$ as $Ord(\sigma)$.

2.3 Collecting of Connecting Pair Sets

For any cycle $\sigma_i, i \neq 0$, let $\mathbf{a} = (a_0, a_1, \ldots, a_{n-1})$ such that $Ord(\mathbf{a}) = Ord(\sigma_i)$, and let $\mathcal{P}_i$ be the set of all the conjugate state pairs of the form

$$\{(a_t, \ldots, a_{n-1}, b_0, \ldots, b_{t-1}), (a_t^*, \ldots, a_{n-1}, b_0, \ldots, b_{t-1})\}, \tag{1}$$

where t is an index such that $a_t \neq p_t$, and a_t^* is an element in $\mathcal{A}$ with the property that $\tau_t(a_t^*) < \tau_t(a_t)$, and $(b_0, \ldots, b_{t-1}) \in \mathcal{A}^t$ such that $(a_t, \ldots, a_{n-1}, b_0, \ldots, b_{t-1}) \in \sigma_i$. We call $\mathcal{P}_i$ a connecting pair set and say that the pair (1) connects σ_i and σ_j if $\mathbf{a} = (a_t^*, \ldots, a_{n-1}, b_0, \ldots, b_{t-1}) \in \sigma_j$. We have

Lemma 2. *For any i, $i \neq 0$, the connecting pair set $\mathcal{P}_i$ is not empty. If $(\mathbf{a}, \mathbf{a}^*)$ is a pair in $\mathcal{P}_i$ connecting σ_i and σ_j, then $Ord(\sigma_j) < Ord(\sigma_i)$.*

Now we denote the collection of the connecting pair sets $\mathcal{P}_i (1 \leqq i < N)$ by $\mathcal{C}$, i.e, $\mathcal{C} = \{\mathcal{P}_i \mid 1 \leqq i < N\}$.

2.4 Modification Set

Let M be a transversal of the collection $\mathcal{C}$, i.e. a set consisting of one and only one element from each set $\mathcal{P}_i$ in the collection $\mathcal{C}$, and let $\mathcal{M}$ be the set of all the states appeared in some connecting pairs in M. We design some ways to modify the values of the function $f(x_0,\ldots,x_{n-1})$ at the states in $\mathcal{M}$ to get a wide variety of recursion functions of m-ary de-Bruijn sequences. We call M a modification set associated with $f(x_0,\ldots,x_{n-1})$.

We say two states $\mathbf{s}=(a_0,\ldots,a_{n-1})$ and $\mathbf{s}^*=(b_0,\ldots,b_{n-1})$ in $\mathcal{M}$ are related to each other of there exists a chain made of connecting pairs in M as below

$$(\mathbf{s}_1,\mathbf{s}_1^*),(\mathbf{s}_2,\mathbf{s}_2^*),\ldots,(\mathbf{s}_r,\mathbf{s}_r^*),$$

such that $\mathbf{s}$ is one of the states $\mathbf{s}_1$ and $\mathbf{s}_1^*$, and $\mathbf{s}^*$ is one of the states $\mathbf{s}_r$ and $\mathbf{s}_r^*$, and any two adjacent pairs $(\mathbf{s}_i,\mathbf{s}_i^*)$ and $(\mathbf{s}_{i+1},\mathbf{s}_{i+1}^*), 1\leqq i<r$, have a common state.

It is clear that this defined relation between the states is an equivalent relation on the set $\mathcal{M}$. Thus $\mathcal{M}$ is partitioned into some equivalent classes, we may assume

$$\mathcal{M}=\{\mathbf{s}_{ij}, 1\leqq i\leqq\lambda, 0\leqq j\leqq l_i\}, \tag{2}$$

where the two states $\mathbf{s}_{i,j}$ and $\mathbf{s}_{i',j'}$ are in the same class if and only if $i=i'$. And we say that the modification set $\mathcal{M}$ is of parameters $(l_1,l_2,\ldots,l_\lambda)$.

Theorem 3. *Let $\mathcal{M}$ be a given modification set which is associated with $f(x_0,\ldots,x_{n-1})$ and is of the parameters $(l_1,\ldots,l_\lambda)$ as shown in (2), and S_{l_i} be the symmetric group on the set $\{0,1,\ldots,l_i-1\}$. For any given $\rho=(\rho_1,\ldots,\rho_\lambda),\rho_i\in S_{l_i}$, let*

$$F_\rho(\mathbf{x})=\begin{cases} f(\mathbf{s}_{i,\rho_i(j+1)}) & \text{if } \mathbf{x}=s_{i,\rho_i(j)}, 1\leqq i\leqq\lambda, 0\leqq j<l_i-1,\\ f(s_{i,l_i}) & \text{if } \mathbf{x}=s_{i,\rho_i(l_i-1)}, 1\leqq i\leqq\lambda,\\ f(s_{i,\rho_i(0)}) & \text{if } \mathbf{x}=s_{i,l_i}, 1\leqq i\leqq\lambda,\\ f(\mathbf{x}) & \text{otherwise,}\end{cases}$$

where $\mathbf{x}=(x_0,\ldots,x_{n-1})$. Then

1. *$F_\rho(\mathbf{x})$ is a recursion function of m-ary de Bruijn sequences of degree n,*
2. *if $\rho\neq\rho^*$, then $F_\rho(\mathbf{x})\neq F_{\rho^*}(\mathbf{x})$, hence $\mid\mathcal{F}_\mathcal{M}\mid=\prod_{i=1}^{\lambda}l_i!$, where $\mathcal{F}_\mathcal{M}=\{F_\rho(\mathbf{x})\mid\rho=(\rho_1,\ldots,\rho_\lambda),\rho_i\in S_{l_i}\}$, in other words, the total number of the recursion functions of m-ary de Bruijn sequences of degree n provided by $\mathcal{M}$ is $\prod_{i=1}^{\lambda}l_i!$.*

Theorem 4. *Let $\mathcal{M}_1$ and $\mathcal{M}_2$ be two different modification sets associated with a given $f(x_0,x_1,\ldots,x_{n-1})$, then we have $\mathcal{F}_{\mathcal{M}_1}\cap\mathcal{F}_{\mathcal{M}_2}=\emptyset$, where the $\mathcal{F}_{\mathcal{M}_2}$ is as in Theorem 3.*

Now let $N_0(n)$ be the number of different modification sets $\mathcal{M}$ associated with $f(x_0,x_1,\ldots,x_{n-1})=x_0$ which can be constructed according to the above method, and let $N_1(n)$ be the total number of the recursion functions of m-ary de Bruijn sequences which can be generated by these modification sets. It is clear that $N_1(n)$ is much larger than $N_0(n)$, but at this moment we do not bother ourselves to figure out how $N_1(n)$ is larger than $N_0(n)$. Just by a very rough estimation we have

Theorem 5. $N_1(n) > N_0(n) > 2^{(m^n/n)-mn}$.

Remark: In the papers [3,7,9,12] some methods of the construction of m-ary de Bruijn sequences for $m \geqq 2$ are discussed. We notice that the m-ary de Bruijn cycles of degree n which can be generated by the algorithm in [3] is only an extremely small portion of those constructed by our above method. In fact, let $N_E(n)$ be the number of the m-ary de Bruijn cycles of degree n which can be generated by the algorithm in [3], it is easy to deduce from the expression for $N_E(n)$ in [3] that $N_E(n) < 2^{m^{n/2}}$. It is clear that $N_E(n) = o(N_1(n))$. In fact the asymptotic order of $N_E(n)$ is much less than that of $N_1(n)$, and moreover, even the asymptotic order of $\log_2 N_E(n)$ is much less than that of $\log_2 N_1(n)$.

3 Example

In this section we provide an example to show our method for construction of 3-ary de Bruijn cycles of degree 4. In the example we take $f(x_0, x_1, x_2, x_3) = x_0$, $\mathbf{p} = (0,0,0,0)$ and $\boldsymbol{\tau} = (\tau.\tau, \tau, \tau)$, where τ is the identity permutation on the set $\mathcal{A} = \{0,1,2\}$, i.e., $\tau_i = i, 0 \leqq i \leqq 2$. There are totally 24 cycles $\sigma_i, 0 \leqq i \leqq 23$, generated by $f(x_0, x_1, x_2, x_3) = x_0$. The following table displays these cycles σ_i and the corresponding connecting pair sets $\mathcal{P}_i, 1 \leqq i \leqq 23$, where the notation $(a_0a_1a_2a_3)$ at the column under σ_i denotes the cycle $[(a_0, a_1, a_2, a_3)(a_1, a_2, a_3, a_0)(a_2, a_3, a_0, a_1)(a_3, a_0, a_1, a_2)(a_0, a_1, a_2, a_3)]$, and the notation $(a_0a_1a_2a_3; a_4)$ denotes the connecting pair $\{(a_0a_1a_2a_3), (a_4a_1a_2a_3)\}$.

i	σ_i	$\mathcal{P}_i$
0	(000)	
01	(1000)	(1000;0)
02	(2000)	(2000;1), (2000;0)
03	(1100)	(1100;0), (1001;0)
04	(2100)	(2100;1), (2100;0), (1002;0)
05	(1200)	(1200;0), (2001;1), (2001;0)
06	(2200)	(2200;1), (2200;0), (2002;1), (2002;0)
07	(1010)	(1010;0)
08	(2010)	(2010;1), (2010;0), (1020;0)
09	(1110)	(1110;0), (1101;0), (1011;0)
10	(2110)	(2110;1), (2110;0), (1102;0), (1021;0)
11	(1210)	(1210;0), (2101;1), (2101;0), (1012;0)
12	(2210)	(2210;1), (2210;0), (2102;1), (2101;0), 1022;0)
13	(2020)	(2020;1), (2020;0)
14	(1120)	(1120;0), (1201;0), (2011;1), (2011;0)
15	(2120)	(2120;1), (2120;0), (1202;0), (2021;1), (2021;0)
16	(1220)	(1220;0), (2201;1), (2201;0), (2012;1), (2012;0)
17	(2220)	(2220;1), (2220;0), (2202;1), (2202;0), (2202;0), (2022;1)(2022;0)
18	(1111)	(1111;0)
19	(2111)	(2111;1), (2111;0), (1112;0), (1121;0), (1211;0)
20	(2211)	(2211;1), (2211;0), (2112;1), (2112;0), (1122;0), (1221;0)
21	(2121)	(2121;1), (2121;0), (1212;0)
22	(2221)	(2221;1), (2221;0), (2212;1), (2212;0), (2122;1), (2122;0), (1222;0)
23	(2222)	(2222;1), (2222;0)

Any traversal M of the collection $\mathcal{C}$ of these $\mathcal{P}_i, 1 \leqq i \leqq 23$, will correspond to a modification set $\mathcal{M}$. For example we consider the modification set $\mathcal{M}_0$ which corresponds to transversal M_0 consisting of the first connecting pair from each of the connecting pair sets $\mathcal{P}_i, 1 \leqq i \leqq 23$. $\mathcal{M}_o$ is of parameters $(2,2,2,2,2,2,2,2,2,1,1,1,1,1)$. In fact it consists of 37 states, which are partitioned into 14 equivalent classes, as shown in the following table, where we enclose the states of each one of the equivalent classes in brackets.

(2000,1000,0000), (2100,1100,0100), (2200,1200,0200),
(2010,1010,0010), (2110,1110,0110), (2210,1210,0210),
(2120,1120,0120), (2220,1220,0220), (2111,1111,0111),
(2020,1020,), (2211,1211), (2121,1121), (2221,1221), (2222,1222).

The modification set $\mathcal{M}_0$ produces $(2!)^9 = 512$ different recursion functions of 3-ary de Bruijn sequences of degree 4 according to Theorem 3. In fact, more than $3^8 2^{12}$ distinct modification sets can be obtained from the transversal of the above collection $\mathcal{C}$, and in general the number of de Bruijn cycles provided by one modification set is far more than 1.

To compare with the algorithm in [3], which produces only 3-ary de Bruijn cycle of degree 4, the total number of 3-ary de Bruijn cycles of degree 4 which can be generated in the above example according to our method is more than $3^8 2^{12}$.

References

1. Z.D. Dai, "On the Construction and Cryptographic Application of de Bruijn Sequences", Presented at *Crypt'88*, 1988.
2. Division of Algebra in the Institute of Mathematics, Department of Mathematics of the University of Science and Technology of China, "On Methods of Constructing Feedback Functions of M-Sequences", ACTA MATHEMATICAE APPLICATAE SINICA, Nov., 1977
3. T. Etzion, "An Algorithm for Constructing m-ary de Bruijn sequences", Journal of Algorithms, Vol. 7, 1986, pp. 331-340.
4. T. Etzion and A. Lempel, "Algorithm for the generation of full length shift-register sequences", IEEE Tras. Inform. Theory IT-30(1984), pp. 480-484.
5. H.M. Fredricksen, "A Survey of Full Length Nonlinear Shift Register Cycle Algorithms", SIAM REVIEW, Vol. 24, No. 2, 1982
6. H.M. Fredricksen and I.J. Kessler, "Lexicographic compositions and de Bruijn sequences", J. Combin Theory 22(1977), pp.17-30
7. H.M. Fredricksen and J. Maiorana, "Necklaces of beads in k colors and k-ary Bruijn sequences", Discrete Math. 23(1978), pp.207-210.
8. C. Larry, W. Doug, "Embedded de Bruijn sequences", Proceedings of the 17th Southeastern International Conference on Combinatorics, Graph Theory and Computing (Boca Raton, Fla. 1986).
9. A. Ralston, "A New memoryless algorithm for de Bruijn sequences", J. Algorithm 2(1981).
10. Z.X. Wan, Z.D. Dai, X.L. Feng, M.L. Liu, "Nonlinear Shift Register Sequences", Science Press, 1978.
11. S. Xie, "Notes on de Bruijn sequences", Discrete Mathematics, 16, 1987, pp. 157-177.
12. R.H. Xiong, "Theories and Algorithms for generating q-ary de Bruijn sequences", Scientia Sinica, A, No.11, 1988, pp. 1386-1397.

Session 9

PSEUDORANDOMNESS

Chair: Bill Caelli
(Queensland University of Technology, Australia)

Information Technology Security Standards - An Australian Perspective

John Snare

Telematic and Security Systems Section
Telecom Australia Research Laboratories

Abstract. From a telecommunications perspective, standards facilitate the implementation of distributed applications. Such systems can be implemented using components produced by different suppliers, at different times, and in ways that involve a minimum of proprietary intellectual property. As such open systems become widely implemented, it is becoming increasingly important to have standards for security services and mechanisms to allow the interests of all interconnected parties to be protected. This paper discusses the role of standards in providing a link between the large body of available theory, and business needs. A standardised approach has the following advantages:

- agreement can be reached on the meaning of security terminology;
- security mechanisms can be subject to international, expert scrutiny before adoption;
- common security mechanisms can be developed in such a way that re-use is possible; and
- the limited amount of available technical expertise can be efficiently used and made accessible to all parts of industry and government.

When an analysis is made of security standardisation activities around the world, it is quickly appreciated that we are in fact well away from realising an optimum approach to security standards development. However, there is still much to be gained from the standardisation process. This paper looks at the range of security standardisation activities, and then focuses on the work being done to develop generic (basic) security building block standards in the International Organisation for Standardisation (ISO)/International Electrotechnical Commission (IEC), Joint Technical Committee 1, Subcommittee 27. The range of Subcommittee 27 activities is summarised, and an update is given of progress to date. This status is then placed in the perspective of related Australian standardisation activity.

Introduction

The development of computer networking from the 1970's has been accompanied by an associated need to provide an ever increasing range of security protections. This networking revolution has drawn into the commercial arena security technologies that had previously only been of interest to the military and government sectors. The mathematics behind various types of puzzle problem, secret sharing, pseudo randomness are now the basis of practical security solutions of direct relevance to people in all walks of life. The security revolution to parallel the network revolution is not yet upon us, but it is only a matter of time.

From a business perspective, there is a gap between the security needs associated with distributed information systems and the security solutions that can be readily implemented. The reasons for this are both interesting and complex. On the one hand, there is now sufficient fundamental security technology upon which the necessary solutions could be built. However on the other hand, there no adequate tools to help design efficient security solutions for distributed systems. There is thus an associated reluctance to invest money in security solutions for which an analytic justification is not possible.

This paper presents standardisation as an intermediate step between the fundamental technology and its deployment. Standards offer the potential to encourage the deployment of secure systems and lower the associated costs. This paper presents a view of current security standards developments from a business perspective. It is not possible in a paper such as this to cover all security standards activities in any depth. This paper therefore indicates the scope of standardisation activities, and then concentrates on the activities of the group working on basic technique standards, as these provide a fundamental basis for all other security standards.

A Distributed IT System Scenario

The pressing need for technology to allow easy incorporation of security in distributed applications can be illustrated by the following business scenario, illustrated in figure 1.

Suppose you have an office LAN, perhaps one of many in a medium size company. Such a LAN would typically contain a file server, print server and probably provide an electronic mail service for its users in a small, closed environment. Having realised the benefits of electronic mail, someone will suggest a connection to other electronic mail systems in the company. What are the security implications of this change?

Before long, you want to give an interstate office access to one of the planning systems on a workstation on your LAN. For cost reasons, a dial-up, public packet switching service seems to be the best option. What are the security implications of this change? Is it safe to use the same system to have staff in remote locations jointly develop a new application? It is only a matter of time before it makes business sense to give your staff access to global electronic mail (either through commercial networks or the internet), and to become involved in trading using EDI.

Finally, in this scenario, suppose you will eventually want to fully internet the LANs, using broad band public switched networks. At the same time you discover that staff have a need to dial in to their workstations from remote locations, usually either from home or when travelling on business.

From a business perspective, such an evolution is both logical and sensible. In many cases, a business could not afford to do other than to fully utilise invest-

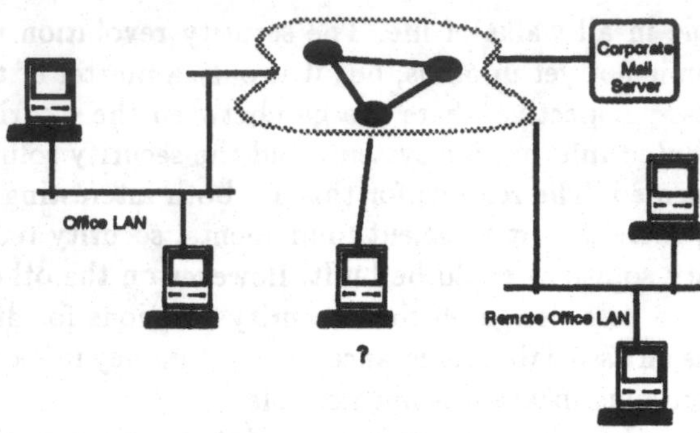

Fig. 1. A Distributed IT System Scenario

ment in IT systems in this way. However, from a security (risk) perspective, this scenario is unacceptable. What is worse, no-one can easily determine what needs to be done to secure this environment. We don't know how to analyse risk in this networked scenario, putting various threats in perspective and allowing priorities to be determined to efficiently reduce risk. We don't know how to describe or measure the security levels achieved in a security implementation. Furthermore, we don't know how to handle networked architectures which develop over time and/or where different people are responsible for different parts.

What is sure though, is that you can't "do" security in just one place, like we did in the past. Any security solution will have to be distributed across the network. At present, the options are to either forbid this type of networking, or to take the risk. What is needed is tools that allows the risk to be managed.

Safe distributed systems will need to have security designed in from the very earliest stages, and managed though all development and operational phases of a project life cycle. No one knows how to do any of these things well yet, but we do know some of the tools that are essential. If these tools can be provided to an organisation, it can then focus on satisfying its security needs, confident that implementation of secure systems will result. This is where standards come in.

Security Requirements

There is considerable experience in the area of securing single site IT systems. It is the distributed element made possible by telecommunications that makes the security problem of the 1990's hard. This paper thus concentrates on security aspects associated with communicating systems.

The security requirements of distributed IT systems can be considered in terms of the data, functions, hardware, software, terminal, and network elements that make up such systems. The protection necessary will be provided by combinations of physical, logical, procedural, and personnel security mechanisms as appropriate to the circumstances and the vulnerability. Emphasis in this paper is on logical security because it is in this area that many developments in information technology are simultaneously generating a wide range of new threats and a range of new mechanisms for protection.

Important logical security services[1] are:

- authentication - corroboration of claims concerning the identity of a remote entity in a telecommunication system;
- access control - protection against unauthorised use of telecommunication system resources (information, processing, peripherals, functionality);
- confidentiality - protection from unauthorised disclosure of information or traffic flows;
- data integrity - proof that data has not been created, altered, or destroyed in an unauthorised manner; and
- non-repudiation - protection against one or both parties involved in communication later denying such involvement (eg. creation or receipt of information, or use of functionality).

These security services must be associated with management functions. Such functions:

- set parameters for use (including cryptographic keys);
- control invocation;
- monitor performance;
- log anomalous events and errors; and
- allow audit and tracing where required.

When considering security requirements in a networked system, it is apparent that some security services (eg. confidentiality) are best applied on an end to end basis between terminal equipments. There are other cases where both networks and end systems need to be involved. For example, it may be important for terminal systems to authenticate themselves on an end to end basis in an application such as Electronic Data Interchange (EDI). However, it is additionally important that terminal and network systems authenticate each other to ensure that charging is correct. Furthermore, although essentially transparent to users, there may be cases where different network sub-systems may need to authenticate each other to allow secure service to be delivered. It can thus be seen, that delivery of complex telecommunication services (such as enhanced "intelligent" voice services, value added information services, or customer network management services) requires logical security to be considered carefully in the context of sophisticated security architectures.

The traditional approach to providing security services has been to place heavy emphasis on the use of access control mechanisms. Thus confidentiality, for example, was provided by implementation of systems such that unauthorised

[1] These services are not independent (for example, access control requires use of an authentication service), and can be provided by a number of security mechanisms

parties could not gain access to a computer, a file in a computer, or a communication channel. The situation has now changed. Not only is there a question as to whether access control can be reliable enough in a distributed system, but there is also realisation the confidentiality is not the only security requirement.

Mechanisms based on applied cryptography have been developed over the last decade to provide security services of higher quality. Such mechanisms, which can be combined and integrated into applications to provide security services, include:

- encipherment - relevant to confidentiality services and other mechanisms;
- digital signatures - relevant to authentication and non-repudiation services;
- access control mechanisms;
- data integrity mechanisms;
- authentication exchange mechanisms;
- traffic padding mechanisms;
- routing control mechanisms; and
- notarisation mechanisms.

Many of the major computer system vendors have now developed sophisticated security architectures for networks of their computers. However, such solutions tend to be both proprietary and expensive, and thus of limited use in the scenario presented in the last section.

Standards as the Basis of a Security Implementor's Tool-kit

Costs can only be reduced to an acceptable level if we learn from experience. The expertise necessary to solve security problems is both scarce and expensive; it must be used efficiently. Problems should be solved well once and the solutions be capable of being re-engineered into many situations. This implies development of a standardised way of doing things, which could be described as an implementor's tool kit. This approach must be accompanied by means to establish confidence in the reliability of the standardised security solutions. This may be accomplished if standards are based on common practice, have endorsement by a respected and trusted authority, and/or have been subjected to wide-scale public scrutiny, and/or are developed by people known to be technically expert.

The types of thing that we can expect a security implementor's tool-kit to contain are:

- **definitions** of a common terminology, so that different people can reliably understand a security problem or solution (a common understanding of the problem space);
- **models and frameworks** that describe, at a high level, what various security functions do and how they should work;
- **architectures** that allow decomposition of complex functions into layers or modules with relatively simple interfaces;

– **mechanisms** that can be used in a variety of implementations, including protocols for communication and procedures that may be invoked in end-systems; and
– **evaluation** methods to assess the behaviour and quality of a design, product, or implemented system.

It would also be nice if common requirements for particular types of application could be identified and common solutions offered. Such applications include the spectrum from computer operating systems and database systems to banking and electronic business systems.

The International Security Standards Scene

There are a large number of international standardisation activities underway that address elements of the "security toolkit" under the auspices of bodies such as the IEEE (Institute of Electrical and Electronic Engineers), ISO (International Organisation for Standardisation), CCITT (International Telecommunications and Telegraphy Consultative Committee), ETSI (European Telecommunications Standards Institute), and the IEC (International Electrotechnical Commission). Figure 2 attempts to summarise the areas of activity.

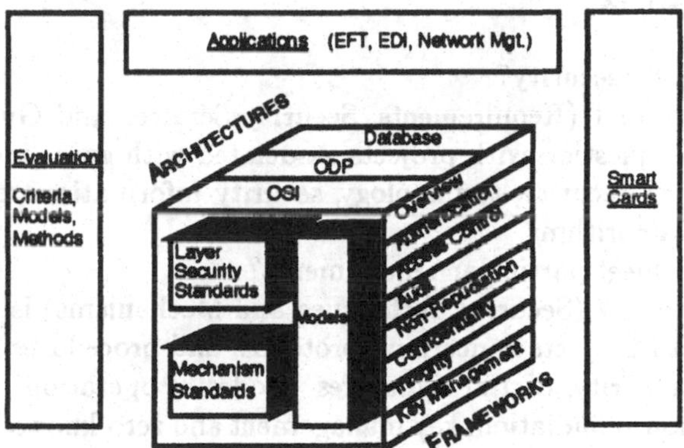

Fig. 2. Overview of Standards Activities

It is beyond the scope of this paper to detail all the associated activities in each area, however it should be appreciated that most of the important issues raised by my scenario are being addressed (at least to some extent). To date, most work has been done concerning the architectures and frameworks at the core of this picture, especially by the group in ISO/IEC that have been developing the Open Systems Interconnection (OSI) standards (ISO/IEC JTC 1/SC 21) and associated security techniques and mechanisms (ISO/IEC JTC 1/SC 27). The future should see increasing activity that builds on this work in the area of application standards. It should also be appreciated that a hierarchy is developing, where applications use common architectures, such as Open Systems Interconnection (OSI). Within OSI, common security frameworks are being developed. And such frameworks use common security mechanisms.

It is at the mechanism level where cryptography appears. This paper is oriented towards the use of cryptography to solve distributed IT system security problems, and thus emphasis is on mechanism standards, and in particular the work of ISO/IEC JTC 1/SC 27, in following sections.

Before proceeding to discuss mechanism standardisation activity, it must be pointed out that standards development is not perfectly coordinated, and thus, for pragmatic reasons, this model is not being perfectly used. We thus find application standards (for example those associated with banking) which also include mechanisms, however it is hoped that out "toolkit" will evolve with time to reduce such anomalies.

Data Security Technique Standards (ISO/IEC JTC 1/SC 27)

Committee ISO/IEC JTC 1/SC 27 is developing standards for general techniques and mechanisms for data security, with emphasis on logical security.

The work of SC 27 can be summarised as addressing, in different working groups, the questions:

- How do you do security?
  Working Group 1 (Requirements, Security Services and Guidelines) is addressing this question with projects associated with guidelines for the management of IT security, terminology, security information objects, and registration of algorithms.
- How do you meet particular requirements?
  Working Group 2 (Security Techniques and Mechanisms) is addressing this question with projects concerning protocols and procedures for authentication, data integrity, digital signatures, modes of operation of ciphers, hash functions, non-repudiation, key management and zero-knowledge techniques.
- How do you know if a solution is any good?
  Working Group 3 (Security Evaluation Criteria) is addressing this question working on issues concerning requirements, functionality and assurance.

The following sections describe some of the more significant of the activities currently being done to address these questions.

Guidelines for the Management of IT Security

Although it is not the task of a standard to describe how to do security, a technical report on this topic is being produced to help not experts get started. This is particularly important, because this topic is not well understood and it is currently very difficult take a systematic approach to the implementation of security. The Technical Report will establish a consistent terminology, and help the reader understand the significance of relationships between assets, threats, vulnerability's impacts, risks and safeguards to name a few. It will then go on to give insight into how systems can be modelled from a security perspective, and on processes that can be adopted by designers concerning security of systems. This work is still in its formative stages, however when it is complete it will significantly and practically contribute to the deployment of better security in distributed systems.

Authentication

Authentication mechanisms are an important feature of many data communication protocols. Currently most protocols use very crude authentication methods such as reliance on identification of access circuits, passwords etc. The emerging generation of communication protocols are increasingly including options to use cryptographic techniques for the purposes of remote party identification. The standards currently under preparation in SC 27 in this area provide a variety of authentication mechanisms that are intended to provide a safe basis for the authentication aspects of new data communication protocols and computer applications.

Although the concept of authentication is apparently quite straight forward, in communications systems there are a number of issues that require careful consideration. The first issue that must be addressed is that of certification of identity. It is important to be able to transfer reliably information through a network concerning claimed identity, and to have confidence that this information can't be tampered with, replayed, or delayed. However it is also important to have a mechanism in place to link authentication credentials with an actual physical or legal entity to answer questions like: "who says I am who I say I am?".

This is especially important in cases where communication is essentially adhoc rather than pre-planned. The approach most developed in the standards arena is to use mechanisms based on public key cryptography (asymmetric cryptosystems) and trusted third parties in the process of creating authentication credentials. Such trusted parties do not need to be directly involved in subsequent instances of authentication, but may become involved again in dispute resolution. The mechanisms currently under study for standardisation are based on both public key algorithms such as RSA and on the newer "zero knowledge" protocols.

Alternative approaches are also possible for applications where public key cryptography is not appropriate and where symmetric algorithms such as DES must be used. In these cases on-line notarisation or key management services provided by trusted third parties are also being standardised. Such systems are both operationally and technically less elegant and more complex than the public key approach, but have a role in cases where security management must be tightly controlled.

The standards under development generally assume the availability of a cryptographic algorithm of the appropriate type and then specify a usage framework along with protocols for the authentication on either or both parties. Such protocols have a number of variants to cope with different application requirements; for example different protocols may be appropriate depending on whether an application is store and forward or interactive. A standard has been published containing a general authentication model. Two more detailed standards are nearing completion describing authentication procedures based on asymmetric and symmetric techniques. These standards are flexible and general and will offer system designers choices concerning the use of random numbers, time-stamps, or both in authentication protocols. For example, the authentication framework designed for use in association with electronic directories (CCITT Recommendation X.509) uses an authentication protocol that is consistent with the proposed SC 27 standard for authentication using asymmetric techniques.

New work has now been proposed for possibly two new authentication standards; the first involving the use of non-reversible functions (as is done in the new GSM digital mobile telephone system), and the second involving the use of zero knowledge techniques (as might be used in pay-TV systems).

Integrity

Logical information integrity can be assured through the use of either symmetric or asymmetric cryptosystems. Symmetric schemes involve either the use of Message Authentication Codes (MAC), or encrypted Manipulation Detection Codes. Both these approaches can protect against tampering by third parties, but on their own offer no protection against receiver tampering. Such protection can be provided by including a trusted third party notary in the communication, but this approach is currently receiving little attention in the standards arena.

Approaches to integrity based on public key cryptosystems are receiving considerable standardisation attention currently. Approaches based on "digital signatures" are attractive because they:

- simultaneously provide integrity protection against both third party and receiver attack;
- provide origin authentication;
- form the basis of non-repudiation services; and
- do not require on-line participation of third parties.

Because public key cryptosystems are relatively inefficient, special approaches must be adopted to handle the protection of large amounts of information. The approach being adopted for integrity standards is to develop a "collision resistant" compression or hash function to produce an unpredictable message summary and then calculate the digital signature based on that summary.. Development of cryptographically secure un-keyed hash functions for a standard is an unexpectedly difficult task. Efforts are in hand to develop hash functions based on symmetric crypto-algorithms, simple finite field arithmetic, and the operations typically used in common public key crypto-systems. However, good progress has only been made on the first of these, based on symmetric cryptosystems.

Protection of the integrity of short data blocks also requires special consideration, and a standard based on public key techniques has recently been approved for publication as a standard to cover this case.

Two standards are currently envisaged in the area of digital signatures, to cover different message sizes. A standard for digital signatures using a variant of the RSA algorithm for small data blocks has now been published. Work is underway concerning a digital signature using a public key algorithm for long (arbitrary length) messages, however it is at an early stage.

Other digital signature schemes exist and may also be standardised in future. Candidates for future consideration include schemes based on zero knowledge protocols and the recently released draft US Federal Information Processing Standard for digital signatures (prepared by the National Institute for Standards and Technology (NIST) and the National Security Agency (NSA) in the US).

Non-repudiation

Non-repudiation security in a telecommunications environment can take a variety of forms. For example it may be necessary to prove information transfer from originator to network, from originator to receiver, from network to receiver. Within the standardisation arena work has recently commenced concerning definition of a general non-repudiation framework along with protocol mechanisms based on the use of symmetric and asymmetric algorithms.

In the case of non-repudiation using asymmetric algorithms, a mechanism for "blind" non-repudiation of receipt is being included that provides for a receiver to promise to accept a message before he can read it. It remains, however, to be seen whether such a mechanism can be defined satisfactorily.

Confidentiality

Both symmetric and asymmetric encryption algorithms are essential component for providing confidentiality and other security services. Unfortunately, the United States has vetoed the development of standards for such algorithms within the ISO data security techniques committees, so this element of the security "tool-kit" cannot be satisfactorily provided at this stage. However certain important, related matters are covered by prospective standards. Most significantly, an international register of cryptographic algorithms is being established, according to standardised procedures. This register will provide an identifier for algorithms that will allow their recognition in security protocols. It also includes provision to record details about the algorithms, such as its properties, intended uses, availability, etc.

Standards have also been produced concerning the modes of use of encryption algorithms for providing confidentiality. Such modes cover encryption on a block by block basis, as well as how data can be encrypted as a chained set of characters or blocks.

Security Management - Key Management

The security of any system that uses cryptography is critically dependent on the secure handling of the associated cryptographic keys, from the time they are created until the time they are destroyed. Key management is the name given to the secure handling of keys throughout their life cycle. Key management techniques for commercial applications will differ markedly from those for military and classified systems, as considerably enhanced flexibility will be needed in conjunction with the need for systems to be able to be operated by non specialist staff. Commercial key management techniques are currently relatively immature. Standards in this area will facilitate rapid development of key management technology. They will also allow similar solutions to be used across a variety of applications and thus reduce costs.

Considerable progress has been made on standards concerning key management mechanisms using symmetric and asymmetric techniques. Working drafts are expected to stabilise in 1993.

The symmetric technique draft standard contains generalisations of protocols using on line key distribution and key translation centres. At the moment, it is inconsistent with the approach taken for key management in banking, and a review is underway to determine whether technical alignment with the banking standard is technically sound.

The asymmetric technique standard (still at an early stage of development) will set out basic requirements for all known types of scheme, and include examples of key management protocols in informative annexes. This standard will hopefully be very flexible, whilst at the same time guiding its users to develop safe (secure) key management protocols. Two types of protocol will be covered:

- key agreement schemes where both parties contribute to key generation; and
- key transfer schemes, where one party generates a key and then transfers it to another party.

The standard will address protocol requirements for key management involving various combinations of end users, certification authorities, and other intermediaries (such as key directories).

The key management standard will also address the topic of how to distribute public keys (thus touching on the topic of key certificates).

At a later date it will also address the requirements of a key certification authority. This topic may not involve any cryptographic mechanisms, and will possibly be a list of procedural requirements. This work is at an early stage, and so this part of the standard may follow at a later date.

Evaluation Criteria

Design and implementation of logically secure networked systems presents interesting questions from a security "strength" perspective. Specifically:

- how do you meaningfully describe the security strength required; and
- how do you measure the security strength realised in implemented systems?

In the past, this question has been addressed by the major western countries for military and government systems through various trusted system evaluation criteria and interpretations. A standardised approach is now considered desirable to cover commercial systems and recognise the international nature of procurement.

The conventional approach is also considered to be inadequate in that evaluations are expensive, configuration specific, and biased towards confidentiality. The process of use of government criteria is essentially adversarial in that systems are procured to a security specification, and then an evaluator hired (with a vested interest in finding fault) to determine whether the supplied goods meet specification. In commercial applications, more of a cooperative approach is seen as desirable. In this case, equipment suppliers can make sustainable claims when products are offered, and after-the-event evaluation avoided as much as possible. Commercial evaluation criteria would also need increased emphasis on security services beyond confidentiality, be flexible in allowing preventative requirements to be traded-off against detection mechanisms, and take into account commercial disaster recovery practice.

Until the meeting of SC 27 held in October 1992, international standardisation work on evaluation criteria had been based on the European "ITSEC" harmonised government criteria. However it has now been proposed to significantly depart from this approach. Both the US and Canada are concerned that the ITSEC was too open-ended for practical use, that the functions were not logically organised, and that issues relating to mechanism effectiveness were too vague. It is also generally accepted that data exchange is very poorly handled in the ITSEC. For all these reasons, it was agreed that a better way had to be found. The "hot" issues are:

- what functions should be included as primary security objectives;
- how to handle practical/common groupings of functions (in "functionality classes"); and
- whether to group common functionality classes, and assurance/effectiveness packages into the US introduced concept of "Protection Profiles".

It seems likely that the functionality part of the evaluation criteria standard will be drafted around the functions of:

- Confidentiality;
- Integrity;
- Availability; and
- Accountability.

Under each heading will be "services" such as access control, etc. The Canadians go further, and suggest that groups of these services can be assembled in an hierarchy; eg Confidentiality with: access control; access control plus audit; access control plus audit plus ... .

Service classes may in turn be assembled into functionality classes.

Summary of Progress to Date

A security developers standards "tool-kit" currently contains standards or drafts covering:

- Guidelines for doing IT security;
- Modes of operation of 64-bit and n-bit block ciphers;
- Procedures for the registration of confidentiality algorithms;
- A general authentication model;
- Authentication based on public key techniques;
- Zero-knowledge techniques;
- Digital signature giving message recovery (for small data blocks);
- Hash functions;
- Non-repudiation;
- Key Management; and
- Evaluation criteria for IT systems.

Appendix 1 summarises the current work programme of ISO JTC 1/SC 27 and project status.

The Australian Security Standards Scene

Australia has been active in the area of security standards for some time, but with activity driven from the application area. In the early 1980's the potential importance of security in electronic banking, especially point of service applications, was recognised. A major effort was undertaken to develop national banking security standards in advance of international standards, which were seen as unlikely to be available in the time frame required by the Australian banking industry. As a result, Standards Australia has published a comprehensive set of standards for electronic banking, with more work in progress. Details of the Australian Banking security standards are given in appendix 2.

The development of mechanism standards by ISO JTC 1/SC 27 has been followed and contributed to by the banking security standards committee in Australia, with a view to utilising SC 27 standards where possible in the Australian banking standards. Although the mechanism standards have not received the attention of the banking application standards, Australia has a continuing, active if relatively low key, involvement in that work.

In parallel with the activity of the Standards Australia banking committees, Australia has a small group who have been tracking OSI security within the OSI standards committee in Australia, and people following developments in smart card standards, particularly concerning operational standards and inter-industry command standards.

A final topic of considerable potential significance is that of standards for EDI security. There is a Standards Australia committee devoted to this topic. Unlike the case for electronic banking a decade ago, there is no clear consensus on the need for provision of security services in EDI. Furthermore, the situation in international standards bodies is uncertain. Accordingly, progress is slow, and it is likely that Australia will not try to set standards in advance of the rest of the world. Internationally, there are several different EDI protocols in use, each with its own value added network service support, and each with differing security concepts being explored. In essence, it needs to be resolved whether it is necessary to integrate security into EDI messages at the finest level of detail (following the traditional banking approach), or whether it is sufficient to adopt a cleaner but less flexible architecture where security is decoupled from detailed message structure using an enveloping technique (following the traditional communications protocol approach). In this case, it is likely to be the demands of the business community that resolves these issues, rather than the standardisation process.

Conclusions

We have a wide range of security technologies that are relevant to distributed IT systems; however we don't know how to apply them. This paper has presented standardisation activities as a way to package and present security in a business relevant way to attempt to close the gap between theory and application. There are, however several impediments to the development of such standards. Firstly, the technical content of standards is not obvious when working near the state of the art. Achieving the necessary quality is not easy when the availability of people with suitable expertise is limited. This is especially true in small countries like Australia. Finally, in some cases, industry is not particularly motivated to contribute to the development of security standards, because they have limited ability to exploit them in "world" products due to export restrictions that apply to security products in many countries.

Despite these difficulties, we know that an increasing number of IT systems are being built that are not as safe as they should be. There is also increasing anecdotal evidence in the trade press that there is substantial and increasing fraud in this area. The challenge is to close the gap between (latent) need and implementable solutions before some business crisis occurs. An increased awareness of standardisation activities, followed by increased effort to develop high quality security standards is an important part of the solution.

Acknowledgment

The permission of the Director of Research, Telecom Australia to publish this paper is hereby acknowledged.

APPENDIX 1 - ISO/IEC JTC 1/SC 27 Standards/Project List

Items with an "ISO" prefix are published ISO standards. Items with a "DIS" (Draft International Standard) prefix are not yet published standards, but have reached a level of technical stability. Items with a "CD" (Committee Draft) prefix have agreed objectives, but are at an early stage of preparation, and subject to possible substantial technical change. Items with a "New" prefix are those for which work has only just begun.

- ISO 8372 - Information Technology - Security Techniques -
  Modes of Operation for a 64-bit Block Cipher Algorithm.

- ISO 9796 - Information Technology - Security Techniques -
  Digital Signature Scheme giving Message Recovery.

- ISO 9797 - Information technology - Security Techniques -
  Data Integrity Mechanism using a Cryptographic Check Function Employing a Block Cipher Algorithm.

- ISO 9798-1 - Information Technology - Security Techniques -
  Entity Authentication Mechanisms, Part 1: General Model.

- CD 9798-2 - Information Technology - Security Techniques -
  Entity Authentication Mechanisms Part 2: Entity Authentication Mechanisms using a Symmetric Algorithm.

- CD 9798-3 - Information Technology - Security Techniques -
  Entity Authentication Mechanisms, Part 3: Entity Authentication Using a Public Key Algorithm.

- New (9798-4) - Information Technology - Security Techniques -
  Entity Authentication Mechanisms, Part 4: Entity Authentication Using Non-reversible Functions.

- New (9798-5) - Information Technology - Security Techniques -
  Entity Authentication Mechanisms, Part 5: Entity Authentication Using Zero Knowledge Techniques.

- ISO 9979 - Information Technology - Security Techniques - **Procedures for the Registration of Cryptographic Algorithms.**

- ISO 10116 - Information Technology - Security Techniques - **Modes of Operation for an n-bit Block Cipher Algorithm.**

- DIS 10118-1 - Information Technology - Security Techniques - **Hash-Functions for Digital Signatures and Authentication Mechanisms, Part 1: General**

- DIS 10118-2 - Information Technology - Security Techniques - **Hash-Functions for Digital Signatures and Authentication Mechanisms, Part 2: Hash Functions using a Symmetric Block Cipher Algorithm.**

- Possible new - Information Technology - Security Techniques - **Hash Functions Part 3: Dedicated Hash Functions.**

- Possible new - Information Technology - Security Techniques - **Hash Functions Part 4: Hash Functions Using Modular Arithmetic.**

- New - Information Technology - Security Techniques - **Key Management, Part 1: Framework.**

- New - Information Technology - Security Techniques - **Key Management, Part 2: Mechanisms for Key Management using Symmetric Techniques.**

- New - Information Technology - Security Techniques - **Key Management, Part 3: Mechanisms for Key Management using Asymmetric Techniques.**

- New - Information Technology - Security Techniques - **Key Management, Part 4: Mechanisms for Cryptographic Separation.**

- New - Information Technology - Security Techniques - **Security Mechanisms using Zero Knowledge Techniques, Part 1: General Model.**

- New - Information Technology - Security Techniques - **Security Mechanisms using Zero Knowledge Techniques, Part 2: Mechanisms Based on Identity and Factorisation.**

– New - Information Technology - Security Techniques -
Non-repudiation Mechanisms, Part 1: General Model.

– New - Information Technology - Security Techniques -
Non-repudiation Mechanisms, Part 2: Non Repudiation Mechanisms using Symmetric Techniques.

– New - Information Technology - Security Techniques -
Non-repudiation Mechanisms, Part 3: Non Repudiation Mechanisms using asymmetric Techniques.

– New - Information Technology - Security Techniques -
Digital Signature with Appendix.

– New - Information Technology - Security Techniques -
Security Information Objects.

– New - Information Technology - Security Techniques -
Guide-lines for the management of IT Security (to be a technical report).

– New - Information Technology - Security Techniques -
Guide-lines for the Use and Management of Trusted Third Party Services.

– New - Information Technology - Security Techniques -
Guide-lines for the Use and Selection of Security Services and Mechanisms for IT Security.

– New - Information Technology - Security Techniques -
Collection and Analysis of Requirements for IT Security Evaluation Criteria (to be a technical report).

– New - Information Technology - Security Techniques -
Evaluation Criteria for IT Security - Part 1 - General Model.

– New - Information Technology - Security Techniques -
Evaluation Criteria for IT Security - Part 2 - Functionality Classes of IT Systems.

– New - Information Technology - Security Techniques -
Evaluation Criteria for IT Security - Part 3 - Assurance of IT systems.

– New - Information Technology - Security Techniques - **Procedures for the Registration of Hash Functions.**

APPENDIX 2 - Standards Australia Electronic Banking Standards

Standards Australia has published, or is preparing, the following electronic banking standards in the AS2805 "Electronic funds transfer - Requirements for interfaces" series.

Part 1	Communications (1991)
Part 2	Message structure, format and content (1986)
Part 3	PIN management and Security (1985)
Part 4	Message Authentication (1985)
Part 5.1	Ciphers - Data encipherment algorithm 1 (DEA-1) (1992)
Part 5.2	Ciphers - Modes of operation for an n-bit block cipher algorithm (1992)
Part 5.3	Ciphers - Data encipherment algorithm 2 (DEA-1) (1992)
Part 6.1	Key Management - Principles (1988)
Part 6.2	Key Management - Transaction keys (1988)
Part 6.3	Key Management - Node to Node (1988)
Part 6.4	Key Management - Terminal to Acquirer (1988)
Part 6.5.1	Key Management - TCU Initialization - Principles (1992)
Part 6.5.2	Key Management - TCU Initialization - Symmetric (in preparation)
Part 6.5.3	Key Management - TCU Initialization - Asymmetric (1992)
Part 6.5.4	Key Management - TCU Initialization - Key Activation (in preparation)
Part 7	POS Message Content (1986)
Part 8	Financial institution message content (1986)
Part 9	Privacy of communications (1991)
Part 10	Secure file transfer (in preparation)
Part 11	Card parameter table (in preparation)

Non–Interactive Generation of Shared Pseudorandom Sequences

Manuel Cerecedo[1]*, Tsutomu Matsumoto[1] and Hideki Imai[2]

[1] Division of Electrical and Computer Engineering, Yokohama National University, 156 Tokiwadai, Hodogaya, Yokohama, 240 Japan
[2] Institute of Industrial Science, University of Tokyo, 7–22–1 Roppongi, Minato–ku, Tokyo, 106 Japan

Abstract. We address the following problem: given a random seed secretly shared among a group of individuals, non–interactively generate pieces corresponding to a much longer shared pseudorandom sequence. Shared randomness is an essential resource in distributed computing and non–interactive ways of generating it can be useful in applications such as Byzantine Agreement, common coin flipping or secure computation protocols.
Our first result is negative: well known cryptographically strong pseudorandom number generators cannot be evaluated without interaction and, in particular, it is shown that constructions that recursively apply a one–way function to a random seed and output at each iteration the simultaneously hard bits in the input of the one–way function are actually incompatible with a homomorphic evaluation.
On the other hand, we show that pseudorandom generators that can be both proven cryptographically strong and sharedly evaluated without interaction do exist. A concrete implementation, under the RSA assumption, is described.

1 Introduction

The straightforward procedure for a group of individuals to compute a common pseudorandom string consists of two stages: first, individually generate independent pseudorandom sequences, that are combined afterwards by means of an interactive protocol (e.g. common coin flipping [21]). Protocols for verifiable secret sharing (VSS, [10, 24, 12]) provide a very simple way of combining the individual pseudorandom strings: each individual verifiably shares its string and those that were correctly shared are linearly combined without interaction, which is possible when the secret sharing scheme is homomorphic. Observe that such procedure is still interactive (two rounds) even if we assume that the VSS sharing takes a single round, as it is the case with 'non–interactive' VSS protocols [12, 23] —notice however that these VSS protocols assume the existence of

* Supported by Japanese Ministry of Education, Science and Culture Scholarship, No. 890864.

'broadcast channels', which in realistic settings would have to be simulated by Byzantine Agreement (BA, e.g. [14]).

Common randomness is a fundamental resource in distributed computing (e.g. there is a constant–rounds reduction of BA to a common coin; see [9]) and, in particular, it is essential to construct secure computation protocols [16, 3]: recall, for example, that the multiplication protocol in [3] requires, during the degree reduction of the polynomial used to share the product, a randomization step which involves the generation of a shared random polynomial. Furthermore, shared pseudorandom sequences can also be used to improve the efficiency of secure computation protocols reducing the need for interaction: in [7], cryptographically strong pseudorandom generators and common coin flipping are used to construct a secure computation protocol requiring only a constant number of rounds of interaction. Since non–interactive secure computation appears to be not possible in the general case (see [15]), improving the efficiency of shared randomness generation may be the best we can do to reduce the communication complexity of secure protocols.

The notion of pseudorandomness we are concerned with for cryptographic purposes is that defined in [6, 26], in terms of polynomial indistinguishability of the pseudorandom sequence from a truly random sequence. Although many, and very general (namely based on any one–way function: [19, 18]) generators have been proposed according to this definition, all of them (with one exception, we will recall below [20]) have followed the central idea in [6] of recursively applying the one–way function and outputting the simultaneously 'hard' (unpredictable) bits in the input. As we will see, it turns out that this construction is incompatible with a homomorphic evaluation of the generators. To make such a homomorphic evaluation possible, we will have to look for different generators.

To ascertain the existence of pseudorandom number generators that can be both proven cryptographically strong and evaluated by a non–interactive protocol we recall generators first discussed by Shamir in [25] and show that a construction related to one of these generators is cryptographically strong under the RSA assumption and can be computed by a non–interactive protocol.

First, to formalize this notion of non–interactive generation of shared pseudorandomness, we define 'pseudorandom generation protocols' as a natural generalization of the concept of cryptographically strong pseudorandom number generation to a shared setting: the generating algorithm is substituted by a protocol and the passive distinguisher, by an arbitrary adversary participating in the generation protocol.

2 Model and Definitions

We consider protocols for a network $\mathcal{N}$ of n computationally bounded probabilistic machines (participants) that secretly shared some information during a preprocessing stage and afterwards communicate only through *public channels* (denoted $i \rightarrow \mathcal{N}$: $\mathsf{message}$): Every message is sent to all the participants but any receiver is not necessarily guaranteed that every other processor received the same message (that is, not broadcast channels in strict sense, as those simulatable by a Byzantine Agreement protocol). Communication is synchronous: In each round, every participant receives the messages that every other participant sent in the previous round, makes some local computation and sends the same output message to every other participant.

We make the assumption that more than half of the participants do not deviate from the protocol; arbitrarily deviating participants are formally dealt with by introducing an ideal adversary $\mathcal{A}$ dynamically corrupting less than half of the participants.

The notation $\{x_i\}_{i\in\mathcal{I}}$, for a given index set $\mathcal{I} = \{i_1, \ldots, i_k\}$, will be used as a shorthand for $\{x_{i_1}, \ldots, x_{i_k}\}$. By $x \in_R X$ it is meant that x is uniformly drawn at random from set X.

An essential component of the protocols below is a *homomorphic* (t,n)–*threshold scheme*[3]: If S is a (finite) set of secrets, a secret $s \in S$ is secretly shared among the n participants in $\mathcal{N}$ if each $i \in \mathcal{N}$ holds a piece —denoted s_i— from a set of pieces P such that, if $\mathcal{A} \subset \mathcal{N}$ is any subset of t or more participants, $\mathcal{B} \subset \mathcal{N}$ any subset of less than t participants, and p denotes the public information accessible to any participant (including, for example, messages broadcasted for verification of the secret pieces), then: (1) $H(s \mid p, \{s_i\}_{i\in\mathcal{A}}) = 0$; (2) $H(s \mid p, \{s_i\}_{i\in\mathcal{B}}) = H(s)$. With H the Shannon entropy function such secret sharing schemes are said to be *perfect*; in computationally bounded settings, as in this paper, the secrecy of s is not unconditional but an analogous definition of *computationally perfect* threshold schemes, in terms of effective entropy [26], can be used instead. Below, by perfect threshold schemes we will mean schemes that are perfect in the computational sense. A (t,n)–threshold scheme is $(\oplus,\otimes)$–homomorphic [5] when, for two given binary functions $\oplus$ and $\otimes$ in S and P respectively, $z = x \oplus y$ iff $z_i = x_i \otimes y_i$, for any $x, y, z \in S$ and every $i \in \mathcal{N}$.

Recall the definition of *cryptographically strong pseudorandom number generator* (from [6, 26]): any polynomial time probabilistic distinguisher $\mathcal{D}$ receiving the string $f(x) \in \{0,1\}^{Q(l)}$ from the deterministic generator with a shorter secret input $x \in_R \{0,1\}^{P(l)}$, with l the security parameter and $P(l), Q(l)$ any given polynomials, has negligible (decreasing faster than the inverse of any polynomial in l) probability of distinguishing $f(x)$ from a $y \in_R \{0,1\}^{Q(l)}$, i.e.

$$|\Pr[\mathcal{D}(f(x)) = 1] - \Pr[\mathcal{D}(y) = 1]|$$

is negligible.

To extend this notion to our shared pseudorandom number generation problem notice that, instead of a passive distinguishing machine $\mathcal{D}$ receiving the output of the generator, we have a generic adversary $\mathcal{A}$ participating in the generation protocol. This adversary may influence the protocol in arbitrary ways and obtain additional information that helps distinguishing the output of the protocol from a random string. We can say that the output is pseudorandom

[3] The discussion can be extended to general homomorphic secret sharing schemes; see [13].

only if even the adversary $\mathcal{A}$ participating in the generation protocol cannot convince a computationally bounded distinguisher that the output is not pseudorandom. As in the context of zero–knowledge interactive proof systems [17] or secure computation protocols [22, 4] this notion can also be formalized in terms of polynomial indistinguishability of probability ensembles; we will follow the same approach and say that the protocol outputs a pseudorandom string if the view of the adversary $\mathcal{A}$ (including all the information accessible to $\mathcal{A}$ during the protocol and the pseudorandom output) in the actual generation protocol and $\mathcal{A}$'s view in a simulated protocol that he may have produced without actually participating in the protocol and with a truly random string substituted for the pseudorandom output are indistinguishable to any computationally bounded machine $\mathcal{D}$.

We first propose a general definition of PRG protocol, independently of the communication model for the network $\mathcal{N}$, and after restrict ourselves to the non–interactive case, in which only public–channels are used.

Definition 1. Let l denote a security parameter, $r \in_R \{0,1\}^{P(l)}$ be (t,n)–secretly shared among the n participants in $\mathcal{N}$, and $\mathcal{A}$ denote an adversary corrupting less than t participants. A protocol Π is a **string generation protocol** with adversary $\mathcal{A}$ if at the end of Π there is a common string s of length $Q(l)$ (longer than r) known to every participant, for any given polynomials $P(l)$ and $Q(l)$. Let $\mathsf{VIEW}_{\mathcal{A}}(\Pi, \mathsf{O}(\Pi))$ denote the probability distribution of the views of adversary $\mathcal{A}$ during string generation protocol Π (including the output $\mathsf{O}(\Pi)$); and let $\mathsf{VIEW}_{\mathcal{A}}(\mathcal{S}, y)$ be the probability distribution of the views of $\mathcal{A}$ in a protocol simulated by a machine $\mathcal{S}$ with access to the adversary's private inputs and output string y. A string generating protocol Π is **pseudorandom** if for any adversary $\mathcal{A}$ there exists a simulator $\mathcal{S}$ working in expected polynomial time such that

$$|\Pr[\mathcal{D}(\mathsf{VIEW}_{\mathcal{A}}(\Pi, \mathsf{O}(\Pi))) = 1] - \Pr[\mathcal{D}(\mathsf{VIEW}_{\mathcal{A}}(\mathcal{S}, y)) = 1]|$$

is negligible (decreases faster than the inverse of any polynomial in l) for any polynomial time probabilistic distinguisher $\mathcal{D}$ and $y \in_R \{0,1\}^{Q(l)}$.

Definition 2. A pseudorandom number generation protocol Π is **non–interactive** if there exist m and k, and a sequence of k secretly shared values $\{s^i \in \{0,1\}^m\}_{1 \le i \le k}$ such that $s = \mathsf{O}(\Pi) = s^1 \circ s^2 \circ \cdots \circ s^k$ ($\circ$ denotes concatenation) and the pieces corresponding to every s^i can be locally generated without communication. That is, the pseudorandom sequence can be reconstructed using only public channels and one round of interaction for every sub–string of m bits (e.g., if $m = 1$, bit by bit; if $m = Q(l)$, the whole sequence has to be reconstructed in one round).

3 Non–interactive Blum–Micali Pseudorandom Bit Generators

A central idea to most of the constructions of pseudorandom generators from one–way functions [6, 19, 18] has been to output the simultaneously hard bits in the input of the one–way function: bits of the input that cannot be guessed better than random by any polynomially bounded observer given access to the output of the one–way function. That is, in terms of indistinguishability, with f a one–way function on $X \subset \{0,1\}^n$ and $b^{k(n)}$ a polynomial–time computable function from domain X into range $\{0,1\}^{k(n)}$; the output bits of $b^{k(n)}$ are *simultaneously hard* if no polynomially bounded observer can distinguish them from random bits:

$$|\Pr[\mathcal{M}(f(x), b^{k(n)}) = 1] - \Pr[\mathcal{M}(f(x), y) = 1]|$$

is negligible for any polynomially bounded distinguisher $\mathcal{M}$ and $y \in_R \{0,1\}^{k(n)}$.

Generators that operate successively applying a one–way permutation on a random seed x and outputting the simultaneously hard bits of the input at each iteration as follows:

$$S_n \stackrel{\text{def}}{=} b^{k(n)}(f(x)) \circ b^{k(n)}(f^2(x)) \circ \cdots \circ b^{k(n)}(f^l(x))$$

where $b^{k(n)}(y)$ denotes the $k(n)$ simultaneously hard bits of y, and $f^2(x)$ denotes $f(f(x))$ are called Blum–Micali pseudorandom bit generators: $[X, f, b^{k(n)}]$.

We first address the problem of whether these generators can be evaluated in a non–interactive way. The following theorem gives a negative answer.

Theorem 1. *Let $X \subset \{0,1\}^n$ be a finite set, $f : X \mapsto X$ a one–way permutation with $k(n)$ simultaneously hard bits in its input, and $b^{k(n)} : X \mapsto \{0,1\}^{k(n)}$ a polynomial–time computable function. Given a network $\mathcal{N}$ and some $x \in_R X$ secretly shared by the participants in $\mathcal{N}$ in a perfect (t,n)–threshold scheme, if there exists a non–interactive protocol Π for $\mathcal{N}$ to evaluate the sequence*

$$S_n = b^{k(n)}(f(x)) \circ b^{k(n)}(f^2(x)) \circ \cdots \circ b^{k(n)}(f^l(x))$$

with $x \in_R X$, then S_n is not pseudorandom.

Proof. We show that if such non–interactive protocol exists the output bits of the function $b^{k(n)}$ cannot be simultaneously hard and thus there exists some polynomial test that distinguishes the previous sequence from a truly random sequence.

First, since the protocol is non–interactive, the participants in $\mathcal{N}$ can independently generate the pieces corresponding to the values $f(x), f^2(x), \ldots, f^l(x)$, that is, there must exist functions $\{f_i\}_{i\in\mathcal{N}}$ such that $f_i(x) = [f(x)]_i, \ldots, f_i^l(x) = [f^l(x)]_i$, where $[f(x)]_i$ denotes i's piece of the value $f(x)$ in a perfect (t,n)–threshold scheme. Since $k(n)$ bits in the input of f are simultaneously hard, there must also be at least $k(n)$ simultaneously hard bits in the input of the functions f_i, for every i.

Also, since the participants in $\mathcal{N}$ can independently generate the pieces corresponding to the values $b^{k(n)}(f(x)), \ldots, b^{k(n)}(f^l(x))$, there must exist functions $b_i^{k(n)}$ such that $b_i^{k(n)}(f(x)) = [b^{k(n)}(f(x))]_i, \ldots, b_i^{k(n)}(f^l(x)) = [b^{k(n)}(f^l(x))]_i$, for every i.

Now, if we assume that the output of $b^{k(n)}$ are the simultaneously hard bits of f, since the threshold scheme is perfect, the output of the $b_i^{k(n)}$ must also be the simultaneously hard bits of f_i, for every i. That is, the sequences

$$S_{n_i} = b_i^{k(n)}(f_i(x)) \circ b_i^{k(n)}(f_i^2(x)) \circ \cdots \circ b_i^{k(n)}(f_i^l(x))$$

for every $i \in \mathcal{N}$ must be pseudorandom. However, the values $b_i^{k(n)}(f_i(x))$ are the secret pieces of $b^{k(n)}(f(x))$ in a threshold scheme and t of them determine the others, which is a contradiction with the pseudorandomness of the sequences S_{n_i}. □

In [20], Impagliazzo and Naor proposed a novel construction of pseudorandom generator, directly proving the impredictability of all the output bits of a one-way function, mapping n bit strings into values mod $2^{l(n)}$, with $l(n) > n$, and based on the intractability of the subset sum problem. Unfortunately, this generator, which outputs the last $l(n) - n$ bits of the function and uses the rest as a new seed, turns also out to be incompatible with a homomorphic evaluation.

4 Constructions Based on the RSA Assumption

In this section we show that shared pseudorandom sequences can indeed be generated by a non-interactive protocol, given a homomorphic secret sharing scheme. We will recall the generators discussed in [25], preceding the work of Blum and Micali. Though Shamir's notion of unpredictable number sequences leaves unsolved the question of whether individual bits are also unpredictable, the particular constructions are suitable for homomorphic sharing schemes and thus they may still be found useful. We will argue that universal families of hash functions can be used to extract, from the output of one of the former generators, cryptographically strong pseudorandom sequences. Only RSA-based generators are described; the discussion can be generalized to other homomorphic one-way permutations.

Assumption 1. The RSA permutation $f : x \mapsto x^e \pmod{N}$ on Z_N, where N (of length n bits) is the product of two large primes p and q and $e > 1$ is relatively prime to $\varphi(N)$, is one-way.

Alexi *et al.* [2] proved that the $O(\log n)$ least significant bits of the input of the RSA permutation are simultaneously hard under Assumption 1. That is, in the sequel $k(n) \geq O(\log n)$

4.1 Pseudorandom Number Generators

The following are variants of pseudorandom generation constructions discussed in [25].

Generator 1. Let f denote the RSA permutation with exponent e and $f^2(x) = f(f(x))$. Given two secret seeds r and s randomly chosen in Z_N, the sequence

$$f(r)\cdot f^l(s),\ f^2(r)\cdot f^{l-1}(s),\ \ldots,\ f^l(r)\cdot f(s)$$

can be proved [25] to be unpredictable in the sense that any element of the sequence cannot be computed given another single element.

Generator 2. Let f_i denote the RSA permutation with exponent e_i, and with $\{e_i\}_{1\leq i\leq l}$ such that for any e_i, $\gcd(\{e_j\}_{j\neq i})/\gcd(e_1,\ldots,e_l) > 1$, and a secret seed r uniformly chosen at random in Z_N, the sequence

$$f_1(r),\ f_2(r),\ldots,\ f_l(r)$$

was considered in [1]; a related sequence had been discussed also in [25]. The problem of computing any element of the sequence given all the others was proved equivalent to the problem of inverting RSA, that is, from a probabilistic polynomial time algorithm A that, given a set of elements, succeeds in computing a new element with non–negligible probability, we can find another probabilistic polynomial time algorithm A' to compute an arbitrary RSA–root and vice versa. A generalized result appears in [11].

Clearly, both of these sequences can be generated given a $(\cdot,\cdot)$–homomorphic secret sharing scheme. The problem we address now is whether a cryptographically strong pseudorandom sequence can be obtained from the output of this last generator. Intuitively, we want to convert the unpredictability of the generator output into uniform random bits. With this purpose, we use ***universal families of hash functions*** [8]: A family of functions $H_{n,k}$ mapping $\{0,1\}^n$ to $\{0,1\}^k$ is a universal family of hash functions if, for every $x, y \in \{0,1\}^n, x \neq y$, then $\Pr[h(x) = h(y)] = 1/2^k$, for $h \in_R H_{n,k}$.

Generator 3. With f_i and r as in the previous generator, and h chosen uniformly at random in a family of universal hash functions $H_{n,k(n)}$, assuming RSA simultaneously hides $k(n)$ input bits (from [2], $k(n) = \log(|N|)$), we define the ensemble S, with

$$S_n \stackrel{\text{def}}{=} h(f_1(r)) \circ h(f_2(r)) \circ \cdots \circ h(f_l(r))$$

We argue that S is computationally indistinguishable from the uniform ensemble U of the same length even if the distinguisher is also given the sequence $\{f_i\}_{1\leq i\leq l}$

Lemma 1. *The two following ensembles are polynomially indistinguishable:*

$$(N,\ \{f_i(r)\}_{1\leq i\leq l},\ h,\ \{h(f_i(x))\}_{1\leq i\leq l})$$

$$(N,\ \{f_i(r)\}_{1\leq i\leq l},\ h,\ y \in_R \{0,1\}^{l\cdot k(n)})$$

for N randomly chosen in the set of acceptable moduli, $r \in_R \{0,1\}^n$ and $h \in_R H_{n,k(n)}$.

Proof. Given that RSA simultaneously hides $k(n)$ bits, that is, the distribution of $y^{1/e} \bmod N$ for $y \in_R Z_N$ is at least as random as the uniform distribution on $k(n)$ bits to any polynomially bounded observer. From the equivalence between computing RSA–roots and unknown elements in S_n for Generator 2, given an arbitrary set of elements in S_n, the distribution of a new $f_j(x)$ must be also at least as random for a polynomially bounded observer as the uniform distribution on $k(n)$ bits; the lemma follows from the definition of universal hashing. □

4.2 $(\cdot,\cdot)$–Homomorphic Threshold Scheme

Following the observation of Feldman [12], Shamir's threshold scheme can be made $(\cdot,\cdot)$–homomorphic in the following way: S and P are both Z_N, pieces are defined as $x_i = x \cdot p_1^i \cdots p_{t-1}^{i^{t-1}} \bmod N$, for $1 \le i \le n$, where the coefficients $\{p_j\}_{1\le j\le t-1}$ are uniformly chosen in Z_N. For any $T \subset \{1,2,\ldots,n\}$, s.t. $|T| = t$, the reconstruction function is given by

$$x = \prod_{i \in T} x_i^{\prod_{j\in T; j\neq i} \frac{j}{j-i}} \bmod N$$

The fact that not all of the $(j-i)$'s have inverses modulo $\varphi(N)$ turns out to be irrelevant since in our context the value x is the original seed of the pseudo–random generator and the value $x^{n!}$ can always be reconstructed because $\prod_{j\in T; j\neq i}(j-i) \mid n!$. The sequence of values to be reconstructed will be

$$x^{n!\cdot e_1}, x^{n!\cdot e_2}, \ldots, x^{n!\cdot e_l} \pmod N$$

In the presence of corrupted participants, pieces to be used in the reconstruction stage have to be verified. We will assume that the random seed was secretly generated during the preprocessing stage by simultaneously running n instances of a VSS protocol and secretly combining the correctly shared values. We also assume that the values $\{p_k^{e_1}\}_{1\le k\le t-1}$ are known to every participant from the preprocessing stage. Despite we will be interested in protocols for Generator 3, in which case the verification of pieces to be used in the reconstruction of the pseudorandom sequence is straightforward, we discuss an alternative verification algorithm, using second order pieces, which we may use for Generator 1, for example.

Verification (simple). To verify pieces of the form $x_i^{e_i}$, when the $t-1$ values $\{p_k^{e_1}\}_{1\le k\le t-1}$ are known to every participant from the preprocessing stage. We check:

$$(x_i^{e_i})^{e_1} \stackrel{?}{\equiv} (x^{e_1})^{e_i} \cdot (p_1^{e_1})^{e_i \cdot i} \cdots (p_{t-1}^{e_1})^{e_i \cdot i^{t-1}} \pmod N$$

Notice that this verification procedure cannot be used for Generator 1, the procedure below is more general.

Verification (with second order pieces). To verify pieces of any form, when each participant has also 'second order' pieces (pieces of the 'first order' pieces).

For example, with Generator 1, after the preprocessing stage, each participant i has the first order pieces of the seeds, r_i and s_i, and the second order pieces $\{r_{ji}, s_{ji}\}_{j\in\mathcal{N}}$, then, for $i, j \in \mathcal{N}$:

$$r_i = r \cdot q^{i}_{r_1} \cdots q^{i^{t-1}}_{r_{t-1}} \bmod N, \quad s_i = s \cdot q^{i}_{s_1} \cdots q^{i^{t-1}}_{s_{t-1}} \bmod N$$

$$r_{ij} = r_i \cdot q^{j}_{r_{i_1}} \cdots q^{i^{t-1}}_{r_{j(t-1)}} \bmod N, \quad s_{ij} = s_i \cdot q^{j}_{s_{i_1}} \cdots q^{i^{t-1}}_{s_{j(t-1)}} \bmod N$$

And, to reconstruct $f^i(r) \cdot f^{l-i+1}(s)$, each participant reveals:

$$f^i(r_i) \cdot f^{l-i+1}(s_i) \text{ and } \{f^i(q_{r_{ij}}) \cdot f^{l-i+1}(q_{s_{ij}})\}_{1\leq j\leq t-1}$$

So that the first order pieces can be verified.

4.3 Shared Pseudorandom Generation Protocol

Given the results of previous sections, the protocol is straightforward; to generate an instance of the generator (steps 1.1 to 1.3 of protocol Π below), the participants can use a trusted individual. If this is not possible, all the computations required to generate N and h (for example, the probabilistic primality testing algorithms) can theoretically be performed by a secure computation protocol, of questionable practical value . The protocol Π is as follows: To generate the random seed, each participant i secretly shares, with a VSS protocol (for example, Feldman's non–interactive protocol [12]; VSS.share, step 2.2), a randomly chosen $r^{(i)}$ and correctly shared values are multiplied (step 2.3). After step 2.3 the values $\{p_k^{e_1}\}_{1\leq k\leq t-1}$ are known to every participant.

To generate the k–th element of the sequence, every participant i just computes $r_i^{e_k} \bmod N$ on its secret piece r_i, step 3.1. The reconstruction and verification is as above; the reconstructed sequence is: $R = r_{(1)} \circ r_{(2)} \circ \cdots \circ r_{(l)}$. See Figure 1.

Theorem 2 *Protocol Π is a non–interactive pseudorandom number generation protocol for any adversary $\mathcal{A}$ corrupting less than $t < n/2$ participants.*

Proof. The sequence generated by the protocol is the one discussed in Lemma 1; intuitively, since the only information revealed by the participants are the values $x^{e_i \cdot n!} \bmod N$ and Generator 3 is cryptographically strong, the sequence R is pseudorandom. To formally prove that this protocol satisfies Definition 1 we have to construct a simulator $\mathcal{S}$ of the adversary's view, with a truly random string instead of the pseudo–random output, and such that the actual and simulated views are polynomially indistinguishable. A static adversary corrupting t participants $i_1, \ldots, i_t$ from the outset is assumed to simplify the proof; the simulator, with access to the private inputs of participants $i_1, \ldots, i_t$, is outlined below.

From step 3.1, for $k \geq 2$, first uniformly choose a random $x \in_R Z_N$ and compute $s_{(k)} = x^{e_k} \bmod N$. As in Lemma 1, there are at least $k(n)$ bits of $s_{(k)}$ that cannot be guessed by $\mathcal{A}$ better than at random; substitute these bits by

Protocol Π

GENERATE INSTANCE:

1.1 $N \in_R M_n$ (set of acceptable RSA moduli)
1.2 $h \in_R H_{n,k(n)}$
1.3 $\{e_i\}_{1 \le i \le l}$ s.t. $\forall e_i,\ \frac{\gcd(\{e_j\}_{j \ne i})}{\gcd(e_1,\ldots,e_l)} > 1$

GENERATE SHARED SEED

2.1 i: $r^{(i)} \in_R Z_N$
2.2 i: Run VSS.share on $r^{(i)}$
2.3 i: $r_i \leftarrow \prod_{j \text{ well shared}} r_i^{(j)}$
$r_i = x \cdot p_1^{i} \cdots p_{t-1}^{i^{t-1}} \bmod N$
$\{p_k^{e_1} \bmod N\}_{1 \le k \le t-1}$

LOCAL COMPUTATION $(k = 2, \ldots, l)$

3.1 i: $r_{(k)i} \leftarrow r_i^{e_k} \bmod N$

RECONSTRUCTION $(k = 2, \ldots, l)$

4.1 $i \to \mathcal{N}$: $r_{(k)i}$
4.2 i: VSS.verify (simple) $\{r_{(k)j}\}_{j \in \mathcal{N}}$
4.3 i: VSS.reconstruct $r^{e_k \cdot n!} \bmod N$
4.4 i: $r_{(k)} \leftarrow h(r^{e_k \cdot n!})$

Fig. 1. Shared pseudorandom generation protocol under the RSA assumption

truly random bits and call the result $s^*_{(k)}$. Given this value and the local values $\{r_i^{e_k} \bmod N\}_{i=i_1 \ldots i_t}$ of the corrupted participants, the simulator can compute the remaining $n - t$ pieces corresponding to the secret $s^*_{(k)}$; we denote these faked pieces by $s^*_{(k)i}$. In step 4.1, corrupted participants broadcast the values $\{r_i^{e_k} \bmod N\}_{i=i_1 \ldots i_t}$ and good participants are simulated by broadcasting the faked pieces $s^*_{(k)i}$. The value that the participants reconstruct from the broadcasted pieces is hashed as in the real protocol.

We argue that the probability distributions of the real and simulated protocols are indistinguishable to any polynomially bounded observer: this follows from the fact that the threshold scheme is perfect and the values $s_{(k)}$ and $s^*_{(k)}$ are polynomially indistinguishable, so that the distribution of the faked values broadcasted during step 4.1 of the simulator and of the real protocol are indistinguishable. Also, from the way we generated the $s^*_{(k)}$ it follows that the output of the simulated protocol is truly random. □

5 Conclusions

We have considered the problem of non-interactively generating a secretly shared pseudorandom sequence from a short secretly shared seed. Cryptographically strong pseudorandom generators based on outputting the hard bits of the one-way function, as in the Blum–Micali construction, appear to require interactive protocols. To construct non-interactive pseudorandom generation protocol we have considered a concrete generator and shown it to be cryptographically strong. The solution proposed, despite being non-interactive, incurs high (polynomial) computational costs and we are left with the problem of finding efficient non-interactive generators.

Acknowledgements

We are grateful to the anonymous referees for several helpful comments on the previous version of this manuscript.

References

1. S. Akl and P. Taylor. "Cryptographic solution to a problem of access control in a hierarchy." *ACM TOCS,* 1, 1983, pp. 239–248.
2. W. Alexi, B. Chor, O. Goldreich and C.P. Schnorr. "RSA and Rabin Functions: Certain Parts are as Hard as the Whole." *SIAM Journal on Computing,* vol. 17, no. 2, April 1988, pp. 194–209.
3. M. Ben–Or, S. Goldwasser and A. Wigderson. "Completeness Theorems for Non–Cryptographic Fault–Tolerant Distributed Computation." *Proc. 20th STOC,* ACM, 1988, pp. 1–10.
4. D. Beaver. "Foundations of Secure Interactive Computing." *Proc. Crypto '91,* Springer–Verlag, LNCS vol. 576, pp. 377–391.
5. J.C. Benaloh. "Secret Sharing Homomorphisms: Keeping Shares of a Secret Secret." *Proc. Crypto '86.* Springer–Verlag, LNCS vol. 293, 1987.
6. M. Blum and S. Micali. "How to Generate Cryptographically Strong Sequences Of Pseudo–Random Bits." *Proc. 22nd FOCS,* IEEE, 1982, pp. 112–117.
7. D. Beaver, S. Micali and P. Rogaway. "The Round Complexity of Secure Protocols." *Proc. 22nd STOC,* ACM, 1990, pp. 503–513.
8. J. Carter and M. Wegman. "Universal Classes of Hash Functions." *Journal of Computer and System Sciences,* 1979, vol. 18, pp. 143–154.
9. B. Chor and C. Dwork. "Randomization in Byzantine Agreement." *Advances in Computing Research,* vol. 5, JAI Press, 1989, pp. 443–497.
10. B. Chor, S. Goldwasser, S. Micali and B. Awerbuch. "Verifiable Secret Sharing and Achieving Simultaneity in the Presence of Faults." *Proc. 26th FOCS,* IEEE, 1985, pp. 383–395.
11. J.–H. Evertse and E. van Heyst. "Which New RSA–Signatures Can Be Computed from Certain Given RSA–Signatures?" *Journal of Cryptology,* vol. 5, no. 1, 1992, pp. 41–52.
12. P. Feldman. "A Practical Scheme for Non–Interactive Verifiable Secret Sharing." *Proc. 28th FOCS,* IEEE, 1987, pp. 427–437.
13. Y. Frankel and Y. Desmedt. "Classification of ideal homomorphic threshold schemes over finite Abelian groups." *Proc. Eurocrypt '92.* To appear in Springer–Verlag, LNCS.

14. P. Feldman and S. Micali. "Optimal Algorithms for Byzantine Agreement." *Proc. 20th STOC,* ACM, 1988, pp. 148–161.
15. Y. Frankel, Y. Desmedt and M. Burmester. "Non–existence of homomorphic general sharing schemes for some key spaces." *Proc. Crypto '92.* To appear in Springer–Verlag, LNCS.
16. O. Goldreich, S. Micali and A. Wigderson. "How to Play Any Mental Game." *Proc. 19th STOC,* ACM, 1987, pp. 218–229.
17. S. Goldwasser, S. Micali and C. Rackoff. The Knowledge Complexity of Interactive Proof Systems. *SIAM Journal on Computing,* vol. 18, no. 1, Feb. 1989, pp. 186–208.
18. J. Håstad. "Pseudo–Random Generators under Uniform Assumptions." *Proc. 22nd STOC,* ACM, 1990, pp. 395–404.
19. R. Impagliazzo, L.A. Levin and M. Luby. "Pseudo–Random Generation from One-way Functions." *Proc. 21st STOC,* ACM, 1989, pp. 12–24.
20. R. Impagliazzo, M. Naor. "Efficient Cryptographic Schemes Provably as Secure as Subset Sum." *Proc. 30th FOCS,* IEEE, 1989, pp. 236–241.
21. S. Micali and T. Rabin. "Collective Coin Tossing without Assumptions nor Broadcasting." *Proc. Crypto '90,* Springer–Verlag, LNCS vol. 537, 1991.
22. S. Micali and P. Rogaway. Secure Computation. *Proc. Crypto '91,* Springer–Verlag, LNCS vol. 576, pp. 392–404.
23. T.P. Pedersen. "Non–Interactive and Information–Theoretic Secure Verifiable Secret Sharing." *Proc. Crypto '91,* Springer–Verlag, LNCS vol. 576, 1992, pp. 129–140.
24. T. Rabin and M. Ben–Or. "Verifiable Secret Sharing and Multi–Party Protocols with Honest Majority." *Proc. 21st STOC,* ACM, 1989, pp. 73–85.
25. A. Shamir. "On the Generation of Cryptographically Strong Pseudorandom Sequences." *ACM Trans. on Computer Systems,* vol. 1, no. 1, Feb. 1983, pp. 38–44.
26. A.C. Yao. "Theory and Applications of Trapdoor Functions." *Proc. 23rd FOCS,* IEEE, 1982, pp. 80–91.

A Generalized Description of DES–based and Benes–based Permutationgenerators

Michael Portz*

RWTH Aachen, Lehrstuhl für Angewandte Mathematik insbesondere Informatik
Ahornstr. 55, D–5100 Aachen, Germany
email: michaelp@terpi.informatik.rwth-aachen.de

Abstract. The construction of pseudorandom permutation generators has been of major interest since Luby and Rackoffs first description. Numerous papers have been dedicated to their simplification. Beside the original DES–based further constructions of pseudorandom permutation generators have been introduced. One of these is based on Benes–networks which are well known tools in the field of parallel processing. Up to now both constructions had apparently nothing much in common but their pseudorandomness. In this paper a new type of construction, the Clos–based permutation generator, is introduced. The DES–based and the Benes–based generators are shown to be special cases of this new type. Viewing both as Clos–based generators gives new insights into known results: the original construction of the Benes–based generators can be improved; some of the already known impossibility–results concerning the DES–based generators are now better understandable; finally it is possible, to formulate new necessary conditions for the pseudorandomness of generated permutations.

1 A Short Introduction

The study of the pseudorandomness of various objects has been of major interest during the last decade. Goldreich, Goldwasser, Micali were the first who pointed out, that not alone the pseudorandomness of numbers, bits or bitstrings alone is of interest, but that also the pseudorandomness of functions should be studied [GoGM1]. They described, how to construct **pseudorandom generators of functions** out of pseudorandom bitgenerators [GoGM2]. Such a function generator is defined to be a sequence of sets of functions. The sequence–index of a set determines inputsize, outputsize and a certain keylength of each function. Each key of the given length specifies exactly one function of the set. Usually inputsize, outputsize and keylength are coupled to the sequence–index via polynomials. A function generator is called pseudorandom, if no polynomially bounded adversary [2] can distinguish a function, which is picked from the function generator using a uniformly distributed key, from

* partially sponsored by the "Graduiertenkolleg Informatik und Technik der RWTH Aachen"

[2] The bounds are set on space and time, especially on the number of calls to the function. Usually the adversary is formalized as turing machine or Boolean circuit.

a function, which is picked uniformly distributed from all functions with the given input–/outputsize. More formal definitions are contained for example in [LuRa2].

The pseudorandomness of 1–1 onto functions was first investigated by Luby and Rackoff in 1986 [LuRa1,LuRa2]. The generators for these special functions are called **permutation generators**. The pseudorandomness and some additional requirements, e.g. invertibility, make the permutation generators very useful for the construction of cryptosystems with provable security properties[LuRa1]. The authors describe a construction for a pseudorandom permutation generator, which uses three different pseudorandom functions. Their construction is based on the algorithmic structure of the DES as indicated by Fig. 2. Various subsequent papers improved Luby and Rackoffs result, in that the necessary number of different pseudorandom functions was decreased [Pata1,Piep1,SaPi1], or that the preconditions were changed[Maur1].

On Eurocrypt '91 Portz described how pseudorandom permutations can be build using one single pseudorandom Boolean function and a certain type of interconnection network, the Benes–network, as basic tools [Port1]. Up to now, the DES–based permutation generators and the Benes–based permutation generators had apparently nothing much in common but their pseudorandomness. This paper introduces a new type of permutation generator, the **Clos–based permutation generator**. DES–based and Benes–based permutation generators are shown to be special cases of this generator[3]. Like the Benes–based this new type uses an interconnection network as starting point, the network introduced by Clos[Clos1].

It must be pointed out, that already at CRYPTO '83 Davio et al showed, that there is a relationship between the algorithmic structure of the DES and the Clos–networks [DDFG1]. The authors state, however, that "This point is ... probably marginal to the present discussion, ..." ([DDFG1], p.189). From this point of view the current paper makes some loose ends meet.

2 Clos–based Permutation Generators

Like the results presented in [Port1] the generalized description is based on techniques used in the theory of interconnection networks, especially on the networks described by Clos [Clos1]. Some of the basic terminology should be introduced. A $N \times N$–**crossbar** is a switching element with N inputs and N outputs. Depending on its **setting** it realizes a permutation on its inputs. Usually it is assumed, that a crossbar can realize every permutation. This assumption is not necessary here. A **Clos–network** with $N_0 \cdot N_1$ inputs consists of columns of crossbars. The first column consists of N_0 $N_1 \times N_1$–crossbars, the second consists of N_1 $N_0 \times N_0$–crossbars, the third consists again of N_0 $N_1 \times N_1$–crossbars and so on. The i–th output of the j–th crossbar in column k is connected to the j–th input of the i–th crossbar in column $k+1$. The setting of each crossbar is desribed by a **control–setting permutation**.

Thus, the problem of describing one permutation on given inputs is reduced to the problem of describing several smaller permutations. This has technical advantages,

[3] Discussions at AUSCRYPT led to the conjecture, that even the Type-2-Transformation described in [ZhMI1] can be viewed as a special case of Clos–based permutation generators. This discussion will be deepened in the final version of the paper.

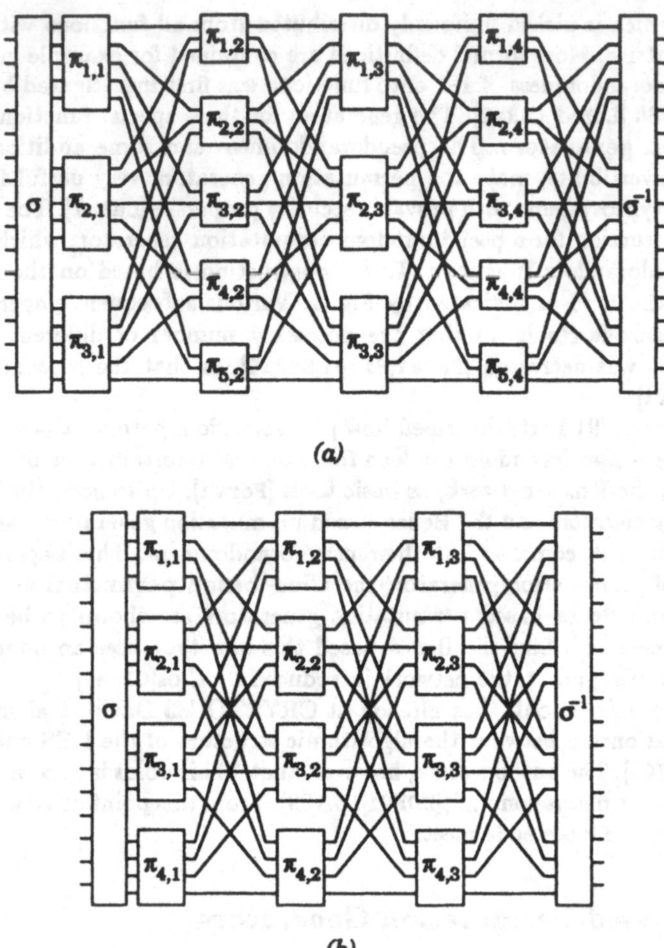

Fig. 1. Two (σ, Π)–Clos–like permutations

if such networks are used to connect e.g. processors in parallel computers. In the following it is shown that this pays in the cryptographic context, too. The Clos–based description of permutation generators is mainly based on these Clos–networks. Note, that in the present context the view of the interconnection network as a circuit is only a model of thought. Constructions relevant for cryptographic purposes have more than 2^{64} inputs and thus are not realizable as circuit. In this paper interconnection networks only serve as a tool to describe and construct permutations.

This section gives two definitions: the Clos-like permutation and the Clos–based permutation generator. Due to the structure of the Clos–network the new permutation generator is not based on general functions, but permutations. Of course these permutations can themselves be based on functions, using a different generating process.

Definition 1. Let s and N be nonnegative integers, N with (arbitrary) factorization $N = N_0 \cdot N_1$. Let Π be a s–tupel $(\Pi_1, \ldots, \Pi_s)$ and let further Π_j for $j \in [1 : s]$ be the tupel $(\pi_{1,j}, \ldots, \pi_{N_{(j+1) \text{ MOD } 2},j})$ whichs entries are permutations

$$\pi_{i,j} : [1 : N_{j \text{ MOD } 2}] \leftrightarrow [1 : N_{j \text{ MOD } 2}], \quad i \in [1 : N_{(j+1) \text{ MOD } 2}] \tag{1}$$

Let σ be a 1-1 onto mapping from $[1 : N]$ to $[1 : N_0] \times [1 : N_1]$ and σ^{-1} its inverse. For $j \in [1 : s]$, $L \in [1 : N_{(j+1) \text{ MOD } 2}]$ and $R \in [1 : N_{j \text{ MOD } 2}]$ the π_j are defined as follows:

$$\pi_j(L, R) = (R, \pi_{R,j}(L)) \tag{2}$$

Then $\pi_{\sigma,\Pi}$ defined by the equation

$$\pi_{\sigma,\Pi}(B) = (\sigma^{-1} \circ \pi_s \circ \ldots \circ \pi_1 \circ \sigma)(B), \quad B \in [1 : N] \tag{3}$$

is the *[(σ, Π)-]Clos-like permutation*. Π is called the *control-setting matrix*. In the following the prefix (σ, Π) is skipped, if both values can be derived from the context.

In the two cases relevant to this paper, the DES–based and the Benes–based permutation generators, the number of columns which is necessary to describe them as Clos–like permutations is 3. This is not too surprising, as it is easy to show and a well known fact, that in Clos–networks 3 columns and an appropriate choice of the permutations in Π are sufficient, to realize every permutation on $[1 : N]$. It is more surprising, that the complexity of finding the permutations $\pi_{i,j}$ is not high. Figure 1 shows two very small examples of Clos–like permutations. In case (a) the input intervall has size $N = 15$, which is factorized to $N_0 = 3$ and $N_1 = 5$, in case (b) these parameters are chosen as $N = 16$ and $N_0 = N_1 = 4$.

Definition 2. Let l be a polynomial, I be a nonfinite subset of $\mathbb{N}$ and $n \in I$. Let further be $N = 2^n$ and s_n, σ_n, N_0 and N_1 be defined according to Def. 1. For $j \in [1 : s_n]$, $k \in \{0,1\}^{l(n)}$ and $i \in [1 : N_{(j+1) \text{ MOD } 2}]$ let $\pi_{(n,k),i,j}$ be a permutation on $[1 : N_{j \text{ MOD } 2}]$. Then define $\Pi_{(n,k)}$ as follows:

$$\Pi_{(n,k),j} = (\pi_{(n,k),1,j}, \ldots, \pi_{(n,k),N_{(j+1) \text{ MOD } 2},j}) \tag{4}$$

$$\Pi_{(n,k)} = (\Pi_{(n,k),1}, \ldots, \Pi_{(n,k),s}). \tag{5}$$

For $\gamma_n = \{\pi \mid \pi \text{ is the } (\sigma_n, \Pi_{(n,k)})\text{–Clos–like permutation for a } k \in \{0,1\}^{l(n)}\}$ the set $\Gamma = \{\gamma_n \mid n \in I\}$ is called a *Clos-based permutation generator*.

If not otherwise stated the input size N is always of the form $N = 2^n$ with $n \in 2\mathbb{N}$. This is in no way a restriction.

3 DES–based Permutation Generators

The construction of Luby and Rackoff is based on the algorithmic structure of the DES, which in turn is derived from the LUCIFER encryption algorithm (see e.g.[Wels1]): each of these algorithms performs a certain number of iterations (Luby/Rackoff: 3, DES: 16, LUCIFER: choosable). The input is of even length (Luby/Rackoff: $2n$, DES: 64, LUCIFER: $2n$) and its left and right halves are treated

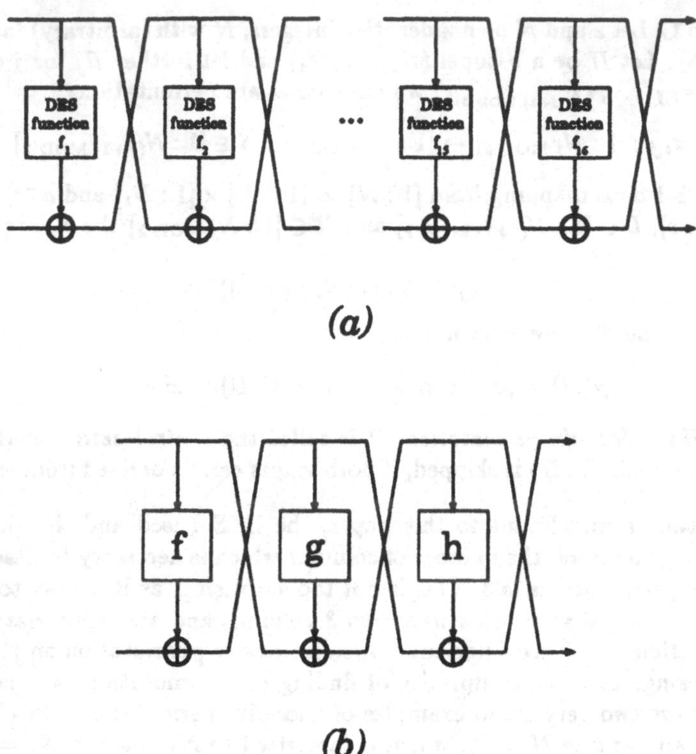

Fig. 2. Structure of DES (*a*) and DES–like permutations (*b*)

seperately in each iteration. The left half is unchanged whereas the right half is XORed with a bitstring, which is derived from the first half, the secret key and the actual number of the iteration. After this computation left and right halves are exchanged. In [ZhMI1] this type of algorithmic structure is called Feistel Type Transformation (FTT) after the inventor of LUCIFER, H. Feistel.

The derivation of the bitstring is usually in each iteration described as a function f and can be pictured as shown in Fig. 2. Luby/Rackoffs construction, DES and LUCIFER differ in this derivation. Luby and Rackoff put the highest restrictions on these functions, they have to be pseudorandom, LUCIFER needs nonlinear functions and for DES the functions f are described using the well known permutations and substitution tables. For this paper it is important, that the computation performed on the right half (XORing with a bitstring) is nothing more than a permutation on all possible right halves, no matter what the precise nature of the functions f is. In fact, it is a very simple permutation, which is moreover selfinverse (Fig. 3).

Luby and Rackoffs main result is, that if one uses pseudorandom functions, the resulting permutation will be pseudorandom as well. The following theorem shows, that Luby and Rackoffs (pseudorandom) permutations can rather simply be transcribed into Clos–like permutations. Note again, that in [DDFG1] a strongly related fact is already pointed out.

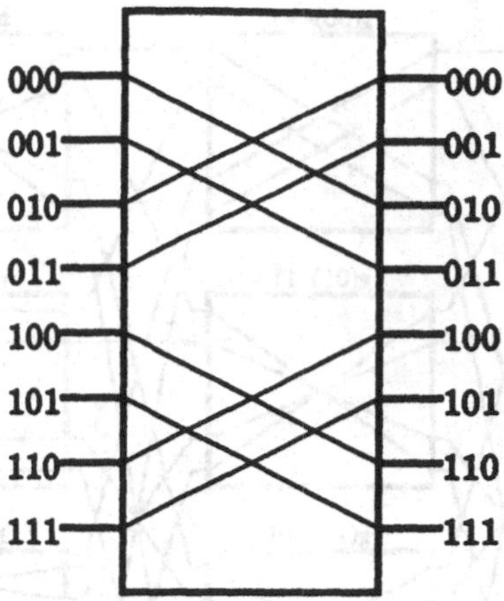

i xor 010

Fig. 3. XORing with a bitstring is a selfinverse permutation

Theorem 3. *Let f, g and h be three different (pseudorandom) functions which generate a permutation π using Luby and Rackoffs construction. Then π can be transcribed into a Clos-like permutation with three columns. The control-setting matrix consists only of permutations, during whichs computation exactly one call to one of the three functions has to be made.*

Proof. Let $N = 2^{2 \cdot n}$ and define σ and σ^{-1} by the equation

$$\sigma((b_1, \ldots, b_{2 \cdot n})) = ((b_1, \ldots, b_n), (b_{n+1}, \ldots, b_{2 \cdot n})) \tag{6}$$

Let further be the permutations $\pi_{i,j}$, $i \in [1 : N_{(j+1) \text{ MOD } 2}]$, $j \in [1 : s]$ for $B \in [1 : N_{j \text{ MOD } 2}]$ be defined by

$$\pi_{i,j}(B) = \begin{cases} B \oplus f(i) & j = 1 \\ B \oplus g(i) & j = 2 \\ B \oplus h(i) & j = 3 \end{cases} \tag{7}$$

The thusly (σ, Π)-like permutation is the same permutation, which the three functions f, g and h generate according to Luby and Rackoffs construction. This is finally proved by showing the identity of the results on same inputs, can simply be derived and therefore is skipped here. □

Figure 4 gives a very small example. Taking a closer look at this proof it shows, that the transformation consists mainly of a renaming of operations. It is just a different view of the same thing. Thus, the following corollary is justified.

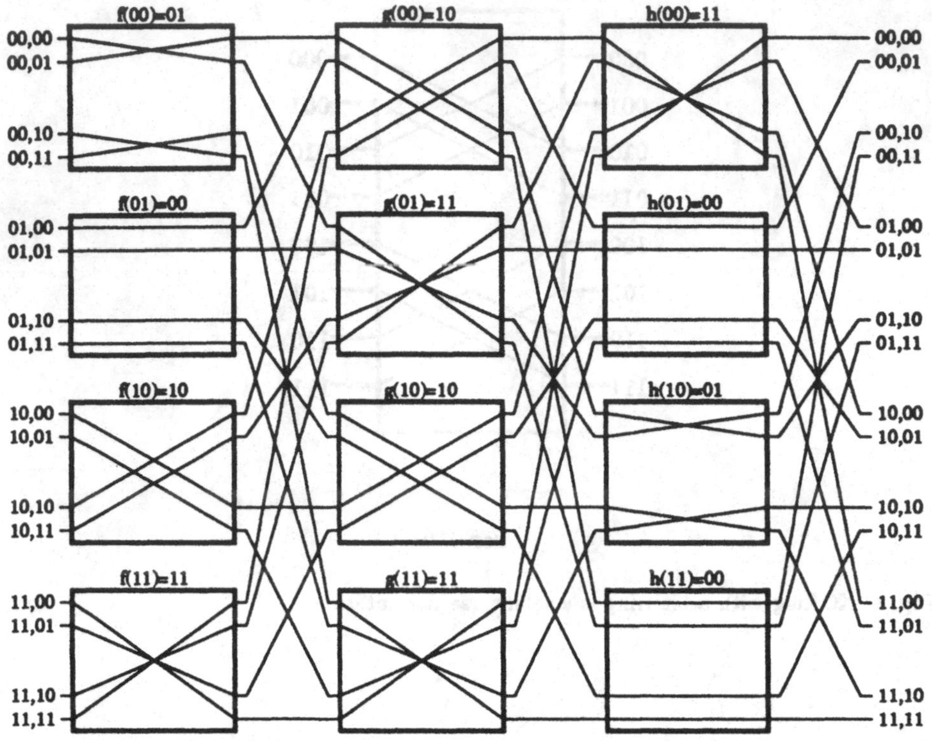

Fig. 4. A DES-like permutation as Clos-like permutation

Corollary 4. *Every DES-based permutation generator can be transformed into a Clos-based permutation generator without any overhead.*

4 Benes-based Permutation Generators

The Benes-based permutation generators introduced in [Port1] are based on Benes-networks. A **Benes-network** is a recursively defined interconnection network. In its most simple form, which is used in [Port1], it consists of 2×2-crossbars, each with a unique index. The actual setting of each crossbar is determined by a **control-setting function**. Such a function computes for each index of a crossbar a Boolean value, namely 0 or 1. W.l.o.g. a 1 should let the crossbar exchange its two inputs and a 0 not. In this section it is shown, that given a control-setting function for a Benes-network (and thus a permutation), there is an algorithm which computes with linear time complexity a control-setting function for a Clos-network.

A Benes-network with 2 inputs simply consists of one 2×2-crossbar. A Benes-network with 2^n inputs consists of two Benes-networks with 2^{n-1} inputs and 2^n additional crossbars. The first 2^{n-1} additional crossbars form the first column of the network, the others the last column. Between them are the two smaller Benes-networks, the upper one and the lower one. The upper (lower) output of the i-th

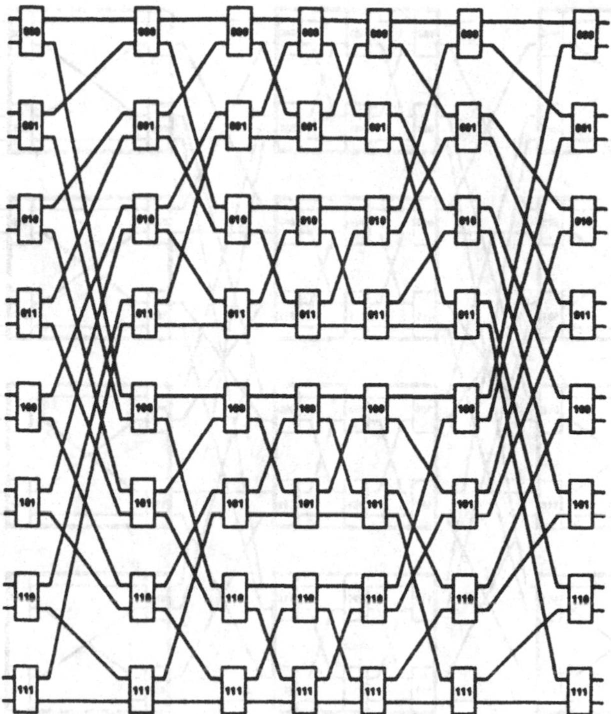

Fig. 5. Benes–network with 16 Inputs

crossbar in the first column is connected to the i–th input of the upper (lower) smaller network. The upper (lower) input of the i–th crossbar in the last column is connected to the i–th output of the upper (lower) smaller network. Furthermore a **Baseline–network** is the restriction of a Benes–network on its first n columns.

Figure 5 shows a Benes–network with 16 inputs. By using a pseudorandom Boolean function as control–setting function Benes–networks generate pseudorandom permutations [Port1]. The following theorem shows, that each permutation which is generated by a Benes–network can simply be transformed to a Clos–like permutation, and thus fits into the generalized description. For that purpose let Be_{h_b} (Ba_{h_b}) be the permutation which is generated by a Benes–network (Baseline–network) using the control–setting function h_b ($b \hat{=}$ Boolean).

Theorem 5. *Let h_b be a control–setting function for a Benes–network with $N = 2^{2 \cdot n}$ inputs and let π be the thusly generated permutation. Then π can be transcribed into a Clos–like permutation with three columns. An evaluation on a single input of any of the permutations of the control–setting matrix has to make exactly n calls to h_b, if it is a permutation from the first or third column and it has to make exactly $2 \cdot n - 1$ calls to h_b if it is a permutation from the second column.*

Proof. The proof uses the following definitions: Let $N = 2^{2 \cdot n}$ and define σ and σ^{-1}

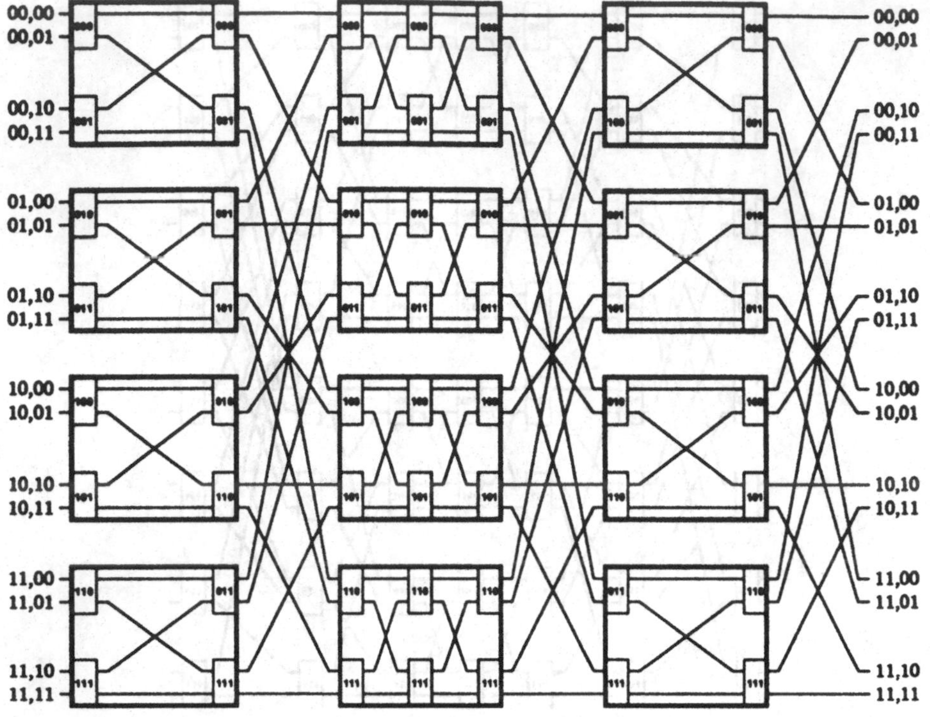

Fig. 6. Benes–like permutation as Clos–like permutation

by the equation

$$\sigma((b_1, \ldots, b_{2 \cdot n})) = ((b_1, \ldots, b_n), (b_{n+1}, \ldots, b_{2 \cdot n})) \tag{8}$$

The control–setting matrix of the Clos–like permutation is described by defining the permutations $\pi_{i,j}$ $i \in [1 : N_{(j+1) \text{ MOD } 2}]$, $j \in [1 : s]$ for $B \in [1 : N_{j \text{ MOD } 2}]$. They are:

$$\pi_{i,j}(B) = \begin{cases} Ba_{h_b,i,j}(B) & j = 1 \\ Be_{h_b,i,j}(B) & j = 2 \\ Ba_{h_b,i,j}(B) & j = 3 \end{cases} \tag{9}$$

This is possible due to certain decomposition properties of Benes–networks. Figure 6 gives a short impression of the proof. The full paper contains a detailed version of the proof. □

A similiar corollary as in Sect. 3 can be derived for the case of Benes–based permutation generators.

Corollary 6. *Every Benes–based permutation generator can be transformed into a Clos–based permutation generator without any overhead.*

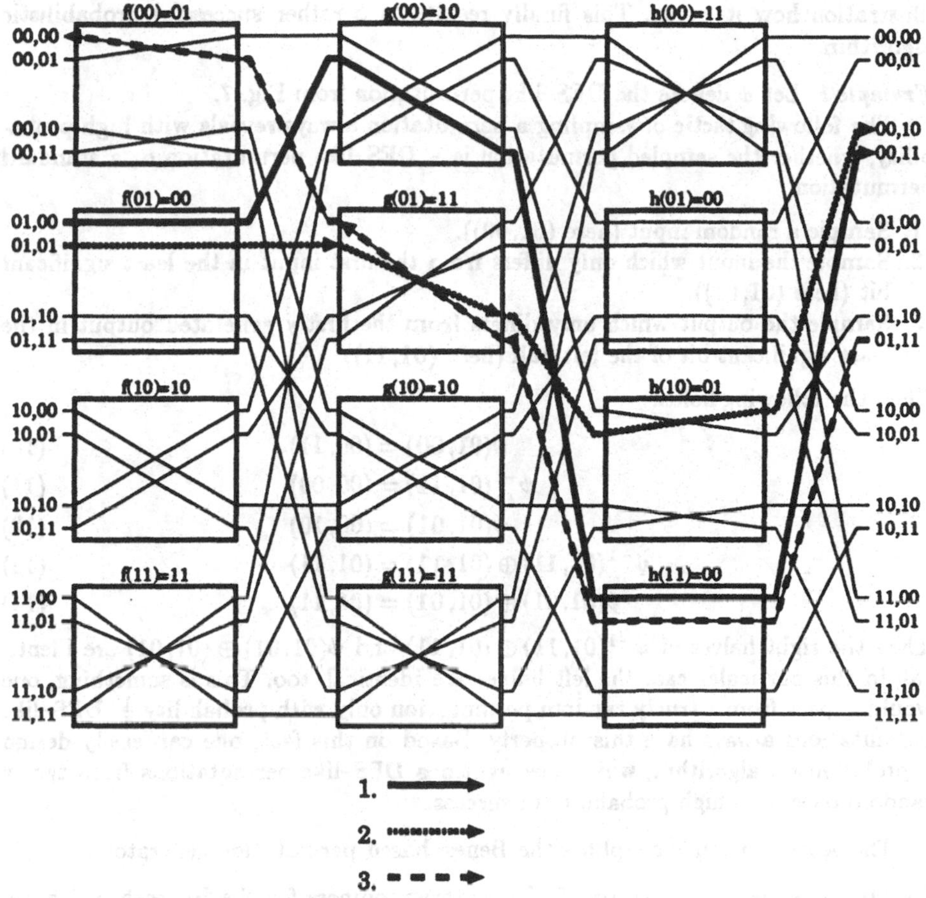

Fig. 7. Why (f, g, h) is not super-pseudorandom

5 Reviewing Permutation Generators

This section gives two sketches of how conventional permutation generators are illuminated, if they are viewed as Clos-based permutation generators. In [LuRa2] it is stated, that the permutation generator with three different pseudorandom functions is not super-pseudoradom. The super-pseudorandomness means resistance against an adversary, who is not only allowed to encrypt but to decrypt, too.

In terms of Clos-networks the original permutation generator is pseudorandom, because it is next to impossible to hit one crossbar in the second column by two or more different cleartexts, if one can only encrypt. The same generator is not super-pseudorandom, because one succeeds rather easily in hitting one crossbar twice, if one is allowed to en- and decrypt[4]. Figure 7 and the following example give an

[4] Please note, that this is only a new illustration of an already known fact.

illustration how it works. This finally results in a rather successful probabilistic algorithm.

Example 1. Let ψ denote the DES–like permutation from Fig. 7.

The following tactic of sampling a permutation always reveals with high probability, whether the sampled permutation is a DES–like permutation or a standard permutation.

1. Sample a random input (here $(01,00)$).
2. Sample the input which only differs from the first input in the least significant bit (here $(01,01)$).
3. Sample the output which only differs from the firstly generated output in the least significant bit of the left half (here $(01,11)$)

Then the following holds:

$$\psi(01,00) = (00,11) \tag{10}$$
$$\psi^{-1}(01,11) = (00,00) \tag{11}$$
$$\psi(01,01) = (00,10) \tag{12}$$
$$\psi^{-1}(01,11) \oplus (01,11) = (01,11) \tag{13}$$
$$\psi(01,01) \oplus (01,01) = (01,11) \tag{14}$$

Then the right halves of $\psi^{-1}(01,11) \oplus (01,11)$ and $\psi(01,01) \oplus (01,01)$ are identical; in this particular case the left halves are identical, too. This is something, one would expect from a truely random permutation only with probability $\frac{1}{4}$. DES–like permutations always have this property. Based on this fact, one can easily design a probabilistic algorithm, which distinguishes DES–like permutations from truely random ones with high probability of success.

The second example simplifies the Benes–based permutation generators.

Example 2. Considering the proof of pseudorandomness for the Benes–based generators there are three properties of them, which guarantee the pseudorandomness of the generator:

1. No polynomially bounded adversary can choose two or more inputs, so that there is a significant chance, that at least two of them enter the same crossbar in the second column, even if he can take samples before.
2. If there is only one input to a crossbar in the second column, this single input is mapped uniformly distributed to the outputs of the crossbar.
3. If there is only one input to a crossbar in the third column, this single input is mapped uniformly distributed to the outputs of the crossbar.

Looking at the proof of Theorem 5, one can see, that the crossbars in the second column are realized by Benes–networks themselves. This is much more than necessary, as e.g. the half as big Baseline–network, which already is used in columns one and three, has the property, that it uniformly distributes any single input to the outputs. Replacing the smaller Benes–networks in Clos–column two by Baseline–networks, already reduces the number of calls to the controll–setting function from $2n-1$ to $\frac{3n}{2}$. Exploiting the degrees of freedom in the original proof, a further reduction to $\frac{3}{2}n - O(\log n)$ columns is possible.

The full version of this paper [Port2] will take a look at other constructions e.g. [Maur1,Pata1,Piep1] and will consider different approaches e.g. [SaPi1,ZhMI1]. Further it will contain a more formal version of the necessary conditions for the construction of pseudorandom permutation generators, formulated in terms of Clos-networks, which already have been stated above informally.

6 Conclusion

This paper introduces the Clos-based permutation generators. It describes, how DES-based and Benes-based permutation generators can be shown to be special cases of Clos-based permutation generators. This allows an immediate comparison of the DES-based and the Benes-based generators. From this comparison arise a better understanding and possible simplifications and improvements of both. Thus the contents of this paper may contribute to the further development on this area and thus to the development and the design of new symmetric cryptosystems and the cryptanalysis of older ones.

7 Acknowledgements

Volker Hatz, Toni Klingler and Andreas Schikarski encouraged me to take the "Clos"er look. Yvo Desmedt pointed out the results of [DDFG1]. Many thanks to all four.

References

[Bene1] V. E. Benes: Mathematical theory of connecting networks and telephone traffic, Academic Press, New York, 1965

[BeHP1] T. Beth, P. Horster, M. Portz: Verbindungsnetzwerke in der Kryptologie

[Clos1] C. Clos: A study of nonblocking switching networks. Bell System Technical Journal, Vol. 32, 1953, S.406–424.

[DDFG1] M. Davio, Y. Desmedt, M. Fosseprez, R. Govaerts, J. Hulsbosch, P. Neutjens, P. Piret, J.-J. Quisquater, J. Vandewalle, and P. Wouters: Analytical characteristics of the DES. Advances in Cryptology. Proc. Crypto 83, Plenum Press, 1984, pp. 171-202

[GoGM1] O. Goldreich, S. Goldwasser, S. Micali: On the cryptographic application of random functions. Advances in Cryptology - Proceedings of CRYPTO '84, Lecture Notes in Computer Science 196, Springer Verlag, 1985

[GoGM2] O. Goldreich, S. Goldwasser, S. Micali: How to construct random functions. Journal of the ACM, Vol. 33, No. 4, Oct. 1986, pp. 792-807

[LuRa1] M. Luby, C. Rackoff: Pseudo-random permutation generators and cryptographic composition. Proceedings of the 18th ACM Symposium on the Theory of Computing, ACM, 1986, pp. 356-363

[LuRa2] M. Luby, C. Rackoff: How to construct pseudo-random permutations from pseudorandom functions. SIAM Journal of Computing, Vol. 17(2), 1988, pp. 373-386

[Maur1] U. Maurer: A simplified and generalized treatment of Luby-Rackoff pseudorandom permutation generators. Abstracts of Eurocrypt '92, Balatonfüred, Hungary, 1992

[Pata1] J. Patarin: How to construct pseudorandom and super pseudorandom permutations from one single pseudorandom function. Abstracts of Eurocrypt '92, Balatonfüred, Hungary, 1992

[Piep1] J. Pieprzyk: How to construct pseudorandom permutations from single pseudorandom functions. Advances in Cryptology - EUROCRYPT '90, Lecture Notes in Computer Science 473, Springer Verlag, 1991

[Port1] M. Portz: On the use of interconnection networks in cryptography. Advances in Cryptology - EUROCRYPT '91, Lecture Notes in Computer Science 547, Springer Verlag, 1991

[Port2] M. Portz: Clos-based Permutation Generators. to appear

[SaPi1] B. Sadeghiyan, J.Pieprzyk: A construction for super pseudorandom permutations from a single pseudorandom function. Abstracts of Eurocrypt '92, Balatonfüred, Hungary, 1992

[Wels1] D. Welsh: Codes and Cryptograhpy. Clarendon Press — Oxford University Press, Oxford UK

[ZhMI1] Y. Zheng, T. Matsumoto, H. Imai: On the construction of block ciphers provably secure and not relying on any unproved hypothesis. Advances in Cryptology – CRYPTO 89, Lecture Notes in Computer Science 435, Springer, 1990

Session 10

ODDS AND ENDS

Chair: Valery Korjik
(Bronch-Bruevitch Technical Communications University, Russia)

Prime Generation with the Demytko–Miller–Trbovich Algorithm

Leisa Condie

Department of Mathematics, Statistics and Computing Science,
University of New England - Armidale, Australia.

Abstract. Prime numbers satisying certain constraints are used in public-key cryptosystems. Consequently, some attention has been paid to algorithms which allow large, cryptographically useful, primes to be created. One algorithm, devised by Miller and Trbovich, was neglected until Nick Demytko of Telecom Research Laboratories published a theoretical examination of it in 1988. At Telecom's invitation the following practical analysis of the method was undertaken. An implementation of the algorithm was written, and nearly one million primes generated, and analysed. This paper gives details of the algorithm, implementation, experiments and results, and draws conclusions about the applicability of the algorithm.

1 Overview of Available Algorithms

1.1 Gordon's Method

In [4] and [5] a method is outlined for finding large primes p, with a specified number of bits, such that

1. p is chosen at random in response to a random seed,
2. $p-1$ has a large prime factor r,
3. $p+1$ has a large prime factor s, and
4. $r-1$ has a large prime factor t.

Any number which satisfies all these criteria is called a *strong prime*.

The algorithm can be boken into three basic steps:

• Choose random seeds a and b and generate the random primes s and t from them.

Given that s (or t) is required to be n bits long, simply ensure the seed has this number of bits and find the first prime larger than the seed. It is suggested that a (or b) be chosen from the range $(2^{n-1}, 2^{n-1}+2^{n-2}-1)$ to ensure it starts with the digits 10. This will give the best chance for finding a prime before the number of bits needs to be increased.

• From t construct r.

* This research was supported by Telecom Australia under contract 7027

A prime is sought which has the form $r \equiv 1 \pmod{2t}$. One approach is to ensure $2t$ is, for instance, $\log_2(n)$ bits shorter than the desired length (n) of r. Starting with unity, successive multiples of $2t$ can then be added until the desired length of r is reached, whereupon primality checks can be undertaken with each subsequent addition of $2t$.

• From r and s construct p.

Let $p_0 = (s^{r-1} - r^{s-1}) \pmod{r}s$. Then for some k

$$p = \begin{cases} p_0 + 2krs, & \text{if } p_0 \text{ odd;} \\ p_0 + (2k+1)rs, & \text{if } p_0 \text{ even.} \end{cases}$$

As p is a few bits longer than $2rs$ it is probably best to have $2rs$ about $\log_2(n)$ bits shorter than the desired length (n) of p. As r and s should be about equal in size, their lengths can thus be ascertained.

An alternate method for p_0 calculation was given in [3]: it was shown that $p_0 + rs = 2ss^{-1} - 1$ where s^{-1} is defined by $ss^{-1} \equiv 1 \pmod{r}$ and $1 \leq s^{-1} < r$. As only a modular inverse is needed, rather than two modular exponentiations, this method is quite efficient.

1.2 Shawe-Taylor Variation

This variation, presented in [7], uses the same basic algorithm as Gordon, however substitutes a different primality test. It is claimed that this test is less prone to erroneously certifying a composite as a prime. It is of comparable efficiency to Gordon's algorithm.

Algorithm-L

Given a number of the form $n = 2vk + 1$ where v is prime and $v > \sqrt{n}$ then

```
success = false
select random b such that 2 ≤ b ≤ n − 2
x = b^{2k}  (mod n)
if x ≠ 1 then
        if gcd(x − 1, n) = 1 then
                if x^v ≡ 1  (mod n) then
                        success = true
if success then
        return n is prime
else
        return n is probably not prime
```

When given a prime, the probability of error for this algorithm is $< 1/v$. The differences caused by the use of this test are:

1. Recursion is used to generate s and t. Specifically, we start with a prime p_i of m bits. A random integer t, of $m-2$ bits (formed from the next $n-2$ bits of the random seed) is taken. A prime is sought of the form $p_{i+1} = 2p_ik + 1$, taking $k = t, t+1, \ldots$ and using Algorithm-L with $v = p_i$.
2. In the construction of r Algorithm-L is used with $v = t$.

3. In the construction of p Algorithm-L is used with $v = r$. Note that the lengths of r and s are different to those specified by Gordon. Here it is required that r be one bit larger than half the length of p rather than a few bits shorter. Also, s needs to be $\log_2(n) + 2$ bits shorter than half the length of p.

1.3 Miller-Trbovich Algorithm

There are basically two methods outlined in [2] for the Miller-Trbovich algorithm [6]. In the first

$$p_{i+1} = h_i p_i + 1$$

is prime provided the following criteria are met:

1. p_i is an odd prime
2. $h_i \leq 4(np_i + 1)$ is even, where $2 \leq n \leq 2p_i + 2$
3. $2^{h_i p_i} \equiv 1 \pmod{p_{i+1}}$
4. $2^{h_i} \not\equiv 1 \pmod{p_{i+1}}$
5. $(2jp_i + 1) \not| \, p_{i+1} \quad \forall \; 1 \leq j \leq n - 1$

Hence, given a small odd prime p_0 and an even number $h_0 \leq 4(p_0 + 1)$, large, strong primes can be quickly built as the length of the prime is doubled at each iteration.

The extended method bears many similarities to Gordon's algorithm. Given two odd primes p_1 and p_2, where $p_1 - 1$ has a large prime factor, then let

$p_0 = (p_2^{p_1-1} - p_1^{p_2-1}) \pmod{p}_1 p_2$
$k = 0$
While p is not prime do
begin

$$p = \begin{cases} p_0 + 2kp_1p_2, & \text{if } p_0 \text{ odd;} \\ p_0 + (2k+1)p_1p_2, & \text{if } p_0 \text{ even.} \end{cases}$$

Now check for primality (†):
$h = (p - 1)/p_1$
Find n such that $h \leq 4(np_1 + 1)$ and check that it is the range $2 \leq n \leq 2p_1 + 2$
Check that:

$2^{hp_1} \equiv 1 \pmod{p}$
$2^h \not\equiv 1 \pmod{p}$
$(2jp_1 + 1) \not| \, p \quad \forall \; 1 \leq j \leq n - 1$

If all of these conditions are met then prime, else continue.
$k = k + 1$

end.

(†) In the implementation of this algorithm, code was added at this stage to trial divide by all primes up to a certain limit. Trial division by all of the primes to 1000, for instance, eliminates 84% of candidates, without requiring the use of computationally expensive tests. Alternatively, the small primes could be multiplied together and the result held as a constant against which a single gcd could be calculated, saving even more time.

2 Implementation

An implementation of the Miller-Trbovich algorithm, now known as the Demytko-Miller-Trbovich algorithm, is presented in [1]. The program is given two *starter primes* - the initial primes from which the larger prime number will be built - and a lower limit on the length of prime it must build. The program will continue to *build* successively larger prime numbers until that limit is equalled or exceeded. The number of algorithm iterations required to build a prime of given length is called the number of *build steps*. The entire process of building a large prime (a *round* of the program) is repeated for a number of different pairs of starter primes or sets of random numbers, depending on the experiment being performed.

The program implementing the algorithm was written in the C language, as was the multiple precision mathematics package used. The implementation is reasonably fast: on an 80486 DX2/50 an eighty digit prime can be produced in one to two seconds. On more modest facilities, such as a 25MHz 80386, the same length prime takes around 12 to 15 seconds to produce.

3 Experiments

The role of the random number generator, and its practical effect on the algorithm, was of particular interest in the experiments undertaken. Three random and pseudo-random number generators were incorporated to find whether the choice of generator was significant. The first generator was the pseudo-random function rand() supplied with the Unix operating system. Rand uses a simple multiplicative congruential generator to return a value in the range 0 to $2^{31} - 1$. As with all such generators, the quality of its output is poor, its spectral properties limited. The second pseudo-random number generator was the public domain package xrand, which also produces 32 bit output. Its output is better than that of rand(): it returns numbers generated by a linear congruential generator, which have been permuted under control of an additive congruential generator. Two permutations take place: the four bytes of one linear congruential generator number are subjected to one of 16 permutations selected by four bits of the additive congruential generator. The permutation is such that a byte of the result may come from each byte of the linear congruential generator number. This effectively destroys the structure within a word. Finally, the sequence of such numbers is permuted within a range of 256 numbers, greatly improving independence. The last generator was a random number board installed into a personal computer, and its output used to build random numbers of an arbitrary length. The board was the Calnet RNG-810 on which there are eight independent Johnson (resistor thermal) noise sources, giving eight random bits per board call.

In the experiments performed a total of 946,677 primes were generated. In descriptions below the first letter of the filename indicates the random number used: z the Unix random function, x the xrand package, and b the random number board. Experiments were classified into one of three types:

1. p and q are constant in each round, and the random numbers vary, or
2. the same set of "random" number inputs are used in each round, with p and q varying, or
3. all of p, q and the random numbers vary.

4 Results

4.1 Methods

The major area of interest with this algorithm was that of distribution, and thus predictability, of the built primes. At the coarsest level of analysis the lengths of the primes were examined for gross features, before a close look at the distribution at each build step was taken. In these *distribution profiles* all the primes of each build step were gathered, the maximum and minimum values determined, and the interval between them broken into ten equal subintervals. A count of the primes in each subinterval was made, and if a particular subinterval was heavily favoured (by containing 30% or more of the primes) that subinterval was further divided into ten equal subdivisions, and the process continued until roughly equal numbers were obtained across a set of ten subintervals, or there were less than 200 primes in a subinterval. The latter condition was included to prevent infinite recursion. A check was also made of the differences between primes of the same build step in successive rounds to see whether patterns emerged, indicating a relationship between them, and implying predictability. This was called *relationship checking.*

4.2 Experiment One

In this experiment the starter primes were kept constant, whilst the random numbers changed across each step and round. Seven xrand files and three random board files were created, containing a total of 172,113 primes. The most striking feature of these files is that in each all rounds comprise build steps of identical sizes. Table 1 shows these lengths.

This suggests the distribution is poor, and examination of the maxima and minima for each build step certainly shows it to be limited. It must be remembered, however, that there are many primes, and even the ranges given would be impossible to exhaust in any feasible period of time. Using the Prime Number Theorem, which gives an estimate of the expected number of primes between 1 and x as $\pi(x) = \frac{x}{\ln(x)}$, the file bo1 provides a typical example, in Table 2. Relative to other experiments, the range of output primes is certainly low, but still large enough to have no reasonable chance of collision (two productions of the same prime) occurring.

It is interesting to compare the file xv1 with bo1, and xv2 with bo2, as they have the same starter primes and experiment type. Large differences in the performance of xrand and the random board would be expected to show here, but no such differences can be found. The Table 3 shows the maxima and

Table 1. Step sizes for experiment 2 files

File	Step 1	Step 2	Step 3	Step 4	Step 5
xo1	10	20	40	80	160
xo2	10	19	39	78	155
xp1	10	20	40	80	160
xp2	38	77	-	-	-
xp3	38	77	153	306	-
xv1	9	18	38	76	-
xv2	10	19	39	78	-
bo1	10	20	39	78	156
bo2	9	19	38	76	152
bo3	10	20	39	78	156

Table 2. Maxima, minima and the expected number of primes for steps of bo1

Step	Maximum	Minimum	Range	Expected
2	$0.3608115 * 10^{19}$	$0.2706538 * 10^{19}$	$9.0 * 10^{17}$	$2.067 * 10^{16}$
3	$0.3905605 * 10^{36}$	$0.2197660 * 10^{36}$	$1.8 * 10^{37}$	$1.956 * 10^{35}$
4	$0.45762 * 10^{76}$	$0.14489 * 10^{76}$	$3.1 * 10^{75}$	$1.789 * 10^{73}$
5	$0.6283061 * 10^{151}$	$0.6299568 * 10^{150}$	$5.6 * 10^{151}$	$1.615 * 10^{149}$

minima of the steps in each file. The implication is that xrand performs very slightly better, however the random board runs were short, and this had some impact on the ranges produced.

Distribution profiles of these files are notably homogenous: the primes of each build step are quite evenly distributed across the ten subintervals. In the files created with xrand the distribution is nearly identical for each step, whereas those created with the random board show it changing with each step, although a clear trend is visible for all.

With the starter primes being held constant it might be expected that successive rounds show relationships: patterns would emerge amongst the built primes. This is not, however, the case, with one striking exception. The file xp2 had a first step of 38 digits, and a second step of 77 digits. The differences between each successive pair of 77 digit numbers showed patterns within themselves. As the length of the starter primes (18 digits each) far exceeded the length of the pseudorandom numbers being applied to them, it is felt that this is the source of the anomoly. The file xp3 was produced to see whether the effect extends to higher steps, and it was found that it did not. Nevertheless, small starter primes (32 bits - around 10 digits - or less) are recommended if such a 32 bit pseudorandom number generator is being used.

4.3 Experiment Two

Varying starter primes were used in this experiment, but the 'random' numbers used were fixed for each round. That is, for a round of four build steps four

Table 3. Comparison of ranges for xrand and random board

xv1 step	minimum	maximum
2	$.270656 * 10^{19}$	$.360855 * 10^{19}$
3	$.219764 * 10^{38}$	$.390650 * 10^{38}$
4	$.144889 * 10^{76}$	$.457822 * 10^{76}$
bo1 step	**minimum**	**maximum**
2	$.270653 * 10^{19}$	$.360811 * 10^{19}$
3	$.219766 * 10^{38}$	$.390560 * 10^{38}$
4	$.144890 * 10^{76}$	$.457620 * 10^{76}$
xv2 step	**minimum**	**maximum**
2	$.820555 * 10^{19}$	$.109401 * 10^{20}$
3	$.201993 * 10^{39}$	$.359058 * 10^{39}$
4	$.122403 * 10^{78}$	$.386768 * 10^{78}$
bo2 step	**minimum**	**maximum**
2	$.998844 * 10^{19}$	$.109378 * 10^{20}$
3	$.299308 * 10^{39}$	$.358911 * 10^{39}$
4	$.268756 * 10^{78}$	$.386453 * 10^{78}$

different numbers were used, but each round in a file would use the same four numbers. This was achieved by repeatedly seeding the generator with the same seed. The purpose was to investigate the extent of the effect that small changes in starter primes (one or both) would have on the final primes. A total of 194,708 primes were produced and analysed.

The starter primes for p were the four digit primes from 9001 to 9133, whilst the starter q primes were all five digit primes from 10001 to 19999. For a given p a round of four build steps was completed for each successive q before the next p was chosen. Table 4 shows the p's that were completed, and the q that an incomplete run finished on.

Table 4. p and q for Experiment Two.

File	Range of p
xm1	9001-9041 + 9043 (q to 15101)
xm2	9001-9103 + 9109 (q to 19571)
xm3	9001-9127 + 9133 (q to 15361)
xm4	9001-9103 + 9109 (q to 17387)
xm5	9001 (q to 13537)
xm6	9001 (q to 13033) and 9001 (q to 10169)
xm7	9001 (q to 11113)

The first attribute of the files examined was the lengths of the primes built. The file xm2 is a typical example of Experiment Two files: at the first step the lengths were only distributed amongst a handful of lengths, with a marked concentration at one particular length (in the example given 10 digits shows a strong concentration of primes). From the second step onwards, however, the lengths become more widely spread, and peak concentrations are far lower - the distribution becomes quite even across several lengths. In xm2 the final step shows the bulk of primes evenly divided between 76, 77, 78, 79, 80 and 81 digits - a very wide range.

The distribution profiles confirm the gross analysis: the range of primes built at each step is wide, and distribution within a length is reasonably even. There were no relationships evident between the primes, and no pattern to their production was found. Even without the effect of the random number generator, small changes to one of the starter primes is enough for unpredictability. Of course, it is not recommended that the system be used without one: it weakens the system for no good reason.

4.4 Experiment Three

This experiment was to round off the above investigations by ensuring that a combination of varying starter primes and the application of pseudo-random numbers did not have adverse effects on the output primes. The worst of the three random number generators - the Unix rand() function - was used. If good results could be obtained with this generator there could be no problems with the stronger generators. An extra test with xrand was made anyway, to double-check this conclusion. A total of 579,656 primes were generated in this section, by far the largest of the experiments, and the same checks were performed upon them as for Experiment One. Starter primes were as for Experiment Two, with two additions: zpp used a 9 digit p, and zpq used 21 digit primes for p and q.

Initially the gross distribution was checked, by looking at the lengths of primes produced, and the results were the same as for Experiment Two: there was good scattering, indicating a good distribution of built primes. This was verified when the maxima and minima were examined for each file. As expected with such results, the distribution profiles showed the ranges increasing with each step, and the built primes were reasonably distributed amongst the subintervals. Relationship checking revealed no anomalies. This was of particular interest with respect to the file zpq, which was using two steps and 21 digit starter primes. In Experiment One xp2, with 18 digit starter primes, revealed relationships between the build steps, due to the large size of the starter primes in comparison to the size of the random number. Here it can be seen that varying the starter primes, even in such a small way as was done here, removes those relationships.

Although the Unix rand() function performed well in these experiments, it is not recommended for use with this algorithm as it is quite weak. The use of large starter primes was also shown to have no noticeably weakening effect, however it is recommended that starter primes of comparable size to the random numbers in use are used. Certainly this experiment shows that the algorithm produces a well distributed, unpredictable set of large prime numbers, which was what it was designed to do.

5 Conclusions

The use of a strong pseudo-random number generator, such as that presented, is encouraged: it is far more portable than the random number board, its spectral properties are good, and the results obtained are comparable to those obtained with the board. Care is required however - the starter primes should not be longer than 32 bits to ensure proper scattering by the generator.

Small input variations become large output variations, and after the first build step even the size of the initial primes cannot be guessed. There do not appear to be distribution anomalies of any kind: a sequence of final primes do not relate to one another in any way, and so the sequence cannot be predicted.

A minimum of three build steps, and preferably four or five, are recommended. This would give good spreading of the output primes, but still take only a small amount of time to generate. If the strong pseudo-number generator is in use this becomes mandatory, as small starter primes are required, and thus four or more steps are needed to build a prime of the desired length.

Generation of large primes is possible in a few minutes on a personal computer, which makes the algorithm suitable for use in situations where only a modest amount of computing power is available.

The algorithm is suitable for use where non-predictable, strong, large primes are required.

References

1. Condie, L.: *Speech Signal Analysis and Investigations in Cryptography.* Ph.D. thesis, Dept. of Computer Science, University College, University of New South Wales, Australian Defence Force Academy, 1991.
2. Demytko, N.: Generating multiprecision integers with guaranteed primality. *IFIP SEC '88: Computer security in the age of information.* 1988.
3. Ganley, M. J.: Note on the generation of p_0 for RSA keysets. *Electronics Letters,* 26(6), 1990.
4. Gordon, J.: Strong RSA keys. *Electronics Letters,* 20(12), 1984.
5. Gordon, J.: Strong primes are easy to find. In *Advances in Cryptology – Eurocrypt 84,* Springer-Verlag, 1984.
6. Miller, W. J., Trbovich, N. G.: RSA public-key data encryption system having large random prime number generating microprocessor or the like. International Patent Classification: H04L9/00, International Publication Number: WO82/02129, 1982.
7. Shawe-Taylor, J.: Generating strong primes. *Electronics Letters,* 22(16), 1986.

Constructions of Feebly-One-Way Families of Permutations

Alain P. L. Hiltgen

Signal and Information Processing Laboratory
Swiss Federal Institute of Technology
CH-8092 Zürich, Switzerland

Abstract. The unrestricted circuit complexity $C(.)$ over the basis of all logic 2-input/1-output gates is considered. It is *proved* that certain explicitly defined families of permutations $\{f_n\}$ are feebly-one-way of order 2, i.e., the functions f_n satisfy the property that, for increasing n, $C(f_n^{-1})$ approaches $2 \cdot C(f_n)$ while $C(f_n)$ tends to infinity. Both these functions and their corresponding complexities are derived by a method that exploits certain graphs called $(n\text{-}1,s)$-stars.

1 Introduction

"A function f is a one-way function if, for any argument x in the domain of f, it is easy to compute the corresponding value $f(x)$, yet, for almost all y in the range of f, it is computationally infeasible to solve the equation $y = f(x)$ for any suitable argument x." This mathematically imprecise, but intuitively clear definition, formulated by Diffie and Hellman [1] in 1976, has been of great practical relevance to cryptography now for more than fifteen years .

Most mathematically precise definitions of one-wayness that have appeared in the literature [2] [3] during this time are based on Turing machine complexity. They describe uniform notions of one-wayness. 'Function' is then taken to mean an infinite family of functions; 'easy' and 'infeasible' are taken to mean computable in polynomial time and not computable in polynomial time, respectively. However, if we use a specific function like exponentiation modulo a fixed prime p_0 to generate public keys from private ones [1], it seems not realistic from a practical cryptographic point of view to suppose that an attacker can evaluate the particular inverse function efficiently (i.e., compute the logarithm modulo p_0) only if he knows an efficient algorithm that solves this problem for all primes p. This practical worry about the existence of special cases with only weak instances has been confirmed in particular for the discrete logarithm by the results in [4].

A more practical precise definition of one-wayness has been formulated by Boppana and Lagarias [5] in 1987. Their non-uniform notion of one-wayness guarantees a computationally difficult (with respect to unrestricted circuit complexity) right-inverse for every member in a one-way family of functions[1]. If the

[1] In this paper we consider only families of permutations so that their inverses are always defined.

circuits, however, are restricted (e.g., constant-depth circuits (cf. [5])), then the complexity results that can be proved again have only little practical implication for cryptography as an attacker is unlikely to observe such restrictions. For real relevance to practical cryptography, unrestricted circuit complexity has to be considered.

In this paper, we use two fundamental lower bounds from circuit-complexity theory to prove that certain explicitly defined families of permutations are'feebly-one-way' with respect to unrestricted circuit complexity. We define an infinite family of permutations f_n to be feebly-one-way if, for increasing n, the complexity ratio $C(f_n^{-1})/C(f_n)$ stays strictly greater than 1 while the complexity $C(f_n)$ tends to infinity. In Section 2, we state some definitions and recall the two lower bounds mentioned above. In Section 3, an explicit family of linear permutations is proved to be feebly-one-way. We prove that, for every member f_n in this family, the complexity of its inverse is about 50% larger than the complexity of f_n itself. This result is generalized in Section 4, yielding explicit families of linear permutations for which the complexity of their respective inverses is about twice their own complexity. The complexity difference between these functions and their inverses increases linearly with n, the length of their input. This seems to be the strongest one-wayness that can presently be proved, since there exist only linear lower bounds on the unrestricted circuit complexity of explicitly defined Boolean functions [6, p.120]. In Section 5, we briefly discuss this problem as well as the implications of our results for practical cryptography.

2 Definitions and Preliminaries

Let $B_{n,m}$ denote the set of $2^{m \cdot 2^n}$ functions $f : \{0,1\}^n \to \{0,1\}^m$ (B_n also stands for $B_{n,1}$). We consider only realizations of such functions by B_2-circuits [6, chapt.1], i.e., by acyclic logical circuits constructed with 2-input/1-output gates, where the n Boolean input variables x_i and the constants **0** and **1** are the only valid inputs to the circuit, and where each gate may compute any of the 16 Boolean functions in 2 variables (B_2-function). Note that neither fan-out nor depth are restricted. The *size* or complexity of a circuit is the number of its gates. The *unrestricted circuit complexity* of a function f is the smallest number of gates in a B_2-circuit computing f. It is denoted by $C(f)$ and simply called *complexity* in what follows.

Let S_{2^n} denote the subset of $B_{n,n}$ containing the $2^n!$ invertible functions, i.e., the set of permutations of $\{0,1\}^n$, and let f_n denote any function from S_{2^n}. In order to distinguish between different notions of one-wayness, we introduce

the measure of feeble one-wayness: $\quad M_F(f_n) = C(f_n^{-1})/C(f_n)$,

and we recall

the measure of practical one-wayness[2]: $\quad M_P(f_n) = \log_2[C(f_n^{-1})]/\log_2[C(f_n)]$.

[2] This was called simply the *measure of one-wayness* in [5].

Now, let $\{f_n\}$ denote a *family of permutations*, i.e., an infinite sequence $f_1, f_2, \ldots$ of functions such that, for each n, the function f_n belongs to S_{2^n}.

Definition 1. A family of permutations $\{f_n\}$ is *feebly-one-way of order k* if $\liminf_{n\to\infty}[C(f_n)] = \infty$ and $\liminf_{n\to\infty}[M_F(f_n)] = k$, with k strictly larger than 1 but not necessarily finite.

The condition $\liminf_{n\to\infty}[C(f_n)] = \infty$ is needed to ensure that a single permutation f_{n_o} with $M_F(f_{n_o})=k >1$ is not artificially boosted to a feebly-one-way family of permutations $\{f_n\}$ by adding, for each $n > n_o$, n-n_o input variables that are fed directly through to the outputs.

Definition 2. A family of permutations $\{f_n\}$ is *practically-one-way of order k'* if $\liminf_{n\to\infty}[C(f_n)] = \infty$ and $\liminf_{n\to\infty}[M_P(f_n)] = k'$, with k' strictly larger than 1 but not necessarily finite.

Note that for finite k and k', these definitions imply that asymptotically $C(f_n^{-1}) \approx k \cdot C(f_n)$ in the case of feeble one-wayness and $C(f_n^{-1}) \approx [C(f_n)]^{k'}$ in the case of practical one-wayness. If $M_P(f_n)$ tends to infinity, the family $\{f_n\}$ is one-way in the classical sense (cf. [5]); however, if $M_F(f_n)$ tends to infinity, the family $\{f_n\}$ is not necessarily practically-one-way as one sees by considering a hypothetical family $\{f_n\}$ with $C(f_n^{-1}) = C(f_n) \cdot \log_2[C(f_n)]$.

We conclude this section by recalling two fundamental lower bounds due to Lamagna and Savage [7] [8].

Theorem 3. *If $f \in B_n$ depends non-idly on each of its n variables, then*

$$C(f) \geq n - 1.$$

Theorem 4. *Let $f=\{f^{(1)}, \ldots, f^{(m)}\} \in B_{n,m}$. If the m component functions $f^{(i)}$ are pairwise different and if they satisfy $C(f^{(i)}) \geq c \geq 1$, then*

$$C(f) \geq c + m - 1.$$

Theorem 3 follows from the fact that, in every circuit realizing such a function, the number of unused nodes (inputs or gate-outputs) must be reduced from n input nodes to 1 output node and that each gate can achieve a reduction by at most 1. Theorem 4 is based on the fact that, in every circuit computing a Boolean function of complexity c, there are at least $c - 1$ gates computing at their outputs Boolean functions of complexity strictly less than c.

In the next sections these two bounds are combined to prove in a very simple way that specific families of permutations are feebly-one-way of a certain order.

3 A Feebly-One-Way Family of Permutations

The family of permutations $\{\varphi_n\}$ investigated in this section is essentially the family of permutations $\{\tau_n\}$ that was shown in [5] to be 'bit-wise one-way for constant-depth circuits'. Each function φ_n is defined for $n \geq 3$ to be the linear function

$$\varphi_n([x_1,\cdots,x_n]) = [y_1,\ldots,y_n]$$

where
$$\begin{array}{ll} y_i(\underline{x}) = x_i \oplus x_{i+1} & 1 \leq i < n \\ y_i(\underline{x}) = x_1 \oplus x_{\lceil n/2 \rceil} \oplus x_n & i = n, \end{array}$$

with $\oplus$ denoting addition modulo 2. It is quite easy to check that the inverse function φ_n^{-1} is the linear function

$$\varphi_n^{-1}([y_1,\ldots,y_n]) = [x_1,\ldots,x_n]$$

where

$$\begin{array}{ll} x_i(\underline{y}) = (y_1 \oplus \cdots \oplus y_{i-1}) \oplus (y_{\lceil n/2 \rceil} \oplus \cdots \oplus y_{n-1}) \oplus y_n & 1 \leq i \leq \lceil n/2 \rceil \\ x_i(\underline{y}) = (y_1 \oplus \cdots \oplus y_{\lceil n/2 \rceil -1}) \oplus (y_i \oplus \cdots \oplus y_{n-1}) \oplus y_n & \lceil n/2 \rceil \leq i \leq n. \end{array}$$

We can now prove the following theorem which gives the exact complexities of φ_n and φ_n^{-1}.

Theorem 5. *For all $n \geq 5$, the functions φ_n satisfy*

$$C(\varphi_n) = n+1 \quad \textit{and} \quad C(\varphi_n^{-1}) = \lfloor \tfrac{3}{2}(n-1) \rfloor.$$

Proof. First we prove that $C(\varphi_n) = n+1$. By considering independent realizations of the component functions, we see immediately that $C(\varphi_n) \leq n+1$. Theorems 3 and 4 further imply that $C(\varphi_n) \geq n$, because each $y_k(\underline{x})$ depends non-idly at least on two different x_i's. In order to prove that $C(\varphi_n) = n+1$, we partition the x_i's into two different sets: $S_1 = \{x_1, x_{\lceil n/2 \rceil}, x_n\}$ and $S_2 = \{x_1,\ldots,x_n\} \setminus S_1$. Note that, for each $n \geq 5$, each one of the component functions $y_k(\underline{x})$ with $k \neq n$ depends on two different x_i's of which at least one is not an element of S_1. Thus, setting the x_i's in S_2 to zero eliminates at least the n-1 output gates computing $y_1(\underline{x}),\ldots,y_{n-1}(\underline{x})$ and leaves the component function $y_n(\underline{x})$ unchanged. Because $C(y_n(\underline{x})) \geq 2$ (by Theorem 3), this yields

$$C(\varphi_n) - (n-1) \geq 2$$

and consequently $$C(\varphi_n) \geq n+1.$$

We conclude that $C(\varphi_n) = n+1$.

Now we prove that $C(\varphi_n^{-1}) = \lfloor \frac{3}{2}(n-1) \rfloor$ by exploiting two useful properties of φ_n^{-1}, namely,

(P1) each $x_i(\underline{y})$ is the modulo 2 sum of at least $\lceil n/2 \rceil$ of the y_k's and this lower bound holds with equality for $i = n$;

(P2) for each $i \neq n$, $x_i(\underline{y}) \;=\; x_{i+1}(\underline{y}) \oplus y_i$.

Starting with $x_n(\underline{y})$ (by Theorem 3 and an obvious realization, $C(x_n(\underline{y})) = \lceil n/2 \rceil - 1$) and using property (P2) to calculate the remaining $x_i(\underline{y})$, always yields a realization for φ_n^{-1} that uses no more than

$$(\lceil n/2 \rceil - 1) + (n-1) \;=\; \lfloor \tfrac{3}{2}(n-1) \rfloor \text{ gates.}$$

Property (P1), however, implies that $\min_i[C(x_i(\underline{y}))] = C(x_n(\underline{y}))$ so that, by Theorem 4, we have $C(\varphi_n) \geq \lfloor \frac{3}{2}(n-1) \rfloor$. We conclude that the realization above is always optimal and hence that $C(\varphi_n^{-1})$ equals $\lfloor \frac{3}{2}(n-1) \rfloor$. □

Defining φ_n to be the identity function (with complexity 0) for n=1 and n=2, we obtain the following corollary as a direct consequence of Theorem 5.

Corollary 6.
$\{\varphi_n\}$ is a family of permutations that is feebly-one-way of order 3/2.

We conclude this section by giving explicitly the smallest example (n=7) for which $C(\varphi_n^{-1})$ is strictly larger than $C(\varphi_n)$.

$$\underline{y} = \varphi_7(\underline{x}) = \begin{bmatrix} 1100000 \\ 0110000 \\ 0011000 \\ 0001100 \\ 0000110 \\ 0000011 \\ 1001001 \end{bmatrix} \cdot \underline{x} \qquad\qquad \underline{x} = \varphi_7^{-1}(\underline{y}) = \begin{bmatrix} 0001111 \\ 1001111 \\ 1101111 \\ 1111111 \\ 1110111 \\ 1110011 \\ 1110001 \end{bmatrix} \cdot \underline{y}$$

$$C(\varphi_7) = 8 \qquad\qquad\qquad C(\varphi_7^{-1}) = 9$$

The linear functions φ_7 and φ_7^{-1} are represented here by their corresponding matrices. The respective complexities follow directly from application of Theorem 5.

4 A Method for Proving Feeble One-Wayness of Order 2

To generalize the results of Section 3, we consider only a small subset H_n of all the invertible functions in S_{2^n}. Let h denote any function from H_n that maps the vector $\underline{x}$=$[x_1,\ldots,x_n]$ to the vector $\underline{y}$=$[y_1,\ldots,y_n]$; then h satisfies the following properties:

(P1') h is invertible;
(P2') $\forall k \neq n, \exists!\, i,\, j$ such that $y_k = x_i \oplus x_j$;
(P3') $\forall i,\, j,\quad y_n \neq x_i$ and $y_n \neq x_i \oplus x_j$.

With each function h in H_n we can now associate a graph $G(h)$ whose vertices are labeled from x_1 to x_n and in which two vertices x_i and x_j are joined by an edge labeled y_k if and only if $x_i \oplus x_j = y_k$. This graph has exactly $V(G)=n$ vertices and $E(G)=n$-1 edges. The values for $V(G)$ and $E(G)$ follow directly from the previous definition and properties (P2') and (P3'). Moreover, $G(h)$ has no cycles, because in any cycle the edge-labels would sum to zero modulo 2, contradicting properties (P2') and (P1') which imply that each edge in $G(h)$ is labeled by a different y_k and that the y_k's are linearly independent. Thus, using elementary graph theory, we conclude that $G(h)$ is a forest with $V(G)$-$E(G)=1$ components, i.e., a tree (cf. [9, p.19]). This tree will be referred to as the respective *h-tree* in what follows.

From properties (P1'), (P2') and (P3') above, we derive the following theorem about the complexities of h and h^{-1}.

Theorem 7. *For each function $h \in H_n$,*

$$n \leq C(h) \leq C(y_n(\underline{x})) + n-1 \quad \textit{and} \quad C(h^{-1}) = \min_i[C(x_i(\underline{y}))] + n-1.$$

Proof. By considering independent realizations of the component functions, we immediately obtain that $C(h) \leq C(y_n(\underline{x})) + n-1$ (by property (P2')). Theorems 3 and 4 further imply that $C(h) \geq n$, because each $y_k(\underline{x})$ depends non-idly at least on two different x_i's (cf. (P2') and (P3')). This yields the desired bounds on the complexity of h.

Considering now h^{-1}, it follows directly from (P1'), (P2') and (P3') that the conditions for the application of Theorem 4 are all satisfied and consequently that

$$C(h^{-1}) \geq \min_i[C(x_i(\underline{y}))] + n-1.$$

In order to prove that equality holds, we make use of the h-trees associated with the functions in H_n. The tree structure of $G(h)$ tells us immediately that it is always possible to start at any vertex x_i and compute the component functions corresponding to the other n-1 vertices by doing exactly the n-1 modulo 2 additions indicated by the respective edges. This means that, by starting at any vertex x_j with $C(x_j(\underline{y}))=\min_i[C(x_i(\underline{y}))]$, one obtains a realization for h^{-1} that requires exactly $\min_i[C(x_i(\underline{y}))] + n-1$ gates. □

We will see that the concept of the associated h-tree is very useful also in the following derivations. Because the functions φ_n from the previous section belong to H_n, we start our investigations by considering their h-trees first. As illustrated by Figure 1, $G(\varphi_n)$ is an (n-1)-arc in which the vertex corresponding to the component function $\bigoplus_{k=1}^n y_k$ (this vertex is denoted by Σ in what follows) appears always as near to the center of the (n-1)-arc as possible ($x_{\lceil n/2\rceil}=\Sigma$). Note that specifying the position of the Σ vertex in the labeled (n-1)-arc completely specifies the component functions $x_i(\underline{y})$ of φ_n^{-1}.

If we depart from the vertex Σ on any path to a monovalent vertex (in this case there are only two), the number of different y_k's in the modulo 2 sum corresponding to any vertex x_i (when interpreted as the component function

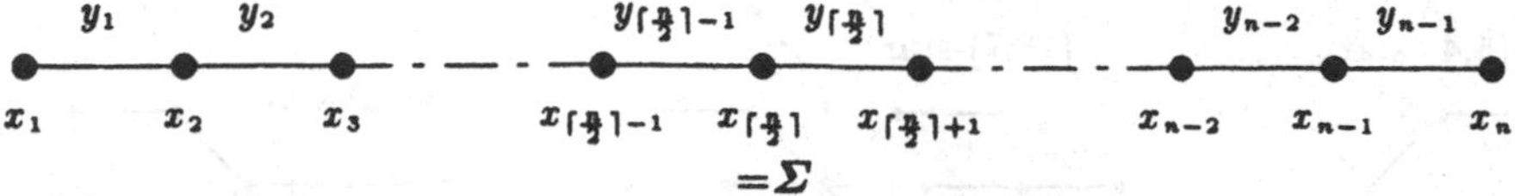

Fig. 1. Representation of the labeled (n-1)-arc, the h-tree associated with the functions φ_n.

$x_i(\underline{y})$ of φ_n^{-1}) is given by $z_i = n - \mathrm{d}(\Sigma, x_i)$, where the distance $\mathrm{d}(\Sigma, x_i)$ equals the number of edges between Σ and x_i. If we wish to maximize $C(h^{-1})$, we can (because of Theorem 7) maximize $\min_i[C(x_i(\underline{y}))] = \min_i[z_i\text{-}1]$ (by Theorem 3 and obvious realizations) or, equivalently, we can minimize $\max_i[\mathrm{d}(\Sigma, x_i)]$. This explains why it is important that Σ is always placed as near as possible to the center of the (n-1)-arc.

From the (n-1)-arc with center Σ, we notice further that each edge y_k ($k=$ 1 .. n-1) partitions the h-tree into a first part where the expressions corresponding to the vertices x_i depend on y_k and a second part where the expressions corresponding to the x_i's are independent of y_k. Because y_n never appears as an edge of the h-tree (cf. (P3')), it is also evident that each component function $x_i(\underline{y})$ depends on y_n. This immediately implies the following property of the monovalent vertices of $G(\varphi_n)$

$$\bigoplus_{\text{monov. } x_i} x_i(\underline{y}) = x_1(\underline{y}) \oplus x_n(\underline{y}) = \bigoplus_{k=1}^{n-1} y_k .$$

Thus,

$$\Sigma(\underline{y}) \oplus x_1(\underline{y}) \oplus x_n(\underline{y}) = y_n .$$

From the last expression it follows equivalently that $y_n(\underline{x})$ is always a modulo 2 sum of three x_i's, the two monovalent vertices and the Σ vertex.

The generalization follows straightforwardly from the previous observations simply by changing the valence s of the vertex Σ (recall that $\Sigma(\underline{y}) = \bigoplus_{k=1}^{n} y_k$). The aim of this approach is to reduce the maximum distance $\mathrm{d}(\Sigma, x_i)$ without significantly increasing the complexity $C(y_n(\underline{x}))$ (cf. Theorem 7).

Definition 8. A tree with n-1 edges is called an *(n-1, s)-star* if the n vertices satisfy the following conditions:

i) one vertex, the center of the (n-1, s)-star, has valence $s \geq 2$;
ii) s vertices have valence 1 and are at a distance $\lceil \frac{n-1}{s} \rceil$ or $\lfloor \frac{n-1}{s} \rfloor$ from the center;
iii) n-s-1 vertices have valence 2.

Figure 2 shows different examples of (n-1, s)-stars. It is evident that (n-1, s)-stars exist only for those values of n and s satisfying $n > s \geq 2$.

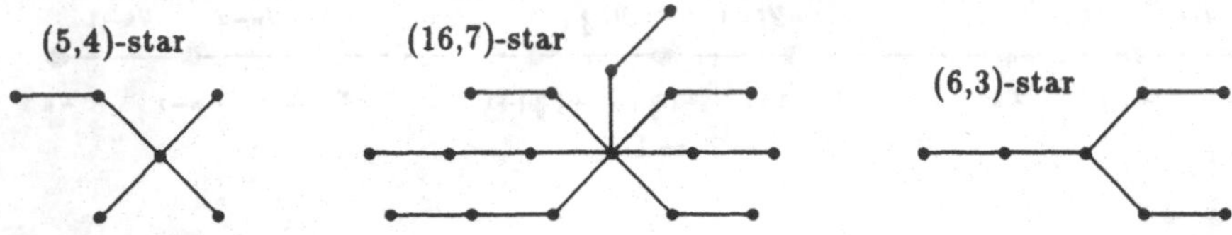

Fig. 2. Representation of different $(n\text{-}1, s)$-stars.

Let $\lambda_{n,s}$ denote any of those functions from H_n for which the corresponding h-tree is (up to permutations of the x_i's and permutations of the y_k's ($\neq y_n$)) the $(n\text{-}1, s)$-star with center Σ . We prove that the functions $\lambda_{n,s}$ satisfy the following theorem.

Theorem 9. *For all $n > s \geq 2$, the functions $\lambda_{n,s}$ satisfy*

$$n \leq C(\lambda_{n,s}) \leq n + s - \begin{cases} 1 \text{ for even } s \\ 2 \text{ for odd } s \end{cases} \quad \text{and} \quad C(\lambda_{n,s}^{-1}) = \lfloor \tfrac{2s-1}{s}(n-1) \rfloor .$$

Proof. The condition $n > s \geq 2$, as mentioned above, is necessary for the existence of the corresponding $(n\text{-}1, s)$-star. Thus, following the observations made for $\varphi_n = \lambda_{n,2}$, we know from Definition 8 that

$$\max_{i=1...n} \mathrm{d}(\Sigma , x_i) = \lceil \tfrac{n-1}{s} \rceil$$

and consequently that

$$\min_{i=1...n} [C(x_i(\underline{y}))] = \lfloor \tfrac{s-1}{s} (n-1) \rfloor .$$

The complexity of $\lambda_{n,s}^{-1}$ follows now directly from the application of Theorem 7. Note that $\lambda_{n,s}^{-1}$ is completely specified by the labeled $(n\text{-}1, s)$-star with center Σ.

To prove the bounds on the complexity of $\lambda_{n,s}$ we have to distinguish two different cases:

i) *s is even*: Similarly to the case s=2, each edge y_k (k= 1 .. n-1) partitions the $(n\text{-}1, s)$-star with center Σ into a first part (with s-1 monovalent vertices) where the expressions corresponding to the vertices x_i depend on y_k and a second part (with one monovalent vertex) where the expressions corresponding to the x_i's are independent of y_k. Again, y_n never appears as an edge of the h-tree so that each component function $x_i(\underline{y})$ depends on y_n. This immediately implies that

$$\bigoplus_{\text{monov. } x_i} x_i(\underline{y}) = \bigoplus_{k=1}^{n-1} y_k \qquad \text{because } s \text{ is even and } s\text{-1 is odd.}$$

Thus,

$$\Sigma(\underline{y}) \oplus \bigoplus_{\text{monov. } x_i} x_i(\underline{y}) = y_n .$$

From the last expression it follows equivalently that, for even values of s, $y_n(\underline{x})$ is always a modulo 2 sum of s+1 different x_i's, the s monovalent vertices and the Σ vertex, with $C(y_n(\underline{x}))=s$ (by Theorem 3 and an obvious realization).

ii) *s is odd*: The derivations follow exactly those of case i), the only exception being that

$$\bigoplus_{\text{monov. } x_i} x_i(\underline{y}) \;=\; y_n \qquad \text{because } s \text{ is odd and } s\text{-1 is even.}$$

Thus, for odd values of s, $y_n(\underline{x})$ equals simply the modulo 2 sum of the s different x_i's corresponding to the monovalent vertices and $C(y_n(\underline{x}))=s-1$.

In both cases, $\lambda_{n,s}$ is completely specified by the labeled (n-1,s)-star and the expression for $y_n(\underline{x})$. The bounds on the complexity of $\lambda_{n,s}$ now follow directly from application of Theorem 7. □

Note that by partitioning the x_i's into two sets as in the proof of Theorem 5, it is possible to prove the following stronger result.

Theorem 10. *For all $s \geq 2$ and all $n \geq 2s+e_s$, where $e_s = 1(0)$ for even(odd) values of s, the functions $\lambda_{n,s}$ satisfy*

$$C(\lambda_{n,s}) \;=\; n+s+e_s-2 \quad \textit{and} \quad C(\lambda_{n,s}^{-1}) \;=\; \lfloor \tfrac{2s-1}{s}(n-1) \rfloor .$$

Proof. Define the sets S_1 and S_2 by $S_1 = \{x_i \mid y_n(\underline{x}) \text{ depends non-idly on } x_i\}$ and $S_2 = \{x_1, \ldots, x_n\} \setminus S_1$. By considering the h-trees associated with the functions $\lambda_{n,s}$, we see immediately that, for each $n \geq 2s+e_s$, each one of the component functions $y_k(\underline{x})$ with $k \neq n$ depends on two different x_i's of which at least one is not an element of S_1. Thus, setting the x_i's in S_2 to zero eliminates at least the n-1 output gates computing $y_1(\underline{x}), \ldots, y_{n-1}(\underline{x})$ and leaves the component function $y_n(\underline{x})$ unchanged. Because $C(y_n(\underline{x})) \geq s+e_s-1$ (by Theorem 3), the new theorem follows directly from Theorem 9. □

Defining $\lambda_{n,s}$ to be the identity function for all $1 \leq n \leq s$ ($s \geq 2$), we obtain the following two corollaries.

Corollary 11. *For any integer $s_0 \geq 2$,*
$\{\lambda_{n,s_0}\}$ is a family of permutations that is feebly-one-way of order $\frac{2s_0-1}{s_0} = 2-\frac{1}{s_0}$.

Corollary 12. *For any integer-valued sequence $s_1, s_2, \ldots,$ satisfying*

$$s_n \geq 2, \quad \liminf_{n\to\infty}[s_n/n] = 0 \quad \textit{and} \quad \liminf_{n\to\infty}[s_n] = \infty,$$

$\{\lambda_{n,s_n}\}$ is a family of permutations that is feebly-one-way of order 2.

Note that the first limit guarantees that Theorem 10 holds for all sufficiently large values of n, while both limits are necessary to guarantee that the measure of feeble-one-wayness $M_F(\lambda_{n,s_n})$ tends to 2 as n tends to infinity.

By considering the derivative of $M_F(\lambda_{n,s})$ with respect to s, we find that the sequence s_n with an almost optimal value of s for each $n \geq 5$ is given by

$$s_n^{\text{opt}} = \begin{cases} 2 & \text{for } n = 5 \\ \text{OddVal}\left(\lfloor 0.5 + \sqrt{\frac{n-1.5}{2}} \rfloor,\ \lfloor 1.5 + \sqrt{\frac{n-1.5}{2}} \rfloor\right) & \text{for } n \geq 6. \end{cases}$$

The behavior of the corresponding measure $M_F(\lambda_{n,s_n^{\text{opt}}})$, for practical values of n, is shown in Figure 3. For larger n, it follows directly from Corollary 12 that $M_F(\lambda_{n,s_n^{\text{opt}}})$ tends to the finite value 2.

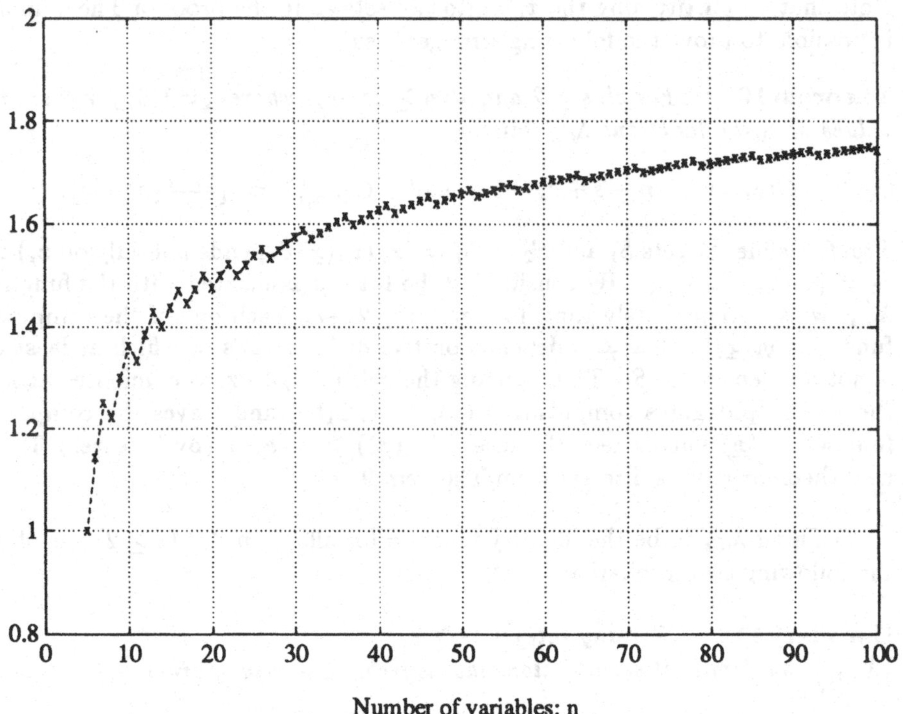

Fig. 3. Behavior of $M_F(\lambda_{n,s_n})$ for $s_n = s_n^{\text{opt}}$.

We conclude this section by an example that illustrates how the function $\lambda_{11,s}$ changes as s is modified. After assigning x_i's to the vertices and y_k's ($\neq y_n$) to the edges of the respective $(10, s)$-stars with center Σ , we can immediately write down the following matrices.

$$\underline{y} = \lambda_{11,2}(\underline{x}) = \begin{bmatrix} 11000000000 \\ 01100000000 \\ 00110000000 \\ 00011000000 \\ 00001100000 \\ 10000010000 \\ 00000011000 \\ 00000001100 \\ 00000000110 \\ 00000000011 \\ 10000100001 \end{bmatrix} \cdot \underline{x} \qquad \underline{x} = \lambda_{11,2}^{-1}(\underline{y}) = \begin{bmatrix} 11111111111 \\ 01111111111 \\ 00111111111 \\ 00011111111 \\ 00001111111 \\ 00000111111 \\ 11111011111 \\ 11111001111 \\ 11111000111 \\ 11111000011 \\ 11111000001 \end{bmatrix} \cdot \underline{y}$$

$$C(\lambda_{11,2}) = 12 \qquad C(\lambda_{11,2}^{-1}) = 15$$

$$\underline{y} = \lambda_{11,3}(\underline{x}) = \begin{bmatrix} 11000000000 \\ 01100000000 \\ 00110000000 \\ 00011000000 \\ 10000100000 \\ 00000110000 \\ 00000011000 \\ 10000000100 \\ 00000000110 \\ 00000000011 \\ 00001001001 \end{bmatrix} \cdot \underline{x} \qquad \underline{x} = \lambda_{11,3}^{-1}(\underline{y}) = \begin{bmatrix} 11111111111 \\ 01111111111 \\ 00111111111 \\ 00011111111 \\ 00001111111 \\ 11110111111 \\ 11110011111 \\ 11110001111 \\ 11111110111 \\ 11111110011 \\ 11111110001 \end{bmatrix} \cdot \underline{y}$$

$$C(\lambda_{11,3}) = 12 \qquad C(\lambda_{11,3}^{-1}) = 16$$

$$\underline{y} = \lambda_{11,5}(\underline{x}) = \begin{bmatrix} 11000000000 \\ 01100000000 \\ 10010000000 \\ 00011000000 \\ 10000100000 \\ 00000110000 \\ 10000001000 \\ 00000001100 \\ 10000000010 \\ 00000000011 \\ 00101010101 \end{bmatrix} \cdot \underline{x} \qquad \underline{x} = \lambda_{11,5}^{-1}(\underline{y}) = \begin{bmatrix} 11111111111 \\ 01111111111 \\ 00111111111 \\ 11011111111 \\ 11001111111 \\ 11110111111 \\ 11110011111 \\ 11111101111 \\ 11111100111 \\ 11111111011 \\ 11111111001 \end{bmatrix} \cdot \underline{y}$$

$$C(\lambda_{11,5}) = 14 \qquad C(\lambda_{11,5}^{-1}) = 18$$

In these representations, x_1 has always been chosen to correspond to the center Σ. The respective complexities follow directly from application of Theorem 10. Note that $M_F(\lambda_{11,s})=C(\lambda_{11,s}^{-1})/C(\lambda_{11,s})$ is indeed maximal for $s = s_{11}^{\text{opt}} = 3$.

5 Comments

To our knowledge, the only previous results concerning *complexity-asymmetric* permutations, i.e., permutations f on $\{0,1\}^n$ with $C(f^{-1}) \neq C(f)$, were given in our earlier work [10] and in an unpublished dissertation by S. Boyack [11]. Both of these state the result that no complexity-asymmetric permutations exist for

$n \leq 3$. In [10], we proved for a particular nonlinear permutation f on $\{0,1\}^4$ that $C(f)$=5 and $C(f^{-1})$=6. This is the smallest example of a complexity-asymmetric permutation. In [11], a construction was described that yields, for all $n = 8d$-1 (d an integer >0), a linear permutation f_n on $\{0,1\}^n$ with $C(f_n) = 10d$-2 and 11d-3 $\leq C(f_n^{-1}) \leq$ 12d-3. These permutations can easily be extended to give a family of permutations that is feebly-one-way of some order between 1.1 and 1.2. It is, however, not possible to give the exact order because the exact value of $C(f_n^{-1})$ cannot be determined by presently known methods.

Thus, the families of permutations described in this paper are the first examples which allow us to prove the existence of feeble one-wayness of order 2. Note that this is not at all a typical worst-case result. Using the complexities given in Theorem 10, it is indeed always possible to determine the exact *order of asymmetry*, $C(\lambda_{n,s_n}^{-1})/C(\lambda_{n,s_n})$, for every single member in one of our families. Moreover, it follows straightforwardly from Theorem 4 that any approximation for a specific λ_{n,s_n}^{-1} that uses less than $C(\lambda_{n,s_n}^{-1})$ gates makes an output error for at least 50% of the inputs.

Finally, it is obvious that feeble one-wayness is still too weak to be used in practice. We believe, however, that the lack of techniques for proving nonlinear lower bounds on the unrestricted circuit complexity is the major reason for the fact that there are at present no proofs for practical one-wayness, let alone classical one-wayness. We hope therefore that our results will encourage other persons to do research in this direction, which aims to provide the foundation for provable security.

Acknowledgements

The author is grateful to J.L. Massey for motivating this research and for many helpful comments. Helpful discussions with J. Ganz are also gratefully acknowledged.

References

1. W. Diffie and M. E. Hellman, "New directions in cryptography," *IEEE Trans. Inform. Theory*, vol. IT-22, no. 6, pp. 644–655, Nov. 1976.
2. G. Brassard, "Relativized cryptography," *IEEE Trans. Inform. Theory*, vol. IT-29, no. 6, pp. 877–894, Nov. 1983.
3. L. A. Levin, "One-way functions and pseudorandom generators," *Proc. 17th Ann. ACM Symp. Theory of Comput.*, pp. 363–365, 1985.
4. S. C. Pohlig and M. E. Hellman, "An improved algorithm for computing logarithms over GF(p) and its cryptographic significance," *IEEE Trans. Inform. Theory*, vol. IT-24, no. 1, pp. 106–110, Jan. 1978.
5. R. B. Boppana and J. C. Lagarias, "One-way functions and circuit complexity," *Information and Computation*, vol. 74, pp. 226–240, 1987.

6. I. Wegener, *The Complexity of Boolean Functions.* New York: Wiley (Stuttgart: Teubner), 1987.
7. E. A. Lamagna and J. E. Savage, "On the logical complexity of symmetric switching functions in monotone and complete bases," *Brown University Techn. Rep., Providence, Rhode Island*, July 1973.
8. J. E. Savage, *The Complexity of Computing.* New York: Wiley, 1976.
9. W. T. Tutte, *Graph Theory*, vol. 21 of *Encyclopedia of Math. and its Appl.* Massachusetts: Addison-Wesley, 1984.
10. A. P. Hiltgen and J. Ganz, "On the existence of specific complexity-asymmetric permutations." Techn. Rep., Signal and Inform. Proc. Lab, ETH-Zürich, 1992.
11. S. W. Boyack, "The robustness of combinatorial measures of boolean matrix complexity." Ph.D. thesis, Massachusetts Inst. of Techn., 1985.

On Bit Correlations Among Preimages of "Many to One" One-Way Functions

— A New Approach to Study on
Randomness and Hardness of One-Way Functions —

KOUICHI SAKURAI[1] TOSHIYA ITOH[2]

[1] Computer & Information Systems Laboratory, Mitsubishi Electric Corporation, 5-1-1 Ofuna, Kamakura 247, Japan (sakurai@isl.melco.co.jp).
[2] Dept. of Information Processing, Tokyo Institute of Technology, 4259 Nagatsuta, Midori-ku, Yokohama 227, Japan (titoh@ip.titech.ac.jp).

Abstract. This paper presents a new measure of the complexity of *many to one* functions. We study bit correlations among the preimages of an element of the range of many to one one-way functions. Especially, we investigate the correlation among the least significant bit of the preimages of 2 to 1 one-way functions based on algebraic problems such as the factorization and the discrete logarithm.

1 Introduction and motivation

A one-way function is a central topic in modern cryptography and computational complexity theory. Intuitively, a function is *one-way* if it is easy to compute but hard to invert. It must be noted that one-way functions are not necessarily 1 to 1. Although 1 to 1 one-way functions have better properties and are easy to deal with, *many to one* one-way functions are a more general primitive in cryptography and some explicit one-way functions based on some number -theoretic assumption like as Rabin encryption functions [Ra79] are not 1 to 1.

However, some important cryptographic results have constructed from 1 to 1 one-way functions or one-way permutations (e.g. the secure signature proposed by Naor and Yung [NY89]). To reduce these results to *many to one* one-way functions, one approach is to transfer the function to a 1 to 1 function by using a number-theoretic structure [BM84] and another is to use additional tools such as universal hash functions [GKL88, ILL89, Ha90]. But, there are no known direct study on the *many to one* one-way function itself.

In this paper, we introduce a new measure to study intrinsic properties of *many to one* one-way functions, bit correlations among the preimages of an element in the image of *many to one* one-way functions. Especially, we investigate the correlations among the least significant bit of the preimages of 2 to 1 one-way functions based on algebraic problems such as the factorization and the discrete logarithm.

A certain bit (or parts) of the preimages of one-way functions has been studied from the point of view of "hard core" [BM84, ACGS88, Ya82, SS90]. However, we should note that these hard core bits are defined for only 1 to 1 one-way functions. Goldreich and Levin [GL89] showed how to construct a hard core predicate from one-way functions. Goldreich and Levin's hard core predicate is useful not only for one-way permutations but also for general one-way functions. But, their hard core predicate is not well defined, i.e. the value of the predicate of preimages of the function is not necessarily well-defined. This is an obstruction of constructing cryptographic objects from one-way functions (e.g. non-interactive zero-knowledge [FLS90]). Our study suggests when one-way functions have well-defined (hard core) predicates.

We construct 2 to 1 functions based on the factorization and the discrete logarithm, in which finding collision is as hard as factoring the modulus, and shows 3 types of them:

Type 1: lsb of two preimages are equal for (almost) exactly half of the elements in the image of the function.

Type 2: lsb of two preimages of any element in the image of the function are always equal.

Type 3: lsb of two preimages of any element of the image of the function are always different.

The property of type 1 function implies that: *The two preimages of any element are not independent because the distinction of the two preimages gives the factorization of the modulus. However, their least significant bit is independent from each other.* The properties of type 2 and type 3 functions suggest that the type 1 is not a natural property of the functions.

Furthermore, we consider the cryptographic application of the properties of many to one one-way function, especially the well-defined hard core predicates.

2 Preliminaries

A length $l(n)$ is a monotone increasing function from N to N such that $l(n)$ is computable in time polynomial in n. A function f with input length $m(n)$ and output length $l(n)$ specifies for each $n \in N$ a function $f_n : \{0,1\}^{m(n)} \to \{0,1\}^{l(n)}$. For simplicity, we write $f(x)$ in place of $f_n(x)$. We call f *polynomial-time computable* if there exists a polynomial-time Turing machine that on input $x \in \{0,1\}^{m(n)}$ computes $f(x) \in \{0,1\}^{l(n)}$.

At first, we recall the definition of (strong) one-way functions.

Definition 1 [Ya82]. A function $f : \{0,1\}^* \to \{0,1\}^*$ is called *one-way* if it is polynomial time computable, but not "polynomial time invertible". More precisely, for every probabilistic polynomial time algorithm A, and for every constant $c > 0$ there exists an N_c such that for all $k > N_c$

$$Prob\left[A\left(f(x), 1^k\right) \in f^{-1}\left(f(x)\right)\right] < k^{-c},$$

where the probability is taken over all x's of length k and the internal coin tosses of A, with uniform probability distribution.

Note that the role of 1^k in the definition above is to a void the *trivial* one-way function, which shrinks the inputs by more than a polynomial amount.

A hash function, which compresses the size of inputs, is a kind of "many to one" function. Damgård defined a *family of collision intractable hash functions*.

Definition 2 [Da87, Ru92]. A *family of collision intractable hash functions* is a set of hash functions $\{h_i | i \in I\}$ for some index set $I \subseteq \{0,1\}^*$, where $h_i : \{0,1\}^{|i|+1} \mapsto \{0,1\}^{|i|}$, with the following properties:

1. There is a probabilistic polynomial time algorithm G such that $G(1^n) \in \{0,1\}^n \cap I$.
2. All functions in the family are computable in polynomial time.
3. The problem of finding a collision $x \neq y$ such that $h_i(x) = h_i(y)$ for a given h_i in the family is computationally infeasible to solve. More precisely, for every probabilistic polynomial time algorithm A, and for every constant $c > 0$ there exists an N_c such that for all $k > N_c$

$$Prob\left[i \leftarrow G(1^n), (x,y) \leftarrow A(i) : h_i(x) = h_i(y) \bigwedge x \neq y\right] < k^{-c},$$

where the probability is taken over the internal coin tosses the algorithm G and C.

In this paper, we consider the single function rather than the family. We define a collision intractable function (CIF) as follows.

Definition 3. A polynomial time computable function $f : \{0,1\}^* \rightarrow \{0,1\}^*$ is called *collision intractable* if it is has no "polynomial time collision finding algorithm". More precisely, for every probabilistic polynomial time algorithm A, and for every constant $c > 0$ there exists an N_c such that for all $k > N_c$

$$Prob\left[A\left(x, f(x), 1^k\right) \in \{f^{-1}(f(x)) \setminus \{x\}\}\right] < k^{-c},$$

where the probability is taken over all x's of length k and the internal coin tosses of A, with uniform probability distribution.

We will denote the domain of the function f by $Dom(f)$, and the image $\{y \in \{0,1\}^* : \exists x \in Dom(f)$ such that $y = f(x)\}$ by $Im(f)$.

3 Bit correlations among preimages of CIF

3.1 A new measure to study many to one functions

We study the properties of the preimages of elements in the range of (many to one) one-way functions. A new measure to study the complexity of many to one one-way functions is introduced.

Definition 4. Let f be a many to one function, and $b : \{0,1\}^* \to \{0,1\}$ be a polynomial time computable predicate for f. The probability of the distribution of the gap of the function f w.r.t b is defined as:
PGap(f,b) =

$$\frac{\left\{\sum_{y\in Im(f)} \left(1 - \frac{|\#\{x\in f^{-1}(y) : b(x)=1\} - \#\{x\in f^{-1}(y) : b(x)=0\}|}{\#f^{-1}(y)}\right)\right\}}{\#Im(f)}$$

Remark. For a 1 to 1 function f and any predicate b of f, PGap$(f,b) = 0$.

Our main research interest is that:

PROBLEM: Study PGap(f,b) for a collision intractable function f and a predicate b for f.

In the following, we consider 2 to 1 one-way functions. Such functions has two preimages x and y for any element in $Im(f)$, i.e. $f(x) = f(y) (x \neq y)$. Our interest is the (un)correlation of the least significant bit (lsb) of x and y.

Definition 5. Let f be a 2 to 1 function. The probability of the distribution of distinct lsb of the preimages of f is

$$\text{PDL}(f) = Prob\left[lsb(x) \neq lsb(y) \wedge f(x) = f(y) \wedge x \neq y\right],$$

where the probability is taken over all elements in the image of f.

Namely, PDL(f) = PGap$(f,$ lsb$)$ if the function f is 2 to 1.

We construct three type of 2 to 1 functions, for which difficulty of finding collision is as hard as the factorization of the modulus, with the property that PDL$(f) = 0, 1/2$, and 1.

3.2 CIF with PDL = 1/2

At first, we consider a variant of the Rabin encryption function [Ra79]. Let p and q be odd primes distinct from each other with same size $|p| = |q|$. For $N = pq$, we consider the function $R(x) = x^2 \pmod N$. The original Rabin encryption function with $Dom(R) = Z_N^*$ is 4 to 1.

To modify the function 2 to 1, we consider the function R which has a restricted domain $Dom(R) = [0, N/2]$.

Proposition 6. *The function* $F(x) = x^2 \pmod N$ *with* $Dom(F) = [0, N/2]$ *is 2 to 1.*

As the extracting square roots is polynomially equivalent to factoring [Ra79], we obtain the following proposition.

Proposition 7. *Under the intractability assumption on the factoring the modulus, the function F is collision intractable.*

The proposition above suggests that two preimages of any element in the image are not independent: the distinction of two preimages of an element induces the factor of the modulus.

Our question is that:

Question A: Are there any correlation between $lsb(x)$ and $lsb(y)$ such that $F(x) = F(y)(x \neq y)$?

A known property of lsb of preimages of the Rabin encryption function is that if the modulus N is ***Blum integer*** then lsb of the preimage, which is the set of all integers of the quadratic residuosity modulo N, of an element in the image is a hard bit [ACGS88]. However, the argument of hard bits is useful for only a variant of Rabin encryption function that is *1 to 1*. No previous results answer to the question above.

One of our results is as follows[3].

Theorem 8. *If N is sufficiently large, then PDL(F)* $\approx 1/2$.

The following proof is based on a direct counting argument.

Proof: Let N be the product of two distinct odd primes p and q. Let D be a set of integers $x \in \mathcal{Z}_N^*$ such that $0 < x < N/2$. Here we define a set C to be

$$\mathcal{C} = \{y_1 \in \mathcal{Z}_N^* \mid y_1 \in \mathcal{D} \bigwedge \exists y_2 \in \mathcal{Z}_N^* [y_2 \in \mathcal{D} \bigwedge y_1^2 \equiv y_2^2 \pmod N \bigwedge \mathrm{lsb}(y_1) = \mathrm{lsb}(y_2)]\}.$$

Then the probability P that we want to evaluate is given by

$$P = \frac{\|\mathcal{C}\|}{\|\mathcal{D}\|} = \frac{\|\mathcal{C}\|}{\varphi(N)/2} = \frac{2 \cdot \|\mathcal{C}\|}{\varphi(N)},$$

where $\|\mathcal{A}\|$ is the ***cardinality*** of a set A and $\varphi(N)$ is the Euler's totient function [Kra86] of N. From the assumption that $y_1^2 \equiv y_2^2 \pmod N$, it follows that $y_1^2 - y_2^2 \equiv (y_1 + y_2)(y_1 - y_2) \equiv 0 \pmod N$ iff

$$(\text{Case 1}) \begin{cases} p \mid y_1 + y_2; \\ q \mid y_1 - y_2; \end{cases} \quad \text{or} \quad (\text{Case 2}) \begin{cases} q \mid y_1 + y_2; \\ p \mid y_1 - y_2. \end{cases}$$

We first consider the (Case 1), i.e., $p \mid y_1 + y_2$ and $q \mid y_1 - y_2$. This implies that for such y_1 and y_2, $\mathrm{lsb}(y_1) = \mathrm{lsb}(y_2)$ iff there exists integers s and t such that

$$(1)\ y_1 + y_2 = 2sp, \qquad (2)\ y_1 - y_2 = 2tq.$$

[3] Kurosawa et al.[KOT90] announced that there exists a 4 move blackbox simulation perfect ZKIP for Quadratic Residuosity(QR) modulo two distinct primes. However, Okamoto [Oka90] pointed out that the proof of soundness of the scheme in [KOT90] depends upon an unproven assumption, i.e., distribution of a hard bit of square roots modulo such a composite. The authors [SI90] showed the assumption is true.

It is immediate to see that $\text{lsb}(y_1) = \text{lsb}(y_2)$ iff for some integers s and t,

$$(3)\ y_1 = sp + tq, \qquad (4)\ y_2 = sp - tq.$$

From the assumption that $0 < y_1, y_2 < N/2$, we have $0 < y_1 + y_2 < N$ and $-N/2 < y_1 - y_2 < N/2$, and it follows from the Equations (1) and (2) that

$$(5)\ 0 < s < \frac{q}{2}, \qquad (6)\ -\frac{p}{4} < t < \frac{p}{4}.$$

In addition, from the assumption that $0 < y_1, y_2 < N/2$ and the Equations (3) and (4), we have

$$(7)\ -\frac{p}{q}s < t < -\frac{p}{q}s + \frac{p}{2}, \qquad (8)\ \frac{p}{q}s - \frac{p}{2} < t < \frac{p}{q}s.$$

Figure 1 shows the set C_1 of all pairs of (s,t) that satisfies the Equations (5)-(8).

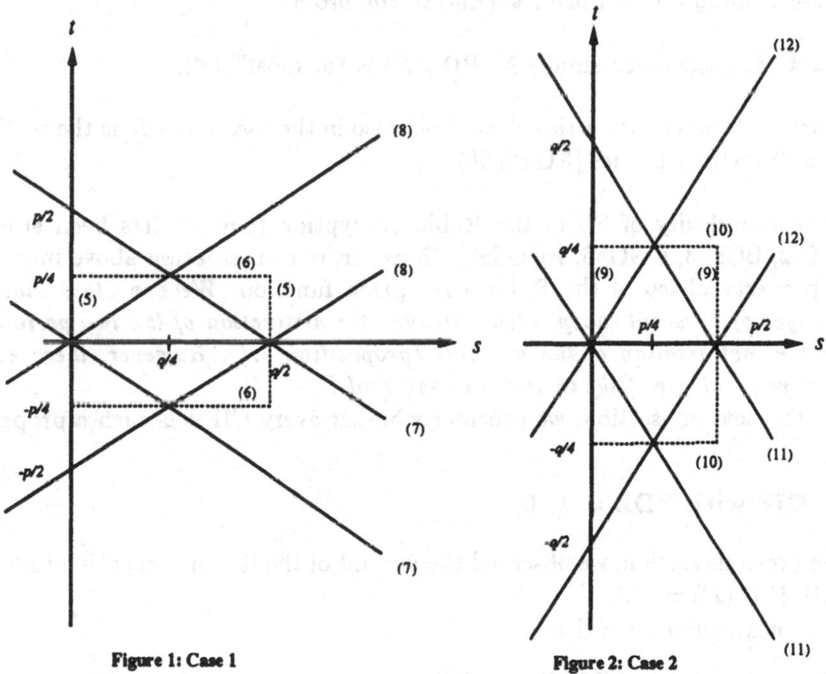

Figure 1: Case 1

Figure 2: Case 2

We then consider the (Case 2), i.e., $q \mid y_1 + y_2$ and $p \mid y_1 - y_2$. This implies that for such y_1 and y_2, $\text{lsb}(y_1) = \text{lsb}(y_2)$ iff there exist integers s and t such that $y_1 + y_2 = 2sq$ and $y_1 - y_2 = 2tp$. In almost the same way as the (Case 1), we have

$$(9)\ 0 < s < \frac{p}{2}, \qquad (10)\ -\frac{q}{4} < t < \frac{q}{4},$$

$$(11)\quad -\frac{q}{p}s < t < -\frac{q}{p}s + \frac{q}{2}, \qquad\qquad (12)\ \frac{q}{p}s - \frac{q}{2} < t < \frac{q}{p}s.$$

Figure 2 (in a way similar to Figure 1) shows the set C_2 of all pairs of (s,t) that satisfies the Equations (9)-(12).

Thus the probability P that for $N = pq$ and any $y_1 \in \mathcal{D}$, there exists a $y_2 \in \mathcal{D}$ such that $y_1 \equiv y_2 \pmod N$ and $\mathrm{lsb}(y_1) = \mathrm{lsb}(y_2)$ is

$$P = \frac{\|C\|}{\|\mathcal{D}\|} = \frac{\|C_1\| + \|C_2\|}{\varphi(N)/2}$$

$\|C_1\| + \|C_2\|$ is approximated by $\{(p-1)/2\} \cdot \{(q-1)/2\}$ under the condition that p and q are the same size and $N = pq$ is sufficiently large. (NOTE: $\frac{(p-1)}{2} \cdot \frac{(q-1)}{2} - (\|C_1\| + \|C_2\|)$ is negligible.) Then,

$$P \approx \frac{2 \cdot \{(p-1)/2\} \cdot \{(q-1)/2\}}{(p-1)\cdot(q-1)} = \frac{1}{2},$$

because p and q are odd primes. (End of the proof)

Remark. For any (even small) N, PDL(F) is "at most" 1/2.

Remark. The same properties above hold also in the case when N is the product of $k(> 2)$ distinct primes [KOSIT90].

The complexity of lsb of the Rabin encryption function has been studied [GMT82, BCS83, CGG86, ACGS88]. However, our observation above implies a new property of lsb of the Rabin encryption function. *We can state that the preimages of F is not independent, because the distinction of the two preimages gives the factorization of the modulus (proposition 3.4). However, there exists an independent part (lsb) in two preimages of F.*

In the next subsection, we consider whether every CIF has such a property.

3.3 CIF with PDL = 1, 0

In the previous section, we observed the variant of the Rabin encryption function F with PDL(F) = 1/2.

Our next question is that:

> **Question B:** Is PDL(f) = 1/2 a natural property of CIF f? Is the following statement true?
>
> "If f is CIF, then PDL(f) is 1/2."

We give negative evidences to this question. In the following, we construct collision intractable one-way functions with PDL = 0 and 1.

We consider the one-way function $f_{g,N}(X) = g^X \pmod N$, where N is a composite of two distinct (special) integers and g satisfies a (specific) condition.

Theorem 9. *Let P and Q be two primes with the form $P = 2p+1, Q = 2q+1$, where p and q are odd primes distinct from each other with same size $|p| = |q|$ and set $N = PQ = (2p+1)(2q+1)$. Suppose that $g \in Z_N{}^*$ has order 2pq. We define a function $G(x) = g^x \pmod N$. Then*

1. G is (almost everywhere) 2 to 1 if $Dom(G) = [0, N]$.
2. G is a CIF under the assumption that the factoring of the modulus is hard.
3. $PDL(G) = 0$,

Proof of theorem 9:

1. $Ord(g) = 2pq$ and $N \approx 4pq$.
2. The same kind of argument inspired by Miller [Mi76] is well known [Ba84, Wo72, SS90]. Suppose that we find the y such that $g^x = g^y \pmod N$ for a given x. Then the Carmichael function [Kra86] of N, $\lambda(N)$ divides $x - y$ because $g^{(x-y)} = 1 \pmod N$. Then there exists the constant $k = 1$, or 2 such that $x - y = k\lambda(N)$. Consider the algorithm below. (1) Choose $a \in Z_N^*$ at random. (2) Compute the odd number u such that $x - y = u2^t$ for some positive integer t, and set $power \leftarrow a^u \pmod N$. (3) Successively square *power* until arriving *at the first* 1. At this point, $power = a^{u2^v} = 1 \pmod N$ for some positive integer $v(< t)$. (4) With probability 1/2, $a^{u2^{v-1}} \neq \pm 1 \pmod N$, which gives the factorization of N. Otherwise, go back to step (1).
3. $lsb(x-y) = 0$ because $x - y = k\lambda(N)$ and $\lambda(N) = l.c.m(P-1)(Q-1) = 2pq$ is even. Then $lsb(x) = lsb(y)$.

Theorem 10. *Let P and Q be two primes with the form P=2p+1, Q=2q+1, where p and q are two distinct primes. Set $N = PQ$. We define a function $H(x) = h^x \pmod N$, where the order $h \in Z_n{}^*$ is pq, over $(0 < x < N/2)$. Then*

1. H is (almost everywhere) 2 to 1 if $Dom(H) = [0, N/2]$.
2. H is a CIF under the assumption that the factoring of the modulus is hard.
3. $PDL(H) = 1$.

Proof of theorem 10:

1. $Ord(h) = pq$ and $N/2 \approx 2pq$
2. Same as Proof of Theorem 3.7 (1)
3. $lsb(x - y) = 1$ because $x - y = \lambda(N)/2$ and $\lambda(N)/2 = pq$ is odd. Then $lsb(x) \neq lsb(y)$.

3.4 PDL on general CIF

In the previous subsection, we observe PDL of CIFs based on algebraic problems. Although we want to study PDL of the general CIF, there are few known results on the the general CIF. The first work on a general construction of CIF was done by Damgård [Da87]. He presents a method for constructing family of collision intractable *hash* function from *claw free pair of permutations* introduced by [GMR88].

Definition 11 [Da87, Ru92]. A *family of claw free permutation pair* is a family of a sets of permutation pair $\{(f_i^0, f_i^1)|i \in I\}$ for some index set $I \subseteq \{0,1\}^*$, where $f_i^l : \{0,1\}^{|i|+1} \mapsto \{0,1\}^{|i|} (l = 0,1)$, with the followings properties:

1. There is a probabilistic polynomial time algorithm G such that $G(1^n) \in \{0,1\}^n \cap I$.
2. All permutations in the family are computable in polynomial time.
3. The problem of finding a claw (x, y) such that $f_i^0(x) = f_i^1(y)$ for a given (f_i^0, f_i^1) in the family is computationally infeasible to solve. More precisely, for every probabilistic polynomial time algorithm A, and for every constant $c > 0$ there exists an N_c such that for all $k > N_c$

$$Prob\left[i \leftarrow G(1^n), (x, y) \leftarrow A(i) : f_i^0(x) = f_i^1(y)\right] < k^{-c},$$

where the probability is taken over the internal coin tosses the algorithm G and C.

Recently, Russell [Ru92] gave conditions for the existence of collision-free hash functions.

Theorem 12 [Ru92]. *There exists a collection of collision-free hash functions iff there exists a collection of claw-free pair of pseudo-permutations.*

It is hard to investigate $PDL(f, b)$ for collision-free hash functions f constructed by Russell. However, he also showed the following.

Theorem 13 [Ru92]. *There exists a collection of collision-free hash functions* $\{f_i\}_{i \in I}$ *and a predicate* b *with* $PDL(f_i, b) = 1$ *for any element* f_i *iff there exists a collection of claw-free pair of permutations.*

4 Well-defined hard bit of many to one one-way function

A certain bit (or parts) of the preimages of one-way functions has been studied from the point of view of "hard core" [BM84, ACGS88, Ya82, SS90]. However, these works are only on 1 to 1 one-way functions. Indeed Goldreich and Levin [GL89] constructed the hard core predicate based on any one-way functions, however, in general the preimage of the hard core predicate is not necessary well-defined. This is an obstruction[4] of constructing a non-interactive zero-knowledge proof which is introduced in [BFM88] from one-way function. Feige, Lapidot, and Shamir constructed a non-interactive zero-knowledge proof from a one-way *permutation*, where the preimage of the hard core predicate is well-defined.

The following is an open problem concerning on the *hardness* of the hard bit of a many to one one-way function.

The lsb of the preimages of an element of the the range of 2 to 1 one-way function with PDL(f) = 1 is *information-theoretically* unpredictable.

---

[4] Goldwasser and Ostrovsky [GO92] overcame this problem by introducing the notion of *invariant signatures*.

We observed (in section 3) the function $F(x) = x^2 \pmod{N = pq}$ with $Dom(F) = [0, N/2]$ satisfies the condition $\mathrm{PDL}(F) = 1/2$. This result suggests that the following question is meaningful.

Question: Is computing the least-significant bit of x from $F(x)$ (with success probability $1/2 + 1/poly(|N|)$) as hard as the factorization of N?

The same kind of problem is affirmatively solved [ACGS88] under the condition that N is a Blum integer [Bl82] and $F(x) = x^2 \pmod N$ with $Dom(F) = QR_N$ (in this case, F is a permutation over QR_N).

If the question above is positively solved, we can construct directly a PRG from the Rabin function [5] even when N is not a Blum integer.

5 Concluding remarks and open problem

In this paper, we have introduced a new measure of the complexity of *many to one* functions, and have investigated the correlation among the least significant bit of the preimages of 2 to 1 one-way functions. Our observation suggests that many to one one-way functions could have well-defined hard core predicates like as one-way permutation. Another application of bit uncorrelations of the preimages of many to one one-way function is bit commitment scheme [BCC88, Na90] secure against infinitely powerful receiver. (This idea was inspired by Naor and Yung [NY89] and also investigated in [KOT90].)

The authors will describe the detail in the final version of this paper.

Acknowledgments:

The authors wish to thank to Dr. Tatsuaki Okamoto. His question [Oka90] on the paper [KOT90] inspired the authors to study this topic. Also, the authors would like to thank to Mr. Masahiro Takei. He kindly helped the authors to write up the figures.

References

[ACGS88] Alexi,W., Chor,B., Goldreich,O., and Schnorr, C.P., "RSA and Rabin functions: Certain Parts are as Hard as the Whole," *SIAM J. on Computing*, Vol.17, No.2, pp.194-209 (April 1988).

[Ba84] Bach, E., "Discrete Logarithms and Factoring," *Report No. UCB/CSD 84/186*, Univ. of California (1984).

[BCC88] Brassard, G., Chaum, D., and Crépeau, C., "Minimum Disclosure Proofs of Knowledge," *Journal of Computer and System Sciences*, Vol.37, No.2, pp.156-189 (1988).

[5] Goldreich el al. [GKL88] constructed a PRG from the Rabin function with general modulus. However, their method is based on the universal hash function.

[BCS83] Ben-Or, M., Chor, B., and Shamir, A., "On the cryptographic security of single RSA bits," *Proceedings of the 15th Annual ACM Symposium on Theory of Computing*, pp.421-430 (1983).

[BFM88] Blum, M., Feldman,P., and Micali, S., "Non-interactive zero-knowledge and its applications," *Proceedings of the 20th Annual ACM Symposium on Theory of Computing*, pp.103-112 (1988).

[Bl82] Blum, M., "Coin Flipping by Telephone," *IEEE Spring COMPCOM* (1982).

[BM84] Blum, M., and Micali, S., "How to Generate Cryptographically Strong Sequences of Pseudo-Random Bits," *SIAM J. on Computing*, Vol.13, pp.850-864 (1984).

[CGG86] Chor, B., Goldreich, O., and Goldwasser, S., "The bit security of modular squaring given partial factorization of the modulus," *Proceedings of CRYPTO'85*, Lecture Notes in Computer Science 218, pp.448-457, Springer-Verlag, Berlin (1986).

[Da87] Damgård,I.B., "Collision Free Hash Functions and Public Key Signature Scheme," *Proceedings of EUROCRYPT87*, Lecture Notes in Computer Science 304, Springer-Verlag, pp.203-216, Berlin (1988).

[FLS90] Feige, U., Lapidot, D., and Shamir A., "Multiple Non-Interactive Zero-Knowledge Proofs Based on a Single Random String," *Proceedings of the 31st Annual Symposium on Foundations of Computer Science*, pp.308-318 (1990).

[GKL88] Goldreich.O., Krawczyk,H. and Luby, M., "On the Existence of Pseudorandom Generator," *Proceedings of the 29st Annual Symposium on Foundations of Computer Science*, pp.12-24 (1988).

[GMT82] Goldwasser, S., Micali, S., and Tong, P., "Why and how to establish a private code on a public network," *Proceedings of the 23st Annual Symposium on Foundations of Computer Science*, pp.134-144 (1982).

[GMR88] Goldwasser, S., Micali, S., and Rivest, L., "A digital signature scheme against adaptive chosen-message attack," *SIAM J. on Computing*, Vol.17, No.2, pp.281-308 (April 1988).

[GL89] Goldreich.O., and Levin, L.A., "A Hard-Core Predicate for all One-Way Functions," *Proceedings of the 21th Annual ACM Symposium on Theory of Computing*, pp.25-32 (May 1989).

[GO92] Goldwasser, S. and Ostrovsky, R. "Non-Interactive Zero-Knowledge Proofs and Invariant Signature are Equivalent," Preproceedings of CRYPTO'92 (1992).

[Ha90] Håstad, J., "Pseudo-Random Generator under Uniform Assumptions," *Proceedings of the 22th Annual ACM Symposium on Theory of Computing*, pp.12-24 (May 1990).

[ILL89] Impagliazzo, R., Levin, L.A., and Luby, M., "Pseudo-random generation from one-way functions," *Proceedings of the 21th Annual ACM Symposium on Theory of Computing*, pp.12-24 (May 1989).

[KOT90] Kurosawa, K., Ogata, W., and Tsujii, S., "4 Move ZKIP," *IEICE Technical Report*, Vol.90, No.125, pp.63-69 (July 1990).

[KOSIT90] Kurosawa, K., Ogata, W., Sakurai, K., Itoh, T., and Tsujii, S., "4-move zero-knowledge interactive proof systems," *IEICE Technical Report*, Vol.90, No.365, pp.7-10 (Dec. 1990).

[Kra86] Kranakis, E., "Primality and cryptography," Wiley-Teubner Series in Computer Science (1986).

[Mi76] Miller, G., "Riemann's Hypothesis and Test for Primality," *Journal of Computer and System Sciences*, Vol.13, pp.300-317 (1976).

[Na90] Naor,M., "Bit Commitments using Pseudo-Randomness," *Proceedings of CRYPTO'89*, Lecture Notes in Computer Science 435, pp.128-136, Springer-Verlag, Berlin (1990).

[NY89] Naor,M., and Yung,M., "Universal One-way Hash functions and their Cryptographic Applications," *Proceedings of the 21th Annual ACM Symposium on Theory of Computing*, pp.33-43 (May 1989).

[Oka90] Okamoto, T., *private communication* (Oct. 1990).

[Ra79] Rabin,M.O., "Digital Signatures and public key functions as intractable as factorization," *Technical Memo TM-212, LCS/MIT (1979).*

[Ru92] Russell A., "Necessary and Sufficient Conditions for Collision-Free Hashing," *Preproceedings of Crypto'92*, (1992).

[SI90] Sakurai,K., and Itoh,T., "On the Distribution of a Hard Bit of Square Roots Modulo a Product of Two Distinct Odd Primes," *Manuscript* (Oct. 1990).

[SS90] Schrift, A.W. and Shamir, A. "The Discrete Log is very Discrete," *Proceedings of the 22th Annual ACM Symposium on Theory of Computing*, pp.405-415 (May 1990).

[Wo72] Woll, H., "Reductions among Number Theoretic Problems," *Information and computation*, 72, pp.167-179 (1987).

[Ya82] Yao, A.C. "Theory and applications of trapdoor functions," *Proceedings of the 23st Annual Symposium on Foundations of Computer Science*, pp.80-91 (1982).

The Fast Cascade Exponentiation Algorithm and its Applications on Cryptography

Sung-Ming Yen and Chi-Sung Laih

Communication Laboratory,
Department of Electrical Engineering,
National Cheng Kung University,
Tainan, Taiwan, Republic of China

Abstract. In this paper, the evaluation of $\prod_{i=1}^{P} M_i^{b_i}$ is considered and a fast algorithm is proposed. Through out this paper, the above evaluation is called the *cascade exponentiation*. The cases of $P = 1, P = 2$, and $P = 3$ of the cascade exponentiations have special importance in cryptographic applications. The performance analysis and comparisons of the proposed algorithm and the best known method will be given.

1 Introduction

Most of the public key cryptosystems use hard computational problems such as the discrete logarithm problem, e.g., the ElGamal cryptosystem [1], and the factorization problem, e.g., the RSA cryptosystem [2],as the guarantees for system security. Almost all the cryptosystems of the above two kinds use modular exponentiations as the fundamental arithmetics. For this reason, many papers discuss more efficient exponentiation evaluation algorithms [3-6]. Some other researchers also try to reduce the overall computation time through another domain, the modular multiplications. Essentially, the computational algorithms of exponentiation and multiplication can be considered independently. In this paper, we consider only the total number of multiplications in the exponentiation operation but not the computational aspect of multiplication. Both in the areas of cryptology and fundamental computer science, the exponentiation evaluation is an important topic from the view-points of application and theoretical research. The fundamental type is the single exponentiation such as M^b . This problem can be optimally solved by addition chain [3]. In this paper, a more general type of exponentiation algorithm called the *cascade exponentiation* algorithm will be discussed. A fast algorithm will be proposed to solve the problem such as evaluating $M_1^{b_1} \times M_2^{b_2} \times M_3^{b_3}$ where $M_1 \neq M_2 \neq M_3$ and $b_1 \neq b_2 \neq b_3$ in general. This fast cascade exponentiation algorithm can be used to speed up very complicated computations that are required for most cryptographic protocols based on the cascade exponentiations, e.g., the RSA cryptosystem [2] and the Diffie-Hellman key distribution scheme [7] for the case of $p = 1$, the Schnorr's [8] and the Sandia [9] interactive identification and digital signature schemes and the DSS of the NIST [10] for the case of p=2, and the ElGamal digital signature scheme [1] for the case of p=3.

2 The Proposed Fast Cascade Exponentiation Algorithm

2.1 Review of the evaluation of exponentiation:

The concept of addition chain has been widely considered as the optimal method to compute the exponentiation requiring the minimum number of multiplications if a shortest addition chain for the exponent can be found [3].

An ***addition chain*** for a given number n is a sequence of increasing numbers $A = \{a_1, a_2, \ldots, a_r\}$ such that: (1) a_1=1, a_r=n, (2) $a_i = a_j + a_k$, for $i > j \geq k$, and (3) the length of the addition chain A is equal to $r - 1$.

Here we give two examples of the evaluation of M^{75} as follows

$M \Rightarrow M^2 \Rightarrow M^3 \Rightarrow M^6 \Rightarrow M^{12} \Rightarrow M^{15} \Rightarrow M^{30} \Rightarrow M^{60} \Rightarrow M^{75}$,

$M \Rightarrow M^2 \Rightarrow M^4 \Rightarrow M^8 \Rightarrow M^9 \Rightarrow M^{18} \Rightarrow M^{36} \Rightarrow M^{37} \Rightarrow M^{74} \Rightarrow M^{75}$.

In the above examples, the former is computed based on one of the shortest addition chains while the later is based on the regular ***square-multiply*** method. It is obvious that using the shortest addition chain, the exponentiation can be optimally evaluated with the minimum number of multiplications. Unfortunately, the problem of finding a shortest addition chain for a given number has been proven to be an NP-complete problem [5]. So, some heuristic algorithms are developed for the construction of near optimal addition chain in a reasonable time for large exponentiation evaluations [6]. A concept similar to the addition chain, called the addition sequence is given below.

The ***addition sequence*** for a given sequence of increasing numbers $B = \{b_1, b_2, \ldots, b_t\}$ is also a sequence of increasing numbers $A = \{a_1, a_2, \ldots, a_r\}$ such that: (1) $a_1 = 1, a_r = b_t$, (2) $a_i = a_j + a_k$, for $i > j \geq k$,(3) $B \subseteq A$, and (4) the length of the addition sequence A is equal to $r - 1$.

Using the concept of addition sequence, the following t exponentiation operations, $M^{b_1}, M^{b_2}, \ldots, M^{b_t}$, can be evaluated in a sequence of multiplications.

The concept of addition chain can be extended to the vector addition chain [6]. In the following discussions, we use upper case letter as the vector and lower case letter as the element in each vector.

The ***vector addition*** chain for a given p-dimensional vector $V = [v_1, v_2, \ldots, v_p]$ is a sequence of increasing vectors $A = \{A_1, A_2, \ldots, A_r\}$ such that
(1) $A_1 = [0,0,...,0,1], A_2 = [0,0,\ldots,1,0], \ldots, A_p = [1,0,\ldots,0,0]$, and $A_r = V$;
(2) $A_i = A_j + A_k$, for $i > j \geq k$;
(3) the length of the vector addition chain A is equal to $r - p$.

Extended from the definitions of addition sequence and ***vector addition chain***, the definition of ***vector addition sequence*** can be obtained and the detail is omitted.

Using the concept of vector addition chain, the cascade exponentiation can also be evaluated. Same as the addition chain for M , if a shortest vector addition chain for $V = [b_1, b_2, b_3]$ can be found then the evaluation of $M_1^{b_1} \times M_2^{b_2} \times M_3^{b_3}$ can be optimally solved with the minimum number of multiplications.

2.2 The proposed cascade exponentiation evaluation algorithm:

Now, a fast cascade exponentiation algorithm is proposed which is based upon the ***square-multiply*** method in vector form and the usage of ***table lookup***. Shamir had developed a simple method [1] for the evaluation of $M_1^{b_1} \times M_2^{b_2} \times M_3^{b_3}$. In Shamir's method, $b_i (i = 1, 2, 3)$ are encoded as binary representations. Suppose b_i is an n bit number and encoded as $((b_{i,n-1}), (b_{i,n-2}) \ldots, (b_{i,1}), (b_{i,0}))_2$. Then the cascade exponentiation is performed in the square-multiply like manner from the most significant bit (MSB) to the least significant bit (LSB).

In Shamir's method, the following numbers $M_1 \times M_2, M_1 \times M_3, M_2 \times M_3$, and $M_1 \times M_2 \times M_3$ have to be evaluated in advance and saved in a table. First, T is initialized to be $M_1^{b_{1,n-1}} \times M_2^{b_{2,n-1}} \times M_3^{b_{3,n-1}}$ which is fetched from the table and the operation $T \leftarrow T^2$ is performed. After the squaring operation, the following multiplication is performed $T \leftarrow T \times (M_1^{b_{1,n-2}} \times M_2^{b_{2,n-2}} \times M_3^{b_{3,n-2}})$. Using this square-multiply like algorithm from the MSB to the LSB, the cascade exponentiation can be performed in a vector like manner.

Based on the above idea for cascade exponentiation, a more efficient cascade exponentiation algorithm is developed below. In this new algorithm we ***dynamically*** scan the binary representations of bi's from the MSB to the LSB by W-bit each time. In general, the parameter W depends on the total number of bases and the bit length of their exponent. Without loosing any generality, in order to simplify the description of the algorithm, we consider the case of $M_1^{b_1} \times M_2^{b_2} \times M_3^{b_3}$ and the parameter W is selected to be 2.

[Algorithm]: Fast cascade exponentiation

Step-1: Evaluate all the following numbers and save them in a table

$$M_1^{((e_{1,1}),(e_{1,0}))_2} \times M_2^{((e_{2,1}),(e_{2,0}))_2} \times M_3^{((e_{3,1}),(e_{3,0}))_2}$$

where $e_{i,j} \in \{0, 1\}$ except the cases of $e_{1,0} = e_{2,0} = e_{3,0} = 0$.

Step-2: Dynamically scan the binary representations of b_i's from MSB to LSB two bits each time.

$$t = n - 3,$$

$$S = M_1^{((b_{1,n-1}),(b_{1,n-2}))_2} \times M_2^{((b_{2,n-1}),(b_{2,n-2}))_2} \times M_3^{((b_{3,n-1}),(b_{3,n-2}))_2},$$

(where S is used to save the result)

while ($t \geq 0$) do

begin

$$\text{if}\left[\begin{pmatrix} b_{1,t}, b_{1,t-1} \\ b_{2,t}, b_{2,t-1} \\ b_{3,t}, b_{3,t-1} \end{pmatrix} = \begin{pmatrix} 0, x_1 \\ 0, x_2 \\ 0, x_3 \end{pmatrix} \text{ where } x_i \in \{0,1\}\right]$$

then $\{S = S^2, \ \ t = t-1\}$

$$\text{else if}\left[\begin{pmatrix} b_{1,t}, b_{1,t-1} \\ b_{2,t}, b_{2,t-1} \\ b_{3,t}, b_{3,t-1} \end{pmatrix} = \begin{pmatrix} x_1, 0 \\ x_2, 0 \\ x_3, 0 \end{pmatrix} \text{ where } x_i \in \{0,1\}\right.$$

$\left.\text{and } x_i's \text{ are not all zero}\right]$ then

$\{S = S^2,$

$S = S \times (M_1^{X_1} \times M_2^{X^2} \times M_3^{X_3}),$

if ($t \neq 0$) then $\{ \ S = S^2 \ \},$

$t = t - 2\}$

else $\{ \ S = S^4,$

$S = S \times (M_1^{((b_{1,t}),(b_{1,t-1}))_2} \times M_2^{((b_{2,t}),(b_{2,t-1}))_2} \times M_3^{((b_{3,t}),(b_{3,t-1}))_2}),$

$t = t - 2\}$

end

In the above algorithm, the exponents b_1, b_2, and b_3 are scanned from the MSB to the LSB two bits each time in the **dynamic approach**. If the bit pattern under scanned is $\begin{pmatrix} 0, x_1 \\ 0, x_2 \\ 0, x_3 \end{pmatrix}$ then the left $\begin{pmatrix} 0 \\ 0 \\ 0 \end{pmatrix}$ is passed through accompanying one square operation and the right $\begin{pmatrix} x_1 \\ x_2 \\ x_3 \end{pmatrix}$ is taken into account during the next scan to save some multiplications. Furthermore, if the bit pattern under scanned is $\begin{pmatrix} x_1, 0 \\ x_2, 0 \\ x_3, 0 \end{pmatrix}$ then we just consider it as $\begin{pmatrix} x_1 \\ x_2 \\ x_3 \end{pmatrix}$ to save some multiplications and table space during the step-1.

Example 1. Evaluate $M_1^{27} \times M_2^{66} \times M_3^{98}$ using the fast cascade exponentiation algorithm.

The binary representations of the exponents and the dynamic scanning process are shown below:

$$27 = (0\ \ 0\ \ 1\ \ 1\ \ 0\ \ 1\ \ 1)_2$$

$$66 = (1\ \ 0\ \ 0\ \ 0\ \ 0\ \ 1\ \ 0)_2$$

$$98 = (\underbrace{1\ \ 1}_{\begin{pmatrix}0\\2\\3\end{pmatrix}}\ \ \underbrace{0\ \ 0}_{\begin{pmatrix}3\\0\\0\end{pmatrix}}\ \ 0\ \ \underbrace{1\ \ 0}_{\begin{pmatrix}3\\2\\2\end{pmatrix}})_2$$

Then the cascade exponentiation is performed as follows (the computation in step-1 is not shown here):

$$(M_2^2 \times M_3^3) \rightarrow \{double\ square\} \rightarrow (M_2^8 \times M_3^{12}) \rightarrow \{\times M_1^3\} \rightarrow$$

$$(M_1^3 \times M_2^8 \times M_3^{12}) \rightarrow \{square\} \rightarrow (M_1^6 \times M_2^{16} \times M_3^{24}) \rightarrow \{double\ square\}$$

$$\rightarrow (M_1^{24} \times M_2^{64} \times M_3^{96}) \rightarrow \{\times(M_1^3 \times M_2^2 \times M_3^2)\} \rightarrow (M_1^{27} \times M_2^{66} \times M_3^{98})$$

3 Performance Analysis

Three special forms of cascade exponentiation are particularly important in cryptographic applications, i.e., $M_1^{b_1}$ (this is called the reduced form), $(M_1^{b_1} \times M_2^{b_2})$, and $(M_1^{b_1} \times M_2^{b_2} \times M_3^{b_3})$. For the reason of simplification, more detailed analysis of $(M_1^{b_1} \times M_2^{b_2})$ will be given first then the similar analysis process can be applied to the other cases straightforward.

3.1 Performance analysis of $M_1^{b_1} \times M_2^{b_2}$:

Let both b_1 and b_2 be n-bit numbers and they are scanned in the algorithm from the MSB to the LSB with W bits each time. If n=512 and W=3 is considered. In step-1, 48 numbers have to be evaluated with 48 multiplications and saved in a table with the aid of the following *vector addition sequence*

{[0,1], [1,0], [0,2], [2,0], [0,3], [0,5], [0,7], [1,1], [1,2], [1,3], [1,4], [1,5], [1,6], [1,7], [2,1], [2,3], [2,5], [2,7], [3,0], [3,1], [3,2], [3,3], [3,4], [3,5], [3,6], [3,7], [4,1], [4,3], [4,5], [4,7], [5,0], [5,1], [5,2], [5,3], [5,4], [5,5], [5,6], [5,7], [6,1], [6,3], [6,5], [6,7], [7,0], [7,1], [7,2], [7,3], [7,4], [7,5], [7,6], [7,7]} which is *optimal* for the required 48 vectors. In the above sequence, {[0,2], [2,0]} are the auxiliary vectors. In step-2, the algorithm takes 509 square operations and takes at most 170 and on the average 151 (which will be examined later) multiplications when the **dynamic scanning process** is taken into account. Analysis shows that the algorithm needs overall at most 727 and on the average 708 (described below) multiplications. The algorithm is then 28.9% and 20.9% faster than the Shamir's method for the worst case and average case respectively. The performance analysis of dynamic scanning process for W=3 is given below

(case-1): $\begin{pmatrix} 0\,X\,X \\ 0\,X\,X \end{pmatrix}$ which will save $\frac{12}{64} \times \frac{1}{3} \times 170$ multiplications on the average where the left $\begin{pmatrix} X \\ X \end{pmatrix}$ are not both zero .

(case-2): $\begin{pmatrix} 0\,0\,X \\ 0\,0\,X \end{pmatrix}$ which will save $\frac{3}{64} \times \frac{2}{3} \times 170$ multiplications on the average where the $\begin{pmatrix} X \\ X \end{pmatrix}$ are not both zero .

(case-3): $\begin{pmatrix} 0\,0\,0 \\ 0\,0\,0 \end{pmatrix}$ which will save $\frac{1}{64} \times 1 \times 170$ multiplications on the average .

So, these three cases can save 19 multiplications totally.

For b_i to be 512 bits, the parameter W is optimally selected to be 3 if the extra memory required is not taken into account.

For the cases of $M_1^{b_1} \times M_2^{b_2} \times M_3^{b_3}$ and $M_1^{b_1}$, similar performance analysis can be carried out and all the results are summarized in Table-1 and Table-3.

3.2 General performance analysis of cascade exponentiation algorithm:

In the following paragraph, we will give the general case performance analysis. Based upon this analysis, the optimal parameter W for any given p and $|\ b_i\ |$ can be easily obtained. In the following analysis, we will denote $|\ b_i\ |$ as n, i.e., $n = 1 + \lfloor log_2 b_i \rfloor$. The performance analysis will be divided into the worst case and average case. Also, the analysis will be divided into the cases of $W \geq 2$ and W=1 because the step-1 of the proposed algorithm is somewhat different for $W \geq 2$ and W=1 to utilize the optimal vector addition sequence. We also use $M(n,p,W)$ to denote the number of multiplications required in the proposed algorithm.

Theorem 1 . *In the proposed algorithm for $n >> W$ and $W \geq 2$, in step-1, it requires $(2^p)^W - (2^p)^{W-1}$ multiplications to obtain the all required $(2^p)^W -$*

$(2^p)^{W-1}$ *numbers using an optimal vector addition sequence.*

Theorem 2 . *The worst case performance of the proposed algorithm for* $n >> W$ *and* $W \geq 2$ *requires* $M(n,p,W) = [(2^p)^W - (2^p)^{W-1}] + [(n-W) + \lceil \frac{n}{w} \rceil - 1)]$.

Theorem 3 . *The average case performance of the proposed algorithm for* $n >> W$ *and* $W \geq 2$ *requires* $M(n,p,W) = [(2^p)^W - (2^p)^{W-1}] + (n - W) + \{\lceil \frac{n}{w} \rceil - 1 - [[\frac{-1+W-W\times 2^p+(2^p)^W}{(2^p)^W \times (2^p-1)}] \times \frac{n}{W^2} + \frac{1}{(2^p)^W} \times \frac{n}{W}]\}$.

Theorem 4 . *The worst case performance of the proposed algorithm for* $n >> W$ *and* W*=1 requires* $M(n,p,W) = 2^p - (p+1) + 2 \times (n-1)$.

Theorem 5 . *The average case performance of the proposed algorithm for* $n >> W$ *and* W*=1 requires* $M(n,p,W) = 2^p - (p+1) + [1 + \frac{2^p-1}{2^p}] \times (n-1)$.

From the above four theorems, the optimal parameter W of each case for any p and $\mid b_i \mid$ can be obtained by solving the following equation

$$\frac{\partial M(n,p,W)}{\partial W} = 0.$$

Of course, the solution of the above equation may not be an integer. Then the optimal solution W is selected to be the integer such that the above equation approaches to zero nearestly. The optimal parameter W can also be obtained simply by enumeration of $M(n,p,W)$ over a small range of possible integers.

3.3 Further improvement:

In the previous performance analysis, it is assumed that the cost of multiplication and the cost of square operation are identical. In fact, for both the theoretical analysis [3] and some special purpose hardwares, square operations can be performed three times faster than multiplications [11]. If this factor is considered in the performance analysis, the proposed fast cascade exponentiation algorithm will be more efficient than what has been evaluated in the previous discussions. The reason is obvious, what the proposed algorithm doing is to decrease the number of multiplications in the evaluation of cascade exponentiation and the cost of multiplication is more expensive than that of square operation. Of course, the performance enhancement depends on the computational architecture used in the hardware implementation. The performance evaluations are listed in Table-2 and Table-4 when the square operation is three times faster than the multiplication for the average case and the worst case respectively.

4 The Conclusions

In this paper, a fast cascade exponentiation, $\prod_{i=1}^{p} M_i^{b_i}$, evaluation algorithm is proposed which is on the average 21%, 20.9%, and 16.5% faster than the Shamir's method for p=1, p=2, and p=3 respectively under the consideration of 512-bit exponents. For some special hardware architectures, square operation can be three times faster than multiplication then the proposed algorithm can be on the average 37.3%, 33.5%, and 25.6% faster than the Shamir's method for p=1, p=2, and p=3 respectively under the consideration of 512-bit exponents. The basic idea used in this algorithm is to store a small table of precomputed values from part of the following cascade exponentiation $\prod_{i=1}^{p} M_i^{b_i}$, $2^w - 1 \leq b_i \leq 0$. From the precomputed table, the required large cascade exponentiation can be performed more efficiently. The algorithm is both simple and regular so it is suitable for software or even hardware implementations. A hardware architecture of the proposed algorithm and a novel dynamic scanning operation circuit are available in [12].

References

1. T. ElGamal, "A public key cryptosystem and signature scheme based on discrete logarithms," IEEE Trans. on Inform. Theory, Vol. IT-31, No. 4, pp.469-472, July 1985.
2. R.L. Rivest, A. Shamir, and L. Adleman, "A method for obtaining digital signatures and public-key cryptosystem," Commun. ACM, Vol. 21, pp.120-126, Feb. 1978.
3. D.E. Knuth, The art of computer programming, Vol. II: Seminumerical algorithms. Reading, Addison Wesley, 1969.
4. Andrew Yao, "On the evaluation of powers," Siam. J. Comput. 5, (1976).
5. P. Downey and B. Leony and R. Sethi, "Computing sequences with addition chains," Siam Journ. Comput. 3 (1981) pp.638-696.
6. J. Bos, M. Coster, "Addition Chain Heuristics," Proceedings CRYPTO'89, Springer-Verlag Lecture Notes in Computer Science, pp.400-407.
7. W. Diffie and M.E, Hellman, "New directions in cryptography," IEEE Trans. on Inform. Theory, vol.IT-22, pp.644-654, 1976.
8. C.P. Schnorr, "Efficient identification and signatures for smart cards," Advances in Cryptology-Proceedings of Crypto'89, Lecture Notes in Computer Science, Vol.435, Springer Verlag, New York, 1990, pp.239-252.
9. E.F. Brickell and K.S. McCurley , "Interactive identification and digital signatures," AT and T Technical Journal, November/December 1991, pp.73-86.
10. "A Proposed Federal Information Processing Standard for Digital Signature Standard (DSS)," Federal Register , Vol.56, No.169, August 31, 1991, pp.42980-42982.
11. Y. Yacobi , "Exponentiating faster with addition chains," Advances in Cryptology-Proceedings of Eurocrypt'90, pp.205-212, 1990.
12. C.S. Laih and S.M. Yen, "The Study of Fast Exponentiation Algorithm," Technical Report of NSC 81-0408-E-006-02 (1992).

Table-1 Average case performance of the proposed fast cascade exponentiation algorithm

Required number of multiplications (average case)	$\|b_i\|=$ 1024	$\|b_i\|=$ 512	$\|b_i\|=$ 256	$\|b_i\|=$ 128	$\|b_i\|=$ 64
(P=1)					
Shamir's method	1534	766	382	190	94
new proposed method	1192	605	308	156	80
optimal value of W=	6	5	4	4	3
R=	4.4	5.1	6	2.9	3.5
performance improved	22.3%	21%	19.4%	17.9%	14.9%
(P=2)					
Shamir's method	1791	895	447	223	111
new proposed method	1372	708	373	191	100
optimal value of W=	3	3	2	2	2
R=	6.3	3.1	8.9	4.4	2.2
performance improved	23.4%	20.9%	16.6%	14.3%	9.9%
(P=3)					
Shamir's method	1922	962	482		
new proposed method	1553	803	428		
optimal value of W=	2	2	2		
R=	8.5	4.2	2.1		
performance improved	19.2%	16.5%	11.2%		
(P=4)					
Shamir's method	1993				
new proposed method	1756				
optimal value of W=	2				
R=	2.1				
performance improved	11.9%				

* The Cascade Exponentiation is defined as $\prod_{i=1}^{p} M_i^{b_i}$

* $|b_i|$: the bit length of b_i

Table-2 Average case performance of the proposed fast cascade exponentiation algorithm using faster squaring operation

Required number of multiplications (average case)	$\|b_i\|=$ 1024	$\|b_i\|=$ 512	$\|b_i\|=$ 256	$\|b_i\|=$ 128	$\|b_i\|=$ 64
(P=1)					
optimal value of W=	6	5	4	4	3
performance improved	39.9%	37.3%	34.3%	31.1%	26.4%
(P=2)					
optimal value of W=	3	3	2	2	2
performance improved	37.7%	33.5%	27%	23%	15.7%
(P=3)					
optimal value of W=	2	2	2		
performance improved	29.7%	25.6%	17.3%		
(P=4)					
optimal value of W=	2				
performance improved	18%				

* In this table it is assumed that

$$\frac{\text{operation time of multiplication}}{\text{operation time of squaring}} = 3$$

Table-3 Worst case performance of the proposed fast cascade exponentiation algorithm

Required number of multiplications (worst case)	$\|b_i\|$= 1024	$\|b_i\|$= 512	$\|b_i\|$= 256	$\|b_i\|$= 128	$\|b_i\|$= 64
(P=1)					
Shamir's method	2046	1022	510	254	126
new proposed method	1220	623	318	163	83
optimal value of W=	6	6	5	4	4
R=	5.3	2.7	3.2	3.9	1.9
performance improved	40.37%	39%	37.6%	35.8%	34.1%
(P=2)					
Shamir's method	2047	1023	511	255	127
new proposed method	1410	727	386	201	105
optimal value of W=	3	3	3	2	2
R=	7.1	3.5	1.8	5.3	2.6
performance improved	31.1%	28.9%	24.5%	21.2%	17.3%
(P=3)					
Shamir's method	2050	1026	514	258	
new proposed method	1589	821	437	245	
optimal value of W=	2	2	2	2	
R=	9.1	4.6	2.3	1.1	
performance improved	22.5%	20%	15%	5%	
(P=4)					
Shamir's method	2057	1033			
new proposed method	1773	1005			
optimal value of W=	2	2			
R=	2.1	1.1			
performance improved	13.8%	2.7%			

* The Cascade Exponentiation is defined as $\prod_{i=1}^{P} M_i^{b_i}$

* $|b_i|$: the bit length of b_i

Table-4 Worst case performance of the proposed fast cascade exponentiation algorithm using faster squaring operation

Required number of multiplications (worst case)	$\|b_i\|$= 1024	$\|b_i\|$= 512	$\|b_i\|$= 256	$\|b_i\|$= 128	$\|b_i\|$= 64
(P=1)					
optimal value of W=	6	6	5	4	4
performance improved	60.3%	58.2%	55.9%	52.4%	48.8%
(P=2)					
optimal value of W=	3	3	3	2	2
performance improved	46.6%	43.2%	36.4%	31.6%	24.7%
(P=3)					
optimal value of W=	2	2	2	2	
performance improved	33.6%	29.9%	22.1%	7.5%	
(P=4)					
optimal value of W=	2	2			
performance improved	20.6%	4%			

* In this table it is assumed that

$$\frac{\text{operation time of multiplication}}{\text{operation time of squaring}} = 3$$

Session 11

PUBLIC KEY CRYPTOGRAPHY I

Chair: Jennifer Seberry
(University of Wollongong, Australia)

The Design of a Conference Key Distribution System

Chin-Chen Chang*, Tzong-Chen Wu**, and C. P. Chen*

* Institute of Computer Science and Information Engineering,
National Chung Cheng University, Chiayi, Taiwan 601, R.O.C.

** Department of Information Management,
National Taiwan Institute of Technology, Taipei, Taiwan 107, R.O.C.

ABSTRACT

In this paper, we propose a conference key distribution system for generating a common secret key for two or more users. In our system, each user possesses a secret key and a public key. Initially, the chairperson constructs a conference key associated with his secret key and the conference members' public keys. Then each member can obtain and authenticate the conference key by using his secret key. Further, we have shown that the security of our proposed system is based on the difficulty of breaking the Diffie-Hellman key distribution system.

Keywords: public key cryptosystem, key distribution system, conference key distribution, authentication

1. Introduction

With the increasing prevalence of communication networks, cryptographic algorithms for achieving data security are becoming more and more important. A key management scheme plays an important role in key-controlled cryptographic algorithms. In 1976, Diffie and Hellman [2] invented the concept of public key cryptography, which is referred to as a key distribution system. The Diffie and Hellman key distribution system is described as follows.

Let x_i be a secret key of user U_i, and let x_j be a secret key of user U_j. Suppose that U_i and U_j want to share a common secret key k_{ij}. First, U_i and U_j compute their public keys y_i and y_j, respectively, such that

$$y_i = \alpha^{x_i} \bmod p, \text{ and } y_j = \alpha^{x_j} \bmod p.$$

Here p is a large prime number and α is a primitive element modulo p, and p and α are known only to U_i and U_j, respectively. Then a common secret key k_{ij} can be computed as

$$\begin{aligned} k_{ij} &= \alpha^{x_i x_j} \bmod p \\ &= y_i^{x_j} \bmod p \\ &= y_j^{x_i} \bmod p. \end{aligned}$$

That is, U_i can obtain the common secret key k_{ij} with his secret key x_i and U_j's public key y_j. Similarly, U_j can obtain the common secret key k_{ij} with his secret key x_j and U_i's public key y_i. Although it is not yet proved that breaking the Diffie-Hellman key distribution system is equivalent to computing discrete logarithms over finite fields, it appears to be difficult for an intruder to compute k_{ij} if he only knows y_i and y_j [2, 3].

There are several public key cryptosystems [1, 3, 4, 5, 10, 11] that can be used to generate a two-user common secret key with authentication. Consider two or more users who want to hold a secure conference. They should derive one common secret key to be shared among them. The derived common secret key is referred to as a conference key. Recently, there have been several conference key distribution systems [6, 7, 8, 9] proposed for generating a conference key. In this paper, we first propose a key distribution system derived from the Diffie-Hellman key distribution system. Extending from the proposed key distribution system, we present a conference key distribution system for two or more users.

In Section 2, we propose a key distribution system KDS. We present a conference key distribution system CKDS in Section 3. Finally, conclusions are given in Section 4.

2. KDS

Let p and α be two positive integers as defined in the previous section. Let x_i be the secret key and y_i be the public key of user U_i, and let x_j be the secret key and y_j be the public key of user U_j, respectively, as derived from the Diffie-Hellman key distribution system. Suppose that U_i wants to communicate with U_j. The KDS for generating a common secret key k shared by U_i and U_j is stated below.

First, U_i constructs a common secret key k and an authentication message A by the following steps:

<u>Step 1</u>: Randomly select a number r between 1 and p-2, such that

$$\gcd(r, p-1) = 1.$$

<u>Step 2</u>: Compute $u = y_j^r \bmod p$.

<u>Step 3</u>: Compute an integer k by solving

$$k \cdot u \equiv 1 \bmod p. \tag{2-1}$$

<u>Step 4</u>: Compute $s = \alpha^r \bmod p$. (2-2)

<u>Step 5</u>: Compute t by solving

$$k = x_i \cdot s + r \cdot t \pmod{p-1}. \qquad (2\text{-}3)$$

Step 6: Construct an authenticating message $A = (s, t)$ and send A to U_j.

After receiving the authenticating message A, U_j can obtain and authenticate the common secret key k by the following steps:

Step 1: Compute $u = s^{x_j} \bmod p$. (2-4)

Step 2: Obtain k by solving

$$k \cdot u \equiv 1 \bmod p. \qquad (2\text{-}5)$$

Step 3: Verify the common secret key k obtained from Step 2.

If $\alpha^k = y_i^s \cdot s^t \bmod p$, (2-6)

then k was indeed generated by U_i; otherwise, k is a forgery.

Now, we shall show that the recipient U_j can obtain and authenticate the common secret key k from the authenticating message A generated by the originator U_i. Since $y_j \neq 0$, we have $\gcd(y_j^r, p) = 1$, for any r. Thus, Eq. (2-1) has a unique solution for k. By Eq. (2-4), we have

$$\begin{aligned} u &= s^{x_j} \bmod p \\ &= (\alpha^r)^{x_j} \bmod p \\ &= (\alpha^{x_j})^r \bmod p \\ &= y_j^r \bmod p. \end{aligned}$$

Thus, U_i and U_j obtain the same common secret key k from Eqs. (2-1) and (2-5), respectively. By raising α (given by Eq. (2-3)) to power k, the value of α^k in Eq. (2-6) is easily obtained.

Let us consider the case of an intruder who tries to reveal k from s and t. It is obvious that the problem of revealing k from s and t is as difficult as breaking the Diffie-Hellman key distribution system without knowing the randomly selected number r [4]. Furthermore, since x_i and r are only known to U_i, an intruder does not have the ability to compute t for forging an authenticating message A to reveal k. The following simple example shows how our KDS works.

Example 1.

Suppose that user U_i wants to communicate with user U_j. Let $\alpha = 6$ and $p = 11$. Let $x_i = 8$ and $x_j = 3$. By the Diffie-Hellman public key system, we have $y_i = 4$ and $y_j = 7$.

User U_i should perform the following tasks:

1. Choose a random number r between 1 and p - 1, say r = 9.
2. Compute $u = y_j^r \bmod p = 7^9 \bmod 11 = 8$.
3. Compute k by solving $ku \equiv 1 \bmod p$. We have $k \cdot 8 \equiv 1 \bmod 11$.
   That is, k = 7.
4. Compute $s = \alpha^r \bmod p = 6^9 \bmod 11 = 2$.
5. Compute t from $k = x_i \cdot s + r \cdot t \pmod{p - 1}$. We have $7 = 8 \cdot 2 + 9 \cdot t \pmod{10}$.
   That is, t = 9.
   The authenticating message A = (s, t) = (2, 9).

User U_j, after receiving A, should perform the following tasks:

1. Compute $u = s^{x_j} \bmod p = 2^3 \bmod 11 = 8$.
2. Obtain k by solving $k \cdot u \equiv 1 \bmod p$. We have $k \cdot 8 \equiv 1 \bmod 11$.
   That is, k = 7.
3. Since $\alpha^k = 6^7 \bmod 11 = 8$, and $y_i^s \cdot s^t \bmod p = 4^2 \cdot 2^9 \bmod 11 = 8$, we have $\alpha^k = y_i^s \cdot s^t \bmod p$. Thus, U_j and U_i share the common secret key k = 8.

Extending from KDS as presented above, we shall propose a conference key distribution system CKDS in the next section.

3. CKDS

Let U_0 be the chairperson and U_j be an authorized member of a conference, where j=1, 2, ..., n, for some n. Let x_0 be the secret key and y_0 be the public key of the chairperson U_0. Again, let x_j be the secret key and y_j be the public key of the authorized member U_j. Our CKDS is stated below.

For generating a conference key k, the chairperson U_0 does the following:

Step 1: Randomly select a random number r and a conference key k between 1 and p-1, such that gcd(r, p-1) = 1.

Step 2: Compute

$$z_i = y_i^r \bmod p, \qquad (3\text{-}1)$$

for i = 1, 2, ..., n.

Step 3: Construct a polynomial h(x) with degree n such that h(0) = k and h(x) passes through (i, z_i) for i = 1, 2, ..., n.

Step 4: Find $w_i = h(n + i)$, for i = 1, 2, ..., n.

Step 5: Compute $s = \alpha^r \bmod p$. (3-2)

Step 6: Compute t by solving

$$k = x_0 \cdot s + r \cdot t \pmod{p-1}. \qquad (3\text{-}3)$$

Step 7: Broadcast s, t, and the w_i's.

After receiving s, t, and the w_i's, the authorized member U_j can individually obtain and authenticate the conference key k by the following steps:

Step 1: Compute $z_j = s^{x_j} \bmod p$. (3-4)

Step 2: Reconstruct the polynomial h(x) from (j, z_j), $(n+1, w_1)$, ..., $(2n, w_n)$ and find $k = h(0)$.

Step 3: Verify the common secret key k obtained from Step 2.

If $\alpha^k = y_0^{s} \cdot s^t \bmod p$, (3-5)

then k was indeed generated by U_0; otherwise, k is a forgery.

Since gcd(r, p - 1) = 1, there is a unique solution k satisfying Eq. (3-3). Here, we show how a conference member, say U_j, can obtain and authenticate the conference key k from the authenticating message A_j generated by the chairperson U_0. We know that the polynomial reconstructed from (j, z_j), $(n+1, w_1)$, ..., $(2n, w_n)$ is identical to h(x) constructed by the chairperson U_0. Consequently, each authorized member U_j, for j=1, 2,..., n, in the conference obtains the same conference key $k = h(0)$ generated by the chairperson U_0. Again, the security of our CKDS is based on the difficulty of breaking the Diffie-Hellman key distribution system, as was our KDS.

4. Conclusions

We have presented a key distribution system KDS. We have also extended KDS to a conference key distribution system, CKDS. It has been shown that the security of KDS and CKDS is based on the difficulty of breaking the Diffie-Hellman key distribution system.

REFERENCES

[1] Denning, D. E. (1982): *Cryptography and Data Security*, Addison-Wesley, Mass., 1982.

[2] Diffie, W. and Hellman, M. (1976): "New directions in cryptography," *IEEE Trans. Information Theory*, Vol. IT-22, 1976, pp. 472-492.

[3] Ehrsam, W. F., Matyas, S. M., Meyer, C. H., and Tuchman, W. L. (1978): "A cryptographic key management scheme for implementing the Data Encryption Standard," *IBM Systems Journal*, Vol. 17, No. 2, 1978, pp. 106-125.

[4] ElGamal, T. (1985): "A public key cryptosystem and a signature scheme based on discrete logarithms," *IEEE Trans. Information Theory*, Vol. IT-31, No. 4, July 1985, pp. 469-472.

[5] Fiat, A. and Shamir, A. (1986): "How to prove yourself : practical solutions to identification and signature problems," *Proceedings of CRYPTO'86*, Lecture Notes in Computer Science, Springer-Verlag, 1987, pp. 186-194.

[6] Ingemarsson, I., Tang, D. T., and Wong, C. K. (1981): "A conference key distribution system," *IEEE Trans. Information Theory*, Vol. IT-22, 1982, pp. 714-720.

[7] Koyama, K. and Ohta, K. (1987): "Identity-based conference key distribution system," *Proceedings of CRYPTO'87*, Lecture Notes in Computer Science, Springer-Verlag, 1988, pp. 175-184.

[8] Koyama, K. and Ohta, K. (1988): "Security of improved identity-based conference key distribution system," *Proceedings of EUROCRYPT'88*, Springer-Verlag, 1989, pp. 11-19.

[9] Okamoto, E. and Tanaka, K. (1989): "Key distribution system based on identification information," *IEEE Journal on Selected Areas in Communications*, Vol. 7, No. 4, May 1989, pp. 481-485.

[10] Rivest, R., Shamir, A., and Adleman, L. (1978): "A method for obtaining digital signatures and public key cryptosystems," *Commun. ACM*, Vol. 21, No. 2, February 1978, pp. 120-126.

[11] Shamir, A. (1984): "Identity-based cryptosystems and signature schemes," *Proceedings of CRYPTO'84*, Lecture Notes in Computer Science, Springer-Verlag, 1985, pp. 47-53.

Remarks on "The Design of a Conference Key Distribution System"

Edward Zuk

Telecom Research Laboratories,
770 Blackburn Road,
Clayton Victoria 3168, Australia

Abstract. Chang, Wu and Chen propose a Conference Key Distribution System (CKDS) in these proceedings. This note describes an attack on their system and presents a possible solution.

1 An Attack on the proposed Conference Key Distribution System

As presented in these proceedings, the Chang, Wu and Chen Conference Key Distribution System (CKDS) can be broken. Observe that in step 5 of the proposed protocol, the chairperson sends as part of the authentication message, Y_j to each of the other n authorised members of the conference. An attacker tapping the communications of the chairman could easily intercept all these messages and construct the following product:

$$\prod_{j=1}^{n} Y_j \equiv \prod_{j=1}^{n} Y/\left(y_j^r\right) \bmod p \tag{1}$$

¿From the definition of Y

$$Y \equiv \left(\prod_{j=1}^{n} y_j^r\right) \bmod p \tag{2}$$

we arrive at

$$\prod_{j=1}^{n} Y_j \equiv Y^n/Y \equiv Y^{n-1} \bmod p \tag{3}$$

As the prime p is public it is possible to find the $(n-1)^{\text{th}}$ roots. If $\gcd(n-1, p-1) = 1$, then the solution is uniquely obtained by evaluating the following.

$$Y \equiv \left(Y^{n-1}\right)^e, \text{ where } e \cdot (n-1) \equiv 1 \bmod p \tag{4}$$

For the more complicated cased when $n-1$ has a common divisor with $p-1$, the square root algorithm given in [KNUTH] can be adapted to calculate the $n-1$ roots. It would then be a simple matter to check which of the $n-1$ roots is the correct key from the authentication information, $c1$ and $c2$.

2 A Possible Solution

A possible solution to this problem could be to use a modulus which is the product of two primes. All knowledge of the two primes would have to be destroyed after the product is formed. This solution however requires proper analysis to ensure that it does not introduce any other weaknesses.

References

[KNUTH] Knuth, D.E.: "The Art of Computer Programming, Volume 2 / Seminumerical Algorithms" Second edition (1981), Addison-Wesley. pp 625-626

Public-key Cryptosystem Based on the Discrete Logarithm Problem

Lein Harn[1] and Shoubao Yang[2]

[1] Computer Science Telecommunications Program
University of Missouri - Kansas City
Kansas City, MO 64110 USA
Tel: (816)235-2367
E-mail: harn@cstp.umkc.edu
[2] Department of Computer Science
University of Science and Technology of China
Hefei, Anhui 230026 PRC

Abstract. In 1985, T. ElGamal proposed a public-key cryptosystem and a signature scheme, in which the difficulty of breaking the system is based on the difficulty of computing a discrete logarithm in a finite group. For the same security level, the size of the ciphertext and the computational time of ElGamal's encryption are double those of the well-known RSA scheme. In this paper, we propose a public-key cryptosystem based on the discrete logarithm, in which the size of the ciphertext and the computational time are the same as those of the RSA scheme, and the security level is the same as the ElGamal cryptosystem.

1 Introduction

A cryptosystem can help users to establish secure communication channels in an open environment. In general, it involves two parts: the key distribution and the message encryption. In 1976, Diffie and Hellman [1] introduced the concept of a public-key distribution system to solve the key distribution problem. Since then, several two-key cryptosystems have been proposed [2,3,4]. Among them, the RSA cryptosystem [2] has been receiving much attention. This is mostly due to the following two reasons:

1. Its security, which is based on the difficulty of factoring a large integer into two large primes, has been thoroughly investigated.
2. Its performance, which requires only one exponentiation for enciphering or deciphering one message block and provides 1:1 transmission efficiency, is optimal compared with other schemes.

In 1985, ElGamal [4] proposed a public-key cryptosystem and a signature scheme, in which the difficulty of breaking the system is based on the difficulty of computing a discrete logarithm in a finite group. However, for the same security level, the size of the ciphertext and the computational time of ElGamal's encryption are double those of the well-known RSA scheme.

A striking property of the ElGamal system is that if we encrypt the same message twice, we will not get the same ciphertext. Also, due to the structure of ElGamal's system, there is no obvious relationship between the enciphering of m_1, m_2, $m_1 m_2$, or any other simple functions of m_1 and m_2. This is not the case for other known schemes, such as the RSA scheme. In other words, ElGamal's system is a probabilistic cryptosystem and can prevent some kind of attacks, such as a probable-plaintext attack.

In 1990, Laih and Lee [5] proposed a modification of ElGamal's encryption scheme. Instead of providing a 1:2 transmission efficiency as in the original ElGamal encryption scheme, their scheme provides an $n : (n+t)$ transmission efficiency, where n is the total number of message blocks and $t = [log_2(n+1)]$. The same order of improvement of the computational time can also be obtained.

In this paper, we generalize Laih and Lee's approach to achieve an optimal performance. The performance of our system, in terms of the size of ciphertext and the computational time, is equivalent to the RSA scheme. The security of our system is the same as the ElGamal cryptosystem.

We will briefly review ElGamal's cryptosystem in Section 2. Our proposed cryptosystem is given in Section 3, including some discussions and security analysis. In Section 4, the randomness of the encryption keys used in our proposed system is analysed.

2 Review of the ElGamal Cryptosystem

Since ElGamal's cryptosystem utilizes the basic idea of the Diffie-Hellman Public Key Distribution System (PKDS) [1], let us briefly review the PKDS first.

We start with a large prime, p, and a primitive element, α, of $GF(p)$, which are all known to the public. In order to provide adequate security for the PKDS, Pohlig and Hellman [6] indicate that p should be selected such that $p-1$ contains at least one large prime factor. They recommend picking $p = 2p' + 1$, where p' is also a large prime.

Suppose A and B want to share a common secret key K_{AB} during a communication session, where A and B have their own secret keys x_A and x_B respectively. A computes $y_A = \alpha^{x_A} \; mod \; p$ and sends it to B. Similarly, B computes $y_B = \alpha^{x_B} \; mod \; p$ and sends it to A. Thus, the common secret session key K_{AB} can be obtained by A and B individually as

$$K_{AB} = \alpha^{x_A x_B} \; mod \; p = y_A^{x_B} \; mod \; p = y_B^{x_A} \; mod \; p$$

However, any intruder has to solve a discrete logarithm problem in order to obtain K_{AB}. For more information of the discrete logarithm problem, interested readers may consult [7] and [8].

In ElGamal's cryptosystem, suppose A wants to send B a message m, where $0 \leq m \leq p-1$. First, A randomly selects a number k from $[0, p-1]$. This k will serve as the secret key x_A in the PKDS. Then A computes $K = y_B^k \; mod \; p$, where $y_B = \alpha^{x_B} \; mod \; p$ is the public key from B, which is sent by B during the initial phase of setting up the communication session. The ciphertext (c_1, c_2) is computed as

$$c_1 = \alpha^k \ mod \ p$$

and

$$c_1 = Km \ mod \ p$$

Since (c_1, c_2) is the transmitted ciphertext, according to the property of the PKDS, B can obtain K as

$$K = c_1^{x_B} \ mod \ p$$

Then message m can be decrypted as

$$m = c_2 K^{-1} \ mod \ p$$

where K^{-1} is the multiplicative inverse of K *mod* p.

Note that the size of the ciphertext is double the size of the message. It requires two modular exponentiations for enciphering one message block and one modular exponentiation and one modular inversion for deciphering one ciphertext block. For security reasons, the random number k should never be used twice [4].

3 Our Proposed Public-key Cryptosystem

Just like any other cryptosystem, there are three phases involved in our proposed cryptosystem: public key distribution, encryption, and decryption.

Phase 1: Public Key Distribution

We start with two large primes, p and p', where $p = 2'p + 1$, and a primitive element, α, of $GF(p)$, which are all known to the public.

Suppose A wants to send some secret information to B. First, A accesses the public-key directory and obtains the public key from B, $y_B = \alpha^{x_B} \ mod \ p$, where x_B is the secret key from B. Then, A randomly selects a number k' from $[0, p-1]$. According to the PKDS, a common secret key K'_{AB} can be obtained by A as

$$K'_{AB} = y_B^{k'} \ mod \ p$$

Then A checks whether K'_{AB} *mod* p' is a primitive element of $GF(p')$ by using the following corollary based on D. E. Knuth's Theorem C [9, pp. 20]:

Corollary 1. *For prime $p' > 2$ and $0 < a < p'$, a is a primitive, if and only if $a^{(p'-1)/q}$ mod $p' \neq 1$ for any prime divisor q of $p' - 1$*

If K'_{AB} *mod* p' is not a primitive element of $GF(p')$, then A re-selects random number k' and re-computes K'_{AB} until it becomes true. The probability of K'_{AB} *mod* p' being a primitive element of $GF(p')$ is almost 0.5 if we further restrict p' as $p' = 2p'' + 1$, where p'' is also a large prime.

Corollary 2. *For prime* $p' = 2p'' + 1, p > 2$ *and* p'' *is also a large prime, one half of the integers in* $[2, p' - 2]$ *are primitive elements of* $GF(p')$.

Proof of Corollary 2. For prime $p' = 2p'' + 1$, $p' > 2$, an integer $a \in [1, p' - 1]$ and $0 < a < p'$, a is a quadratic nonresidue modulo p', if and only if

$$a^{(p'-1)/2} \; mod \; p' = p' - 1$$

[10]

Since one half of the integers in $[1, p' - 1]$ are quadratic nonresidue modulo p', according to Corollary 1, they are all primitive elements of $GF(p')$, except $p' - 1$.

This k', denoted as k, will serve as the "secret session key" from A , and the corresponding K'_{AB}, denoted as K_{AB}, will become the "common secret session key" shared by A and B. The corresponding public session key of A, $y_A = \alpha^k$ *mod* p, is then computed and transmitted to B.

Phase 2: Encryption

For each message block, m_i, in a sequence of message blocks, $\{m_1, m_2, \ldots, m_i, \ldots\}$, A computes two encryption keys, $K_{i,1}$ and $K_{i,2}$, iteratively as

$$K_{i,1} = K_{i-1,1} K_{AB} \; mod \; p' = K_{AB}{}^i \; mod \; p',$$

and

$$K_{i,2} = \alpha^{K_{i,1}} \; mod \; p,$$

where $K_{0,1} = 1$.

The corresponding ciphertext block C_i is computed as

$$C_i = K_{i,2} m_i \; mod \; p.$$

The sequence of ciphertext blocks $\{C_1, C_2, \ldots, C_i, \ldots\}$ is transmitted to B.

Phase 3: Decryption

Once receiving the public session key y_A from A, B can compute the common secret session key K_{AB} as

$$K_{AB} = y_A^{x_B} \ mod \ p,$$

and encryption keys, $K_{i,1}$ and $K_{i,2}$, as well. For each received ciphertext C_i, the corresponding message m_i is computed as

$$m_i = C_i K_{i,2}{}^{-1} \ mod \ p,$$

where $K_{i,2}{}^{-1}$ is the multiplicative inverse of $K_{i,2} \ mod \ p$.

However, $K_{i,2}{}^{-1}$ can be computed without knowing $K_{i,2}$ thus speeding up the computation. This is due to Fermat's Theorem [10, pp. 42]. Since $gcd(\alpha, p) = 1$, we have

$$1 = \alpha^{p-1} \ mod \ p = \alpha^{K_{i,1}+(p-1-K_{i,1})} \ mod \ p = K_{i,2}\alpha^{p-1-K_{i,1}} \ mod \ p.$$

Hence

$$K_{i,2}{}^{-1} = \alpha^{p-1-K_{i,1}} \ mod \ p.$$

By this way, the computation of $K_{i,2}{}^{-1}$ is reduced significantly.

Discussion

1. In our proposed cryptosystem, two modular multiplications and one modular exponentiation are required for enciphering or deciphering each message block. For the same level of security, the size of the modulus p used in our scheme should be about the same size of the modulus n used in the RSA scheme [4]. Since modular exponentiation determines the computational time of both schemes, the computational time of our scheme is equivalent to the RSA scheme.
2. In our proposed cryptosystem, for every message block m_i, there is a corresponding ciphertext block C_i. The transmission efficiency is 1:1.
3. In our system, the unique secret session key k chosen by A, and the unique common secret session key K_{AB}, shared by A and B, are used throughout the session to generate the encryption keys, $K_{i,1}$ and $K_{i,2}$, for $i = 1, 2, \ldots$. These keys, K_{AB}, $K_{i,1}$ and $K_{i,2}$, will differ from one session to the next, if the value of k is different. Hence k should be selected randomly to insure the security. Note that since $K_{AB} \ mod \ p'$ is a primitive element of $GF(p')$, we can obtain a period of length $p' - 1$ for the encryption keys, $K_{i,1}$ and $K_{i,2}$. This is one less than the maximum length p'. So for all practical applications, such a period of length is as long as what we want. In other words, within a session, even we use the same k to generate the encryption keys, $K_{i,1}$ and $K_{i,2}$, these keys will differ for each message block. In the next section, we will describe the randomness tests of $K_{i,2}$.

Security Analysis

We propose three possible attacks, but for each attack, the difficulty of breaking our system is equivalent to breaking the Diffie-Hellman PKDS.

1. In the ciphertext-only attack, the intruder needs to derive $K_{i,2}$ in order to obtain m_i. Since $K_{i,2} = \alpha^{K_{i,1}}\ mod\ p$, $K_{i,1} = K_{AB}{}^{i}\ mod\ p'$ and K_{AB} is the common secret session key based on the PKDS, obtaining $K_{i,2}$ is equivalent to computing a discrete logarithm.
2. In the chosen-plaintext attack, the intruder can obtain an encryption key, $K_{i,2} = \alpha^{K_{i,1}}\ mod\ p$, but not $K_{i,1}$, unless he can solve the discrete logarithm problem.
3. In the chosen-plaintext attack, since

$$K_{i,2} = \alpha^{K^{i}_{AB}\ mod\ p'}\ mod\ p$$

and

$$K_{i+1,2} = \alpha^{K^{i+1}_{AB}\ mod\ p'}\ mod\ p$$

knowing $K_{i,2}$ does not lead to obtaining $K_{i+1,2}$. Similarly, knowing $K_{i+1,2}$ does not yield $K_{i,2}$, unless the discrete logarithm problem is solved. In other words, it is computationally infeasible to convert a known encryption key (or a sequence of known encryption keys) to an unknown encryption key (or a sequence of unknown encryption keys) without knowing the common secret session key K_{AB}.

4 The Randomness Test of Encryption Keys $K_{i,2}$

As pointed out in the previous section, a common secret session key K_{AB} is shared by A and B during a communication session. For every message m_i in a sequence of message blocks, the encryption keys $K_{i,1}$ and $K_{i,2}$ are generated iteratively as

$$K_{i,1} = K_{i-1,1}K_{AB}\ mod\ p' = K^{i}_{AB}\ mod\ p'$$

and

$$K_{i,2} = \alpha^{K_{i,1}}\ mod\ p = \alpha^{K^{i}_{AB}\ mod\ p'}\ mod\ p.$$

That means A and B are using the following pseudo-random number generator

$$K_{i,2} = \alpha^{K_{i,1}}\ mod\ p = \alpha^{K^{i}_{AB}\ mod\ p'}\ mod\ p,\ for\ i = 1, 2, \ldots$$

to generate a sequence of random numbers, $K_{i,2}$, $i = 1, 2, \ldots$ with the common secret session key, K_{AB}, as a seed.

In the security analysis, we have pointed out that knowing one random number does not reveal the other random numbers. However, we feel it is still necessary to test these bit strings for the appearance of randomness based on the statistical analysis.

Three conventional methods are used for the randomness testing. They are frequency test, serial test, and binary derivative test [11]. The first two are based on "Chi-square" test [9].

The rationale for using chi-square tests is to establish the degree of conformity to two criteria of randomness:

1. Every number in the range of interest shall have an equal chance of being chosen (Frequency Test).
2. The occurrence of any number shall in no way affect the occurrence of any other number (Serial Test).

The frequency test used here consists of selecting 50 samples of 1000 numbers each, classifying them into 100 equal intervals on [0,1] (99 degree of freedom) and computing 50 values of chi-squared based on an expectation of 100. The result falls into 71%-72%, which is acceptable for randomness.

The serial test consists of selecting 50 samples of 1000 numbers each, cross-classifying each overlapping sequential pair in the cells of a square matrix each of whose sides is divided into 10 equal intervals on [0,1] (99 degree of freedom), and computing 100 values of chi-squared based on an expectation of 100. The result falls into 41%-42%, which is also acceptable for randomness.

As to the binary derivative test, we select 50 strings each consisting of 1000 bits and calculate 100 derivatives of each basic sequence. The result also indicates that the proposed random number generator produces "uncorrelated" random sequence.

From the results of these three randomness tests, it leads us to believe that a "good" random number generator has been proposed in this paper.

5 Conclusion

We have proposed a public-key cryptosystem based on the discrete logarithm in this paper. The performance of our system, in terms of transmission efficiency and computational time, is equivalent to the RSA scheme, while the security of the proposed system remains the same as the ElGamal cryptosystem. The randomness tests provide us with extra evidence to believe that this proposed cryptosystem is secure.

References

[1] W. Diffie and M. E. Hellman.: New Directions in Cryptography, IEEE Trans. Inform. Theory, vol. IT-22, pp. 644-654, Nov. 1976.

[2] R. L. Rivest, A. Shamir and L. Adelman.: A Method for Obtaining Digital Signatures and Public-Key Cryptosystem, Commun. of ACM, vol. 21, No. 2, pp.120-126, Feb. 1978.

[3] R. C. Merkle and M. E. Hellman.: Hiding Information and Signatures in Trapdoor Knapsack, IEEE Trans. Inform. Theory, vol. IT-24, pp. 525-530, Sep. 1978.

[4] T. ElGamal.: A Public Key Cryptosystem and a Signature Scheme Based on Discrete Logarithms, IEEE Trans. Inform. Theory, vol. IT-31, pp. 469-472, July 1985.
[5] C. S. Laih and J. Y. Lee.: Efficient Probabilistic Public-Key Cryptosystem Based on the Diffie-Hellman Problem, Electronics Letters, vol. 26, No. 5, pp. 326-327, March 1990.
[6] S. Pohlig and M. Hellman.: An Improved Algorithm for Computing Logarithms over GF(p) and Its Cryptographic Significance, IEEE Trans. Inform. Theory, vol. IT-24, pp. 106-110, 1978.
[7] K. S. McCurley.: The discrete logarithm problem, in Proc. of Symposia in Applied Mathematics, vol. 42, pp. 49-74, American Mathematical Society, Providence, 1990.
[8] A. M. Odlyzko.: Discrete logarithms in finite fields and their cryptographic significance, Advances in Cryptology, Eurocrypt '84, pp. 224-314, Springer-Verlag, 1985.
[9] D. E. Knuth.: The Art of Computer Programming, vol. II: Seminumerical Algorithms, Addison-Wesley Publishing Company, 1969.
[10] D. E. R. Denning.: Cryptography and Data Security, Addison-Wesley, 1982.
[11] J. M. Carroll.: The binary derivative test: noise filter, crypto aid, and random-number seed selector, SIMULATION, Sept. 1989.

Session 12
PUBLIC KEY CRYPTOGRAPHY II

Chair: John Snare
(Telecom Research Laboratories, Australia)

Elliptic Curves over F_p Suitable for Cryptosystems

Atsuko Miyaji

Matsushita Electric Industrial Co. , LTD.
1006, KADOMA, KADOMA-SHI, OSAKA, 571 JAPAN

Abstract. Koblitz ([5]) and Miller ([6]) proposed a method by which the group of points on an elliptic curve over a finite field can be used for the public key cryptosystems instead of a finite field. To realize signature or identification schemes by a smart card, we need less data size stored in a smart card and less computation amount by it. In this paper, we show how to construct such elliptic curves while keeping security high.

1 Introduction

Public key cryptosystems based on the discrete logarithm problem on an elliptic curve (EDLP) can offer small key length cryptosystems. If an elliptic curve is chosen to avoid the Menezes-Okamoto-Vanstone reduction ([9]), then the only known attacks on EDLP are the Pollard ρ−method ([11])and the Pohlig-Hellman method ([10]). So up to the present, such elliptic curve cryptosystems on E/F_q are secure if $\#E(F_q)$ is divisible by a prime only more than 30 digits ([3]).

If we use an elliptic curve E/F_q for digital signature or identification by a smart card ([12]), data size and computation amount of signature generation should be as small as possible. We may publish only the x-coordinate $x(P)$ of a public key P and one bit necessary to recover the y-coordinate $y(P)$ of P since the public key of an elliptic curve point is 2 times as large as the definition field F_q. Then we can reduce the data size to one half. But it will cause the computation amount to recover $y(P)$.

In this paper, we investigate an elliptic curve suitable for cryptosystems, in the sense that it requires less data size and less computation, while maintaining the security. We also show the advantage of our elliptic curve in the case of the Schnorr's digital signature scheme on an elliptic curve.

This paper is organized as follows. Section 2 summarizes the addition formula of an elliptic curve ([13]). Section 3 describes the Schnorr signature on an elliptic curve, and show the data size and the computation amount for two cases, the basic version and the reducing-data version. Section 4 discusses the elliptic curve which gives cryptosystems that reduce both of data sizes and the computation amount.

2 Addition formula of Elliptic curve

Cryptosystems on an elliptic curve E/F_q, for example the Diffie-Hellman key distribution and ElGamal cryptosystems, require the computation of kP ($P \in$

$E(F_q)$). We will discuss the computation amount of kP. For simplicity, we neglect addition, subtraction and multiplication by a small constant in F_q because they are much faster than multiplication and division in F_q.

Let K be a finite field F_q of characteristic $\neq 2, 3$. An elliptic curve over K is given as follows,

$$E : y^2 = x^3 + ax + b \ \ (a, b \in K, 4a^3 + 27b^2 \neq 0).$$

Then the set of K-rational points on E (with a special element $\mathcal{O}$ at infinity), denoted $E(K)$, is a finite abelian group, where $E(K) = \{(x, y) \in K^2 | y^2 = x^3 + ax + b\} \cup \{\mathcal{O}\}$. For the curve E, the addition formulas in the affine coordinate are the following. Let $P = (x_1, y_1)$, $Q = (x_2, y_2)$ and $P + Q = (x_3, y_3)$ be points on $E(K)$.

- **Curve addition formula in the affine coordinates ($P \neq \pm Q$)**

$$\begin{aligned} x_3 &= \lambda^2 - x_1 - x_2, \\ y_3 &= \lambda(x_1 - x_3) - y_1, \\ \lambda &= \frac{y_2 - y_1}{x_2 - x_1}; \end{aligned} \tag{1}$$

- **Curve doubling formula in the affine coordinates ($P = Q$)**

$$\begin{aligned} x_3 &= \lambda^2 - 2x_1, \\ y_3 &= \lambda(x_1 - x_3) - y_1, \\ \lambda &= \frac{3x_1^2 + a}{2y_1}. \end{aligned} \tag{2}$$

The formula (1) requires two multiplications and one division in K, while the formula (2) requires three multiplications and one division in K. The computation amount of division in K is more than that of multiplication in K. So we often use the projective coordinates to avoid divisions in K. The addition formulas in the projective coordinates are the following. Let $P = (X_1, Y_1, Z_1)$, $Q = (X_2, Y_2, Z_2)$ and $P + Q = (X_3, Y_3, Z_3)$.

- **Curve addition formula in the projective coordinates ($P \neq \pm Q$)**

$$\begin{aligned} X_3 &= vA, \\ Y_3 &= u(v^2 X_1 Z_2 - A) - v^3 Y_1 Z_2, \\ Z_3 &= v^3 Z_1 Z_2, \end{aligned} \tag{3}$$

where $u = Y_2 Z_1 - Y_1 Z_2, v = X_2 Z_1 - X_1 Z_2, t = X_2 Z_1 + X_1 Z_2, A = u^2 Z_1 Z_2 - v^2 t$;

- **Curve doubling formula in the projective coordinates ($P = Q$)**

$$\begin{aligned} X_3 &= 2hs, \\ Y_3 &= w(4B - h) - 8{Y_1}^2 s^2, \\ Z_3 &= 8s^3, \end{aligned} \tag{4}$$

where $w = a{Z_1}^2 + 3{X_1}^2, s = Y_1 Z_1, B = X_1 Y_1 s, h = w^2 - 8B$. The formula (3) requires 15 multiplications, while the formula (4) requires 12 multiplications.

For the use of cryptosystems, we may set $z(P) = Z_1$ to one in the formula (3). Then the formula (3) requires 12 multiplications.

Subtractions are as expensive as additions over elliptic curves. So the computation amount of kP by the addition-subtraction method ([2, 8]) is less than that by the binary method, while both methods need memory storage only for P. We assume to compute kP by the addition-subtraction method. The computation by the addition-subtraction requires n times of curve doubling and $\frac{n}{3}$ times of curve adding on the average, where $n = |K|$. Computation of kP in the projective coordinate requires one division and two multiplications in the final stage. Since n is larger than about 100, the computations in the projective coordinates are faster than that in the affine coordinates if the ratio of the computation amount of division in K to that of multiplication in K is larger than 9. In order to compare the computation amount of Schnorr signature scheme on a finite field and on an elliptic curve, we assume to compute kP in the projective coordinate by the addition-subtraction method and compute the power residue by the binary method.

3 Elliptic curve cryptosystems

If $E(K)$ and a basepoint $P \in E(K)$ are carefully chosen, then the only known attacks on the cryptosystems are the square root attacks. EDLP on such E to the base P is secure up to the present ([3]), if the order of P, $ord(P)$, is divisible by more than a 30-digit prime. Here we summarize the Schnorr signature on such an elliptic curve and establish a basis for evaluation of the elliptic curve proposed in the next chapter.

Let $M \in \mathbb{Z}$ be a message. User A sends the message M to user B with her or his signature of M.

- Initialization
  - system parameter
    - $E : y^2 = x^3 + ax + b$ $(a, b \in F_p\ ; p$ is a prime of $n(\geq 97)$ bits).
    - $P \in E(F_p)$: a basepoint (chosen as the above).
    - $l = ord(P)$ (l is $m(\geq 97)$ bits).
  - a one-way hash function $h : \mathbb{Z}_l \times \mathbb{Z} \to \{0, \cdots, 2^t - 1\}$, where t is the security parameter.
- Key generation

  User A randomly chooses an integer s , a secret key, and makes public the point $P_A = -sP$ as a public key.
- Signature generation

  1 Pick a random number $k \in \{1, ..., l\}$ and compute

$$R = kP = (r_x, r_y). \tag{5}$$

  Here $r_x = x(R)$ and $r_y = y(R)$.

  2 Compute $e := h(r_x, M) \in \{0, \cdots, 2^t - 1\}$.

  3 Compute $y \equiv k + se \pmod{l}$ and output the signature (e, y).

– Signature verification

1 Compute $\overline{R} = yP + eP_A = (\overline{r_x}, \overline{r_y})$ and check that $e = h(\overline{r_x}, M)$.

As we described in Section 2, the computation of kP requires m curve doublings and $\frac{m}{3}$ curve additions on the average, where k is a m-bit number. Extending the addition-subtraction method to the computation in the verification, we can calculate $yP + eP_A$ in m curve doublings and $\frac{1}{3}(m-t) + \frac{5}{9}t$ curve additions on the average with precomputations of $\pm(P \pm P_A)$, which require about the same computaion amount as one curve addition.

Here we set $n, m = 128$. Then the known attacks on such an elliptic curve cryptosystems requires at least 2^{64} elliptic curve operations. This is roughly equal to that of the original Schnorr on F_p (p is 512 bits). If lower security is required, then n, m can be replaced by a smaller number like 97. For the security parameter, here we set $t = 128$.

We will present two versions of Schnorr signature on an elliptic curve. One is the basic Schnorr signature on an elliptic curve described above, called Basic EC version. Another is called Reducing data EC version. In this version, only $x(P_A)$ and the least significant bit of $y(P_A)$ are published as a public key to reduce the data size. The same is done for the basepoint P. On the other hand, the original Schnorr signature scheme on F_p, called Finite field version (p is 512 bits, the security parameter t=128) roughly has the same security as that on the above elliptic curves. So the size of the definition field of Finite field version is four times as large as that of Basic and Reducing data EC versions.

We compare Basic EC version, Reducing data EC version and Finite field version, with respect to data size. Table 1 shows the comparison.

• **Basic EC version**

The system key is a, p, P, and l (640 bits). The secret key is s (128 bits). They are stored in a smart card. So the data size stored in a smart card is 768 bits. The public key is (P_A) (256 bits) and the signature is e and y (256 bits).

• **Reducing data EC version**

In this version, we have to publish one more parameter "b" of E as a system key to recover a point by the x-coordinate of the point and the least significant bit of the y-coordinate of the point. It requires power residue to recover the y-coordinate of P and increases computation for signature. The system key is a, b, p, $x(P)$, the least significant bit of $y(P)$ and l (641 bits). The secret key is s (128 bits). So the data size stored in a smart card is 769 bits. It is almost equal to that of Basic EC version. The public key is ($x(P_A)$ and the least significant bit of $y(P_A)$) (129 bits) and the signature is e and y (256 bits).

• **Finite field version**

The system key of Finite field version is a set of the definition field, the basepoint and the order of basepoint (1164 bits), where the size of the definition field is 512 bits and the order of basepoint is 140 bits. The secret key is 140 bits. So the data size stored in a smart card is 1304 bits.

The size of the definition fields of both EC versions is reduced to 25% of Finite field version. But the stored data size is not so reduced (59%). This is because an elliptic curve point has 2 coordinates and we need a parameter to decide E.

Let us compare the three cases with respect to the computation amount. We assume the computation method that we described in Section 2. Table 2 shows the comparison of the computation amount of signature generation and verification. Here we assume $m(n) = (n/t)^2 m(t)$, where $m(n)$ denotes the amount of work to perform one modular multiplication whose modulus size is n bits. We assume the ratio of the computation amount of division in K to that of multiplication in K to 10. We see the computation amount of signature generation of Reducing data EC version is reduced to 67% of Finite field version. It is not so reduced as the size of the definition field. This is because the computation amount of one elliptic curve addition is much more than that of one multiplication in the same definition field and we need to recover a basepoint.

We see that both EC versions seem to be better than Finite field version for both points of the data size and the computation amount. But actually they are not so efficient considering the less size of the definition field of E. For the stored data size, the ratio of the stored data size to the definition field for both EC versions is 6. On the other hand, for Finite field version, the ratio is 2.5. For the computation amount, one elliptic curve addition requires about 12 multiplications. If we require higher security, for example $t = 160$, then we will have to construct an elliptic curve over at least a 160-bit finite field. Then the advantage for EC versions shown in Table 1 and 2 decreases.

	System Key	Secret Key	Public Key	Signature size
Basic EC version	640	128	256	256
Reducing data EC version	641	128	129	256
Finite field version	1164	140	512	268

Table 1. Comparison of data size(in bits)

	Signature Generation	Signature Verification
Basic EC version	129	151
Reducing data EC version	141	175
Finite field version	210	242

Table 2. Comparison of the computation amount(number of 512-bit modular multiplications)

In the next section, we construct an elliptic curve cryptosystem, which has (1)the less ratio of the stored data size to the definition field than 6;
(2)the same public key size as Reducing data EC version;
(3)the less computation amount than that of Basic EC version.
It will be also best implementation for the higher security parameter.

4 Elliptic curves suitable for Cryptosystems

If $E(F_p)$ and the basepoint $P \in E(F_p)$ are appropriately chosen, then the only known attacks on the cryptosystems are the square root attacks. We first discuss a method to construct such elliptic curves and then investigate what elliptic curve among them is suitable for implementation with respect to less data size (key length) and less computation amount.

4.1 Decision of the class of elliptic curves

One method to avoid the recent attack is to construct EDLP on E/F_p with p elements ([7]). We describe a modified method to decide the class of such elliptic curves. There are two phases for the decision of E/F_p with p elements.

The first phase is to find an appropriate prime p. Such p is a form of $p = db^2 + db + \frac{d+1}{4}$ (b is an integer) for $d \in \{3, 11, 19, 43, 67, 163\}$. Such integers d enable us to construct easily the j-invariant j_d of E/F_p with p elements for the prime p, which is uniquely determined by d. Table 3 lists integers d and the j-invariant j_d.

d	j_d
3	0
11	$(-2^5)^3$
19	$(-2^5 * 3)^3$
43	$(-2^6 * 3 * 5)^3$
67	$(-2^5 * 3 * 5 * 11)^3$
163	$(-2^6 * 3 * 5 * 23 * 29)^3$

Table 3. Integers d and j-invariant j_d

Once the prime $p = db^2 + db + \frac{d+1}{4}$ and j_d are given, then the next phase is to decide the class of E/F_p with p elements. There is a little difference between the case of $d = 3$ and others. First we investigate the case of $d \in \{11, 19, 43, 67, 163\}$. Then the elliptic curves over F_p with the j-invariant j_d are given as follows.

$$E_{c,d} : y^2 = x^3 + 3c^2 a_d x + 2c^3 a_d,\ a_d = \frac{j_d}{1728 - j_d} \quad (\forall c \in F_p^*).$$

For each d, we can classify $\{E_{c,d}|c \in F_p^*\}$ into two equivalence classes of twists, namely

$$\mathcal{E}_d = \{E_{c,d}|c \in F_p^*, \left(\frac{c}{p}\right) = 1\} \text{ and } \mathcal{E}'_d = \{E_{c,d}|c \in F_p^*, \left(\frac{c}{p}\right) = -1\},$$

where $\left(\frac{c}{p}\right)$ denotes the Legendre symbol. Then only one of the two classes gives the elliptic curves with p elements. A general condition to decide the class was investigated ([1]). In our case, the condition can be simplified as follows.

Theorem 1. *Let p be a prime represented by $p = db^2 + db + \frac{d+1}{4}$ (b is an integer) for $d \in \{11, 19, 43, 67, 163\}$. Then the class which gives elliptic curves with p elements is determined as:*

$$\mathcal{E}_d \text{ if } \left(\frac{\alpha_d}{p}\right) = -1,$$

$$\mathcal{E}'_d \text{ if } \left(\frac{\alpha_d}{p}\right) = 1,$$

where α_d is an integer determined by d. Table 4 shows the values of α_d.

d	α_d
11	3 * 7
19	3
43	2 * 5 * 7
67	3*5*7*11*31
163	2*3*5*7*11*19*23*29*127

Table 4. Integers d and α_d

Now we get the following procedure to decide the class of elliptic curves with p elements.

Procedure 1

1 Search a large prime p such that $p = db^2 + db + \frac{d+1}{4}$ (b is an integer) for $d \in \{11, 19, 43, 67, 163\}$.

2 Calculate $\left(\frac{\alpha_d}{p}\right)$. If $\left(\frac{\alpha_d}{p}\right) = -1$, then $\mathcal{E}_d$ is the class. Else if $\left(\frac{\alpha_d}{p}\right) = 1$, then $\mathcal{E}'_d$ is the class.

Next we will investigate the case of $d = 3$. Then the elliptic curves over F_p ($p = 3b^2 + 3b + 1$) with the j-invariant j_d are given as follows.

$$E_\xi : y^2 = x^3 + \xi \quad (\forall c \in F_p^*). \tag{6}$$

In this case, we can classify $\{E_\xi | c \in F_p^*\}$ into six equivalence classes of twists, namely

$$\mathcal{E}_{3,i} = \{E_\xi | xi \in F_p^*, \left(\frac{\xi}{p}\right)_6 = (-\omega)^i\} \ (0 \le i \le 5, \omega = \frac{-1+\sqrt{-3}}{2}),$$

where $\left(\frac{\xi}{p}\right)_6$ denotes the sixth power residue symbol. Then exactly one of the six classes gives the elliptic curves with p elements. We have a next formula on the number of rational points of the elliptic curves (6).

Theorem 2 ([4]). *If $p \equiv 1 \pmod 3$, let $p = \pi\bar{\pi}$ with $\pi \in \mathbb{Z}[\omega]$ and $\pi \equiv 2 \pmod 3$. Then*

$$\#E_\xi(F_p) = p + 1 + \overline{\left(\frac{4\xi}{\pi}\right)_6}\pi + \left(\frac{4\xi}{\pi}\right)_6\bar{\pi}. \tag{7}$$

Using the formula (7), the condition to decide the class can be given as follows.

Theorem 3. *Let p be a prime represented by $p = 3b^2 + 3b + 1$ (b is an integer). Then the class which gives elliptic curves with p elements is determined as:*

$$\mathcal{E}_{3,1} \text{ if } b \equiv 0, 2, 4 \pmod 6,$$

$$\mathcal{E}_{3,5} \text{ if } b \equiv 1, 3, 5 \pmod 6.$$

Proof. We prove only the case of $b \equiv 1 \pmod 6$. As for the other cases, we can do the same way. Let $\pi = (2b+1)\omega + (b+1)$. Then $p = \pi\bar{\pi}$ and $\pi \equiv 2 \pmod 3$. Since $\left(\frac{4}{\pi}\right)_6 = \omega$, we get that $\#E_\xi(F_p) = p$ if and only if

$$\overline{\left(\frac{\xi}{\pi}\right)_6}\omega^2\pi + \left(\frac{\xi}{\pi}\right)_6\omega\bar{\pi} = -1,$$

that is, $tr(\omega\left(\frac{\xi}{\pi}\right)_6\bar{\pi}) = -1$. So we get $\left(\frac{\xi}{\pi}\right)_6 = -\omega^2$. This means that the class which gives elliptic curves with p elements is $\mathcal{E}_{3,5}$.

Now we get the following procedure to decide the class of elliptic curves with p elements.

Procedure 2

1 Search a large prime p such that $p = 3b^2 + 3b + 1$ (b is an integer).

2 If $b \equiv 0, 2, 4 \pmod 6$, then $\mathcal{E}_{3,1}$ is the class. Else if $b \equiv 1, 3, 5 \pmod 6$, then $\mathcal{E}_{3,5}$ is the class.

We have seen that the time to decide the class of E/F_p with p elements depends on the time finding $p = db^2 + db + \frac{d+1}{4}$ for $d \in \{3, 11, 19, 43, 67, 163\}$. We can easily find such a prime. In fact we were convinced experimentally that finding a prime $p = db^2 + db + \frac{d+1}{4}$ in the range of $30 \sim 90$ digits is as easy as finding a prime in that range. So we can easily decide the class of E/F_p with p elements which gives secure cryptosystems.

4.2 Selection of an elliptic curve and a basepoint

Elliptic curve cryptosystems require the computation of kP, where $P = (X_1, Y_1, 1)$ is a fixed point called basepoint. It is accomplished by repeated doubling, adding and subtracting of P. If we can select a basepoint P with a small x-coordinate X_1 or a small y-coordinate Y_1, the amount of computation of kP will be reduced. Especially in the case of signature and identification by a smart card, reducing of total data size stored in a smart card and the computation amount by a smart card is important. If fewer parameters represent an elliptic curve and a basepoint, the data stored in a smart card is reduced. Furthermore we wish to recover P easily from the parameters.

In the last section, we have decided the class of elliptic curves which gives the secure cryptosystems. Note that any elliptic curve E/F_p of the class and any basepoint $P \in E(F_p)$ give cryptosystems with the same security. We will discuss how to select E of the class and P in E suitable for cryptosystems, in the sense that it reduces computation amount of kP and necessary data size to be stored. We will classify d into two cases, $d = 3$ and others.

• **Proposed scheme A**

First we deal with the case of $d \in \{11, 19, 43, 67, 163\}$. For a given $p = db^2 + db + \frac{d+1}{4}$, we know which class, $\mathcal{E}_d$ or $\mathcal{E}'_d$, gives an elliptic curve with p elements in Section 4.1. Without loss of generality, we will discuss the case of $\mathcal{E}_d$. Let $y_0 = x_0^3 + 3a_d x_0 + 2a_d$ for $x_0 \in F_p$. Then we get one elliptic curve in $\mathcal{E}_d$ and the basepoint following (8).

$$\mathcal{E}_d \ni E_{y_0,d},\ E_{y_0,d} \ni P = (y_0 x_0, y_0^2) \text{ if } \left(\frac{y_0}{p}\right) = 1 \tag{8}$$

If y_0 satisfies the condition of (8) for $x_0 = 0$, then we get $\mathcal{E}_d \ni E_{y_0,d}$ and $E_{y_0,d} \ni P = (0, 4{a_d}^2)$. In fact such y_0 satisfies the condition of (8) if and only if

$$\left(\frac{y_0}{p}\right) = \left(\frac{2a_d}{p}\right) = 1.$$

Except for $d = 19$, there exists $p = db^2 + db + \frac{d+1}{4}$ which satisfies $\left(\frac{2a_d}{p}\right) = 1$. Combining the condition on p to decide a class (i.e. $\left(\frac{a_d}{p}\right) = -1$ or 1), we obtain that such an elliptic curve over F_p exits if and only if $\left(\frac{\beta_d}{p}\right) = -1$ in both cases, $\mathcal{E}_d$ and $\mathcal{E}'_d$. Table 5 shows the value of β_d.

We were also convinced experimentally that, for $\forall p = db^2 + db + \frac{d+1}{4}$ ($d \in \{11, 43, 67, 163\}$), such an elliptic curve exists with a probability of about one half. Here is one example for a 128-digit prime in the case of $d = 11$.

$$E : y^2 = x^3 + 12a^3x + 16a^4;\ E(F_p) \ni P = (0, 4a^2),$$
$$p = 1701\ 41183\ 46046\ 92395\ 60785\ 96622\ 40717\ 16369,$$
$$a = \ 527\ 15357\ 39869\ 82616\ 07887\ 30307\ 87012\ 55349.$$

d	β_d
11	3 * 7
43	3 * 7
67	7*31
163	7*11*19*127

Table 5. Integers d and β_d

Let us use this elliptic curve $E_{y_0,d}$ and basepoint $P = (0, 4a_d^2)$ for Schnorr signature, where $E_{y_0,d} = E$ and $a = a_d$. We further assume that the public key P_A is represented by $x(P_A)$ and the least significant bit of $y(P_A)$. The computation of kP requires the addition to the basepoint P, which is calculated in 9 modular multiplications. So the computation of kP requires $1932m(128)$. We can recover the basepoint in one modular multiplication, only if we store a_d. Since $ord(P)$ equals p, the system key is a_d and p (256 bits). Table 6 shows the data size and Table 7 shows the computation amount. The data size stored in a smart card is reduced to one half of that of Reducing data EC version and Basic EC version. The public key size is the same as that of Reducing data EC version.

The computation amount of the signature generation is reduced by 6% (resp. 14%) of that of Basic EC version (resp. Reducing data EC version). The computation amount of the signature verification is reduced by 10 % of that of Reducing data EC version. It is increased by 5 % of that of Basic EC version. This is because we need one power residue to recover one's public key in the signature verification. If we publish P_A instead of $x(P_A)$ and the least significant bit of $y(P_A)$ as a public key, then the computation amount of the signature verification is reduced by 3% of that of Basic EC version. Even in this case, the public key size is only 50% of Finite field version.

We can choose a prime p and an elliptic curve E/F_p as follows.

$$E : y^2 = x^3 + 12a^3x + 16a^4; \quad E(F_p) \ni P = (0, 4a^2),$$
$$p = 2^{128} - 89\,25388\,84800\,47273\,94087$$
$$a = 1887\,65172\,00252\,43003\,83780\,59753\,00282\,08521$$

The form of p simplifies the arithmetic modulo p and we can store p with only 73 bits. Of course, the particular form of p provides no disadvantage on the security for now.

- **Proposed scheme B**

Next we deal with the case of $d = 3$. For a given $p = 3b^2 + 3b + 1$, we know which class, $\mathcal{E}_{3,1}$ or $\mathcal{E}_{3,5}$, gives the elliptic curve with p elements in Section 4.1. We only discuss the case of $\mathcal{E}_{3,1}$. As for the other case, we can do in the same way.

An elliptic curve E/F_p with p elements and a basepoint P is given as follows,

$$E_\xi : y^2 = x^3 + \xi y_0{}^3; \quad E_\xi(F_p) \ni P = (x_0 y_0, y_0{}^2),$$
$$(\forall \xi \text{ such that } \left(\frac{\xi}{p}\right)_6 = -\omega, \ \forall y_0 = x_0{}^3 + \xi \in F_p{}^{*2}).$$

In this case, there doesn't exist an elliptic curve with the point whose x-coordinate equals 0 because of $\xi \notin F_p{}^{*2}$. But we can select a small ξ such that $\left(\frac{\xi}{p}\right)_6 = -\omega$ and a small x_0 such that $y_0 = x_0{}^3 + \xi \in F_p{}^{*2}$. Here is one example for a 128-digit prime.

$$E \ : y^2 = x^3 + 3 * 4^3; \quad E(F_p) \ni P = (4, 16),$$
$$p = 1701\ 41183\ 46046\ 92480\ 63157\ 20930\ 49376\ 39647$$
$$(x_0 = 1, \ \xi = 3)$$

Let us use the elliptic curve E_ξ and the basepoint $P = (x_0 y_0, y_0{}^2)$ for Schnorr signature. We further assume that one's public data P_A is represented by $x(P_A)$ and the least significant bit of $y(P_A)$. Then the addition to $P = (x_0 y_0, y_0{}^2) = (X_1, Y_1)$ is accomplished in 9 modular multiplications because we can neglect the multiplications by a small constants X_1 and Y_1. Furthermore the simple equation of E reduces the computation amount of doubling. It is accomplished in 10 modular multiplications. As for the computation amount of kP, it requires $1676m(128)$. As for the recovering the basepoint, we can recover it in a negligible computation amount only if we store x_0 and ξ whose data size is enough small. As for the data size, the data size of x_0 and ξ is neglected and $ord(P)$ equals p. So the size of system parameters x_0, ξ and p of Schnorr signature scheme on such E_ξ is about the same as that of the definition field. Table 6 shows the data size and Table 7 shows the computation amount.

We see that the elliptic curves and the basepoints in the case of $d = 3$ give good properties for the cryptosystems, especially in the application of digital signature and identification by a smart card. The data size stored in a smart card is reduced to one third of that of Reducing data EC version and Basic EC version. The public key size is the same as that of Reducing data EC version. The computation amount of the signature generation is reduced by 19% (resp. 26%) of that of Basic EC version (resp. Reducing data EC version). The computation amount of the signature verification is reduced by 6% (resp. 19%) of Basic EC version (resp. Reducing data EC version). If we publish P_A as a public key, then the computation amount of the signature verification is reduced by 14% of that of Basic EC version.

In the same way as Proposed scheme A, we can choose a prime p and an elliptic curve E/F_p as follows.

$$E \ : y^2 = x^3 + 3 * 4^3; \quad E(F_p) \ni P = (4, 16),$$
$$p = 2^{128} - 86\ 61755\ 49264\ 58706\ 00985$$
$$(x_0 = 1, \ \xi = 3)$$

The form of p simplifies the arithmetic modulo p and we can store p with only 73 bits.

	System Key	Secret Key	Public Key	Signature size
Proposed scheme A	256(201)	128	129	256
Proposed scheme B	131 (76)	128	129	256

Table 6. Data size of the Proposed schemes(in bits)

	Signature Generation	Signature Verification
Proposed scheme A	121	158
Proposed scheme B	105	142

Table 7. Computation amount of the Proposed schemes (number of 512-bit modular multiplications)

5 Conclusion

Elliptic curve cryptosystems often require the computation of kP, where P is a fixed basepoint. We have proposed the elliptic curves and basepoints suitable for cryptosystems, in the sense that they require less data size and less computation amount for kP. Especially if we use the Proposed version B in Schnorr signature scheme by a smart card, we have seen that
(1) the data size stored in a smart card is reduced to one third of that of Basic EC version and Reducing data EC version;
(2) the data size of public key is reduced to one half of that of Basic EC version and is the same as Reducing data EC version;
(3) the computation amount of the signature generation is reduced by 19% (resp. 26%) of that of Basic EC version (resp. Reducing data EC version);
(4)The computation amount of the signature verification is reduced by 6% (resp. 19%) of Basic EC version (resp. Reducing data EC version);
(5)In the case where we publish the point P_A as a public key, the computation amount of the signature verification is reduced by 14% of that of Basic EC version.

Acknowledgements
The author would like to thank Alfred Menezes and Tatsuaki Okamoto for helpful conversations. The author wishes to thank Makoto Tatebayashi for helpful advice.

References

1. A. O. L. Atkin and F. Morain, "Elliptic curves and primality proving", *Research Report 1256, INRIA*, Juin 1990. Submitted to Math. Comp.
2. M. J. Coster, "Some algorithms on addition chains and their complexity", Center for Mathematics and Computer Science Report CS-R9024.
3. G. Harper, A. Menezes and S. Vanstone, "Public-key cryptosystems with very small key lengths", *Abstracts for Eurocrypt 92*, 1992.
4. K. Ireland and M. Rosen, *A classical introduction to modern number theory*, GTM 84, Springer-Verlag, New-York, 1982.
5. N. Koblitz, "Elliptic curve cryptosystems", *Mathematics of Computation*, 48(1987), 203-209.
6. V. S. Miller, "Use of elliptic curves in cryptography", *Advances in Cryptology-Proceedings of Crypto'85*, Lecture Notes in Computer Science, 218 (1986), Springer-Verlag, 417-426.
7. A. Miyaji, "On ordinary elliptic curves", *Abstract of proceedings of ASIACRYPT'91*, 1991.
8. F. Morain and J. Olivos, "Speeding up the computations on an elliptic curve using addition-subtraction chains", Theoretical Informatics and Applications Vol. 24, No. 6 (1990), p531-544
9. A. Menezes, T. Okamoto and S. Vanstone "Reducing elliptic curve logarithms to logarithms in a finite field", *Proceedings of the 22nd Annual ACM Symposium on the Theory of Computing*, 80-89, 1991.
10. S. Pohlig and M. Hellman, "An improved algorithm for computing logarithm over $GF(p)$ and its cryptographic significance", *IEEE Trans. Inf. Theory*, IT-24(1978), 106-110.
11. J. Pollard, "Monte Carlo methods for index computation(mod p)", *Mathematics of Computation*, 32 (1978), 918-924.
12. C. P. Schnorr, "Efficient Signature Generation by Smart Cards", *Journal of Cryptology*, Vol.4 (1991), No.3, 161-174.
13. J. H. Silverman, *"The Arithmetic of Elliptic Curves"*, GTM106, Springer-Verlag, New York, 1986.

The Probability Distribution of the Diffie-Hellman Key

Christian P. Waldvogel

James L. Massey

Signal and Information Processing Laboratory
Swiss Federal Institute of Technology
CH-8092 Zurich, Switzerland

Abstract. The probability distribution of the key generated by the Diffie-Hellman Public Key-Distribution system is derived. For different prime factorizations of $p-1$, where p is the prime modulus of the Diffie-Hellman system, the probabilities of the most and the least likely Diffie-Hellman key are found. A lower bound for the entropy of the Diffie-Hellman key is also derived. For the case $p-1=2q$, with q prime, it is shown that the key distribution is very close to the uniform distribution and the key entropy is virtually the maximum possible. A tight upper bound on the probability of the most likely key is also derived, from which the form of the prime factorization of $p-1$ maximizing the probability of the most likely Diffie-Hellman key is found. The conditions for generating equally likely Diffie-Hellman keys for any prime factorization of $p-1$ is given.

1 Introduction

Let $\mathbf{Z}_m$ denote the commutative monoid of elements $\mathbf{Z}_m = \{0, 1, \ldots m-1\}$ whose rule is multiplication modulo m, and let $\mathbf{Z}_m^*$ denote the multiplicative group consisting of all the invertible elements of $\mathbf{Z}_m$ (i.e., $\mathbf{Z}_m^* = \{x : x \in \mathbf{Z}_m and \gcd(x,m) = 1\}$). The symbol $m(.)$ denotes the multiplicative order of its argument, i.e., given β in the multiplicative group $\mathbf{Z}_m^*$, $m(\beta) = i$ such that i is the minimum positive integer for which $\beta^i = 1$. For the prime p we have $\mathbf{Z}_p^* = \{1, 2, \ldots p-1\}$ and in that case we write α to denote a generator of the multiplicative group $\mathbf{Z}_p^*$, i.e., an element of multiplicative order $p-1$.

In the Diffie-Hellman Public Key-Distribution system, each user chooses uniformly at random a private key x in $\mathbf{Z}_{p-1}$ and generates the public key $y = \alpha^x$ in $\mathbf{Z}_p^*$, which he or she then submits to the public directory. When Alice and Bob wish to generate a secret Diffie-Hellman key, Alice requests Bob's public key y_B from the directory; with her private key x_A, she generates the Diffie-Hellman key ${y_B}^{x_A} = \alpha^{x_A x_B}$. Similarly, Bob requests Alice's public key y_A and generates the same Diffie-Hellman key ${y_A}^{x_B} = \alpha^{x_A x_B}$ by using his private key x_B.

In this paper, we are concerned with the probability distribution and the entropy of the Diffie-Hellman key. With the help of the algebraic results in section 2, we derive in section 3 the probability distribution and the entropy of the Diffie-Hellman key. In section 4, we compare, for different prime factorizations of $p-1$, the entropy of the Diffie-Hellman key and the probabilities of the most and the least likely key. Special consideration is given to the case $p-1=2q$, where q is prime. We also derive a tight upper bound on the probability of the most likely key, from which the form of the prime factorization of $p-1$ maximizing the probability of the most likely Diffie-Hellman key is found. Some concluding remarks are given in section 5.

2. Algebraic Preliminaries

Let α be a generator of the multiplicative group Z_p^* and let t be an element in Z_{p-1}. Let $R(t)$ denote the set of pairs (x_A, x_B) in $Z_{p-1} \times Z_{p-1}$ satisfying $\alpha^{x_A x_B} = \alpha^t$, i.e.,

$$R(t) \triangleq \left\{(x_A, x_B) : \alpha^{x_A x_B} = \alpha^t \text{ and } x_A \in Z_{p-1},\ x_B \in Z_{p-1}\right\} \quad . \tag{1}$$

Because $m(\alpha) = p - 1$, it follows that the arithmetic in the exponent of α is arithmetic in Z_{p-1} so that (1) can be written as

$$R(t) = \left\{(x_A, x_B) : x_A x_B = t \text{ in } Z_{p-1} \text{ and } x_A \in Z_{p-1},\ x_B \in Z_{p-1}\right\} \quad . \tag{2}$$

We make use of the following special case of Theorem 57 (p. 51) in [Hardy 60]:

Lemma 1: The equation $x_A x_B = t$ in Z_{p-1} has solutions for the unknown x_B if and only if, as integers, $\gcd(x_A, p-1)$ divides t; moreover, when $\gcd(x_A, p-1)$ divides t, there are exactly $\gcd(x_A, p-1)$ solutions for x_B.

Making use of Lemma 1, we can write the cardinality $\#[R(t)]$ of the set $R(t)$ as

$$\#\left[R(t)\right] = \sum_{x_A \in S(t)} \gcd(x_A, p-1) \tag{3}$$

where

$$S(t) \triangleq \left\{u : \gcd(u, p-1) \text{ divides } t \text{ and } u \in Z_{p-1}\right\} \quad . \tag{4}$$

Lemma 2: Let t be an element in Z_{p-1}. Then for any integer u, $\gcd(u, p-1)$ divides t if and only if $\gcd(u, p-1)$ divides $\gcd(t, p-1)$.

Proof: Suppose that $\gcd(u, p-1)$ divides t. Since $\gcd(u, p-1)$ also divides $p-1$, it follows that $\gcd(u, p-1)$ divides $\gcd(t, p-1)$. Conversely, suppose that $\gcd(u, p-1)$ divides $\gcd(t, p-1)$. Because $\gcd(t, p-1)$ divides t, we conclude that $\gcd(u, p-1)$ divides t. □

Making use of Lemma 2 in (4) gives

$$S(t) = \left\{u : \gcd(u, p-1) \text{ divides } \gcd(t, p-1) \text{ and } u \in Z_{p-1}\right\} \quad . \tag{5}$$

From Theorem 1.15 (ii) (p. 7) in [Lidl 86] we have

Lemma 3: The element α^t in Z_p^* has multiplicative order

$$m(\alpha^t) = \frac{p-1}{\gcd(t, p-1)} \quad .$$

By Lemma 3, we can substitute $\frac{p-1}{m(\alpha^t)}$ for $\gcd(t, p-1)$ in (5) to obtain

$$S(t) = \left\{u : \gcd(u, p-1) \text{ divides } \frac{p-1}{m(\alpha^t)} \text{ and } u \in Z_{p-1}\right\} \quad . \tag{6}$$

Let $p-1 = p_1^{e_1} \dots p_K^{e_K}$ such that the factors $p_1, \dots, p_K$ are distinct primes and the exponents $e_1, \dots, e_K$ are positive integers. Because $m(\alpha^t)$ must divide the order $p-1$ of the multiplicative group Z_p^* by Lagrange's Theorem, $m(\alpha^t)$ must be of the form

$$m(\alpha^t) = p_1^{\hat{e}_1} \dots p_K^{\hat{e}_K} \quad , \tag{7}$$

for some integers $\hat{e}_1, \dots, \hat{e}_K$ such that $0 \le \hat{e}_i \le e_i$ for all i. For such $m(\alpha^t)$, we have

$$\frac{p-1}{m(\alpha^t)} = p_1^{e_1-\hat{e}_1} \dots p_K^{e_K-\hat{e}_k} \quad ,$$

which when substituted in (6) gives

$$\mathrm{S}(t) = \left\{ u : \gcd(u, p-1) \text{ divides } p_1^{e_1-\hat{e}_1} \dots p_K^{e_K-\hat{e}_k} \text{ and } u \in Z_{p-1} \right\} \tag{8}$$

where $\hat{e}_1, \dots, \hat{e}_K$ are defined by (7). Letting $p_1^{c_1} \dots p_K^{c_K}$ denote the prime factorization of $\gcd(u, p-1)$, we can write (8) as

$$\mathrm{S}(t) = \left\{ u : \gcd(u, p-1) = p_1^{c_1} \dots p_K^{c_K},\ p_1^{c_1} \dots p_K^{c_K} \text{ divides } p_1^{e_1-\hat{e}_1} \dots p_K^{e_K-\hat{e}_k} \text{ and } u \in Z_{p-1} \right\}$$

which reduces to

$$\mathrm{S}(t) = \left\{ u : \gcd(u, p-1) = p_1^{c_1} \dots p_K^{c_K}, 0 \le c_i \le e_i - \hat{e}_i \text{ for all } i\ (1 \le i \le K) \text{ and } u \in Z_{p-1} \right\} .$$

Making use of this last expression in (3) gives

$$\begin{aligned}
\#\left[\mathrm{R}(t)\right] &= \sum_{c_1=0}^{e_1-\hat{e}_1} \dots \sum_{c_K=0}^{e_K-\hat{e}_K} \left(\sum_{b \in \mathrm{T}(c_1,\dots,c_K)} p_1^{c_1} \dots p_K^{c_K} \right) \\
&= \sum_{c_1=0}^{e_1-\hat{e}_1} \dots \sum_{c_K=0}^{e_K-\hat{e}_K} p_1^{c_1} \dots p_K^{c_K} \left(\sum_{b \in \mathrm{T}(c_1,\dots,c_K)} 1 \right) \\
&= \sum_{c_1=0}^{e_1-\hat{e}_1} \dots \sum_{c_K=0}^{e_K-\hat{e}_K} p_1^{c_1} \dots p_K^{c_K} \quad \#\left[\mathrm{T}(c_1, \dots, c_K)\right] \quad ,
\end{aligned} \tag{9}$$

where $\#\left[\mathrm{T}(c_1, \dots, c_K)\right]$ denotes the cardinality of the set

$$\mathrm{T}(c_1, \dots, c_K) = \left\{ u : \gcd(u, p-1) = p_1^{c_1} \dots p_K^{c_K} \text{ and } u \in Z_{p-1} \right\} \quad . \tag{10}$$

Lemma 4: Let $c_1, \dots, c_K$ be some integers such that $0 \le c_i \le e_i$ for all i. Then the integer b belongs to the set $\mathrm{T}(c_1, \dots, c_K)$ given in (10) if and only if $b \in Z_{p-1}$ and $m(\alpha^b) = p_1^{e_1-c_1} \dots p_K^{e_K-c_k}$.

Proof: Because $p_1^{e_1} \dots p_K^{e_K}$ denotes the prime factorization of $p-1$, we have

$$p_1^{c_1} \dots p_K^{c_K} = \frac{p_1^{e_1} \dots p_K^{e_K}}{p_1^{e_1-c_1} \dots p_K^{e_K-c_k}} = \frac{p-1}{p_1^{e_1-c_1} \dots p_K^{e_K-c_k}} \quad .$$

By substituting this last expression in (10) we obtain

$$\mathrm{T}(c_1, \dots, c_K) = \left\{ u : \gcd(u, p-1) = \frac{p-1}{p_1^{e_1-c_1} \dots p_K^{e_K-c_k}} \text{ and } u \in Z_{p-1} \right\} \quad .$$

from which it follows that b belongs to $\mathrm{T}(c_1,\ldots,c_K)$ if and only if

$$\gcd(b,p-1)=\frac{p-1}{p_1^{e_1-c_1}\cdots p_K^{e_K-c_k}} \tag{11}$$

and $b\in \mathbf{Z}_{p-1}$. Since by Lemma 3 we see that (11) holds if and only if $m(\alpha^b)=p_1^{e_1-c_1}\ldots p_K^{e_K-c_k}$, we conclude that b belongs to $\mathrm{T}(c_1,\ldots,c_K)$ if and only if $m(\alpha^b)=p_1^{e_1-c_1}\ldots p_K^{e_K-c_k}$ and $b\in\mathbf{Z}_{p-1}$. □

By Lemma 4, we see that $\#[\mathrm{T}(c_1,\ldots,c_K)]$ is equal to the number of elements in the multiplicative group $\mathbf{Z}_p^*$ with multiplicative order $p_1^{e_1-c_1}\ldots p_K^{e_K-c_k}$. Since, by Theorem 1.15 (p.7) in [Lidl 86], the number of such elements is given by $\varphi(p_1^{e_1-c_1}\ldots p_K^{e_K-c_k})$, where $\varphi(.)$ denotes Euler's Totient function, we conclude that $\#[\mathrm{T}(c_1,\ldots,c_K)]=\varphi(p_1^{e_1-c_1}\ldots p_K^{e_K-c_k})$. By substituting this last expression in (9), we obtain

$$\#\left[\mathbf{R}(t)\right]=\sum_{c_1=0}^{e_1-\hat{e}_1}\cdots\sum_{c_K=0}^{e_K-\hat{e}_K} p_1^{c_1}\ldots p_K^{c_K}\ \ \varphi(p_1^{e_1-c_1}\ldots p_K^{e_K-c_k})\ ,$$

which, upon using the property that $\varphi(a\cdot b)=\varphi(a)\,\varphi(b)$ when $\gcd(a,b)=1$, reduces to

$$\begin{aligned}\#\left[\mathbf{R}(t)\right] &= \sum_{c_1=0}^{e_1-\hat{e}_1}\cdots\sum_{c_K=0}^{e_K-\hat{e}_K} p_1^{c_1}\ldots p_K^{c_K}\ \ \varphi(p_1^{e_1-c_1})\ldots\varphi(p_K^{e_K-c_k})\\ &= \left(\sum_{c_1=0}^{e_1-\hat{e}_1} p_1^{c_1}\ \varphi(p_1^{e_1-c_1})\right)\cdots\left(\sum_{c_K=0}^{e_K-\hat{e}_K} p_K^{c_K}\ \varphi(p_K^{e_K-c_K})\right)\\ &= \prod_{i=1}^{K}\left(\sum_{c_i=0}^{e_i-\hat{e}_i} p_i^{c_i}\ \varphi(p_i^{e_i-c_i})\right)\\ &= \prod_{i=1}^{K}\ n(i)\end{aligned} \tag{12}$$

where for all i $(1\le i\le K)$

$$\begin{aligned} n(i) &= \sum_{c_i=0}^{e_i-\hat{e}_i} p_i^{c_i}\ \varphi(p_i^{e_i-c_i})\\ &= \begin{cases} p_i^{e_i} + \displaystyle\sum_{c_i=0}^{e_i-\hat{e}_i-1} p_i^{c_i}\ \varphi(p_i^{e_i-c_i}) & \text{if } \hat{e}_i=0\ ,\\ \displaystyle\sum_{c_i=0}^{e_i-\hat{e}_i} p_i^{c_i}\ \varphi(p_i^{e_i-c_i}) & \text{if } \hat{e}_i>0\ .\end{cases}\end{aligned} \tag{13}$$

With the help of the Kronecker-Delta function $\delta(\cdot)$, which is defined by $\delta(e)=1$ if $e=0$ and $\delta(e)=0$ otherwise, we can write (13) as

$$n(i)=p_i^{e_i}\ \ \delta(\hat{e}_i)+\sum_{c_i=0}^{e_i-\hat{e}_i-\delta(\hat{e}_i)} p_i^{c_i}\ \varphi(p_i^{e_i-c_i}) \tag{14}$$

where the exponent $e_i - c_i$ is positive for $c_i = 0, 1, \ldots, e_i - \hat{e}_i - \delta(\hat{e}_i)$. Using the property of Euler's Totient function that $\varphi(p^e) = p^{e-1}(p-1)$ for every prime p and every positive exponent e, we can simplify the sum in (14) to obtain

$$\begin{aligned} n(i) &= p_i^{e_i}\ \delta(\hat{e}_i) + \sum_{c_i=0}^{e_i-\hat{e}_i-\delta(\hat{e}_i)} p_i^{c_i}\ p_i^{e_i-c_i-1}\ (p_i-1) \\ &= p_i^{e_i}\ \delta(\hat{e}_i) + p_i^{e_i-1}\ (p_i-1)\left(\sum_{c_i=0}^{e_i-\hat{e}_i-\delta(\hat{e}_i)} 1\right) \\ &= p_i^{e_i}\ \delta(\hat{e}_i) + p_i^{e_i-1}\ (p_i-1)(e_i-\hat{e}_i-\delta(\hat{e}_i)+1) \\ &= p_i^{e_i-1}\ \left((p_i-1)(e_i-\hat{e}_i+1)+\delta(\hat{e}_i)\right) . \end{aligned}$$

Substituting this last expression for $n(i)$ in (12) we obtain

$$\#\left[\mathbf{R}(t)\right] = \prod_{i=1}^{K} p_i^{e_i-1}\ \left((p_i-1)(e_i-\hat{e}_i+1)+\delta(\hat{e}_i)\right)$$

where $\hat{e}_1, \ldots, \hat{e}_K$ are defined by (7). We have proven:

Theorem 1: Let $p_1^{e_1} \ldots p_K^{e_K}$ denote the prime factorization of $p-1$ and let t be an element in $\mathbf{Z}_{p-1}$. Let α be a generator of the multiplicative group $\mathbf{Z}_p^*$ and let $m(\alpha^t) = p_1^{\hat{e}_1} \ldots p_K^{\hat{e}_K}$ $(0 \leq \hat{e}_i \leq e_i)$ be the multiplicative order of α^t. Then the number $\#[\mathbf{R}(t)]$ of pairs (x_A, x_B) in $\mathbf{Z}_{p-1} \times \mathbf{Z}_{p-1}$ satisfying $\alpha^{x_A x_B} = \alpha^t$ is given by

$$\#\left[\mathbf{R}(t)\right] = \prod_{i=1}^{K} p_i^{e_i-1}\ \left((p_i-1)(e_i-\hat{e}_i+1)+\delta(\hat{e}_i)\right) , \tag{15}$$

where $\delta(.)$ denotes the Kronecker-Delta function.

Example 1: Let $p-1 = 2q$, with q prime. Substituting $p_1 = 2$, $p_2 = q$ and $e_1 = e_2 = 1$ in Theorem 1 gives

$$\begin{aligned} \#\left[\mathbf{R}(t)\right] &= \left(2-\hat{e}_1+\delta(\hat{e}_1)\right)\left((q-1)(2-\hat{e}_2) + \delta(\hat{e}_2)\right) \\ &= \begin{cases} 6q-3 & \text{if } m(\alpha^t) = 1 , \\ 3q-3 & \text{if } m(\alpha^t) = q , \\ 2q-1 & \text{if } m(\alpha^t) = 2 , \\ q-1 & \text{if } m(\alpha^t) = 2q . \end{cases} \end{aligned}$$

Letting t be an element in $\mathbf{Z}_{p-1}$, we write $\mathbf{R}^*(t)$ to denote the set of pairs (x_A, x_B) in $\mathbf{Z}_{p-1}^* \times \mathbf{Z}_{p-1}^*$ satisfying $\alpha^{x_A x_B} = \alpha^t$, i.e.,

$$\mathbf{R}^*(t) \triangleq \left\{(x_A, x_B) : \alpha^{x_A x_B} = \alpha^t \text{ and } x_A \in \mathbf{Z}_{p-1}^*,\ x_B \in \mathbf{Z}_{p-1}^*\right\} \tag{16}$$

which reduces to

$$\mathbf{R}^*(t) = \left\{(x_A, x_B) : x_A x_B = t \text{ in } \mathbf{Z}_{p-1} \text{ and } x_A \in \mathbf{Z}_{p-1}^*,\ x_B \in \mathbf{Z}_{p-1}^*\right\} . \tag{17}$$

When $t \notin \mathbf{Z}_{p-1}^*$ it follows, from the closure property of the multiplicative group $\mathbf{Z}_{p-1}^*$, that no pair (x_A, x_B) in $\mathbf{Z}_{p-1}^* \times \mathbf{Z}_{p-1}^*$ exists such that $x_A x_B = t$, which implies that for such t we have $\#[\mathbf{R}^*(t)] = 0$. When $t \in \mathbf{Z}_{p-1}^*$ the choice of x_A in (17) uniquely determines x_B, i.e., $x_B = x_A^{-1}\, t$, where x_A^{-1} denotes the multiplicative inverse of x_A in $\mathbf{Z}_{p-1}^*$. Thus for such t (17) reduces to

$$\mathbf{R}^*(t) = \left\{(x_A, x_A^{-1}\, t) : x_A \in \mathbf{Z}_{p-1}^*\right\} \quad ,$$

which implies that $\#[\mathbf{R}^*(t)] = \#\left[\mathbf{Z}_{p-1}^*\right] = \varphi(p-1)$, where $\varphi(.)$ denotes Euler's Totient function. We have proven:

Theorem 2: Let p be a prime number, let α be a generator of the multiplicative group $\mathbf{Z}_p^*$ and let t be an element in $\mathbf{Z}_{p-1}$. Then the number $\#[\mathbf{R}^*(t)]$ of pairs (x_A, x_B) in $\mathbf{Z}_{p-1}^* \times \mathbf{Z}_{p-1}^*$ satisfying $\alpha^{x_A x_B} = \alpha^t$ is given by

$$\#\left[\mathbf{R}^*(t)\right] = \begin{cases} 0 & \text{if } t \notin \mathbf{Z}_{p-1}^* \\ \varphi(p-1) & \text{if } t \in \mathbf{Z}_{p-1}^* \end{cases} \quad ,$$

where $\varphi(.)$ denotes Euler's Totient function.

Example 2: Given $p - 1 = 2q$ with q prime, it follows by Theorem 2 that

$$\#\left[\mathbf{R}^*(t)\right] = \begin{cases} 0 & \text{if } t \notin \mathbf{Z}_{p-1}^* \\ q-1 & \text{if } t \in \mathbf{Z}_{p-1}^* \end{cases} \quad .$$

3 Probability Distribution of the Diffie-Hellman Key

We now use the results of the previous section to derive the probability distribution of the Diffie-Hellman key generated by Alice and Bob. Let the discrete random variable X_A over $\mathbf{Z}_{p-1}$ denote Alice's private key so that $\mathcal{P}(X_A = x_A)$ gives the probability of Alice choosing x_A as her private key, and let the discrete random variable X_B over $\mathbf{Z}_{p-1}$ denote Bob's private key so that $\mathcal{P}(X_B = x_B)$ gives the probability of Bob choosing x_B as his private key. Then $\alpha^{X_A X_B}$ is a discrete random variable such that $\mathcal{P}(\alpha^{X_A X_B} = \alpha^t)$ gives the probability of Alice and Bob generating the Diffie-Hellman key α^t.

We make two assumptions:

Assumption 1: All users choose their private key independently; in particular,

$$\mathcal{P}(X_A = x_A\,,\ X_B = x_B) \;=\; \mathcal{P}(X_A = x_A)\;\,\mathcal{P}(X_B = x_B)$$

for all x_A and x_B in $\mathbf{Z}_{p-1}$.

Assumption 2: Each user chooses her or his private key uniformly at random in $\mathbf{Z}_{p-1}$; in particular, for all x in $\mathbf{Z}_{p-1}$,

$$\mathcal{P}(X_A = x) \;=\; \mathcal{P}(X_B = x) \;=\; \frac{1}{p-1} \quad .$$

Letting the set $\mathbf{R}(t)$ be defined by (1), i.e., $\mathbf{R}(t)$ contains the set of pairs (x_A, x_B) in $\mathbf{Z}_{p-1} \times \mathbf{Z}_{p-1}$ satisfying $\alpha^{x_A x_B} = \alpha^t$, we have

$$\mathcal{P}(\alpha^{X_A X_B} = \alpha^t) = \sum_{(x_A, x_B) \in \mathbf{R}(t)} \mathcal{P}(X_A = x_A \, , \, X_B = x_B) \, . \tag{18}$$

Using Assumption 1 (i.e., X_A and X_B are statistically independent) in (18) gives

$$\mathcal{P}(\alpha^{X_A X_B} = \alpha^t) = \sum_{(x_A, x_B) \in \mathbf{R}(t)} \mathcal{P}(X_A = x_A) \, \mathcal{P}(X_B = x_B)$$

which by Assumption 2 (i.e. $\mathcal{P}(X_A = x_A) = \mathcal{P}(X_B = x_B) = \frac{1}{p-1}$) reduces to

$$\mathcal{P}(\alpha^{X_A X_B} = \alpha^t) = \sum_{(x_A, x_B) \in \mathbf{R}(t)} \left(\frac{1}{p-1}\right)\left(\frac{1}{p-1}\right) = \frac{1}{(p-1)^2} \, \# \left[\mathbf{R}(t)\right] \, .$$

Substituting the expression for $\# \left[\mathbf{R}(t)\right]$ given in Theorem 1, we can write this last equation as

$$\begin{aligned} \mathcal{P}(\alpha^{X_A X_B} = \alpha^t) &= \frac{1}{(p-1)^2} \prod_{i=1}^{K} p_i^{e_i - 1} \left((p_i - 1)(e_i - \hat{e}_i + 1) + \delta(\hat{e}_i) \right) \\ &= \left(\frac{1}{p-1}\right) \prod_{i=1}^{K} \left(\left(\frac{p_i - 1}{p_i}\right)(e_i - \hat{e}_i + 1) + \frac{\delta(\hat{e}_i)}{p_i} \right) \, . \end{aligned}$$

We have proven:

Theorem 3: Let $p_1^{e_1} \ldots p_K^{e_K}$ denote the prime factorization of $p-1$ and let $p_1^{\hat{e}_1} \ldots p_K^{\hat{e}_K}$ $(0 \leq \hat{e}_i \leq e_i)$ be the multiplicative order of α^t in the multiplicative group $\mathbf{Z}_p^*$. Then the probability $\mathcal{P}(\alpha^{X_A X_B} = \alpha^t)$ of Alice and Bob generating the Diffie-Hellman key α^t, under Assumptions 1 and 2, is given by

$$\mathcal{P}(\alpha^{X_A X_B} = \alpha^t) = \left(\frac{1}{p-1}\right) \prod_{i=1}^{K} \left(\left(\frac{p_i - 1}{p_i}\right)(e_i - \hat{e}_i + 1) + \frac{\delta(\hat{e}_i)}{p_i} \right) \, , \tag{19}$$

where $\delta(.)$ denotes the Kronecker-Delta function.

Let

$$\mathcal{P}_{min} \triangleq \min_{t \in \mathbf{Z}_{p-1}} \left[\mathcal{P}(\alpha^{X_A X_B} = \alpha^t)\right]$$

and

$$\mathcal{P}_{max} \triangleq \max_{t \in \mathbf{Z}_{p-1}} \left[\mathcal{P}(\alpha^{X_A X_B} = \alpha^t)\right]$$

denote the probabilities of the least and the most likely Diffie-Hellman key, respectively. From (19) we see that $\mathcal{P}(\alpha^{X_A X_B} = \alpha^t)$ is minimum when $\hat{e}_1 = e_1, \ldots, \hat{e}_K = e_K$, i.e., when the multiplicative order of α^t is $p-1$. Substituting e_i for $\hat{e}_i$ in (19) gives

Corollary 1:

$$\mathcal{P}_{min} = \left(\frac{1}{p-1}\right) \prod_{i=1}^{K} \left(\frac{p_i - 1}{p_i}\right) \, .$$

Because $\frac{p_i - 1}{p_i} \approx 1$ for almost all primes p_i, it follows that $\mathcal{P}_{min} \approx \frac{1}{p-1}$, i.e., $\mathcal{P}_{min}$ is smaller than the average key probability by only a small factor. From (19) we see that $\mathcal{P}(\alpha^{X_A X_B} = \alpha^t)$ is maximum when $\hat{e}_1 = 0, \ldots, \hat{e}_K = 0$, i.e., when the multiplicative order of α^t is one. Substituting 0 for $\hat{e}_i$ in (19) gives

Corollary 2: $$\mathcal{P}_{max} = \left(\frac{1}{p-1}\right) \prod_{i=1}^{K} \left(e_i\,(\frac{p_i-1}{p_i})+1\right) .$$

Let
$$H(\alpha^{X_A X_B}) \triangleq - \sum_{t\in \mathbf{Z}_{p-1}} \mathcal{P}(\alpha^{X_A X_B} = \alpha^t)\; log_2\left(\mathcal{P}(\alpha^{X_A X_B} = \alpha^t)\right)$$

denote the entropy (measured in bits) of the discrete random variable $\alpha^{X_A X_B}$. Since $\mathcal{P}_{max} \geq \mathcal{P}(\alpha^{X_A X_B} = \alpha^t)$ for all t in $\mathbf{Z}_{p-1}$, we can write

$$\begin{aligned} H(\alpha^{X_A X_B}) &\geq - \sum_{t\in \mathbf{Z}_{p-1}} \mathcal{P}(\alpha^{X_A X_B} = \alpha^t)\; log_2(\mathcal{P}_{max}) \\ &= -log_2(\mathcal{P}_{max}) \sum_{t\in \mathbf{Z}_{p-1}} \mathcal{P}(\alpha^{X_A X_B} = \alpha^t) \\ &= -log_2(\mathcal{P}_{max}) \ . \end{aligned} \tag{20}$$

Replacing $\mathcal{P}_{max}$ in (20) by the expression from Corollary 2 gives

$$\begin{aligned} H(\alpha^{X_A X_B}) &\geq -log_2\left(\left(\frac{1}{p-1}\right) \prod_{i=1}^{K} \left(e_i\,(\frac{p_i-1}{p_i})+1\right)\right) \\ &= log_2(p-1) - \sum_{i=1}^{K} log_2\left(e_i\,(\frac{p_i-1}{p_i})+1\right) \ . \end{aligned} \tag{21}$$

Because $\frac{p_i-1}{p_i} < 1$ for all i we obtain

Corollary 3: $$H(\alpha^{X_A X_B}) > log_2(p-1) - \sum_{i=1}^{K} log_2(e_i+1) \ .$$

We now modify Assumption 2 by restricting the way each user randomly chooses her or his private key in $\mathbf{Z}_{p-1}$:

Assumption 3: Each user chooses her or his private key uniformly at random in the set of all *invertible elements* of $\mathbf{Z}_{p-1}$; in particular,

$$\mathcal{P}(X_A = x) = \mathcal{P}(X_B = x) = \begin{cases} 0 & \text{if } x \notin \mathbf{Z}^*_{p-1} \\ \frac{1}{\varphi(p-1)} & \text{if } x \in \mathbf{Z}^*_{p-1} \end{cases} .$$

Letting the set $\mathbf{R}^*(t)$ be defined by (16) we have

$$\mathcal{P}(\alpha^{X_A X_B} = \alpha^t) = \sum_{(x_A,x_B)\in \mathbf{R}^*(t)} \mathcal{P}(X_A = x_A\ ,\ X_B = x_B) \ . \tag{22}$$

Using Assumption 1 (i.e., X_A and X_B are statistically independent) in (22) gives

$$\mathcal{P}(\alpha^{X_A X_B} = \alpha^t) = \sum_{(x_A,x_B)\in \mathbf{R}^*(t)} \mathcal{P}(X_A = x_A)\ \mathcal{P}(X_B = x_B)$$

which by Assumption 3 reduces to

$$\mathcal{P}(\alpha^{X_A X_B} = \alpha^t) = \sum_{(x_A, x_B) \in \mathbf{R}^*(t)} \left(\frac{1}{\varphi(p-1)}\right)\left(\frac{1}{\varphi(p-1)}\right) = \frac{1}{\varphi(p-1)^2} \; \# \left[\mathbf{R}^*(t)\right] .$$

Substituting the expression for $\#[\mathbf{R}^*(t)]$ given in Theorem 2 in this last equation, we obtain:

Theorem 4: Let p be a prime number, let α be a generator of the multiplicative group $\mathbf{Z}_p^*$ and let t be an element in $\mathbf{Z}_{p-1}$. Then the probability $\mathcal{P}(\alpha^{X_A X_B} = \alpha^t)$ of Alice and Bob generating the Diffie-Hellman key α^t, under Assumptions 1 and 3, is given by

$$\mathcal{P}(\alpha^{X_A X_B} = \alpha^t) = \begin{cases} 0 & \text{if } t \notin \mathbf{Z}_{p-1}^* \\ \frac{1}{\varphi(p-1)} & \text{if } t \in \mathbf{Z}_{p-1}^* \end{cases} ,$$

where $\varphi(.)$ denotes Euler's Totient function.

Theorem 4 shows that when each user chooses independently and uniformly at random her or his private key in the set of all the invertible elements of $\mathbf{Z}_{p-1}$, the Diffie-Hellman keys are uniformly distributed over $\mathbf{Z}_{p-1}^*$.

4 Key Distribution for different prime factorizations of $p-1$

Consider the case when the prime modulus p of the Diffie-Hellman Public Key-Distribution system is a strong prime, i.e., $p-1 = 2q$ where q is prime. For such a prime factorization of $p-1$, it follows from Corollaries 1 and 2 that

$$\mathcal{P}_{min} = \left(\frac{1}{p-1}\right)\left(\frac{1}{2}\right)\left(\frac{q-1}{q}\right)$$

and

$$\mathcal{P}_{max} = \left(\frac{1}{p-1}\right)\left(\frac{3}{2}\right)\left(\frac{q-1}{q} + 1\right)$$

which, under the assumption that p is large, gives

$$\mathcal{P}_{min} \approx \left(\frac{1}{p-1}\right)\left(\frac{1}{2}\right) \tag{23}$$

and

$$\mathcal{P}_{max} \approx \left(\frac{1}{p-1}\right)(3) . \tag{24}$$

From (23) we see that the probability of the least likely Diffie-Hellman key is about one-half the average key probability, while from (24) we see that the probability of the most likely Diffie-Hellman key is about three times the average key probability. These results indicate that, when p is a strong prime, the Diffie-Hellman keys are virtually equally likely. Furthermore, for such a strong prime p, it follows by Corollary 3 that

$$H(\alpha^{X_A X_B}) > log_2(p-1) - 2 ,$$

which shows that the entropy $H(\alpha^{X_A X_B})$ of the Diffie-Hellman key is less than two bits below the maximum possible entropy, $log_2(p-1)$, of a random variable with $p-1$ possible values. When p

is large, this implies that the entropy of the Diffie-Hellman key is virtually the maximum possible.

Given the order of magnitude of p, we look for the worst possible prime factorization of $p-1$, i.e., the prime factorization of $p-1 = \prod_{i=1}^{K} p_i^{e_i}$ that yields the largest probability $\mathcal{P}_{max}$. From Corollary 2 we see that a good approximation to $\mathcal{P}_{max}$ is given by

$$\mathcal{P}_{max} \approx \left(\frac{1}{p-1}\right) \prod_{i=1}^{K} e_i \ , \tag{25}$$

from which it follows that a large exponent product $\prod_{i=1}^{K} e_i$ yields a large probability $\mathcal{P}_{max}$.

Lemma 5: Let $\rho_1, \ldots, \rho_K$ and η be distinct real numbers each greater than one. Then the positive real numbers $\varepsilon_1, \ldots, \varepsilon_K$ that maximize the product $\prod_{i=1}^{K} \varepsilon_i$ subject to the constraint $\eta = \prod_{i=1}^{K} \rho_i^{\varepsilon_i}$ are given by $\varepsilon_i = \frac{1}{K} \ log_{\rho_i}(\eta)$ for $i = 1, \ldots, K$.

Proof: For the case $K = 1$, the unique and thus the optimum value for ε_1 is $log_{\rho_1}(\eta)$ as stated in the lemma. We now consider the case $K \geq 2$. Assuming that an oracle has given us the optimum values for all but two of the unknowns, say ε_s and ε_t, we compute the values for ε_s and ε_t that maximize the product $\varepsilon_s \ \varepsilon_t$. We can write the constraint $\eta = \prod_{i=1}^{K} \rho_i^{\varepsilon_i}$ as

$$\begin{aligned} ln(\eta) &= \sum_{i=1}^{K} \varepsilon_i \ ln(\rho_i) \\ &= \varepsilon_s \ ln(\rho_s) + \varepsilon_t \ ln(\rho_t) + \sum_{i=1:i\neq s,t}^{K} \varepsilon_i \ ln(\rho_i) \ , \end{aligned} \tag{26}$$

where $ln(\cdot)$ denotes the natural logarithm, or equivalently

$$\varepsilon_s \ ln(\rho_s) + \varepsilon_t \ ln(\rho_t) = \mu \tag{27}$$

where

$$\mu = ln(\eta) - \sum_{i=1:i\neq s,t}^{K} \varepsilon_i \ ln(\rho_i) \ .$$

From (27) we see that

$$\varepsilon_t = \frac{\mu - \varepsilon_s \ ln(\rho_s)}{ln(\rho_t)} \tag{28}$$

which, when substituted in the product $\varepsilon_s \ \varepsilon_t$ gives

$$\varepsilon_s \ \varepsilon_t = \frac{\varepsilon_s \ \mu - \varepsilon_s^{\ 2} \ ln(\rho_s)}{ln(\rho_t)} \ .$$

Thus maximizing $\varepsilon_s \ \varepsilon_t$ reduces to the simple problem of maximizing the quadratic expression $\varepsilon_s \ \mu - \varepsilon_s^{\ 2} \ ln(\rho_s)$ with respect to ε_s. The maximizing value of ε_s satisfies

$$\varepsilon_s \ ln(\rho_s) = \frac{\mu}{2} \ . \tag{29}$$

Substituting (29) in (28) gives

$$\varepsilon_t \ ln(\rho_t) = \frac{\mu}{2} \ . \tag{30}$$

Equation (29) and (30) show that $\varepsilon_s \; ln(\rho_s) = \varepsilon_t \; ln(\rho_t)$ must hold for every s and t $(1 \le s < t \le K)$ and hence,

$$\varepsilon_1 \; ln(\rho_1) = \varepsilon_2 \; ln(\rho_2) = \ldots = \varepsilon_K \; ln(\rho_K) \; . \tag{31}$$

Equation (31) and (26) imply that $ln(\eta) = K \; \varepsilon_i \; ln(\rho_i)$ for all i $(1 \le i \le K)$ from which we conclude that $\varepsilon_i = \frac{1}{K} \; log_{\rho_i}(\eta)$. □

Because $p - 1 = \prod_{i=1}^{K} \; p_i^{e_i}$ where $p, p_1, \ldots, p_K$ are prime numbers and $e_1, \ldots, e_K$ are positive integers, it follows by taking $\eta = p - 1$ and $\rho_i = p_i$ in Lemma 5 that

$$\begin{aligned} \prod_{i=1}^{K} e_i &\le \prod_{i=1}^{K} \frac{1}{K} \; log_{p_i}(p-1) \\ &= \left(\frac{ln(p-1)}{K}\right)^K \prod_{i=1}^{K} \frac{1}{ln\,(p_i)} \end{aligned} \tag{32}$$

where near equality holds when $e_i \approx \frac{1}{K} \; log_{p_i}(p-1)$ for all i. Making use of (32) in (25) gives

$$\mathcal{P}_{max} \lessapprox \left(\frac{1}{p-1}\right)\left(\frac{ln(p-1)}{K}\right)^K \prod_{i=1}^{K} \frac{1}{ln\,(p_i)} \; , \tag{33}$$

with the approximation sign holding when $e_i \approx \frac{1}{K} \; log_{p_i}(p-1)$ for all i. It remains to find primes $p_1, \ldots, p_K$ and the positive integer K that maximize the right side of (33). From the product in (33), we see that these optimum primes are the first K prime numbers. Letting q_i denotes the ith smallest prime (i.e., $q_1 = 2, q_2 = 3, q_3 = 5$, etc.), we see from (33) that

$$\mathcal{P}_{max} \lessapprox \left(\frac{1}{p-1}\right)\left(\frac{ln(p-1)}{K}\right)^K \prod_{i=1}^{K} \frac{1}{ln\,(q_i)} \triangleq f(K) \; , \tag{34}$$

with the approximation sign holding when $e_i \approx \frac{1}{K} \; log_{q_i}(p-1)$ and $p_i = q_i$ for all i. From the definition of $f(K)$ in (34), it follows that

$$\begin{aligned} \frac{f(K)}{f(K+1)} &= \left(1 + \frac{1}{K}\right)^K (K+1) \; \frac{ln(q_{K+1})}{ln(p-1)} \\ &\approx e \; (K+1) \; \frac{ln(q_{K+1})}{ln(p-1)} \; , \end{aligned} \tag{35}$$

where e denotes Euler's number. By Tchebycheff's Theorem, the $(K+1)^{th}$ prime number q_{K+1} satisfies

$$K + 1 \approx \frac{q_{K+1}}{ln(q_{K+1})} \; , \tag{36}$$

which substituted for $K+1$ in (35) gives

$$\frac{f(K)}{f(K+1)} \approx e \; q_{K+1} \; \frac{1}{ln(p-1)} \; ,$$

or equivalently

$$q_{K+1} \approx \frac{ln(p-1)}{e} \; \frac{f(K)}{f(K+1)} \; .$$

Replacing this last approximation for q_{K+1} in (36) gives

$$K+1 \;\approx\; \frac{\frac{ln(p-1)}{e}\,\frac{f(K)}{f(K+1)}}{ln\left(\frac{ln(p-1)}{e}\,\frac{f(K)}{f(K+1)}\right)} \;. \tag{37}$$

Because $\frac{f(K)}{f(K+1)}$ is monotonically increasing, it follows that $f(K) \lessapprox f(\kappa)$ where κ is the real number maximizing the function $f(.)$, i.e., the real number for which $\frac{f(\kappa)}{f(\kappa+1)} = 1$; moreover, the approximation sign in $f(K) \lessapprox f(\kappa)$ holds when $K \approx \kappa$. Substituting $\frac{f(\kappa)}{f(\kappa+1)} = 1$ and κ for K in (37) gives

$$\kappa \;\approx\; \frac{\frac{ln(p-1)}{e}}{ln\left(\frac{ln(p-1)}{e}\right)} - 1 \;\approx\; \frac{ln(p-1)}{e\;(ln\;ln(p-1)-1)} - 1 \;. \tag{38}$$

We have proven:

Corollary 4: Let q_i denote the ith prime number and let $p-1 = \prod_{i=1}^{K} p_i^{e_i}$. Then an approximate upper bound for the probability $\mathcal{P}_{max}$ is given by

$$\mathcal{P}_{max} \lessapprox \left(\frac{1}{p-1}\right)\left(\frac{ln(p-1)}{\kappa}\right)^{\kappa} \prod_{i=1}^{round(\kappa)} \frac{1}{ln\,(q_i)} \tag{39}$$

with κ being defined by (38) and where $round(\kappa)$ denotes the nearest positive integer to the real number κ. Moreover, the approximation holds when $K \approx \kappa$, $p_i = q_i$ and $e_i \approx \frac{1}{\kappa}\, log_{q_i}(p-1)$ for all i $(1 \leq i \leq K)$.

Example 3: Given the prime $p = 2^7\,3^4 + 1 = 10369$, Corollary 4 gives us the approximate upper bound $\mathcal{P}_{max} \lessapprox 2.4 \cdot 10^{-3}$, while Corollary 2 gives us the exact value $\mathcal{P}_{max} = \frac{11}{6912} \approx 1.59 \cdot 10^{-3}$. These last two results show that for $p = 10369$ the upper bound on $\mathcal{P}_{max}$ is tight, from which it follows that for such an order of magnitude of p, the prime $p = 10369$ virtually maximizes the probability of the most likely Diffie-Hellman key.

Example 4: Letting $p \approx 10^{100} \approx 2^{332}$, Corollary 4 gives us $\mathcal{P}_{max} \lessapprox 1.7 \cdot 10^{-88}$. Thus, for the worst possible prime factorization of $p-1$, i.e., the prime factorization of $p-1$ (should such prime p exist) for which the approximation in (39) holds, the most likely Diffie-Hellman key is about 10^{12} times the average key probability 10^{-100}. This shows a weakness in the Diffie-Hellman system with respect to the key probability distribution, because one of the keys is much more likely than the others. Nevertheless, the probability $\mathcal{P}_{max}$ for such a prime p is still too large to substantially lighten the task of the cryptanalyst seeking to find the Diffie-Hellman key.

5 Conclusion

Let p be the prime modulus of the Diffie-Hellman system. We have shown that large strong primes, i.e., primes p such that $p-1$ is equal to twice a large prime, are "secure" because they yield a small probability for the most likely key and a virtually maximum possible entropy. We have also given the form of the worst possible prime factorizations of $p-1$, i.e., the prime factorizations of $p-1$ that maximize the probability of the most likely Diffie-Hellman key. Finally, we have shown that the independent and uniformly random choice of the private key in the set of all invertible elements in Z_{p-1} guarantees the generation of equally likely Diffie-Hellman keys.

To conclude, it is interesting to note that strong primes p which are "secure" with respect to the key probability distribution are also "secure" with respect to the computational complexity because of the relative difficulty of taking logarithms in the multiplicative group Z_p^* for such primes. However the worst possible prime factorization of $p-1$ with respect to the key probability distribution is not the worst one with respect to the computational complexity, because the Fermat primes p (i.e., $p = 2^e + 1$ where e belongs to the set $\{1,2,4,8,16\}$) are the primes for which the fastest known algorithm exists for extracting discrete logarithms over Z_p^*, cf. [Pohlig 78].

References

[Lidl 86] Rudolf Lidl and Harald Niederreiter, *Introduction to finite fields and their applications*, Cambridge University Press, 1986.

[Hardy 60] G.H.Hardy and E.M.Wright, *An Introduction to the Theory of Numbers (Fourth Edition)*, Oxford University Press, 1960.

[Pohlig 78] Stephen Pohlig and Martin Hellman, *An Improved Algorithm for Computing Logarithms over GF(p) and its Cryptographic Significance*, IEEE Transactions on Information Theory, Vol. IT-24(1), pp. 106-110, January 1978.

A Modular Exponentiation Unit Based on Systolic Arrays

Jörg Sauerbrey

Lehrstuhl für Datenverarbeitung
Technische Universität München
P. O. Box 20 24 20
W-8000 München 2
Germany

sy@ldv.e-technik.tu-muenchen.de

1. Introduction

A lot of cryptographic methods and protocols rely on the fast evaluation of large powers modulo *m*. One of the famous members of this class of methods is RSA. Modular exponentiation is usually based on modular multiplication. A lot of research work has been done to implement fast modular multiplication (e.g. [AliMar91, LipPos90, Morita90, ShBeVu90, OrSvAn90, Eldrid91, EldWal]). Recently there have been some proposals for using systolic arrays for performing modular multiplication of long integers [KocHun91, Even90, IwMaIm92a, IwMaIm92b, Walter]. Systolic arrays are well suited to be implemented using fast clocked VLSI. Here we describe the architecure of a modular exponentiation-unit using systolic multipliers based on Montgomery's algorithm [Montgo85]. The estimated throughput of such a chip is competitive with existing designs (see tables in [Bricke90, OrSvAn90, IwMaIm92a]).

The multiplier described here is built of two identical systolic arrays, which are used for simultaneous multiplication and Montgomery reduction. The arrays are improvements of the systolic multiplier proposed by [Atrubi65]. The main difference to the Atrubin-array is, that the array can handle multiple bits of the operands simultaneously. The first array multiplies two numbers in a serial mode, LSB first. Using a special input control a second array of the same type can be used for performing the modular reduction of the product proposed by Montgomery. Both arrays build a serial systolic modular multiplier with selectable performance. Two of these multipliers (a multiplier-pair) can work in a pipelined mode for a further increase of the throughput. Because of the serial operation mode one or two of these multipliers or multiplier-pairs can be easily connected with some shift-registers to build an exponentiation-unit using one of the efficient exponentiation methods ([Knuth81], [ZhMaYu88, pp. 346], [BrGoMc92], [SauDie92]).

2. Why Systolic Arrays?

A systolic array consists of a set of interconnected cells, each capable of performing the same simple operation. They work together and synchronously to perform a task. Within a systolic array information flows between the cells in a pipelined mode. Only these cells on the array boundaries may be the systems I/O ports. In each clock cycle every cell takes the data of its neighbours, performs an operation, and passes the result to its neighbour cells ([KunLei78], [Kung82]).

For the following reasons, this concept is well suited for VLSI-Implementations:

- simple, regular communication an control structures (regularity)
- identical cells (modularity)
- only local interconnnections (high clockrate)
- pipelined operation (high throughput)

3. A Systolic Array for Multiplication

The systolic multiplier described here is a modification of the multiplier in [Atrubi65] and [Knuth81, pp. 277]. The new multiplier consists of completely identical cells and can process not only binary digits, but also numbers in an arbitrary base b representation, where b is a power of 2.

The multiplier consists of a linear arrangement of central clocked cells. The number of required cells is determined by the maximum size of the operands and the base b used by a cell for the operand processing. Each cell processes one base b digit per clock cycle, which means that the multiplier processes ld (b) binary digits of the operands per clock cycle (ld is the logarithm to the base 2).

The maximum number of binary digits of the operands is denoted l_{bin}. The digit number l of the same operands in base b representation is:

$$l = \left\lceil \frac{l_{bin}}{\mathrm{ld}\,(b)} \right\rceil \tag{1}$$

The number of required cells k is

$$k = \left\lceil \frac{l}{2} \right\rceil + 1 \tag{2}$$

Such a systolic array can process all operands with a number of digits less or equal than l. The clock cycles t which are needed for one multiplication depends on the actual size of the operands $l_{act} \leq l$ and not on the cell number k of the array. The number of required clock cycles t is:

$$t = 2l_{act} \tag{3}$$

Fig. 1 shows the interconnection and the input of the array when performing the operation $p = x \cdot y + q$.

The operands x and y are fed serially to the array with least significant digit first during l clock cycles. For the next l clock cycles zeroes are fed to the array. Simultaneous with x and y the operand q is fed serially. The least significant digit of the result appears at the output p after the propagation delay time of a cell. The next digits follow during the next clockcycles, so that the multiplication is finished after $2l$ clock cycles. The input z is supplied with a control signal indicating the start of a multiplication.

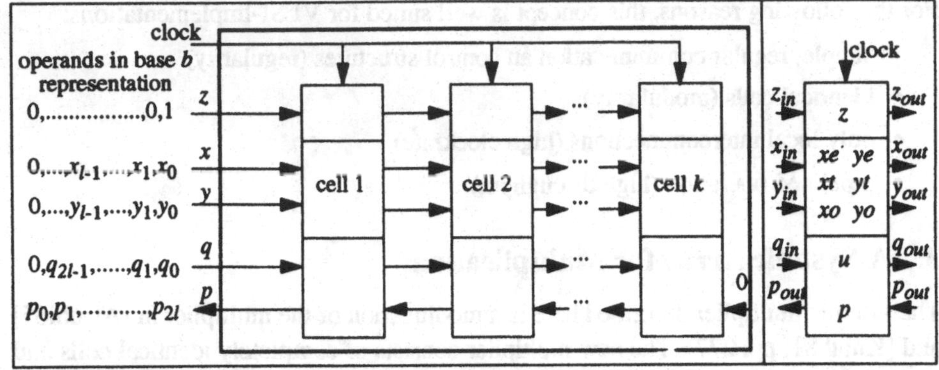

Fig. 1: Construction of a cell and cell-interconnection for building a systolic array performing the operation $p = x \cdot y + q$

Algorithm 1 describes the behaviour of a cell depending on the cell input.

```
cell_behavior(zin, xin, yin, qin, pin, clock)
{ A register transfer (←) is performed with the rising
  slope of the clock }
{ b is the base in which the numbers are represented }
LOOP
 CASE z,zin OF          { state of the cell }
  0,0: z ← 0; zout = 0
  0,1: z ← 1; zout = 0
  1,0 OR 1,1: z ← 2; zout = 0
  2,0 OR 2,1: z ← 3; zout = 0
  3,0 OR 3,1: z ← 0; zout = 1
 ENDCASE

 CASE z,zin OF          { occupation of registers }
  0,1: xe← xin; ye ← yin;
  1,0 OR 1,1: xo ← xin; yo ← yin;
  ELSE xt ← xin; yt ← yin;
 ENDCASE
 xout = xt; yout = yt;

 CASE z,zin OF          {building of intermediate products}
  0,1: qout = 0; s = xin*yin;
  1,0: qout = xin*yin; s = xin*ye+xe*yin;
  ELSE qout = xin*yo+xo*yin; s = xin*ye+xe*yin;
 ENDCASE
 s = s+p+u+qin;
 pout = s mod b;
 u ← s div b;
 p ← pin;
ENDLOOP
```

Algorithm 1: Behaviour of a cell

As an example for the operation of the array, figure 2 shows the interconnection and occupation of 3 cells performing the multiplication $9535 \cdot 8341 = 79531435$ during the 8 clock cycles needed.

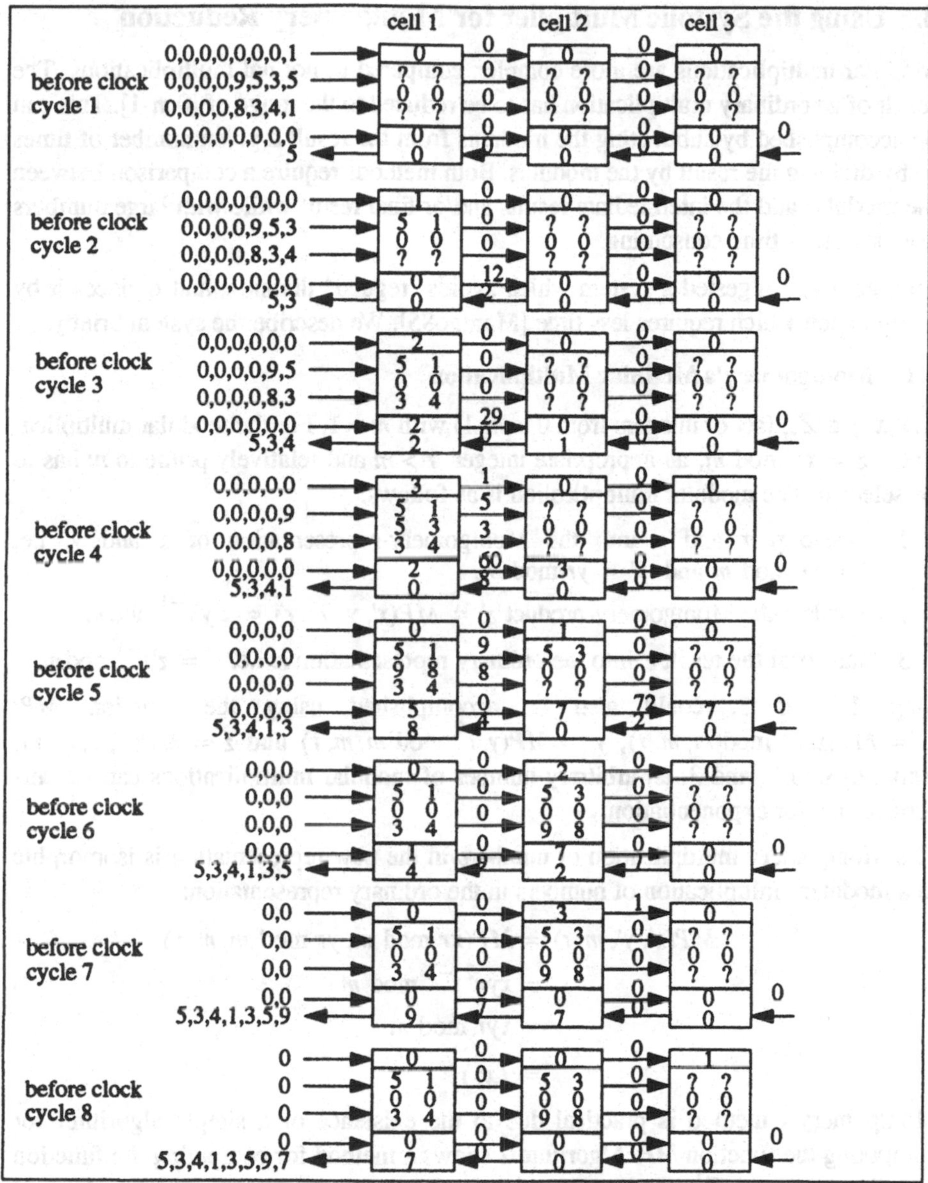

Fig. 2: Interconnection and occupation of the cells performing the multiplication 9535·8341 = 79531435

The number of required clockcycles for one multiplication and with it the throughput of the systolic multiplier can be effected by the choice of the base b. The complexity of one cell increases with an increasing base b value on one hand, but on the other hand less cells and less clock cycles are needed for the processing of operands of the same size (equation (1), (2) and (3)).

4. Using the Systolic Multiplier for Montgomery Reduction

Modular multiplications are more complex compared to normal multiplications. The result of an ordinary multiplication has to be reduced to the range of $[0,m-1]$. This can be accomplished by subtracting the modulus from the result a given number of times or by dividing the result by the modulus. Both methods require a comparison between the modulus and the intermediate results and/or final result while with large numbers this process is time consuming.

Montgomery suggested a system which avoids 'regular' division, and replaces it by an operation which requires less time [Montgo85]. We describe the system briefly.

4.1 Montgomery's Modular Multiplication

Let $x, y \in Z_m$ (set of integers from 0 to m-1) with $m > 1$. For the modular multiplication $z = xy \bmod m$, an appropriate integer $r > m$ and relatively prime to m has to be selected. The modular multiplication is as follows:

1. Transform x and y into the Montgomery-representation of x and y, i.e. $x' = xr \bmod m$ and $y' = yr \bmod m$.
2. Calculate the Montgomery-product $z' = MP(x', y', m, r) = x'y'r^{-1} \bmod m$.
3. Transform the result z' into the ordinary representation z, i.e. $z = z'r^{-1} \bmod m$.

Step 1. and 3. could also be accomplished using the function MP: $x' = MP(x, r^2 \bmod m, m, r)$, $y' = MP(y, r^2 \bmod m, m, r)$ und $z = MP(z', 1, m, r)$. Between step 1. and 3. an arbitrary number of modular multiplications can be performed, i.e. for exponentiation.

The Montgomery-multiplication of numbers in the new representation is isomorphic to a modular multiplication of numbers in the ordinary representation:

$$\begin{aligned} MP(x', y', m, r) &= MP(xr \bmod m, yr \bmod m, m, r) \\ &= xyr^2r^{-1} \bmod m \\ &= xyr \bmod m \\ &= (xy)' \end{aligned}$$

Montgomery's method is practical due to the existence of a simple algorithm for computing the function MP. Algorithm 2 shows a method for computing the function $MP(x, y, m, r) \equiv xyr^{-1} \pmod{m}$. The base used to represent the numbers is denoted by b ($b \geq 2$). The value l and the derived value r are determined using the maximum size of the operands x, y and m.

```
MP(x,y,m,r)              {x,y<b^(l-1); gcd(r,m)=0; r=b^l; m<b^(l-2)}
1. m'0 ← -m^-1 mod b
2. p   ← x*y
3. for i=0 to l-1 do
4. begin
5.   vi ← pi*m'0 mod b {pi is the digit of p with value b^i}
6.   p ← p+vi*m*b^i
7. end
8. return p/r
{ MP(x,y,m,r)≡x*y*r^-1(mod m)    und   MP(x,y,m,r)<2*m<b^(l-1)}
```

Algorithm 2: Montgomery multiplication

Algorithm 2 is correct because:

1. **$MP(x, y, m, r)$ is an integer because for every i from 0 to l-1 the multiple of m, which is added to p in the loop (line 3 to 7), yields a new p where the digit p_i equals 0. Thus the quotient p/r in line 8 is an integer.**

2. **$MP(x, y, m, r) \equiv xyr^{-1}$ (mod$m$) because the algorithm returns $(xy + vm)/r$.**

3. **From $0 \le xy < b^{l-1}b^{l-1} = b^{l-2}b^l < mr$ and $0 \le v < b^l = r$ follows**
   $MP(x, y, m, r) = (xy + vm)/r < (mr + rm)/r = 2m < 2b^{l-2} \le b^{l-1}$.

Algorithm 2 is similar to an algorithm described in [DusKal90]. The condition $x, y < b^{l-1}$, $r = b^l$ and $m < b^{l-2}$ for $MP(x, y, m, r) < b^{l-1}$ has been pointed out by [IwMaIm92b] for the case $b = 2$.

The maximum return value of algorithm 2 is not greater than that of each of the inputs x, y. Thus, the output can be directly used as an input for another application of algorithm 2, to perform an modulo exponentiation. Thus modular exponentiations can be performed easily using algorithm 2 under the assumption that m'_0 and r^2 mod m have been calculated in advance. These calculations have to be done only if the modulus has been changed. At the end of an exponentiation a 'normalisation' of the result to the range of [0,m-1] has to be done by a conditional subtraction of m.

4.2 Systolic Montgomery Reduction

The systolic multiplier described in section 3 can perform a Montgomery-reduction (lines 3 to 7 of algorithm 2), as follows:

If the digits p_i of the product are available serially, the corresponding digit v_i can be calculated by multiplying the digit p_i with the constant m_0' using a single-digit multiplier. The systolic multiplier can start to calculate the product $v \cdot m$ and add it to the product $x \cdot y$ (Montgomery reduction), while more product digits p_i are provided at the same time (pipelining).

Fig. 3 shows the interconnection and the input of the systolic array when performing the Montgomery reduction.

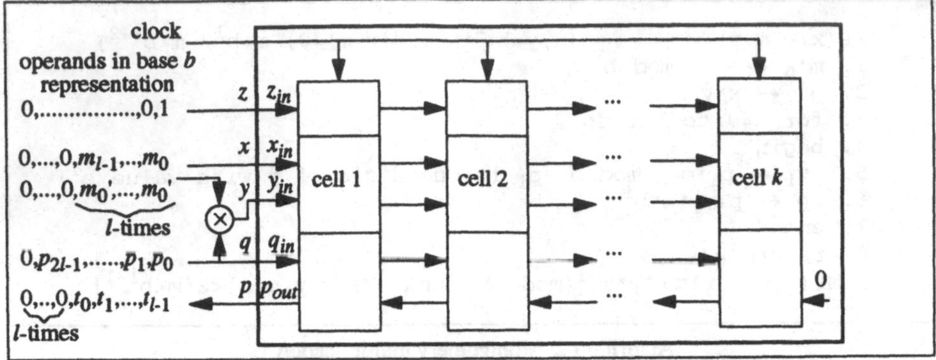

Fig. 3: Interconnection and input of the systolic array performing the Montgomery reduction

Only the l least significant digits of p has to be set to zero and therefore m_0' has to be provided just for l clock cycles. The input is then set to zero. The division p/r (line 8 in algorithm 2) is performed by just taking the l-1 most significant digits of the result t.

The systolic array suggested by [Even90] has a similar characteristic than the architecture proposed here. It works in conjunction with an Atrubin-multiplier and performs a Montgomery-reduction as well. But the array can process just one bit of the operands per clock cycle ($b = 2$) and is limited to the processing of operands of a fixed size.

5. The Modular Multiplier Based on two Systolic Arrays

Both systolic multipliers described in section 3. and 4. can be connected to build a modular multiplier capable to perform the Montgomery-multiplication using algorithm 2 (see figure 4). One systolic multiplier calculates the product $x \cdot y$ (line 2 in algorithm 2), while the second systolic multiplier performs the Montgomery reduction in a pipelined mode (loop 3-7 in algorithm 2). For this the operands for the second systolic multiplier are delayed by one clock cycle with flip-flops.

Since the modular multiplier processes ld (b) bit of the l_{bin}-digit binary operands per clock, it requires $2l_{bin}$ / ld (b) clock cycles for a complete modular multiplication using the Montgomery method. As the least significant bit of the result is already available after l_{bin} / ld (b) clock cycles, a second modular multiplier can start a pipelined modular multiplication using the result bits of the first modular multiplier. Thus using two modular multipliers we need l_{bin} / ld (b) clock cycles per modular multiplication, if several multiplications have to be performed in a sequence (exponentiation).

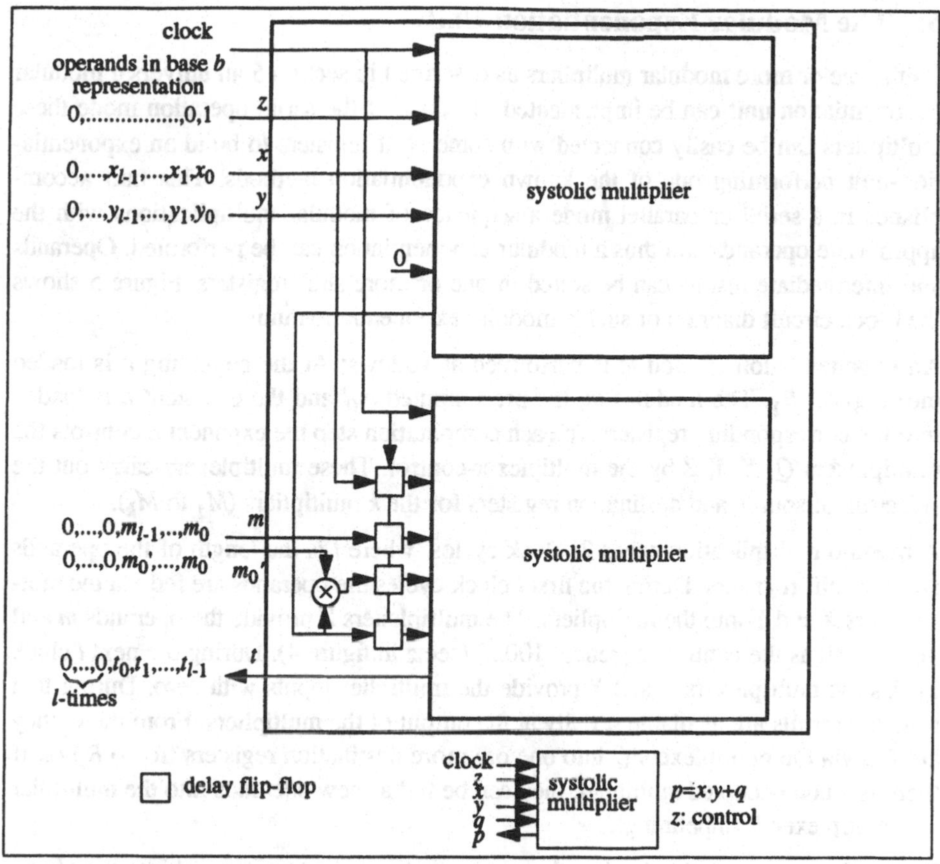

Fig. 4: Modular multiplier for the Montgomery-multiplication $t \equiv (x \cdot y / b^l)$ (mod m)

To estimate the propagation delay and the complexity (area demand) of a cell of the systolic multipliers we used the cell catalog of a 0.8µm-CMOS-Gatc-Array-Process from Siemens as a reference [SiSCE692]. The complexity and the propagation delay of a cell determines the complexity and the throughput of the proposed modular multiplier. Taking a number representation with base $b = 4$ (2 bits of the operands per clock cycle) results in a cell complexity of 174 gates and a propagation delay of 6.025ns per cell. Taking an operand size of 512 bits, a modular multiplier with $b = 4$ consists of about $174 \cdot 512 \cdot 2 \approx 178000$ gates. Due to the regularity of the architecture a full-custom design using the actual technologies for integrating two pipelined modular multipliers on a chip including the I/O- and control-logic should be feasible. Then, assuming a longer sequence of modular multiplications (i.e. for a modular exponentiation), one 512 bit modular multiplication will take $512 \cdot 6.025\text{ns} \approx 3.1\mu\text{s}$.

6. The Modular Exponentiation Unit

Using one or more modular mulipliers as described in section 5 an universal modular exponentiation unit can be implemented. Because of the serial operation mode these multipliers can be easily connected with some shift-registers to build an exponentiation-unit performing one of the known exponentiation methods. This unit accomplishes in a serial or parallel mode a sequence of modular multiplications with the appropriate operands and thus a modular exponentiation can be performed. Operands and intermediate results can be stored in one or more shift registers. Figure 5 shows the block circuit diagram of such a modular exponentiation unit.

An exponentiation c^n mod m is performed as follows: At the beginning c is loaded into register R_1. The modulus m, the precomputed m_0' and the exponent n is loaded into the corresponding registers. In each computation step the exponent n controls the multiplexers Q, X, Y, Z by the multiplexer-control. These multiplexers carry out the selection of source and destination registers for the k multipliers (M_1 to M_k).

A modulo multiplication needs $2l$ clock cycles, where l is the length of the operands and the shift registers. During the first l clock cycles the operands are fed via the multiplexers X and Y into the multipliers. The multiplexers Z provide the operands m and m_0' as well as the control sequence '100...' (see z in figure 4). During the next l clock cycles the multiplexers X and Y provide the multiplier inputs with zero. During that time the results are available serially at the output of the multipliers. From there, they are fed via the multiplexers Q into one ore more destination registers (R_1 to R_j) or, if there is a non occupied multiplier, they can be fed as new operands into the multiplier via multiplexer P (pipelining).

Before the exponentiation begins, the operand c has to be transformed into the Montgomery-representation. This can be accomplished by a multiplication with the precomputed value r^2 mod m (see section 4). This value only depends on the modulus m. At the end of the exponentiation the result has to be transformed back into the ordinary representation. This can be accomplished by a multiplication with the value 1 (see section 4).

Finaly the 'normalisation' of the result to the range of [0,m-1] is performed by a serial subtraction of m with subtractor S in a piplined mode. After the subtraction the final result is the non-negative value of the result and the result minus m.

The multiplexer-control of the unit determines, which exponentiation method is used (e.g. ([Knuth81], [ZhMaYu88, pp. 346], [BrGoMc92], [SauDie92]). The selection of a certain exponentiation method determines the specific implementation of the unit (number of required modular multipliers, registers etc.)

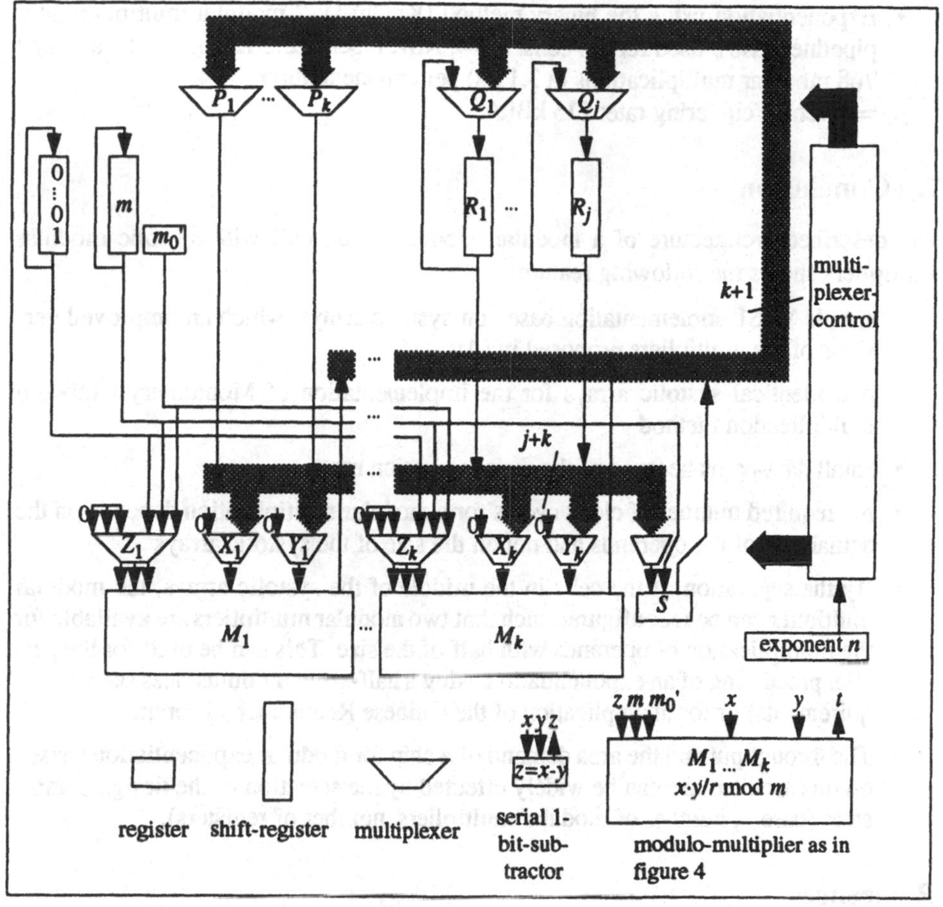

Fig. 5: block circuit diagram of the modular exponentiation unit

The throughput and area demand of a modular exponentiation chip based on this architecture can be effected by the selection of various design parameters:

- The base b used for the cells determines the number of bits per clock cycle processed by the modular multiplier.
- Instead of one modular multiplier a pair of two modular multipliers working in a pipelined mode can be used for reducing the required number of clock cycles form $2l$ to l.
- Several modular multipliers or modular multiplier pairs can be integrated on one chip to make use of possible parallel processing for the exponentiation (see [BrGoMc92].
- Several registers can be integrated to perform certain efficient exponentiation methods (see [SauDie92] and [BrGoMc92]).

The following configuration is an example for the selection of design parameters and the resulting average ciphering rate for RSA.

- Exponentiation using the binary method [Knuth81]; 2 modular multipliers in a pipeline; base b used for the cells: $b = 4$; RSA blocksize: 512 Bit; on the average 768 modular multiplications (à 3.1 µs) per exponentiation.
  ⇒ average ciphering rate: 215 kBit/s

7. Conclusion

The described architecture of a modular exponentiation unit with systolic modular multipliers shows the following features:

- simple VLSI-implementation based on systolic arrays, which are improved versions of the multipliers proposed in [Atrubi65]
- two identical systolic arrays for the implementation of Montomery's modulo multiplication method
- small data-paths because of the serial operation mode
- the required number of clock cycles for a modular multiplication depends on the actual size of the operands and not on the size of the systolic arrays
- By the separation of the cells in the middle of the systolic arrays, the modular multiplier can be reconfigured such that two modular multipliers are available for the multiplication of operands with half of the size. This can be used for the parallel processing of an exponentiation using a half-sized modulus (less security requirements) or for an application of the Chinese Remainder Theorem.
- The throughput and the area demand of a chip for modular exponentiations based on this architecture can be widely effected by the selection of the design parameters (base b, number of modular multipliers, number of registers).

References

[AliMar91] Alia, Giuseppe; Martinelli, Enrico: "A VLSI Modulo m Multiplier", IEEE Transactions on Computers, Vol. 40, No. 7, pp. 873-878, July 1991

[Atrubi65] Atrubin, A.J.: "A One-Dimensional Real-Time Iterative Multiplier", IEEE Transactions on Computers, Vol. 14, pp. 394-399, 1965

[Bricke89] Brickel, Ernest F.: "A Survey of Hardware Implementations of RSA", in Brassard, G. (Ed.): "Advances in Cryptology - Crypto '89", Proceedings (Lecture Notes in Computer Science 435), pp. 368-370, Springer, 1989

[BrGoMc92] Brickell, E.; Gordon, D.M.; McCurley, K.; et.al.: "Fast Exponentiation with Precomputation", appears in Proceedings of EUROCRYPT'92, Springer, 1992

[DusKal90] Dusse, Stephen R.; Kaliski, Burton S.: "A Cryptographic Library for the Motorola DSP56000", in Damgard, I.B. (Ed.): "Advances in Cryptology - EUROCRYPT '90", Proceedings (Lecture Notes in Computer Science 473), pp. 230-244, Springer, 1990

[Eldrid91] Eldridge, Stephen E.: "A Faster Modular Multiplication Algorithm", Intern. J. Computer Math., Vol. 40, pp. 63-68

[EldWal] Eldridge, Stephen E.; Walter, Colin D.: "Hardware Implementations of Montgomery's Modular Multiplication Algorithm", IEEE Transactions on Computers, to appear

[Even90] Even, Shimon: "Systolic Modular Multiplication", in Menezes, A.J.; Vanstone, S.A.(Eds.): "Advances in Cryptology - Crypto'90", Proceedings (Lecture Notes in Computer Science 537), pp. 619-624, Springer, 1990

[IwMaIm92a] Iwamura, K.; Matsumoto, T.; Imai, H.: "High-Speed Implementation Methods for RSA Scheme", appears in Proceedings of EURORYPT'92, 1992

[IwMaIm92b] Iwamura, K.; Matsumoto, T.; Imai, H.: "Modular Exponentiation Using Montgomery Method and the Systolic-Array", IEICE Technical Report, Vol. 92, No. 134, pp. 49-54, ISEC92-7, 1992

[Knuth81] Knuth, Donald E.: "The Art of Computer Programming, Vol. 2: Seminumerical Algorithms", Second Edition, Addison-Wesley, Reading, Massachusetts, 1981

[KocHun91] Koc, C. K.; Hung, C. Y.: "Bit-Level Systolic Arrays for Modular Multiplication", Journal of VLSI Signal Processing, Vol. 3, pp. 215-223, Kluwer Academic Publishers, Boston, 1991

[KunLei78] Kung, H. T.; Leierson, C. E.: "Systolic Arrays (for VLSI)" in Proc. Sparse Matrix Symp. SIAM, pp. 256-282, 1978

[Kung82] Kung, H. T.: "Why Systolic Architectures?", Computer, Vol. 15, No. 1, pp. 37-46, IEEE, January 1982

[LipPos90] Lippitsch, P.; Posch, K.C.; Posch, R.: "Multiplication As Parallel As Possible", Institute for Information Processing Graz, Report 290, October 1990

[Montgo85] Montgomery, P. L.: "Modular Multiplication Without Trial Division", Mathematics of Computation, Vol. 44, No. 170, pp. 519-521, April 1985

[Morita90] Morita, Hikaru: "A Fast Modular-Multiplication Module for Smart Cards", Proceedings of AUSCRYPT '90 (Lecture Notes in Computer Science 453), pp. 406-409, Springer, January 1990

[OrSvAn90] Orup, H.; Svendsen, E.; Andreasen, E.: "VICTOR - and efficient RSA hardware implementation", in Damgard, I.B. (Ed.): "Advances in Cryptology - EUROCRYPT '90", Proceedings (Lecture Notes in Computer Science 473), pp. 245-252, Springer, 1990

[SauDie92] Sauerbrey, Jörg; Dietel, Andreas: "Resource Requirements for the Application of Addition Chains in Modulo Exponentiation", appears in Proceedings of EUROCRYPT'92, Springer, 1992

[ShBeVu90] Shand, M.; Bertin, P.; Vuillemin, J.: "Hardware speedups in long integer multiplication", in Proceedings of the Second ACM Symposium on Parallel Algorithms and Architectures, Crete, July 1990

[SiSCE692] Siemens: "Semicustom ICs; CMOS Family SCxE6; Sea-of-Gates Gate Arrays", V1.0, Version May '92

[Walter] Walter, Colin D.: "Systolic Modular Multiplication", IEEE Transactions on Computers, to appear

[ZhMaYu88] Zhan, C.N.; Martin, H.L.; Yun, D.Y.: "Parallel Algorithms and Systolic Array Designs for RSA Cryptosystem", International Conference on Systolic Arrays, Proceedings, pp. 341-350, May 1988

A Comparison of Key Distribution Patterns Constructed from Circle Geometries

Christine M. O'Keefe*

Department of Pure Mathematics, The University of Adelaide,
GPO Box 498, Adelaide SA 5001, AUSTRALIA

Abstract. A key distribution pattern is a combinatorial structure which provides a secure method of distributing secret keys among a number of participants in a cryptographic network. Inversive and Laguerre planes have been used to construct key distribution patterns with storage requirements lower than the trivial distribution system. In this paper we review these and introduce key distribution patterns arising from Minkowski planes, the third of the so-called *circle geometries*. In addition, we give a comparison of the storage requirements of the key distribution patterns associated with each of the circle geometries.

1 Introduction to Key Distribution Patterns

The problem of secure generation and distribution of cryptographic keys is one of prime importance, since a cryptosystem is useless if its key has been compromised.

Suppose we have a network of $v \geq 3$ nodes, $P_1, \ldots, P_v$; such that each node must be able to communicate with any other node in a secure way. We suppose that each pair of nodes uses a symmetric (secret key) cryptosystem; so that each pair of nodes $\{P_i, P_j\}$ must possess a common, secret key K_{ij} which they use for encryption and decryption of messages sent between them.

The model that we will consider proposes a Key Distribution Centre (KDC) which is responsible for generating, distributing and storing the secret keys K_{ij}. The KDC performs these tasks once only for the life of the keys, and before any messages are sent between the nodes. The advantage of this model over the alternative of an on-line KDC which generates keys only when required is clear; the on-line KDC necessarily creates a bottleneck in the network and would probably have poor response time.

The simplest way of ensuring that each pair of nodes has a common, secret key would be for the KDC to issue a key to each pair of nodes. We call this the ***trivial*** system for key distribution, and note that the KDC needs to generate and store $v(v-1)/2$ keys while each node in the network stores $v - 1$ keys. The object is to find ways of generating and distributing secret keys so that the storage required at the KDC and at each node is lower than the storage required by the trivial system.

* The author acknowledges the support of the Australian Research Council.

In 1985, Blom [4] considered the natural solution of supplying each user with a relatively small amount of secret data from which the keys are derived. As Blom noted, attention must be paid to the fact that dependencies between keys will exist, so that a group of cooperating users might be able to decrease their uncertainty about keys to which they should not have access. Blom presented a scheme based on MDS codes and analysed the storage requirements of the system.

For other work on this problem, see also Jansen [6], Matsumoto and Imai [7] and Yung, Blundo, De Santis, Vaccaro, Herzberg and Kutten [15].

Later, in 1988, Mitchell and Piper [9] generalised the ideas of Blom and Jansen without assuming an algebraic structure on the secret data. They proposed a combinatorial model in which each node is issued with a set of ***subkeys***, and the key used by a pair of nodes $\{P_i, P_j\}$ is found as a one-way function of the set of subkeys common to P_i and P_j. In particular, we think of the set of nodes as the set $\mathcal{P}$ of ***points*** and the set of subkeys as the set $\mathcal{B}$ of ***blocks*** of an incidence structure $(\mathcal{P}, \mathcal{B}, \mathcal{I})$. In this situation, $\mathcal{I}$ is an ***incidence relation*** between points and blocks, defined so that a point is ***incident*** with a block if the corresponding node possesses the corresponding subkey. The key to be used for communication between nodes P_i and P_j is generated from the subkeys in $(P_i) \cap (P_j)$ (where (P) denotes the set of blocks incident with the point P).

Further, suppose we wish to protect the key of a pair of users from attack by a given number w of other participants acting in collusion, in the sense that P_i, P_j should have a common subkey not held by any of a group of w other participants. We make the following definition, following Mitchell and Piper [9] and identifying nodes with points and subkeys with blocks:

Let $v \geq 3$ and let w be an integer with $1 \leq w \leq v-2$. A ***w-KDP*** on v points is a finite incidence structure $\mathcal{K}$ with v points such that, for any pair of points P_i, P_j we have

$$(P_i) \cap (P_j) \not\subseteq \bigcup_{i=1}^{w} (Q_i) \qquad \text{for any points } Q_1, \ldots, Q_w \in \mathcal{P} \setminus \{P_i, P_j\} \ . \tag{1}$$

Condition 1 ensures that P_i and P_j share at least one subkey not in any of $(Q_1), \ldots, (Q_w)$.

It is desirable in an application that each of a pair of nodes be able to determine the common key ***non-interactively***, that is, with no interaction between the nodes. This is achieved by a key distribution pattern as follows. The blocks used in the key distribution pattern are really just names for the subkeys, and each node is issued with the values of only those subkeys which it posesses. Then the key distribution pattern is made public information. When a node wishes to communicate with another node it uses the public information to determine the names of the subkeys that it has in common with with that other node, then uses the (private) values of those subkeys to determine the common secret key, using some one-way function.

The trivial system for key distribution proposed above has the trivial $2-(v,2,1)$ design as its $(v-2)$-KDP. (Recall that a $t-(v,k,\lambda)$ design is an incidence structure of v points and k points per block such that each set of t points lies on a unique block, [5].) To see this, let $\mathcal{P} = \{P_1, \ldots, P_v\}$ and let $\mathcal{B} = \{x_{ij} \mid 1 \le i < j \le v\}$ be such that P_i is incident with x_{jk} if and only if either $i = j$ or $i = k$. Then $(P_i) \cap (P_j) = \{x_{ij}\}$; so this is a $(v-2)$-KDP since the $v-2$ nodes distinct from a given pair $\{P_i, P_j\}$ of nodes, in collusion, do not possess the common subkey x_{ij} of P_i, P_j.

We now turn to the question of the storage required by a w-KDP. Given the assumption that each key to be used in the network consists of n bits, the question is, how many bits must each subkey contain? For the rest of this section, we follow Quinn [12]. We define the *length* $l(x)$ of a subkey x to be the number of bits it contains, and a *length mapping* for a w-KDP $\mathcal{K}$ is a mapping $l: \mathcal{B} \mapsto N$, where $N = \{1, 2, \ldots\}$. Thus a length mapping simply assigns to each subkey x the integer $l(x)$, which is its length. We let L_s denote the constant length mapping $L_s: \mathcal{B} \mapsto \{s\}$.

Let $\mathcal{K}$ be a w-KDP, let P_i, P_j be points and let $Q_1, \ldots, Q_w \in \mathcal{P} \backslash \{P_i, P_j\}$. It would be a desirable property that, even if $Q_1, \ldots, Q_w$ pool their subkey sets, their chance of guessing the n-bit key of $\{P_i, P_j\}$ is no greater than that of someone who knows none of the subkeys in $(P_i) \cap (P_j)$ (in fact this idea is the analogue of the idea of *perfect* for secret sharing schemes, see [13]). So it would be desirable that at least n bits not contained in the subkeys in $\bigcup_{i=1}^{w}(Q_i)$ will contribute to the key of the pair $\{P_i, P_j\}$, for each pair of points P_i, P_j. Thus we make the following definition:

Let $\mathcal{K} = (\mathcal{P}, \mathcal{B}, \mathcal{I})$ be an incidence structure. Then $l: \mathcal{B} \mapsto N$ is a *w-secure length mapping for $\mathcal{K}$* if for each pair $P_i, P_j \in \mathcal{P}$ and for each set of w points $Q_1, \ldots, Q_w \in \mathcal{P} \backslash \{P_i, P_j\}$, we have

$$\sum_{x \in ((P_i) \cap (P_j)) \backslash \bigcup_{i=1}^{w}(Q_i)} l(x) \ge n \ . \tag{2}$$

In Quinn [12, 2.2.1] it is shown that L_n is a w-secure length mapping for any w-KDP. Thus, for any w-KDP, we can choose the lengths of the subkeys to be n. However in many cases it is possible to reduce the lengths of the subkeys and so reduce the storage in the network. We now make these ideas more precise.

Let $\mathcal{K}$ be a w-KDP and let l be a w-secure length mapping for $\mathcal{K}$. For each $P \in \mathcal{P}$, the *node storage* ρ_P of $(\mathcal{K}, l)$ at P is

$$\rho_P = \sum_{x \in (P)} l(x),$$

the number of bits stored at P in the w-KDP. The *average node storage* $\overline{\rho}$ of $\mathcal{K}$ is the average of the node storages. Finally, the *total node storage* β of $(\mathcal{K}, l)$ is the total number of bits in the subkeys of $\mathcal{K}$, that is,

$$\beta = \sum_{x \in \mathcal{B}} l(x).$$

Let $\mathcal{K}$ be a w-KDP, let l be a w-secure length mapping for $\mathcal{K}$, and let $\overline{\rho}$ and β be the average node storage and the total storage respectively. Then l is an *optimal* w-secure length mapping for $\mathcal{K}$ if there is no w-secure length mapping $l' \neq l$ such that either the average node storage of $(\mathcal{K}, l')$ is less than $\overline{\rho}$ or the total node storage of $(\mathcal{K}, l')$ is less than β. Thus l has the lowest average node storage and total node storage among the w-secure length mappings for $\mathcal{K}$.

In the following we sometimes consider only optimal constant w-secure length mappings for w-KDPs. As Quinn [12] points out, this is likely to be the most important w-secure length mapping in a practical situation. For example, the one-way function used to combine the subkeys to get a key might require inputs of constant length.

It is not difficult to show that L_n is an optimal w-secure length mapping for the trivial w-KDP, [12, 2.3.2]. It follows that the trivial w-KDP on v nodes has

$$\overline{\rho} = (v-1)n \quad \text{and} \quad \beta = \frac{v(v-1)n}{2}.$$

This is the standard against which any other w-KDP is compared.

2 Introduction to the Circle Geometries

In this section, we define the three circle geometries, that is, the inversive, Laguerre and Minkowski planes. For more details on the material in this section, see [5, 3, 14].

First, a *(finite) inversive plane* is an incidence structure of *points* and *circles* such that

1. any three points are incident with exactly one circle
2. if P, Q are points and C is a circle incident with P but not incident with Q then there is a unique circle incident with P and Q and having only P in common with C
3. there are four points not incident with a common circle, and each circle is incident with at least one point.

Given an inversive plane, there is an integer s such that each circle has $s+1$ points. We call s the *order* of the inversive plane, which we then denote by $\mathcal{I}(s)$. It follows that $\mathcal{I}(s)$ has s^2+1 points, $s(s^2+1)$ circles, $s(s+1)$ circles on a point and $s+1$ circles on a pair of points.

Let $\mathcal{O}$ be an ovoid in $PG(3,q)$ where $q > 2$ is a power of a prime. The set of points and secant plane sections of $\mathcal{O}$ (the secant planes are the planes which meet $\mathcal{O}$ in more than one point) are the points and circles of an inversive plane $\mathcal{I}(\mathcal{O})$ of order q. An inversive plane isomorphic to an $\mathcal{I}(\mathcal{O})$ is called *egglike*.

All known (finite) inversive planes are egglike. Further, each known inversive plane of odd order s is isomorphic to $\mathcal{I}(\mathcal{O})$ where $\mathcal{O}$ is an elliptic quadric in $PG(3,s)$. For q even, every inversive plane of even order is known to be egglike [5], however there are two classes of ovoids known and the complete classification of ovoids is not known.

Next, a *(finite) Laguerre plane* is an incidence structure of *points, lines,* and *circles* satisfying:

1. each point is on a unique line, and a line and a circle have a unique common point
2. any three points, no two collinear, lie on a unique circle
3. if P and Q are two non-collinear points and if $\mathcal{C}$ is a circle containing P but not Q then there is exactly one circle $\mathcal{C}'$ incident with P, Q and having only P in common with $\mathcal{C}$
4. there exist a point P and a circle $\mathcal{C}$ not containing P, and each circle contains at least three points.

Given a Laguerre plane, there is an integer s such that each circle has $s+1$ points. We call s the *order* of the Laguerre plane, which we then denote by $\mathcal{L}(s)$. It follows easily that a Laguerre plane $\mathcal{L}(s)$ has $s(s+1)$ points, $s+1$ mutually disjoint lines and s^3 circles. Each line has s points, there is a unique line and s^2 circles on a point and there are s circles on a pair of points.

Let $\mathcal{O}$ be an oval in a plane $PG(2,q)$ embedded in $PG(3,q)$, where q is a power of a prime, let $P \in PG(3,q)\backslash PG(2,q)$ and let T denote the cone which projects $\mathcal{O}$ from P. The set of points $\mathcal{P} = T\backslash P$, the set of lines through P and a point of $\mathcal{O}$ and the set of intersections of T with the planes in $PG(3,q)$ not through P form a Laguerre plane $\mathcal{L}(\mathcal{O})$ of order q. A Laguerre plane isomorphic to a $\mathcal{L}(\mathcal{O})$ is called *ovoidal.*

All known (finite) Laguerre planes are ovoidal, and if q is odd then every known Laguerre plane arises from a conic $\mathcal{O}$. However when q is even there are many examples of ovals, and a complete classification is not known.

Finally, a *(finite) Minkowski plane* is an incidence structure of *points, lines,* and *circles*, such that the lines fall into two classes $\mathcal{L}_1$ and $\mathcal{L}_2$ and such that:

1. each point is on a unique line of each class $\mathcal{L}_1$ and $\mathcal{L}_2$, each line of $\mathcal{L}_1$ intersects each line of $\mathcal{L}_2$ in a unique point and a line and a circle have a unique common point
2. any three points, no two of which are collinear, are incident with exactly one circle
3. if P and Q are two non-collinear points and if $\mathcal{C}$ is a circle containing P but not containing Q then there is exactly one circle $\mathcal{C}'$ containing P and Q and having only P in common with $\mathcal{C}$
4. there is a circle containing at least three points.

Given a Minkowski plane, there is an integer s such that each circle has $s+1$ points. We call s the *order* of the Minkowski plane, which we then denote by $\mathcal{M}(s)$. In a Minkowski plane $\mathcal{M}(s)$ there are $(s+1)^2$ points, $s+1$ lines in each system (hence $2(s+1)$ lines) and $s(s^2-1)$ circles. Each line has $s+1$ points, there are 2 lines and $s(s-1)$ circles on a point and $s-1$ circles on a non-collinear pair of points.

The set of points, the two families of generators and the non-tangent plane sections of a hyperbolic quadric $\mathcal{H}$ in $PG(3,q)$, for q a power of a prime, form a Minkowski plane $\mathcal{M}(\mathcal{H})$ of order q. A Minkowski plane isomorphic to some $\mathcal{M}(\mathcal{H})$ is called *classical.*

Every (finite) Minkowski plane of even order is classical but there exist non-classical Minkowski planes of odd order.

In summary, for each value of s equal to a power of a prime, there is an inversive plane of order s, a Laguerre plane of order s and a Minkowski plane of order s. It is currently an unsolved problem in finite geometry to decide whether there are any inversive, Laguerre or Minkowski planes of non-prime power order.

3 Constructions of w-KDPs from the Circle Geometries

The following construction of w-KDPs from inversive planes is implicit in the work of Mitchell [8] and appears explicitly in [12].

Construction 3.1 *[12, 3.1.7, 2.4.3] For each $1 \leq w \leq s$, an inversive plane of order s is a w-KDP $\mathcal{K}_1$ on s^2+1 points with $s(s^2+1)$ subkeys and $s(s+1)$ subkeys at each node. An optimal constant w-secure length mapping for $\mathcal{K}_1$ is $L_{\lceil n/(s-w+1)\rceil}$. This leads to storages*

$$\overline{\rho} = s(s+1)\left\lceil\frac{n}{(s-w+1)}\right\rceil \quad \textit{and} \quad \beta = s(s^2+1)\left\lceil\frac{n}{(s-w+1)}\right\rceil.$$

Mitchell and Piper [9] also pointed out some immediate geometrical consequences of their definition. Recall that if $P \in \mathcal{P}$, then the *external structure* $\mathcal{K}^P$ of $\mathcal{K}$ at P is the incidence structure with point set $\mathcal{P}\backslash\{P\}$ and block set $\{B \in \mathcal{B} \mid P \notin B\}$. It is not difficult to show (see [9]) that if $w \geq 1$ then an incidence structure $\mathcal{K}$ is a $(w+1)$-KDP if and only if $\mathcal{K}^P$ is a w-KDP for each $P \in \mathcal{P}$. Quinn applied this to the w-KDP $\mathcal{K}_1$ as follows:

Construction 3.2 *[12, 3.3.4, 2.4.3] Let P be any point of an inversive plane $\mathcal{I}(s)$ of order s. Then for each $1 \leq w \leq s-1$, it follows that $\mathcal{K}_2 = \mathcal{I}(s)^P$ is a w-KDP on s^2 points with $s^2(s-1)$ subkeys and s^2-1 subkeys at each node. The map $L_{\lceil n/(s-w)\rceil}$ is an optimal constant w-secure length mapping for $\mathcal{K}_2$, giving storages:*

$$\overline{\rho} = (s^2-1)\left\lceil\frac{n}{(s-w)}\right\rceil \quad \textit{and} \quad \beta = s^2(s-1)\left\lceil\frac{n}{(s-w)}\right\rceil.$$

The next two constructions of w-KDPs from Laguerre planes, in a different form, appeared in [12, 3.5.6, 3.5.8], but only in the classical case. The general constructions are found in [10].

Construction 3.3 *[10] Choose u collinear points $R_1, \ldots, R_u$ in $\mathcal{L}(s)$, where $2 \leq u \leq s$. Let $\mathcal{K}_3 = (\mathcal{P}, \mathcal{B}, \mathcal{I})$ be the incidence structure with points the points of $\mathcal{L}(s)$ not on the line containing the points R_i, blocks the circles of $\mathcal{L}(s)$ containing R_i for some i together with the pairs $\{P_i, P_j\}$ where P_i, P_j are collinear in $\mathcal{L}(s)$. Then for each $1 \leq w \leq u-1$, $\mathcal{K}_3$ is a w-KDP on s^2 points with $us^2+s^2(s-1)/2$ subkeys and $us+s-1$ subkeys at each node. Each pair of points in $\mathcal{K}_3$ lies on either 1 or u common blocks.*

The mapping $l: \mathcal{B} \mapsto N$ defined by:

$$l(x) = \begin{cases} \lceil n/(u-w) \rceil & \text{if } x \text{ is a circle in } \mathcal{L}(s); \\ n & \text{otherwise} \end{cases}$$

is a w-secure length mapping for $\mathcal{K}_3$ leading to storages

$$\overline{\rho} = us \left\lceil \frac{n}{u-w} \right\rceil + (s-1)n \quad \text{and} \quad \beta = us^2 \left\lceil \frac{n}{u-w} \right\rceil + \frac{s^2(s-1)n}{2}$$

Construction 3.4 *[10] Choose u collinear points $R_1, \ldots, R_u$ in $\mathcal{L}(s)$, where $2 \leq u \leq s$. Let $\pi_1, \ldots, \pi_t$ be $t \geq 2$ permutations of the set $\mathcal{P}$ of points of $\mathcal{L}(s)$ not lying on the line containing the points R_i and with the additional property that*

> *for each ordered pair (i,j) with $1 \leq i,j \leq t$ and for each pair P_1, P_2 of points in $\mathcal{P}$, if $\pi_i(P_1)$ is collinear with $\pi_i(P_2)$ then $\pi_j(P_1)$ is not collinear with $\pi_j(P_2)$.*

Let $\mathcal{K}_4 = (\mathcal{P}, \mathcal{B}, \mathcal{I})$ be the incidence structure as follows. The points are the points of $\mathcal{P}$, the blocks are the sets $\pi_i^{-1}(C_j)$ for some $i \in \{1, \ldots, t\}$ and where C_j is a circle through the point R_j for some $j \in \{1, \ldots, u\}$. (Note that the blocks might not all be distinct, as sets of points.) Then for each $1 \leq w \leq u-1$, $\mathcal{K}_4$ is a w-KDP on s^2 points with tus^2 subkeys and tus subkeys at each node. Each pair of points lies on either $(t-1)u$ or tu common blocks.

The mapping $L_{\lceil n/(t-1)(u-w) \rceil}$ is a w-secure length mapping for $\mathcal{K}_4$ leading to storages

$$\overline{\rho} = uts \left\lceil \frac{n}{(t-1)(u-w)} \right\rceil \quad \text{and} \quad \beta = uts^2 \left\lceil \frac{n}{(t-1)(u-w)} \right\rceil .$$

It is a corollary of [12, 3.4.7] that the set $\mathcal{P}$ of points of a Laguerre plane not on a fixed line certainly admits at least 2 and at most $s+1$ such permutations. Further, if the Laguerre plane is $\mathcal{L}(\mathcal{O})$ where $\mathcal{O}$ is an elliptic quadric then $\mathcal{P}$ admits $s+1$ such permutations, see [12, 3.5.7].

It is pointed out in [10] that if the w-KDP $\mathcal{K}_4$ has no repeated blocks, then it is in fact a $((t-1)u-1)$-KDP, and in general $\mathcal{K}_4$ will be a w-KDP for some maximum value w with $u-1 \leq w \leq ((t-1)u-1)$. However for simplicity we will refer to it as a $(u-1)$-KDP. This comment will also apply to $\mathcal{K}_6$ below.

These constructions for w-KDPs $\mathcal{K}_3, \mathcal{K}_4$ on Laguerre planes suggest constructions of w-KDPs on Minkowski planes. While the w-KDPs on Laguerre planes are examples of two general construction methods outlined in [12, 3.4.4, 3.4.5], the w-KDPs $\mathcal{K}_5, \mathcal{K}_6$ defined on the Minkowski planes are not. The details of these constructions are given in [11].

Construction 3.5 *[11] Let $R_1, \ldots, R_u$ be collinear points in $\mathcal{M}(s)$, where $4 \leq u \leq s+1$. Let $\mathcal{K}_5 = (\mathcal{P}, \mathcal{B}, \mathcal{I})$ be the incidence structure with points the points of $\mathcal{M}(s)$ not on the line containing the points R_i, blocks the circles of $\mathcal{M}(s)$ containing R_i for some i together with the pairs $\{P_i, P_j\}$ where P_i, P_j are collinear in*

$\mathcal{M}(s)$. Then $\mathcal{K}_5$ is a $(u-3)$-KDP on s^2+s points with $u(s^2-s)+(s^2+s)(2s-1)/2$ subkeys. There are us points with $(u-1)(s-1)+2s-1$ subkeys and s^2+s-us points with $u(s-1)+2s-1$ subkeys. Further, each pair of points in $\mathcal{K}_5$ lies on $1, u-2, u-1$ or u common blocks.

The mapping $l: \mathcal{B} \mapsto N$ defined by:

$$l(x) = \begin{cases} \lceil n/(u-w-2) \rceil & \text{if } x \text{ is a circle in } \mathcal{M}(s); \\ n & \text{otherwise} \end{cases}$$

is a w-secure length mapping for $\mathcal{K}_5$ leading to storages

$$\overline{\rho} = \frac{u(u-1)s(s-1)+(s^2+s-us)u(s-1)}{s^2+s} \left\lceil \frac{n}{u-w-2} \right\rceil + (2s-1)n$$

$$\beta = u(s^2-s) \left\lceil \frac{n}{u-w} \right\rceil + (s^2+s)(2s-1)n/2.$$

Construction 3.6 *[11] Let l_0 and m_0 be two lines in a Minkowski plane $\mathcal{M}(s)$, each belonging to a different class of lines. Choose u points $R_1, \ldots, R_u$ where $6 \le u \le 2s+1$, lying on $l_0 \cup m_0$. Let $\pi_1, \ldots, \pi_t$ be $t \ge 2$ permutations of the set $\mathcal{P}$ of points of $\mathcal{M}(s)$ not in $l_0 \cup m_0$ and with the additional property that*

> *for each ordered pair (i,j) with $1 \le i,j \le t$ and for each pair P_1, P_2 of points in $\mathcal{P}$, if $\pi_i(P_1)$ is collinear with $\pi_i(P_2)$ then $\pi_j(P_1)$ is not collinear with $\pi_j(P_2)$.*

Let $\mathcal{K}_6 = (\mathcal{P}, \mathcal{B}, \mathcal{I})$ be the incidence structure as follows. The points are the points of $\mathcal{P}$ and the blocks are all the sets $\pi_i^{-1}(C_j)$ for some $i \in \{1, \ldots, t\}$ and where C_j is a circle through the point R_j for some $j \in \{1, \ldots, u\}$. (Note that the blocks may not be distinct.) Then for each $1 \le w \le u-5$, $\mathcal{K}_6$ is a w-KDP on s^2 points with $tu(s^2-s)$ subkeys, and each pair of points lies on at least $(t-1)(u-4)$ and at most tu common blocks. For $k = 1,2$ let τ_k be the number of points in $\mathcal{P}$ collinear with exactly k of the points $R_1, \ldots, R_u$. Then for $k = 1,2$, τ_k nodes have exactly $t(u-k)(s-1)$ subkeys and there are $s^2-\tau_1-\tau_2$ nodes with $tu(s-1)$ subkeys.

The map $L_{\lceil n/(t-1)(u-w-4) \rceil}$ is a w-secure length mapping for $\mathcal{K}_6$, leading to storages

$$\overline{\rho} = \frac{t(s-1)(us^2-\tau_1-2\tau_2)}{s^2} \left\lceil \frac{n}{(t-1)(u-w-4)} \right\rceil$$

$$\beta = tu(s^2-s) \left\lceil \frac{n}{(t-1)(u-w-4)} \right\rceil.$$

In [11], it is shown that a classical Minkowski plane $\mathcal{M}(s)$, with $s \ge 3$, admits at least $t = \lfloor (s+1)/2 \rfloor$ such permutations. Further, it is a corollary of [12, 3.4.7] that $\mathcal{M}(s)$ admits at most $s+1$ such permutations. Section 4 addresses the question of which value of t should be chosen in a particular example.

4 Comparison of the w-KDPs Arising from the Circle Geometries

We now compare the total and node storages of the six w-KDPs constructed from the circle geometries, s described in the last section.

The first rows of Tables 1 and 2 (above the bar) give the restrictions on the parameters of the w-KDPs and the rows below the bar give properties of the w-KDPs. We have included the relevant parameters in the name of each w-KDP, for future reference.

Table 1.

	$\mathcal{K}_1(s)$	$\mathcal{K}_2(s)$	$\mathcal{K}_3(s,u)$	$\mathcal{K}_4(s,u,t)$
u			$2 \leq u \leq s$	$2 \leq u \leq s$
t				$2 \leq t \leq s+1$
w	s	$s-1$	$u-1$	$u-1$
v	s^2+1	s^2	s^2	s^2
$\overline{\rho}$	$(s^2+s)\left\lceil\frac{n}{(s-w+1)}\right\rceil$	$(s^2-1)\left\lceil\frac{n}{(s-w)}\right\rceil$	$us\left\lceil\frac{n}{(u-w)}\right\rceil$ $+(s-1)n$	$tus\left\lceil\frac{n}{(t-1)(u-w)}\right\rceil$
β	$(s^3+s)\left\lceil\frac{n}{(s-w+1)}\right\rceil$	$(s^3-s)\left\lceil\frac{n}{(s-w)}\right\rceil$	$us^2\left\lceil\frac{n}{(u-w)}\right\rceil$ $+\frac{(s^3-s)n}{2}$	$tus^2\left\lceil\frac{n}{(t-1)(u-w)}\right\rceil$

Table 2.

	$\mathcal{K}_5(s,u)$	$\mathcal{K}_6(s,u,t)$
u	$4 \leq u \leq s+1$	$6 \leq u \leq 2s+1$
t		$2 \leq t \leq (s+1)/2$
w	$u-3$	$u-5$
v	s^2+s	s^2
$\overline{\rho}$	$\frac{u(u-1)s(s-1)+(s^2+s-us)u(s-1)}{s^2+s}\left\lceil\frac{n}{u-w-2}\right\rceil$ $+(2s-1)n$	$\frac{t(s-1)(us^2-\tau_1-2\tau_2)}{s^2}\left\lceil\frac{n}{(t-1)(u-w-4)}\right\rceil$
β	$u(s^2-s)\left\lceil\frac{n}{u-w}\right\rceil$ $+(s^2+s)(2s-1)n/2$	$tu(s^2-s)\left\lceil\frac{n}{(t-1)(u-w-4)}\right\rceil$

To compare, recall that the trivial KDP on s^2 points will have $w = s-2$ and storages

$$\overline{\rho} = (s^2-1)n \quad \text{and} \quad \beta = s^2(s^2-1)n/2.$$

If small values of w are acceptable, we can achieve a much lower average node storage and total storage using any of the geometrical w-KDPs $\mathcal{K}_1, \ldots, \mathcal{K}_6$. In any case, the w-KDPs $\mathcal{K}_3, \ldots, \mathcal{K}_6$ have average node storage and total storage only of the order of s which improves on the trivial w-KDP and $\mathcal{K}_1, \mathcal{K}_2$ with average node storage of order s^2. Also, the w-KDPs $\mathcal{K}_4$ and $\mathcal{K}_6$ have total storage of order s^2 while the trivial w-KDP has total storage of order s^4 and $\mathcal{K}_1, \mathcal{K}_2, \mathcal{K}_3, \mathcal{K}_5$ have total storage of order s^3.

Of course a detailed analysis in each case is required to determine which is the best w-KDP to use in each particular situation, since the storages can be reduced in general as a function of the security w and the parameters u and t.

In addition, the following should be taken into account in each particular situation. If our network has v nodes, then we can provide a secure key distribution pattern by using a w-KDP on at least v nodes. This might be necessary in some situations, as good w-KDPs are not known to exist for every possible v. As a penalty, the total and node storages generally increase.

To be more specific, suppose we wish to construct a w-KDP on v points using one of $\mathcal{K}_1, \ldots, \mathcal{K}_6$. Since the underlying geometries are known to exist only for prime power orders, we construct the w-KDP on a prime power number of nodes, at least as large as v. If there is a prime power value of s such that $r^2 + r < v \leq s^2$, where r is the next prime power smaller than s, then we would construct a w-KDP using $\mathcal{K}_1(s), \mathcal{K}_2(s), \mathcal{K}_3(s,u), \mathcal{K}_4(s,u,t), \mathcal{K}_5(s,u)$ or $\mathcal{K}_6(s,u,t)$ and the average node and total storages can be read from the table. However, suppose that $v = s^2 + 1$, for a prime power s. Then we would use a w-KDP either $\mathcal{K}_1(s), \mathcal{K}_2(r), \mathcal{K}_3(r,u), \mathcal{K}_4(r,u,t), \mathcal{K}_5(s,u)$ or $\mathcal{K}_6(r,u,t)$ where r is the next prime power greater than s. A new table would need to be drawn up to compare node and total storages in this case. Finally, to construct a w-KDP on $s^2 + 2 \leq v \leq s^2 + s$ points, s a power of a prime, we would use either $\mathcal{K}_1(r), \mathcal{K}_2(r), \mathcal{K}_3(r,u), \mathcal{K}_4(r,u,t), \mathcal{K}_5(s,u)$ or $\mathcal{K}_6(r,u,t)$, where r is the next prime power greater than s. Again, some further analysis would be required to identify the best w-KDP in this situation. The main thing to note is that $\mathcal{K}_5$ can have an advantage over the other w-KDPs on certain numbers of nodes.

References

1. W. Benz: Über die Grundlagen der Geometrie der Kreise in der pseudoeuklidischen (Minkowskischen) Geometrie. J. Reine Angew. Math. **232** (1968) 41–76
2. T. Beth, D. Jungnickel, Lenz: Design Theory. Cambridge University Press, Cambridge, 1986
3. A. Beutelspacher: Einfuhrung in die endliche Geometrie II: Projective Raume. Bibliographisches Institut. Mannheim 1983
4. R. Blom: An optimal class of symmetric key generation systems. Advances in Cryptology: Proceedings of Eurocrypt '84, Springer-Verlag, Berlin. Lecture Notes in Computer Science **209** (1985) 335–338
5. P. Dembowski: Finite Geometries. Springer-Verlag, Berlin Heidelberg, 1968

6. C.J.A. Jansen: On the key storage requirements for secure terminals. Comput. Security **5** (1986) 145–149
7. T. Matsumoto and H. Imai: On the KEY PREDISTRIBUTION SYSTEM: A practical solution to the Key Distribution Problem. Advances in Cryptology: Proceedings of Crypto '87, Springer-Verlag, Berlin. Lecture Notes in Computer Science **293** (1987) 185–193
8. C.J. Mitchell: Combinatorial techniques for key storage reduction in secure networks. Technical memo, Hewlett-Packard Laboratories, Bristol, 1988
9. C.J. Mitchell and F.C. Piper: Key storage in secure networks. Discrete Applied Mathematics **21** (1988) 215–228
10. C.M. O'Keefe: Applications of finite geometries to information security. submitted
11. C.M. O'Keefe: Geometrical constructions for key distribution patterns. in preparation
12. K.A.S. Quinn: Combinatorial Structures with Applications to Information Theory PhD Thesis. RHBNC, University of London, 1991
13. G.J. Simmons: Contemporary Cryptology. ed., IEEE Press, New York, 1992
14. J.A. Thas: Circle geometries and generalized quadrangles. Finite Geometries, Dekker, New York, 1985, 327–352
15. M. Yung, C. Blundo, A. De Santis, U. Vaccaro, X. Herzberg, X. Kutten: Perfectly secure key distribution for dynamic conferences. Advances in Cryptology: Proceedings of Crypto '92, to appear

Rump Session

Chair: Josef Pieprzyk
(University of Wollongong, Australia)

A Block Cipher Method using Combinations of Different Methods under the Control of the User Key

Mike REZNY and Eddie TRIMARCHI

Queensland University of Technology

Abstract. In this paper, we describe a 64-bit multi-round block cipher, suitable for software implementation, in which three different encryption methods are combined in a sequence determined by the user key. In this way, whilst the design is public knowledge, the actual encryption method selected by the user key is kept secret. This method has been implemented using the three block ciphers: Khufu, Loki, and a cipher by Lai and Massey. The performance and cryptanalysis results using the CRYPT-XB package for this example are provided.

1 Introduction

All existing multi-round block ciphers, to our knowledge, use the same encryption method for each round. These ciphers can be implemented more economically in hardware by designing a single round which is reused as many times as necessary, thereby significantly reducing the component count for the chip.

As this cipher was to be implemented entirely in software, we wanted to exploit to the fullest any advantages that this may offer over a hardware implementation. At the same time, we wanted to avoid operations, such as permutations, which are more efficiently implemented in hardware.

It was our intention to design a block cipher that was not merely another multi-round cipher with a novel encryption round. A plethora of these ciphers already exists. We were able to contemplate a cipher using different encryption methods in each round. This would enable combinations of methods to be chosen that could complement each other so as to reduce the weaknesses of any one method on its own.

It is possible to envisage an encryption method whereby a group of users selects a combination of rounds for their private use. A cryptanalyst would then have to determine the combination of cipher rounds used before an attack could be made on the cipher itself. As better encryption rounds are developed they could easily be incorporated into this system.

The user key could be extended to include information that would determine the combination of rounds used in the cipher. This system would then be suitable for use as an encryption standard. This is the basis for the block cipher described in this paper.

2 Description of Method

Our block cipher uses three existing encryption methods: LOKI [1], a method by Lai and Massey [2], referred to from now on as LM, and Khufu [3]. These methods were chosen as they were all efficiently implemented in existing software and each used significantly different encryption methods. This last feature is an important consideration in attempting to ensure that the three methods together did not suffer from weaknesses that could enable easier cryptanalysis.

One of the design considerations was that the cipher execution time should be independent of the combination of rounds chosen. Otherwise, it may be possible to determine the rounds used, based on an analysis of the execution times of the cipher. This has been achieved in this cipher by keeping the number of rounds used of each method constant. The user key is used to determine their order.

We define a cipher set to consist of 8 rounds of Khufu, 2 rounds of LOKI and 1 round of LM. The cipher consists of two cipher sets, as illustrated in figure 1 and thus consists of 16 rounds of Khufu, 4 rounds of LOKI and 2 rounds of LM. The user key determines the order in which the LOKI and LM rounds are interspersed with the Khufu rounds.

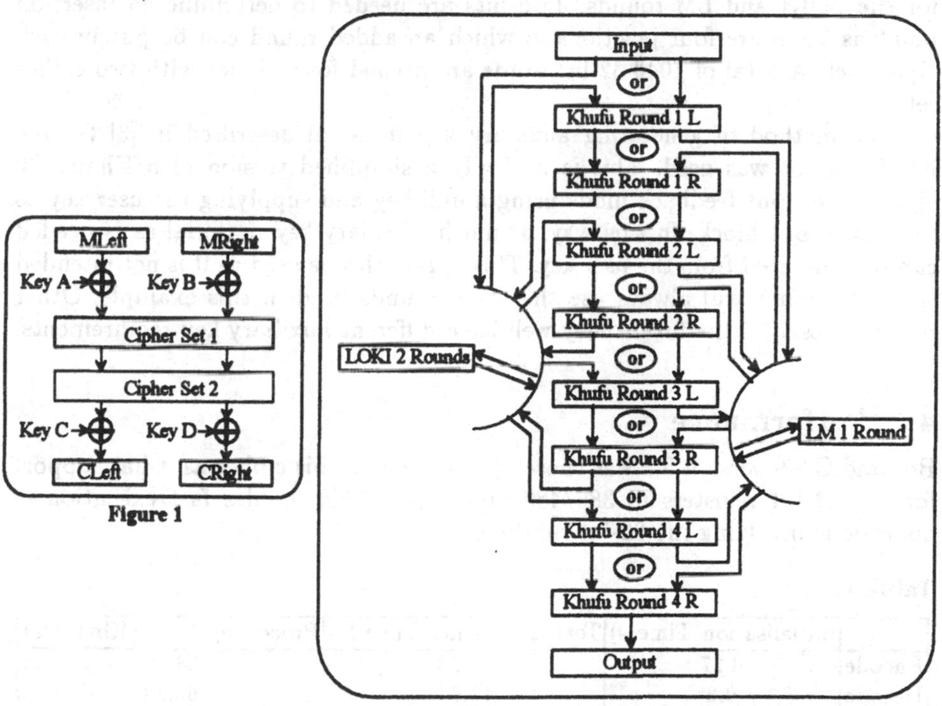

Figure 1

Figure 2

The auxiliary key is used to determine whether the two rounds of LOKI are inserted before the 1st, 3rd, 5th or 7th round of Khufu in each cipher set. In a similar manner, the auxiliary key is used to determine whether the one round of LM is inserted before the 2nd, 4th, 6th or 8th round of Khufu. This gives a total of 16 different permutations of these three methods per cipher set. Figure

2 shows the block diagram of the basic structure for one cipher set.

It should be emphasised that the LOKI block shown above is used only once in each cipher set. The position in which it is used is determined by the auxiliary key. A similar condition also applies to the LM block.

A LOKI round uses information from one sub-block as input to a function whose output is then XORed with the other sub-block. It is necessary to perform two rounds of LOKI for both sub-blocks to be affected. However, one round of LM affects both sub-blocks symmetrically.

3 Auxiliary Key Generation

We chose to set the size of the user supplied key to be the same size as the blocks used in the cipher, ie. 64 bits and generate as much auxiliary key material as needed from this key.

In the main block, shown in figure 1, 64 bits of key material are needed for the XORs, Khufu requires 1024 32-bit words to generate two sets of S boxes, the LOKI method requires 64 bits per cipher set, the LM method requires 160 bits and another 32 bits of key material are needed in determining the insertion points for the LOKI and LM rounds. Two bits are needed to determine an insertion point as there are four positions in which an added round can be put in each cipher set. A total of 1043 32-bit words are needed for a cipher with two cipher sets.

The method of generating auxiliary key material described in [3] for the Khufu cipher was used. This is basically a simplified version of a Khafre [3] cipher in output feedback mode using a null key and supplying the user key as the first input block. In this way, as much auxiliary key material as is needed can be generated from the user key. This approach was used as it is not intended that the cipher will always use the three rounds used in this example. Other rounds chosen in the future may well have different auxiliary key requirements.

4 Performance

Borland C++ version 3.1 was chosen to implement this cipher as it has support for the 32 bit registers in 386/486 processors. This enables fast execution of operations involving the 32-bit sub-blocks.

Table 1.

	Initialisation Time(s)	Total Execution Time (s)	Processing Rate (Kbytes/s)
Encode	0.27	12.96	54.01
Decode	0.2	11.61	60.29

The following encryption and decryption times were recorded on 486DX 33 Mhz IBM compatible PC. These figures represent the averages of three runs to encode and decode a 700 Kbyte file with all disc caching software disabled to ensure the timings were not favourably biased. The computation time to generate the auxiliary key and initialise the S boxes is independent of the file size. The time to encode this file, with a software version of DES, was 520 seconds, giving a processing rate of 1.35 Kbytes/s.

The cipher was tested using the CRYPT-XB [4] software package. The results, listed in Table 2, show that the cipher has no weaknesses based on the results of these tests. In particular, the cipher does not have the complementation property.

Table 2.

Applied test	Result	ChiSqr	α
Affine	Failed	n/a	n/a
Complementation	Failed	n/a	n/a
Plaintext Avalanche (1000)	Passed	$D^* = 1.397$	n/a
Frequency (Patterned)	Passed	2.081	0.8938
Binary Derivative (Patterned)	Passed	28.49	0.4917
Linear Complexity (Patterned)	Passed	8.21	0.8296
Frequency (Random)	Passed	25.08	0.5143
Binary Derivative (Random)	Passes	37.09	0.0566
Linear Complexity (Random)	Passed	15.21	0.1247

5 Conclusions

The method described has been efficiently implemented and encodes about 40 times faster than a software version of DES. It promises extra security by firstly, forcing an attacker to find the combination of methods used before an attack on the cipher itself can be mounted, and secondly, by choosing methods that complement each other. The first property will only be of advantage if no easy way of determining the combinations used can be found. The second property has a potential problem in that methods may be chosen which, instead of enhancing security, may actually provide reduced security. Further research is needed in both of these areas.

The cipher can be easily extended to a stream cipher. The auxiliary key generator supplies extra bits for each encoded block which are used to determine the insertion points for the extra rounds. In this way the actual cipher method changes for each block.

References

1. Brown, L., Pieprzyk, J. and Seberry, J.: LOKI – a cryptographic primitive for authentication and secrecy applications. Advances in Cryptology, AUSCRYPT 90, Lecture Notes in Computer Science, **453**, J. Seberry and J. Pieprzyk eds., Springer-Verlag, (1990) 229–236
2. Lai, X. and Massey, J.: A proposal for a new block encryption standard. Advances in Cryptology, EOROCRYPT 90, Lecture Notes in Computer Science, **473**, I.B. Damgård ed., Springer-Verlag, (1990) 389–404
3. Merkle, R.: Fast software encryption functions. Advances in Cryptolgy, CRYPTO 90, Lecture Notes in Computer Science, **537**, A.J. Menezes and S.A. Vanstone eds., Springer-Verlag, (1991) 476–501
4. Caelli, W., Dawson, E., Gustafson, H. and Nielsen, L.: CRYPT-XB Statistical Package Manual for Block Ciphers. Queensland University of Technology, (1992)

An Attack on Two Hash Functions by Zheng-Matsumoto-Imai

Bart Preneel*, René Govaerts, and Joos Vandewalle

Katholieke Universiteit Leuven, Laboratorium ESAT-COSIC,
Kardinaal Mercierlaan 94, B-3001 Heverlee, Belgium

Abstract. In [ZMI89, ZMI90] two constructions for a collision resistant hash function were proposed. The first scheme is based on a block cipher, and the second scheme uses modular arithmetic. It is shown in this paper that both proposals have serious weaknesses.

1 Introduction

For an informal definition of a collision resistant hash function the reader is referred to [PGV92]. The following model will be used to described iterated hash functions:

$$H_i = f(X_i, H_{i-1}) \quad i = 1, 2, \ldots t\,.$$

Here f is the round function, X_i are the t message blocks, H_i are the chaining variables, H_0 is equal to the initial value, that should be specified together with the scheme, and H_t is the hashcode. It was shown by I. Damgård [Dam89] that if the round function f is a collision resistant function, h is a collision resistant hash function. The authors of [ZMI89, ZMI90] claim that their constructions yield a collision resistant round function. It will be demonstrated that in both cases the round function is not collision resistant, and that in some cases collisions for h can be constructed.

2 The Hash Function Based on a Block Cipher

The round function f compresses a 224-bit input to a 128-bit output and is based on xDES^1. This block cipher is one of the extensions of DES [Fi46] that has been proposed in [ZMI89b]. xDES^1 is a three round Feistel cipher with block length 128 bits, key size 168 bits and with the F function equal to DES. One round is defined as follows:

$$C1_{i+1} = C2_i \quad \text{and} \quad C2_{i+1} = C1_i \oplus \text{DES}(K_i, C2_i) \quad i = 0, 1, 2.$$

The variables $C1_i$ and $C2_i$ are 64-bit blocks, and K_i are 56-bit keys. The block cipher is then written as

$$C2_3 \,\|\, C1_3 = \text{xDES}^1(K_1\|K_2\|K_3, C1_0\|C2_0)\,.$$

* NFWO aspirant navorser, sponsored by the National Fund for Scientific Research (Belgium).

Here $C1_0$ and $C2_0$ are the first and second part of the plaintext, and $C2_3$ and $C1_3$ are the first and second part of the ciphertext. The collision resistant function consists of 2 xDES^1 operations:

$$f(Y1\|Y2) = \text{xDES}^1\left(\text{chop}_{72}\left(\text{xDES}^1(\beta\|Y1,\alpha)\right)\|Y2,\alpha\right).$$

Here $Y1$ and $Y2$ are 112-bit blocks, α is a 128-bit constant, β is a 56-bit initialization variable and chop$_r$ drops the r least significant (or rightmost) bits of its argument. The complete hash function has the following form: $H_i = f(H_{i-1}\|X_i)$, where H_{i-1} is a 128-bit block, and X_i is a 96-bit block. The rate of this scheme is equal to 4, which means that 4 DES encryptions are required to hash 64 bits.

The scheme has two weaknesses, that allow to produce collisions for the round function f. First only 56 bits are kept from the first xDES^1 encryption, and hence a birthday attack will require only 2^{29} operations to produce a collision for the intermediate value and hence for the function f. The second problem is that if $\beta = K_1$ and $Y_1 = K_2\|K_3$, one can use the key collision search algorithm described in [QD89] to produce key collisions for the DES plaintext equal to the second part of α. This yields a collision for f in about 2^{33} operations.

The scheme can be strengthened however by distributing β equally over K_1, K_2, and K_3, and by increasing the size of β [Zhe92]. It will be shown that independently of the size of β, the security level can not be larger than 44 bits. If the size of β is equal to v bits (in the original proposal $v = 56$), the number of fixed bits of β that enter the key port of a single DES block is equal to $v/3$ (it will be assumed that v is divisible by 3). It can be shown that the rate of this scheme is then equal to $R = \frac{6\cdot 64}{208-2v}$. The number of bits of Y_1 that enter the key port will be denoted with y, hence $y + v/3 = 56$. Two attacks are now considered.

For the fixed value of the right part of α and of the first $v/3$ bits of β, one can calculate and store a set of 2^z different ciphertexts. The probability that a collision will be found in this set is approximately equal to 2^{2z-65}. If $y > 32$, implying $v < 72$, a value of $z = 33$ is clearly sufficient to obtain a collision. If on the other hand $y \leq 32$, one will take $z = y$, and the probability of success is smaller than one. One can however repeat this procedure, (e.g., if one attacks a DES block different from the first one, a different value can be chosen for the value of the bits of Y_1 that enter the first DES), and the expected number of operations for a single collision is equal to 2^{65-y}, while the required storage is equal to 2^y. An extension of the Quisquater algorithm [QD89] could be used to eliminate the storage. If the security level S is expressed in bits, it follows that $S = \max\{65 - y, 33\}$. With the relation between y and v, one obtains $S = \max\{9 + v/3, 33\}$.

A second attack follows from the observation that only v bits are kept from the output of the first xDES^1 operation (hence the chop operation is chopping $128 - v$ bits). It is clear that finding a collision for the remaining v bits requires only $2^{v/2+1}$ operations, or $S \leq v/2 + 1$ bits. This attack is more efficient than the first attack if $v < 64$ bits.

The relation between S and v can be summarized as follows: if $v < 64$ then $S = v/2 + 1$, if $64 \leq v < 72$ then S=33, and if $72 \leq v < 104$ then $S = v/3 + 9$.

One can conclude that producing a collision for the proposed round function requires less than 2^{44} operations. Depending on the allocation of the bits of X_i and H_{i-1} to Y_1 and Y_2, it might also be feasible to produce a collision for the hash function with a fixed initial value: it is certainly possible to produce a collision for the hash function if there is a single DES block where all key bits are selected from X_i.

3 The Hash Function Based on Modular Arithmetic

In this case the round function f consists of 2 modular squarings with an n-bit modulus (with $n = 500$):

$$f(Y1\|Y2) = \left(chop'_{450}\left((\beta\|Y1)^2 \bmod N\right)\|Y2\right)^2 \bmod N\,,$$

where $\text{chop}'_r(x)$ drops the r most significant bits of x, $Y1$ and $Y2$ are 450-bit blocks, and β is a 50-bit initialization variable. The complete hash function has the following form: $H_i = f(H_{i-1}\|X_i)$, where H_{i-1} is a 500-bit block, and X_i is a 400-bit block. The security of this scheme is based on the fact that $O(\log N)$ bits of squaring modulo N is hard if N is a Blum integer, i.e., $N = pq$ with $p \equiv q \equiv 3 \bmod 4$. From this it is wrongly concluded that finding two integers such that their squares agree at the 50 least significant positions is hard (a trivial collision for x is $x' = -x$). As only 50 bits of the first squaring are used as input to the second squaring, it follows that collisions can be found with a birthday attack in 2^{26} operations. It can be shown that one can find a second preimage and hence a collision for f even if $k = n/4$ bits are selected, or $3n/4$ bits are chopped. The algorithm is the same as the one presented in [Gir87] to break a related scheme with redundancy in the least significant positions.

References

[Dam89] I.B. Damgård, "A design principle for hash functions," *Advances in Cryptology, Proc. Crypto'89, LNCS 435*, G. Brassard, Ed., Springer-Verlag, 1990, pp. 416–427.

[Fi46] *"Data Encryption Standard,"* Federal Information Processing Standard (FIPS), Publication 46, National Bureau of Standards, U.S. Department of Commerce, Washington D.C., January 1977.

[Gir87] M. Girault, "Hash-functions using modulo-n operations," *Advances in Cryptology, Proc. Eurocrypt'87, LNCS 304*, D. Chaum and W.L. Price, Eds., Springer-Verlag, 1988, pp. 217–226.

[PGV92] B. Preneel, R. Govaerts, and J. Vandewalle, "On the power of memory in the design of collision resistant hash functions," these proceedings.

[QD89] J.-J. Quisquater and J.-P. Delescaille, "How easy is collision search ? Application to DES," *Advances in Cryptology, Proc. Eurocrypt'89, LNCS 434*, J.-J. Quisquater and J. Vandewalle, Eds., Springer-Verlag, 1990, pp. 429–434.

[ZMI89] Y. Zheng, T. Matsumoto, and H. Imai, "Duality between two cryptographic primitives," *Papers of technical group for information security, IEICE of Japan*, March 16, 1989, pp. 47–57.

[ZMI89b] Y. Zheng, T. Matsumoto, and H. Imai, "On the construction of block ciphers provably secure and not relying on any unproved hypothesis," *Advances in Cryptology, Proc. Crypto'89, LNCS 435*, G. Brassard, Ed., Springer-Verlag, 1990, pp. 461–480.

[ZMI90] Y. Zheng, T. Matsumoto, and H. Imai, "Duality between two cryptographic primitives," *Proc. 8th International Conference on Applied Algebra, Algebraic Algorithms and Error-Correcting Codes, LNCS 508*, S. Sakata, Ed., Springer-Verlag, 1991, pp. 379–390.

[Zhe92] Y. Zheng, personal communication, 1992.

Primality Testing with Lucas Functions *

Rudolf Lidl[1] and Winfried B. Müller[2]

[1] Department of Mathematics, University of Tasmania, Hobart, Tas. 7001, Australia
[2] Institut für Mathematik, Universität Klagenfurt, A-9020 Klagenfurt, Austria

Abstract. A generalization of Fermat's Little Theorem is derived by using Lucas functions. This generalization yields new classes of pseudoprimes and can be used to improve some well-known primality tests.

1 Dickson Pseudoprimes

Fermat's Little Theorem plays an important role in many primality tests and in motivating the concept of pseudoprimes. A version of this theorem states that for any prime n and an arbitrary integer b there holds

$$b^n \equiv b \bmod n. \tag{1}$$

If an integer n is not prime it is still possible but not very likely that (1) holds. Odd composite numbers n which satisfy (1) for every $b \in \mathbb{Z}$ are called *Carmichael numbers*. The smallest Carmichael number is $n = 561 = 3 \cdot 11 \cdot 17$. Recently Alford, Granville and Pomerance [1] proved that there are infinitely many Carmichael numbers. As a consequence infinitely many odd composite integers cannot be disclosed as non-prime by Fermat's Little Theorem.

Now, let b, c be integers whereby $b > 0$. If α, β denote the roots of the polynomial $x^2 - bx + c$ then the *Lucas function* $V_n(b,c)$ is defined by

$$V_n(b,c) := \alpha^n + \beta^n \quad \text{for} \quad n \geq 0.$$

From an algebraic point of view the Lucas function $V_n(b,c)$ can also be written as a polynomial over $\mathbb{Z}$ in the indeterminate b. By using *Waring's formula* (cf. [7]) we obtain

$$V_n(b,c) = \alpha^n + \beta^n = \sum_{i=0}^{[n/2]} \frac{n}{n-i} \binom{n-i}{i} (-c)^i \, b^{n-2i} =: g_n(b,c), \tag{2}$$

where $[n/2]$ denotes the greatest integer $\leq \frac{n}{2}$. The polynomial $g_n(x,c)$ is called the *Dickson polynomial of parameter c and degree n*.

From (2) it can be verified immediately that for any prime n and an arbitrary $b \in \mathbb{Z}$ there holds

$$V_n(b,c) = g_n(b,c) \equiv b \bmod n. \tag{3}$$

* This work was supported by the University of Tasmania and the Forschungskommission of the University of Klagenfurt.

Odd composite integers n with $V_n(1,-1) \equiv 1 \bmod n$ are called *Lucas pseudoprimes* [11] or *Fibonacci pseudoprimes* [3]. Filipponi [4] verified that there exist 852 Fibonacci pseudoprimes $n \leq 10^8$.

Composite odd integers n analogous to Carmichael numbers, i.e. numbers n such that congruence (3) is satisfied for a fixed c and every $b \in \mathbb{Z}$, are called *strong Dickson pseudoprimes of the kind c* (in short: strong c-Dickson pseudoprimes).

Obviously, the strong 0-Dickson pseudoprimes are exactly the Carmichael numbers. The strong (-1)-Dickson pseudoprimes are called *strong Fibonacci pseudoprimes* (cf. [6]).

In [8] strong Fibonacci pseudoprimes were characterized as odd composite integers n with

(i) *n is a Carmichael number,*
(ii) $2(p_i+1)|(n-1)$ *or* $2(p_i+1)|(n-p_i)$ *for every prime p_i dividing n.*

Numerical tests on odd composite integers n up to 10^{100} suggest that in contrast to Carmichael numbers for certain parameters $c \neq 0$ strong c-Dickson pseudoprimes seem to be very scarce. Recently, PINCH [9] used the above characterization to identify the smallest strong Fibonacci pseudoprimes

$$\begin{aligned} n &= 443372888629441 \\ &= 17 \cdot 31 \cdot 41 \cdot 43 \cdot 89 \cdot 97 \cdot 167 \cdot 331. \end{aligned}$$

So far we know of only one other strong Fibonacci pseudoprime, namely

$$\begin{aligned} n &= 5893983289990395334700037072001 \\ &= 29 \cdot 31 \cdot 37 \cdot 43 \cdot 53 \cdot 67 \cdot 79 \cdot 89 \cdot 97 \cdot 151 \cdot 181 \cdot 191 \cdot 419 \cdot 881 \cdot 883 \end{aligned}$$

found by MORAIN AND GUILAUME and communicated by PINCH.

A reformulation of a result by KOWOL [5] gives the following characterization of strong c-Dickson pseudoprimes:

An odd composite number n is a strong c-Dickson pseudoprime ($gcd(c,n)=1$) if and only if n is square-free and for every prime p_i dividing n

(i) $(p_i-1)|(n-1)$ *for* $c \not\equiv 1 \bmod p_i$
$(p_i-1)|(n-1)$ *or* $(p_i-1)|(n+1)$ *for* $c \equiv 1 \bmod p_i$,

(ii) $e_i(p_i+1)|(n-1)$ *or* $e_i(p_i+1)|(n-p_i)$,
where e_i denotes the multiplicative order of c mod p_i.

2 Superstrong Dickson Pseudoprimes

In order to use congruences of the type (3) for primality testing one is interested in parameters c for which there exist as few as possible strong c–Dickson pseudoprimes. Kowol's characterization describes such parameters c. If $n = p_1 \cdot p_2 \cdots p_r$ is the factorization of n into primes, a parameter c which is a primitive element mod p_i for $i = 1, 2, ..., r$ is an optimal parameter. But as for practical applications we do not know the factorization of n we can only try to choose a parameter c

which has an order as high as possible mod p_i for as many i as possible. As many primes p_i have the primitive element 2, 3, 5 or 7 the parameter $c = 210 = 2\cdot3\cdot5\cdot7$ turns out to be particulary good for disclosing odd composite integers.

Another consequence of the above characterization of strong c–Dickson pseudoprimes is that any strong 1-Dickson pseudoprime is also a strong (-1)-Dickson pseudoprime. Hence, in general tests with the parameter -1 are better than tests with the parameter 1.

An odd composite integer n is called a ***superstrong Dickson pseudoprime*** if it is a strong c-Dickson pseudoprime for all integers c with $gcd(c, n) = 1$. Obviously, a superstrong Dickson pseudoprime cannot be disclosed as composite by any test (3).

As the order of e_i mod p_i has to be a divisor of $p_i - 1$ and $e_i = p_i - 1$ if c is a primitive element mod p_i we can state:

An odd composite number n is a superstrong Dickson pseudoprime if and only if

(i) *n is a Carmichael number,*

(ii) $(p_i^2 - 1)|(n - 1)$ *or* $(p_i^2 - 1)|(n - p_i)$ *for all prime divisors p_i of n.*

Using this characterization it is easy to check that the smallest strong (-1)-Dickson pseudoprime $n = 443372888629441 = 17 \cdot 31 \cdot 41 \cdot 43 \cdot 89 \cdot 97 \cdot 167 \cdot 331$ is also a superstrong Dickson pseudoprime.

3 Consequences for Primality Testing

As superstrong Dickson pseudoprimes are not identified by any test (3) one has to eliminate these composite numbers in applications of tests (3) for primality testing. A similar problem has been solved e.g. in modifying Fermat's Little Theorem for the Rabin–Miller probabilistic primality test. A realistic improvement of existing primality tests (e.g. BAILLIE AND WAGSTAFF [2], DI PORTO AND FILIPPONI [3]) is to replace tests of the type Little Fermat (1) by tests of the type general Lucas functions–Dickson polynomials (3).

The numerical results collected so far indicate that only few odd composite numbers pass a combination of tests (3) with different parameters c and different values b. For example, only three tests of the type (3) are sufficient to disclose all odd composite numbers up to 10^8. One can choose e.g. the tests

$$V_n(1,-1) \equiv 1 \bmod n, V_n(1,3) \equiv 1 \bmod n, V_n(2,-1) \equiv 2 \bmod n$$

or the tests

$$V_n(1,-1) \equiv 1 \bmod n, V_n(1,-5) \equiv 1 \bmod n, V_n(2,-1) \equiv 2 \bmod n.$$

As for the efficient evaluation of Dickson polynomials we refer to an algorithm due to POSTL [10] which for the evaluation of the polynomial $g_n(x, c)$ takes at most twice the time of the evaluation of the power x^n.

References

[1] Alford, W.R., Granville, A. and Pomerance, C.: There are infinitely many Carmichael numbers. Preprint (April 1992).

[2] Baillie, R., Wagstaff jr., S.S.: Lucas pseudoprimes. Math.Comp. **35**, 1391 - 1417 (1980).

[3] Di Porto, A., Filipponi, P.: A Probabilistic Primality Test Based on the Properties of Certain Generalized Lucas Numbers. In: Advances in Cryptology - Eurocrypt'88, Lecture Notes in Computer Science **330**, Springer–Verlag, New York–Berlin–Heidelberg, pp. 211 - 223 (1988).

[4] Filipponi, P.: Table of Fibonacci Pseudoprimes to 10^8. Note Recensioni Notizie **37**, No. 1–2, 33 - 38 (1988).

[5] Kowol, G.: On Strong Dickson Pseudoprimes. Applicable Algebra in Engineering, Communication and Computing (AAECC) **3**, 129 - 138 (1992).

[6] Lidl, R., Müller, W.B., Oswald A.: Some Remarks on Strong Fibonacci Pseudoprimes. Applicable Algebra in Engineering, Communication and Computing (AAECC) **1**, 59 - 65 (1990).

[7] Lidl, R., Niederreiter, H.: Finite Fields. Addison Wesley, Reading, 1983. (Now published by Cambridge University Press, Cambridge.)

[8] Müller, W.B., Oswald A.: Dickson Pseudoprimes and Primality Testing. In: Advances in Cryptology - Eurocrypt'91, Lecture Notes in Computer Science **547**, Springer–Verlag, New York–Berlin–Heidelberg, pp. 512 - 516 (1991).

[9] Pinch, R.G.E.: The Carmichael Numbers up to 10^{15}. Preprint (April 1992).

[10] Postl, H.: Fast Evaluation of Dickson Polynomials. In: Contributions to General Algebra **6**, Hölder-Pichler-Tempsky Verlag Wien and B.G.Teubner Stuttgart, pp. 223 - 225 (1988).

[11] Singmaster, D.: Some Lucas pseudoprimes. Abstracts Amer.Math.Soc. **4**, No. 83T-10-146, p. 197 (1983).

Author Index

Printing: Weihert-Druck GmbH, Darmstadt
Binding: Buchbinderei Schäffer, Grünstadt

Lecture Notes in Computer Science

For information about Vols. 1–645
please contact your bookseller or Springer-Verlag

Vol. 646: J. Biskup, R. Hull (Eds.), Database Theory – ICDT '92. Proceedings, 1992. IX, 449 pages. 1992.

Vol. 647: A. Segall, S. Zaks (Eds.), Distributed Algorithms. X, 380 pages. 1992.

Vol. 648: Y. Deswarte, G. Eizenberg, J.-J. Quisquater (Eds.), Computer Security – ESORICS 92. Proceedings. XI, 451 pages. 1992.

Vol. 649: A. Pettorossi (Ed.), Meta-Programming in Logic. Proceedings, 1992. XII, 535 pages. 1992.

Vol. 650: T. Ibaraki, Y. Inagaki, K. Iwama, T. Nishizeki, M. Yamashita (Eds.), Algorithms and Computation. Proceedings, 1992. XI, 510 pages. 1992.

Vol. 651: R. Koymans, Specifying Message Passing and Time-Critical Systems with Temporal Logic. IX, 164 pages. 1992.

Vol. 652: R. Shyamasundar (Ed.), Foundations of Software Technology and Theoretical Computer Science. Proceedings, 1992. XIII, 405 pages. 1992.

Vol. 653: A. Bensoussan, J.-P. Verjus (Eds.), Future Tendencies in Computer Science, Control and Applied Mathematics. Proceedings, 1992. XV, 371 pages. 1992.

Vol. 654: A. Nakamura, M. Nivat, A. Saoudi, P. S. P. Wang, K. Inoue (Eds.), Parallel Image Analysis. Proceedings, 1992. VIII, 312 pages. 1992.

Vol. 655: M. Bidoit, C. Choppy (Eds.), Recent Trends in Data Type Specification. X, 344 pages. 1993.

Vol. 656: M. Rusinowitch, J. L. Rémy (Eds.), Conditional Term Rewriting Systems. Proceedings, 1992. XI, 501 pages. 1993.

Vol. 657: E. W. Mayr (Ed.), Graph-Theoretic Concepts in Computer Science. Proceedings, 1992. VIII, 350 pages. 1993.

Vol. 658: R. A. Rueppel (Ed.), Advances in Cryptology – EUROCRYPT '92. Proceedings, 1992. X, 493 pages. 1993.

Vol. 659: G. Brewka, K. P. Jantke, P. H. Schmitt (Eds.), Nonmonotonic and Inductive Logic. Proceedings, 1991. VIII, 332 pages. 1993. (Subseries LNAI).

Vol. 660: E. Lamma, P. Mello (Eds.), Extensions of Logic Programming. Proceedings, 1992. VIII, 417 pages. 1993. (Subseries LNAI).

Vol. 661: S. J. Hanson, W. Remmele, R. L. Rivest (Eds.), Machine Learning: From Theory to Applications. VIII, 271 pages. 1993.

Vol. 662: M. Nitzberg, D. Mumford, T. Shiota, Filtering, Segmentation and Depth. VIII, 143 pages. 1993.

Vol. 663: G. v. Bochmann, D. K. Probst (Eds.), Computer Aided Verification. Proceedings, 1992. IX, 422 pages. 1993.

Vol. 664: M. Bezem, J. F. Groote (Eds.), Typed Lambda Calculi and Applications. Proceedings, 1993. VIII, 433 pages. 1993.

Vol. 665: P. Enjalbert, A. Finkel, K. W. Wagner (Eds.), STACS 93. Proceedings, 1993. XIV, 724 pages. 1993.

Vol. 666: J. W. de Bakker, W.-P. de Roever, G. Rozenberg (Eds.), Semantics: Foundations and Applications. Proceedings, 1992. VIII, 659 pages. 1993.

Vol. 667: P. B. Brazdil (Ed.), Machine Learning: ECML – 93. Proceedings, 1993. XII, 471 pages. 1993. (Subseries LNAI).

Vol. 668: M.-C. Gaudel, J.-P. Jouannaud (Eds.), TAPSOFT '93: Theory and Practice of Software Development. Proceedings, 1993. XII, 762 pages. 1993.

Vol. 669: R. S. Bird, C. C. Morgan, J. C. P. Woodcock (Eds.), Mathematics of Program Construction. Proceedings, 1992. VIII, 378 pages. 1993.

Vol. 670: J. C. P. Woodcock, P. G. Larsen (Eds.), FME '93: Industrial-Strength Formal Methods. Proceedings, 1993. XI, 689 pages. 1993.

Vol. 671: H. J. Ohlbach (Ed.), GWAI-92: Advances in Artificial Intelligence. Proceedings, 1992. XI, 397 pages. 1993. (Subseries LNAI).

Vol. 672: A. Barak, S. Guday, R. G. Wheeler, The MOSIX Distributed Operating System. X, 221 pages. 1993.

Vol. 673: G. Cohen, T. Mora, O. Moreno (Eds.), Applied Algebra, Algebraic Algorithms and Error-Correcting Codes. Proceedings, 1993. X, 355 pages 1993.

Vol. 674: G. Rozenberg (Ed.), Advances in Petri Nets 1993. VII, 457 pages. 1993.

Vol. 675: A. Mulkers, Live Data Structures in Logic Programs. VIII, 220 pages. 1993.

Vol. 676: Th. H. Reiss, Recognizing Planar Objects Using Invariant Image Features. X, 180 pages. 1993.

Vol. 677: H. Abdulrab, J.-P. Pécuchet (Eds.), Word Equations and Related Topics. Proceedings, 1991. VII, 214 pages. 1993.

Vol. 678: F. Meyer auf der Heide, B. Monien, A. L. Rosenberg (Eds.), Parallel Architectures and Their Efficient Use. Proceedings, 1992. XII, 227 pages. 1993.

Vol. 679: C. Fermüller, A. Leitsch, T. Tammet, N. Zamov, Resolution Methods for the Decision Problem. VIII, 205 pages. 1993. (Subseries LNAI).

Vol. 680: B. Hoffmann, B. Krieg-Brückner (Eds.), Program Development by Specification and Transformation. XV, 623 pages. 1993.

Vol. 681: H. Wansing, The Logic of Information Structures. IX, 163 pages. 1993. (Subseries LNAI).

Vol. 682: B. Bouchon-Meunier, L. Valverde, R. R. Yager (Eds.), IPMU '92 – Advanced Methods in Artificial Intelligence. Proceedings, 1992. IX, 367 pages. 1993.

Vol. 683: G.J. Milne, L. Pierre (Eds.), Correct Hardware Design and Verification Methods. Proceedings, 1993. VIII, 270 Pages. 1993.

Vol. 684: A. Apostolico, M. Crochemore, Z. Galil, U. Manber (Eds.), Combinatorial Pattern Matching. Proceedings, 1993. VIII, 265 pages. 1993.

Vol. 685: C. Rolland, F. Bodart, C. Cauvet (Eds.), Advanced Information Systems Engineering. Proceedings, 1993. XI, 650 pages. 1993.

Vol. 686: J. Mira, J. Cabestany, A. Prieto (Eds.), New Trends in Neural Computation. Proceedings, 1993. XVII, 746 pages. 1993.

Vol. 687: H. H. Barrett, A. F. Gmitro (Eds.), Information Processing in Medical Imaging. Proceedings, 1993. XVI, 567 pages. 1993.

Vol. 688: M. Gauthier (Ed.), Ada-Europe '93. Proceedings, 1993. VIII, 353 pages. 1993.

Vol. 689: J. Komorowski, Z. W. Ras (Eds.), Methodologies for Intelligent Systems. Proceedings, 1993. XI, 653 pages. 1993. (Subseries LNAI).

Vol. 690: C. Kirchner (Ed.), Rewriting Techniques and Applications. Proceedings, 1993. XI, 488 pages. 1993.

Vol. 691: M. Ajmone Marsan (Ed.), Application and Theory of Petri Nets 1993. Proceedings, 1993. IX, 591 pages. 1993.

Vol. 692: D. Abel, B.C. Ooi (Eds.), Advances in Spatial Databases. Proceedings, 1993. XIII, 529 pages. 1993.

Vol. 693: P. E. Lauer (Ed.), Functional Programming, Concurrency, Simulation and Automated Reasoning. Proceedings, 1991/1992. XI, 398 pages. 1993.

Vol. 694: A. Bode, M. Reeve, G. Wolf (Eds.), PARLE '93. Parallel Architectures and Languages Europe. Proceedings, 1993. XVII, 770 pages. 1993.

Vol. 695: E. P. Klement, W. Slany (Eds.), Fuzzy Logic in Artificial Intelligence. Proceedings, 1993. VIII, 192 pages. 1993. (Subseries LNAI).

Vol. 696: M. Worboys, A. F. Grundy (Eds.), Advances in Databases. Proceedings, 1993. X, 276 pages. 1993.

Vol. 697: C. Courcoubetis (Ed.), Computer Aided Verification. Proceedings, 1993. IX, 504 pages. 1993.

Vol. 698: A. Voronkov (Ed.), Logic Programming and Automated Reasoning. Proceedings, 1993. XIII, 386 pages. 1993. (Subseries LNAI).

Vol. 699: G. W. Mineau, B. Moulin, J. F. Sowa (Eds.), Conceptual Graphs for Knowledge Representation. Proceedings, 1993. IX, 451 pages. 1993. (Subseries LNAI).

Vol. 700: A. Lingas, R. Karlsson, S. Carlsson (Eds.), Automata, Languages and Programming. Proceedings, 1993. XII, 697 pages. 1993.

Vol. 701: P. Atzeni (Ed.), LOGIDATA+: Deductive Databases with Complex Objects. VIII, 273 pages. 1993.

Vol. 702: E. Börger, G. Jäger, H. Kleine Büning, S. Martini, M. M. Richter (Eds.), Computer Science Logic. Proceedings, 1992. VIII, 439 pages. 1993.

Vol. 703: M. de Berg, Ray Shooting, Depth Orders and Hidden Surface Removal. X, 201 pages. 1993.

Vol. 704: F. N. Paulisch, The Design of an Extendible Graph Editor. XV, 184 pages. 1993.

Vol. 705: H. Grünbacher, R. W. Hartenstein (Eds.), Field-Programmable Gate Arrays. Proceedings, 1992. VIII, 218 pages. 1993.

Vol. 706: H. D. Rombach, V. R. Basili, R. W. Selby (Eds.), Experimental Software Engineering Issues. Proceedings, 1992. XVIII, 261 pages. 1993.

Vol. 707: O. M. Nierstrasz (Ed.), ECOOP '93 – Object-Oriented Programming. Proceedings, 1993. XI, 531 pages. 1993.

Vol. 708: C. Laugier (Ed.), Geometric Reasoning for Perception and Action. Proceedings, 1991. VIII, 281 pages. 1993.

Vol. 709: F. Dehne, J.-R. Sack, N. Santoro, S. Whitesides (Eds.), Algorithms and Data Structures. Proceedings, 1993. XII, 634 pages. 1993.

Vol. 710: Z. Ésik (Ed.), Fundamentals of Computation Theory. Proceedings, 1993. IX, 471 pages. 1993.

Vol. 711: A. M. Borzyszkowski, S. Sokołowski (Eds.), Mathematical Foundations of Computer Science 1993. Proceedings, 1993. XIII, 782 pages. 1993.

Vol. 712: P. V. Rangan (Ed.), Network and Operating System Support for Digital Audio and Video. Proceedings, 1992. X, 416 pages. 1993.

Vol. 713: G. Gottlob, A. Leitsch, D. Mundici (Eds.), Computational Logic and Proof Theory. Proceedings, 1993. XI, 348 pages. 1993.

Vol. 714: M. Bruynooghe, J. Penjam (Eds.), Programming Language Implementation and Logic Programming. Proceedings, 1993. XI, 421 pages. 1993.

Vol. 715: E. Best (Ed.), CONCUR'93. Proceedings, 1993. IX, 541 pages. 1993.

Vol. 716: A. U. Frank, I. Campari (Eds.), Spatial Information Theory. Proceedings, 1993. XI, 478 pages. 1993.

Vol. 717: I. Sommerville, M. Paul (Eds.), Software Engineering – ESEC '93. Proceedings, 1993. XII, 516 pages. 1993.

Vol. 718: J. Seberry, Y. Zheng (Eds.), Advances in Cryptology – AUSCRYPT '92. Proceedings, 1992. XIII, 543 pages. 1993.

Vol. 719: D. Chetverikov, W.G. Kropatsch (Eds.), Computer Analysis of Images and Patterns. Proceedings, 1993. XVI, 857 pages. 1993.

Vol. 720: V.Mařík, J. Lažanský, R.R. Wagner (Eds.), Database and Expert Systems Applications. Proceedings, 1993. XV, 768 pages. 1993.